FOUNDATIONS OF DIFFERENTIAL GEOMETRY

VOLUME II

FOUNDATIONS OF DIFFERENTIAL GEOMETRY

VOLUME II

SHOSHICHI KOBAYASHI

University of California, Berkeley, California

and

KATSUMI NOMIZU

Brown University, Providence, Rhode Island

Wiley Classics Library Edition Published 1996

A WILEY-INTERSCIENCE PUBLICATION

JOHN WILEY & SONS, INC.

New York • Chichester • Brisbane • Toronto • Singapore

Copyright © 1969 by John Wiley & Sons, Inc.

Wiley Classics Library Edition Published 1996.

All rights reserved.

Reproduction or translation of any part of this work beyond that permitted by Sections 107 or 108 of the 1976 United States Copyright Act without the permission of the copyright owner is unlawful. Requests for permission or further information should be addressed to the Permissions Department, John Wiley & Sons, Inc.

Library of Congress Catalog Card Number: 68-19209

ISBN 0-471-15732-5

Printed and bound in the United States of America

10 9 8 7 6 5 4 3 2

PREFACE

This is a continuation of Volume I of the *Foundations of Differential Geometry*. The chapter numbers are continued from Volume I and the same notations are preserved as much as possible. The main text, Chapters VII–XII, deals with the topics that have been promised in the Preface of Volume I. The Notes include material supplementary to Volume I as well. The Bibliography duplicates, for the sake of convenience of readers, all the references in the Bibliography of Volume I in the same numbering and continues to references for Volume II.

The content of each chapter is now briefly described.

Chapter VII gives the fundamental results and some classical theorems concerning geometry of an n-dimensional submanifold M immersed in an $(n + p)$-dimensional Riemannian manifold N, in particular, $\mathbf{R}^{n+p}$. In §1, the natural connections in the orthogonal bundle and the normal bundle over M are derived from the Riemannian connection in the orthogonal bundle over N. In §2, where $N = \mathbf{R}^{n+p}$, we show that these connections are induced from the canonical connections in the Stiefel manifolds $V(n, p)$ and $V(p, n)$, both over the Grassmann manifold $G(n, p)$, respectively, by means of the bundle maps associated to the generalized Gauss map of M into $G(n, p)$. In §§3 and 4, we use the formalism of covariant differentiation $\nabla_X Y$ to study the relationship between the invariants of M and N and obtain the classical formulas of Weingarten, Gauss and Codazzi. We prove a result of Chern-Kuiper which generalizes the theorem of Tompkins. §§5, 6, and 7 are concerned with the classical notions and theorems on hypersurfaces in a Euclidean space, including a result of Thomas-Cartan-Fialkow on Einstein hypersurfaces and results on the type number and the so-called fundamental theorem. In the last §8, we discuss auto-parallel submanifolds and totally geodesic submanifolds of a manifold with an affine connection and prove, in particular, that these two notions coincide in the case where the connection in the ambient space has no torsion. The content of Chapter VII is supplemented by Notes 14, 15, 16, 17, 18, 21, and 27.

Chapter VIII is devoted to the study of variational problems on geodesics. In §1, we define Jacobi fields and conjugate points for a manifold with an affine connection and discuss their geometric meaning. In §2 and 3, we make a further study of these notions in a Riemannian manifold and prove the classical result on the distance between consecutive conjugate points on a geodesic when the sectional curvature (or, more generally, each of the eigenvalues of the Ricci tensor) is greater than a certain positive number everywhere. In §4, we prove Rauch's comparison theorem. In §5, we study the first and second variations of the length integral, considered as a function on the space of all piecewise differentiable curves, and obtain, among others, a proof of Myers's Theorem. The Index Theorem of Morse is proved in §6. In §7, we prove basic properties of cut loci. Although the results of §7 are not used elsewhere in this book, they are basic in the study of manifolds with positive curvature. In §8, we prove a theorem of Hadamard and Cartan which says that for a complete Riemannian space with non-positive curvature the exponential map is a covering map. Applications are made to a homogeneous Riemannian manifold with non-positive sectional curvature and negative definite Ricci tensor. In §9, we prove a theorem to the effect that on a simply connected complete Riemannian manifold with non-positive sectional curvature every compact group of isometries has a fixed point. Applications are given to the case of a homogeneous Riemannian space. Results of §§8 and 9 are used in §11 of Chapter XI. Note 22 supplements the content of this chapter.

In Chapter IX, we provide differential geometric foundations for almost complex manifolds and Hermitian metrics, in particular, complex manifolds and Kaehler metrics. The results in this chapter are essentially of local character. After purely algebraic preliminaries in §1, we discuss in §2 the notion of an almost complex structure, its torsion and integrability as well as complex tangent spaces, operators ∂ and $\bar{\partial}$ for complex differential forms on an almost complex manifold. Many examples are given, including complex Lie groups, complex parallelizable spaces, complex Grassmann manifolds, Hopf manifolds and their generalizations, and a result of Kirchhoff on almost complex structures on spheres. In §3, we discuss connections in the bundle of complex linear frames of an almost complex manifold and relate their torsions with the torsion of the almost complex structure. In §4, Hermitian metrics and the bundle of unitary frames are discussed. The most interesting case is that of a Kaehler metric, whose basic properties are proved here. In §5, we build a bridge between intrinsic notations and complex tensor notations for Kaehlerian geometry. In §6, many examples of Kaehler manifolds are discussed, including the Fubini-Study metric in the complex projective space and the Bergman metric in the open unit ball in C^n. In §7, we give basic local properties of holomorphic sectional curvature and prove that a simply connected and complete Kaehler manifold of constant holomorphic sectional curvature c is a complex projective space, a complex Euclidean space or an open unit ball in a complex Euclidean space according as $c > 0$, $= 0$ or < 0. In §8, we discuss the de Rham decomposition of a Kaehler manifold and the notion of non-degeneracy. §9 is concerned with holomorphic sectional curvature and the Ricci tensor of a complex submanifold of a Kaehler manifold. In the last §10, we study the existence and properties of Hermitian connection in a Hermitian vector bundle following Chern, Nakano, and Singer. This chapter is supplemented by §6 of Chapter X, §10 of Chapter XI (where examples are discussed from the viewpoint of symmetric spaces), and Notes 13, 18, 23, 24, and 26.

In Chapter X, we discuss the existence and properties of invariant affine connections and invariant almost complex structures on homogeneous spaces (especially, reductive homogeneous spaces). In §1, the results of Wang in §11 of Chapter II are specialized to the situation where P is a K-invariant G-structure on a homogeneous space $M = K/H$, and K-invariant connections in P are studied. In §2, we specialize further to the case where K/H is reductive and obtain the canonical connection and the natural torsion-free connection of Nomizu. In §3, we study homogeneous spaces with invariant (possibly indefinite) Riemannian metrics. As an example we provide a differential geometric proof of Weyl's theorem that a Lie group G is compact if the Killing-Cartan form of its Lie algebra is negative-definite. In §§4 and 5, results of Nomizu and Kostant on the holonomy group and reducibility of an invariant affine connection are proved. In §6, following Koszul we give algebraic formulations for an invariant almost complex structure on a homogeneous space and for its integrability. This chapter serves as a basis for Chapter XI and is supplemented by Notes 24 and 25.

In Chapter XI, we present the basic results in the theory of affine, Riemannian, and Hermitian symmetric spaces. We lay emphasis on the affine case a little more than the standard treatment of the subject. In §1, we consider affine symmetric spaces, thus giving a geometric motivation to the group-theoretic notion of symmetric space which is introduced in §2. In §3, we reverse the process in §1; thus we begin with a symmetric space G/H and introduce the canonical affine connection on G/H, making G/H an affine symmetric space. The curvature of the canonical connection is given an algebraic expression. In §4, we study totally geodesic submanifolds of a symmetric space G/H (with canonical connection) from both geometric and algebraic viewpoints. The symmetric Lie algebra introduced in §3 is to a symmetric space what the Lie algebra is to a Lie group. In §5, two results on Lie algebras, namely, Levi's theorem and the decomposition of a semi-simple Lie algebra into a direct sum of simple ideals, are extended to the case of symmetric Lie algebras. The global versions of these results are also given. In §6, we consider Riemannian symmetric spaces and the corresponding symmetric spaces. The symmetric Lie algebra corresponding to a Riemannian symmetric space is called an orthogonal symmetric Lie algebra. In §7, where orthogonal symmetric Lie algebras are studied, the decomposition theorems proved in §5 are made more precise. In §8, the duality between the orthogonal symmetric Lie algebras of compact type and those of non-compact type are studied together with geometric interpretations. In §9, we discuss geometric properties and an algebraic characterization of Hermitian symmetric spaces. Many examples of classical spaces are studied in §10 from viewpoints of symmetric spaces, including real space forms originally defined in Chapter V and complex space forms discussed in Chapter IX. In the last §11, we show, assuming Weyl's existence theorem of a compact real form of a complex simple Lie algebra, that the classification of irreducible orthogonal symmetric Lie algebras is equivalent to the classification of real simple Lie algebras.

In Chapter XII, we present differential geometric aspects of characteristic classes. If G is the structure group of a principal bundle P over M, then using the curvature of a connection in P we can associate to each $\text{Ad}(G)$-invariant homogeneous polynomial f of degree k on the Lie algebra of G a closed $2k$-form on the base space M in a natural manner. The cohomology class represented by this closed $2k$-form is independent of the choice of connection and is called the characteristic class determined by f. In §1, following Chern we prove this basic result of Weil. In §2, we study the algebra of $\text{Ad}(G)$-invariant polynomials on the Lie algebra of G and determine the algebra explicitly when G is a classical group. In §3, adopting the axiomatic definition of Chern classes by Hirzebruch, we express the Chern classes of a complex vector bundle in terms of the curvature form of a connection in the bundle. The formula for the Chern character in terms of the curvature form is also given. In §4, using Hirzebruch's definition of the Pontrjagin classes of a real vector bundle, we derive differential geometric formulas for the Pontrjagin classes. In §5, we characterize the real Euler class of a vector bundle in a simple axiomatic manner and derive the general Gauss-Bonnet formula. This chapter, particularly §5, is supplemented by Notes 20 and 21.

We wish to note specifically that we do not go into the following subjects: the theory of (2-dimensional) minimal surfaces; the theory of global convex surfaces developed by A. D. Aleksandrov and his school; Finsler geometry and its generalizations; the general theory of conformal and projective connections; a deeper study of differential systems. On the subjects of complex manifolds, homogeneous spaces (especially symmetric homogeneous spaces), vector bundles, G-structures and so on, our treatment is limited to the foundational material in differential geometric aspects that does not require deeper knowledge from algebra, analysis or topology. Neither do we treat the harmonic theory nor a generalized Morse theory, although these theories have many important applications to Riemannian geometry. The Bibliography of Volume II contains some basic references in these areas. In particular, for the global theory of compact Kaehler manifolds which requires the theory of harmonic integrals, the reader is advised to read Weil's book: *Introduction à l'Étude des Variétés Kählériennes*.

During the preparations of this volume, we have been most encouraged by the reactions to Volume I of many readers who wanted to find self-contained and complete proofs of the standard results in the field. We sincerely hope that the present volume will continue to meet the needs of these readers.

We should also like to acknowledge the grants of the National Science Foundation which supported part of the work included in this book.

SHOSHICHI KOBAYASHI
KATSUMI NOMIZU

September, 1968

CONTENTS OF VOLUME II

Chapter VII
Submanifolds

1. Frame bundles of a submanifold 1
2. The Gauss map 6
3. Covariant differentiation and second fundamental form . 10
4. Equations of Gauss and Codazzi 22
5. Hypersurfaces in a Euclidean space 29
6. Type number and rigidity 42
7. Fundamental theorem for hypersurfaces 47
8. Auto-parallel submanifolds and totally geodesic submanifolds 53

Chapter VIII
Variations of the Length Integral

1. Jacobi fields 63
2. Jacobi fields in a Riemannian manifold 68
3. Conjugate points 71
4. Comparison theorem 76
5. The first and second variations of the length integral . 79
6. Index theorem of Morse 88
7. Cut loci 96
8. Spaces of non-positive curvature 102
9. Center of gravity and fixed points of isometries . . 108

Chapter IX
Complex Manifolds

1. Algebraic preliminaries 114
2. Almost complex manifolds and complex manifolds . . 121
3. Connections in almost complex manifolds . . . 141
4. Hermitian metrics and Kaehler metrics 146
5. Kaehler metrics in local coordinate systems . . . 155
6. Examples of Kaehler manifolds 159
7. Holomorphic sectional curvature 165
8. De Rham decomposition of Kaehler manifolds . . 171

9.	Curvature of Kaehler submanifolds	175
10.	Hermitian connections in Hermitian vector bundles. .	178

Chapter X
Homogeneous Spaces

1.	Invariant affine connections	186
2.	Invariant connections on reductive homogeneous spaces .	190
3.	Invariant indefinite Riemannian metrics . . .	200
4.	Holonomy groups of invariant connections . . .	204
5.	The de Rham decomposition and irreducibility . .	210
6.	Invariant almost complex structures	216

Chapter XI
Symmetric Spaces

1.	Affine locally symmetric spaces	222
2.	Symmetric spaces.	225
3.	The canonical connection on a symmetric space . .	230
4.	Totally geodesic submanifolds	234
5.	Structure of symmetric Lie algebras	238
6.	Riemannian symmetric spaces	243
7.	Structure of orthogonal symmetric Lie algebras . .	246
8.	Duality	253
9.	Hermitian symmetric spaces	259
10.	Examples	264
11.	An outline of the classification theory	289

Chapter XII
Characteristic Classes

1.	Weil homomorphism	293
2.	Invariant polynomials	298
3.	Chern classes	305
4.	Pontrjagin classes	312
5.	Euler classes	314

Appendices

8.	Integrable real analytic almost complex structures . .	321
9.	Some definitions and facts on Lie algebras . . .	325

Notes

12.	Connections and holonomy groups (Supplement to Note 1)	331
13.	The automorphism group of a geometric structure (Supplement to Note 9) .	332
14.	The Laplacian .	337
15.	Surfaces of constant curvature in $\mathbf{R}^3$	343
16.	Index of nullity .	347
17.	Type number and rigidity of imbedding	349
18.	Isometric imbeddings .	354
19.	Equivalence problems for Riemannian manifolds	357
20.	Gauss-Bonnet theorem.	358
21.	Total curvature .	361
22.	Topology of Riemannian manifolds with positive curvature	364
23.	Topology of Kaehler manifolds with positive curvature	368
24.	Structure theorems on homogeneous complex manifolds	373
25.	Invariant connections on homogeneous spaces.	375
26.	Complex submanifolds.	378
27.	Minimal submanifolds .	379
28.	Contact structure and related structures	381
	Bibliography	387
	Summary of Basic Notations.	455
	Index for Volumes I and II	459
	Errata for *Foundations of Differential Geometry*, Volume I	469
	Errata for *Foundations of Differential Geometry*, Volume II	470

CONTENTS OF VOLUME I

- **I.** Differentiable Manifolds
- **II.** Theory of Connections
- **III.** Linear and Affine Connections
- **IV.** Riemannian Connections
- **V.** Curvature and Space Forms
- **VI.** Transformations

CHAPTER VII

Submanifolds

1. Frame bundles of a submanifold

Let $e_1, \ldots, e_{n+p}$ be the natural basis for $\mathbf{R}^{n+p}$. We shall denote by $\mathbf{R}^n$ and $\mathbf{R}^p$ the subspaces of $\mathbf{R}^{n+p}$ spanned by $e_1, \ldots, e_n$ and $e_{n+1}, \ldots, e_{n+p}$, respectively. Similarly, we identify $O(n)$ (resp. $O(p)$) with the subgroup of $O(n+p)$ consisting of all elements which induce the identity transformation on the subspace $\mathbf{R}^p$ (resp. $\mathbf{R}^n$) of $\mathbf{R}^{n+p}$. In other words,

$$O(n) \approx \begin{pmatrix} O(n) & 0 \\ 0 & I_p \end{pmatrix} \quad \text{and} \quad O(p) \approx \begin{pmatrix} I_n & 0 \\ 0 & O(p) \end{pmatrix},$$

where I_n and I_p denote the identity matrices of order n and p, respectively. Let $\mathfrak{o}(n+p)$, $\mathfrak{o}(n)$ and $\mathfrak{o}(p)$ be the Lie algebras of $O(n+p)$, $O(n)$ and $O(p)$, respectively, and let $\mathfrak{g}(n,p)$ be the orthogonal complement to $\mathfrak{o}(n) + \mathfrak{o}(p)$ in $\mathfrak{o}(n+p)$ with respect to the Killing-Cartan form of $\mathfrak{o}(n+p)$ (cf. Volume I, p. 155 and also Appendix 9). Then $\mathfrak{g}(n,p)$ consists of matrices of the form

$$\begin{pmatrix} 0 & A \\ -{}^t A & 0 \end{pmatrix},$$

where A is a matrix with n rows and p columns and ${}^t A$ denotes the transpose of A.

Let N be a Riemannian manifold of dimension $n+p$ and let f be an immersion of an n-dimensional differentiable manifold M into N. We denote by g the metric of N as well as the metric induced on M (cf. Example 1.2 of Chapter IV). For any point x of M we shall denote $f(x) \in N$ by the same letter x if there is no danger of confusion. Thus the tangent space $T_x(M)$ is a subspace of the tangent space $T_x(N)$. Let $T_x(M)^\perp$ be the orthogonal complement of $T_x(M)$ in $T_x(N)$; it is called the *normal space* to M at x.

Let $O(M)$ and $O(N)$ be the bundles of orthonormal frames over M and N, respectively. Then $O(N) \mid M = \{v \in O(N); \pi(v) \in M\}$, where $\pi: O(N) \to N$ is the projection, is a principal fibre bundle over M with structure group $O(n + p)$. A frame $v \in O(N) \mid M$ at $x \in M$ is said to be *adapted* if v is of the form $(Y_1, \ldots, Y_n, Y_{n+1}, \ldots, Y_{n+p})$ with $Y_1, \ldots, Y_n$ tangent to M (and hence $Y_{n+1}, \ldots, Y_{n+p}$ normal to M). Thus, considered as a linear isomorphism $\mathbf{R}^{n+p} \to T_x(N)$, v is adapted if and only if v maps the subspace $\mathbf{R}^n$ onto $T_x(M)$ (and hence the subspace $\mathbf{R}^p$ onto $T_x(M)^\perp$). It is easy to verify that the set of adapted frames forms a principal fibre bundle over M with group $O(n) \times O(p)$; it is a subbundle of $O(N) \mid M$ in a natural manner. We shall denote the bundle of adapted frames by $O(N, M)$. We define a homomorphism $h': O(N, M) \to O(M)$ corresponding to the natural homomorphism $O(n) \times O(p) \to O(n)$ as follows:

$$h'(v) = (Y_1, \ldots, Y_n) \quad \text{for } v = (Y_1, \ldots, Y_{n+p}) \in O(N, M).$$

If we consider v as a linear transformation $\mathbf{R}^{n+p} \to T_x(N)$, then $h'(v)$ is the restriction of v to the subspace $\mathbf{R}^n$. Hence, $O(M)$ is naturally isomorphic to $O(N, M)/O(p)$. Similarly, denoting by $h''(v)$ the restriction of $v \in O(N, M)$ to the subspace $\mathbf{R}^p$ of $\mathbf{R}^{n+p}$, we obtain a homomorphism $h'': O(N, M) \to O(N, M)/O(n)$ corresponding to the natural homomorphism $O(n) \times O(p) \to O(p)$. By a *normal frame* at $x \in M$, we mean an orthonormal basis $(Z_1, \ldots, Z_p)$ for the normal space $T_x(M)^\perp$. If $(Y_1, \ldots, Y_n, Y_{n+1}, \ldots, Y_{n+p})$ is an adapted frame at x, then $(Y_{n+1}, \ldots, Y_{n+p})$ is a normal frame at x. Since every normal frame is thus obtained and since two adapted frames give rise to the same normal frame if and only if they are congruent modulo $O(n)$, the bundle $O(N, M)/O(n)$ can be considered as the bundle of normal frames over M. Then $h'': O(N, M) \to O(N, M)/O(n)$ maps an adapted frame $v = (Y_1, \ldots, Y_{n+p})$ upon the normal frame $(Y_{n+1}, \ldots, Y_{n+p})$. We denote by $T(M)^\perp$ the set $\bigcup_{x \in M} T_x(M)^\perp$. It is then a vector bundle over M associated to the bundle of normal frames $O(N, M)/O(n)$ by letting the structure group $O(p)$ act naturally on the standard fibre $\mathbf{R}^p$ (cf. §1 of Chapter III). We shall call this vector bundle the *normal bundle* of M (for the given immersion f into N). The following diagrams illustrate these bundles:

$$\begin{array}{ccccc}
O(M) = O(N,M)/O(p) & \xleftarrow{h'} & O(N,M) & \xrightarrow{h''} & O(N,M)/O(n) \\
{\scriptstyle O(n)}\Big\downarrow{\scriptstyle \pi'} & & {\scriptstyle O(n) \times O(p)}\Big\downarrow{\scriptstyle \pi} & & {\scriptstyle O(p)}\Big\downarrow{\scriptstyle \pi''} \\
M & \leftrightarrow & M & \leftrightarrow & M
\end{array}$$

$$\begin{array}{ccccc}
 & O(N,M) & \xrightarrow{i} & O(N)\mid M & \xrightarrow{j} & O(N) \\
{\scriptstyle O(n) \times O(p)}\Big\downarrow{\scriptstyle \pi} & & {\scriptstyle O(n+p)}\Big\downarrow{\scriptstyle \pi} & & {\scriptstyle O(n+p)}\Big\downarrow{\scriptstyle \pi} \\
M & \leftrightarrow & M & \rightarrow & N,
\end{array}$$

where both i and j are injections.

Let θ and φ be the canonical forms of M and N, respectively (cf. §2 of Chapter III); θ is an $\mathbf{R}^n$-valued 1-form on $O(M)$ and φ is an $\mathbf{R}^{n+p}$-valued 1-form on $O(N)$. Then we have

PROPOSITION 1.1. *$h'^*(\theta)$ coincides with the restriction of φ to $O(N, M)$. In particular, the restriction of the $\mathbf{R}^{n+p}$-valued form φ to $O(N, M)$ is $\mathbf{R}^n$-valued.*

Proof. By definition of φ we have

$$\varphi(Y) = v^{-1}(\pi(Y)) \quad \text{for } Y \in T_v(O(N,M)).$$

Since $\pi(Y) \in T_x(M)$, where $x = \pi(v)$, and since v^{-1} maps $T_x(M)$ onto $\mathbf{R}^n$, $\varphi(Y)$ is in $\mathbf{R}^n$. Since $h'(v) = v \mid \mathbf{R}^n$ and since $\pi' \circ h'(Y) = \pi(Y)$, we have

$$\varphi(Y) = v^{-1}(\pi(Y)) = h'(v)^{-1}(\pi' \circ h'(Y)) = \theta(h'(Y))$$
$$= (h'^*(\theta))(Y).$$

QED.

Let ψ be the Riemannian connection form on $O(N)$. Its restriction to $O(N) \mid M$, that is, $j^*\psi$, defines a connection in the bundle $O(N) \mid M$. But its restriction to $O(N, M)$, that is, $i^*j^*\psi$, is not, in general, a connection form on $O(N, M)$.

PROPOSITION 1.2. *Let ψ be the Riemannian connection form on $O(N)$ and let ω be the $\mathfrak{o}(n) + \mathfrak{o}(p)$-component of $i^*j^*\psi$ with respect to the decomposition $\mathfrak{o}(n+p) = \mathfrak{o}(n) + \mathfrak{o}(p) + \mathfrak{g}(n,p)$. Then ω defines a connection in the bundle $O(N, M)$.*

Proof. Since $\mathrm{ad}(O(n) \times O(p))$ maps $\mathfrak{g}(n, p)$ onto itself, we see from Proposition 6.4 of Chapter II that the form ω defines a connection in $O(N, M)$. QED.

PROPOSITION 1.3. *The homomorphism $h'\colon O(N, M) \to O(M) = O(N, M)/O(p)$ maps the connection in $O(N, M)$ defined by ω into the Riemannian connection of M. The Riemannian connection form ω' on $O(M)$ is determined by*
$$h'^*(\omega') = \omega_{\mathfrak{o}(n)},$$
where $\omega_{\mathfrak{o}(n)}$ denotes the $\mathfrak{o}(n)$-component of the $\mathfrak{o}(n) + \mathfrak{o}(p)$-valued form ω.

Proof. By Proposition 6.1 of Chapter II we know that h' maps the connection defined by ω into the connection in $O(M)$ defined by a form ω' such that $h'^*(\omega') = \omega_{\mathfrak{o}(n)}$. To show that ω' defines the Riemannian connection of M we have only to show that the torsion form of ω' is zero. Restricting the first structure equation of ψ to $O(N, M)$, we obtain
$$d(i^*j^*\varphi) = -(i^*j^*\psi) \wedge (i^*j^*\varphi).$$
Since $i^*j^*\varphi$ is equal to $h'^*(\theta)$ and is $\mathbf{R}^n$-valued by Proposition 1.1, comparing the $\mathbf{R}^n$-components of the both sides we obtain
$$d(h'^*(\theta)) = -h'^*(\omega') \wedge h'^*(\theta).$$
Since h' maps $O(N, M)$ onto $O(M)$, this implies $d\theta = -\omega' \wedge \theta$. QED.

Similarly, by Proposition 6.1 of Chapter II we see that there is a unique connection form ω'' on the bundle $O(N, M)/O(n)$ such that
$$h''^*(\omega'') = \omega_{\mathfrak{o}(p)},$$
where $\omega_{\mathfrak{o}(p)}$ denotes the $\mathfrak{o}(p)$-component of the $\mathfrak{o}(n) + \mathfrak{o}(p)$-valued form ω. Geometrically speaking, ω'' defines the parallel displacement of the normal space $T_x(M)^\perp$ onto the normal space $T_y(M)^\perp$ along any curve τ in M from x to y.

The bundles $O(N, M)$, $O(M) = O(N, M)/O(p)$, and $O(N, M)/O(n)$, and their connection form ω, ω', and ω'' are related as follows:

PROPOSITION 1.4. *The mapping $(h', h'')\colon O(N, M) \to O(M) \times (O(N, M)/O(n))$ induces a bundle isomorphism $O(N, M) \approx O(M) + (O(N, M)/O(n))$. The connection form ω coincides with $h'^*(\omega') + h''^*(\omega'')$.*

The proof is trivial (see p. 82 of Volume I).

Finally, we say a few words about the special case of a hypersurface. By a hypersurface in an $(n + 1)$-dimensional manifold N we mean a (generally connected) n-dimensional manifold M with an immersion f. For each $x \in M$, there is a coordinate neighborhood U of x in M and a differentiable field, say, ξ, of unit normal vectors defined on U. Such a ξ can be easily constructed by choosing a coordinate system $x^1, \ldots, x^n$ around x in U and a coordinate system $y^1, \ldots, y^{n+1}$ around $x \ (=f(x))$ in N; in fact, a unit normal vector field on U is determined uniquely up to sign. For a fixed choice of ξ on U, it is obvious that ξ is parallel along all closed curves in U (with respect to the connection in the normal bundle). Assume that N is orientable and is oriented. Then we can choose a differentiable field of unit normal vectors over M if and only if M is also orientable. Indeed, for a fixed orientation on M, there is a unique choice of the field of unit normal vectors ξ such that, for an oriented basis $\{X_1, \ldots, X_n\}$ of $T_x(M)$ at each $x \in M$, $\{\xi_x, X_1, \ldots, X_n\}$ is an oriented basis of $T_x(N)$. Conversely, if a field ξ of unit normal vectors exists globally on M, then a basis $\{X_1, \ldots, X_n\}$ of $T_x(M)$ such that $\{\xi_x, X_1, \ldots, X_n\}$ is an oriented basis of $T_x(N)$ determines an orientation of M. If we forget about the particular orientations of N and M, then again a differentiable field of unit vectors on M is unique up to sign. For a choice of ξ, it is obvious that ξ is parallel along all curves in M.

Without assuming that N and M are orientable, let us choose a unit normal vector ξ_0 at a point x_0 on M. The parallel displacement along all closed curves at x_0 on M will map ξ_0 either upon ξ_0 or upon $-\xi_0$. In other words, the holonomy group of the linear connection in the normal bundle is a subgroup of the group $\{1, -1\}$. (This is also clear from the fact that the bundle $O(N, M)/O(n)$ of normal frames over M has structure group $O(1) = \{1, -1\}$.) If the holonomy group is trivial (and this is the case if M is simply connected), then ξ_0 is invariant by parallel displacement along all closed curves at x_0. In this case, we may define a differentiable unit normal vector field ξ on M by translating ξ_0 parallelly to each point x of M, the result being independent of the choice of a curve from x_0 to x in M. We may thus conclude that *if M is a simply connected, connected hypersurface immersed in a Riemannian manifold N, then M admits a differentiable field of unit normal vectors defined on M.*

2. The Gauss map

We consider $\mathbf{R}^{n+p}$ as an $(n+p)$-dimensional vector space with an inner product such that the natural basis is orthonormal. Let $G(n, p)$ be the set of n-dimensional subspaces of $\mathbf{R}^{n+p}$. We shall make $G(n, p)$ into a manifold as follows. The group $O(n+p)$ acting on $\mathbf{R}^{n+p}$ acts transitively on $G(n, p)$ in a natural manner. The elements of $O(n+p)$ which map the particular subspace $\mathbf{R}^n$ onto itself form the subgroup $O(n) \times O(p)$. Thus we have

$$G(n, p) = O(n+p)/O(p) \times O(p).$$

The manifold $G(n, p)$ is called the *Grassmann manifold* of n-planes in $\mathbf{R}^{n+p}$.

By an *n-frame* in $\mathbf{R}^{n+p}$ we mean an ordered set of n orthonormal vectors in $\mathbf{R}^{n+p}$. The group $O(n+p)$ acts transitively also on the set $V(n, p)$ of n-frames in $\mathbf{R}^{n+p}$ in a natural manner. The elements of $O(n+p)$ which fix the n-frame $(e_1, \ldots, e_n)$ form the subgroup $O(p)$. Thus we have

$$V(n, p) = O(n+p)/O(p).$$

The manifold $V(n, p)$ is called the *Stiefel manifold* of n-frames in $\mathbf{R}^{n+p}$.

Similarly, the isotropy subgroup of $O(n+p)$ at the point of $V(p, n)$ represented by the p-frame $(e_{n+1}, \ldots, e_{n+p})$ is $O(n)$. Hence $V(p, n)$ may be identified with $O(n+p)/O(n)$.

We have now the following three principal fibre bundles over $G(n, p)$:

$$\begin{aligned} E &= O(n+p) &\text{over}\quad& G(n, p) \text{ with group } O(n) \times O(p), \\ E' &= V(n, p) &\text{over}\quad& G(n, p) \text{ with group } O(n), \\ E'' &= V(p, n) &\text{over}\quad& G(n, p) \text{ with group } O(p). \end{aligned}$$

The projection in the bundle E' has the following geometric meaning. An n-frame projects upon the n-plane spanned by it. Similarly, the projection in the bundle E'' maps a p-frame into the n-plane normal to the p-plane spanned by it.

Let γ be the canonical 1-form of $O(n+p)$ with values in $\mathfrak{o}(n+p)$ (cf. §4 of Chapter I). Let ω_E be the $\mathfrak{o}(n) + \mathfrak{o}(p)$-component of γ with respect to the decomposition $\mathfrak{o}(n+p) = \mathfrak{o}(n) + \mathfrak{o}(p) + \mathfrak{g}(n, p)$ defined in §1. By Theorem 11.1 of Chapter

II, the form ω_E defines a connection in E, which will be called the *canonical connection* in E and will be denoted by Γ_E.

The natural projection of $O(n+p)$ onto $V(n, p)$ (resp. onto $V(p, n)$) defines a bundle homomorphism of E onto E' (resp. E'') which will be denoted by f' (resp. f''). By Proposition 6.1 of Chapter II, there exists a unique connection in E', denoted by $\Gamma_{E'}$ (resp. a unique connection in E'', denoted by $\Gamma_{E''}$) such that f' maps Γ_E into $\Gamma_{E'}$ (resp. f'' maps Γ_E into $\Gamma_{E''}$). By the same proposition, the connection forms $\omega_{E'}$ and $\omega_{E''}$ of $\Gamma_{E'}$ and $\Gamma_{E''}$ are determined by

$$f'^*(\omega_{E'}) = \gamma_{\mathfrak{o}(n)} \quad \text{and} \quad f''^*(\omega_{E''}) = \gamma_{\mathfrak{o}(p)},$$

where $\gamma_{\mathfrak{o}(n)}$ and $\gamma_{\mathfrak{o}(n)}$ are the $\mathfrak{o}(n)$- and $\mathfrak{o}(p)$-components of γ, respectively. We call $\Gamma_{E'}$ and $\Gamma_{E''}$ the *canonical connections* in E' and E'', respectively.

In §6 of Chapter II, we defined $E' + E''$ to be the restriction of $E' \times E''$ to the diagonal of $G(n, p) \times G(n, p)$. We have (cf. Proposition 6.3 of Chapter II).

PROPOSITION 2.1. *The mapping* $(f', f'') : E \to E' \times E''$ *induces a bundle isomorphism* $E \approx E' + E''$. *The canonical connection forms are related by*

$$\omega_E = f'^*(\omega_{E'}) + f''^*(\omega_{E''}).$$

Proof. The first statement is trivial. The second statement follows from

$$\omega_E = \gamma_{\mathfrak{o}(n)} + \gamma_{\mathfrak{o}(p)}.$$

QED.

Let M be an n-dimensional manifold immersed in the $(n+p)$-dimensional Euclidean space $\mathbf{R}^{n+p}$. Considering $\mathbf{R}^{n+p}$ as a flat Riemannian manifold we apply results of §1. In particular, we have the following three principal fibre bundles over M:

$P = O(\mathbf{R}^{n+p}, M)$ over M with group $O(n) \times O(p)$,
$P' = O(M) = O(\mathbf{R}^{n+p}, M)/O(p)$ over M with group $O(n)$,
$P'' = O(\mathbf{R}^{n+p}, M)/O(n)$ over M with group $O(p)$.

The connections in P, P', and P'' defined by ω, ω', and ω'' in §1 will be denoted by Γ_P, $\Gamma_{P'}$, and $\Gamma_{P''}$, respectively.

We now define a bundle map $g: P \to E$. The bundle $O(\mathbf{R}^{n+p})$ of orthonormal frames over $\mathbf{R}^{n+p}$ is trivial in a natural manner, that is, $O(\mathbf{R}^{n+p}) = \mathbf{R}^{n+p} \times O(n+p)$. Let $\rho: O(\mathbf{R}^{n+p}) \to O(n+p)$

be the natural projection. Since $P = O(\mathbf{R}^{n+p}, M) \subset O(\mathbf{R}^{n+p})$, every adapted frame $v \in P$ can be considered as an element of $O(\mathbf{R}^{n+p})$. We define

$$g(v) = \rho(v) \qquad \text{for } v \in P.$$

Evidently, g is a bundle map of P into $E = O(n + p)$, that is, g commutes with the right translations by $O(n) \times O(p)$. The bundle map g induces bundles maps $g' \colon P' \to E'$ and $g'' \colon P'' \to E''$ in a natural manner. It induces also a mapping $\tilde{g} \colon M \to G(n, p)$. We summarize various bundles and maps in the following commutative diagram:

$$\begin{array}{ccc} P' & \xrightarrow{g'} & E' \\ h' \uparrow & & \uparrow f' \\ P' + P'' = P & \xrightarrow{g} & E = E' + E'' \\ h'' \downarrow & & \downarrow f'' \\ P'' & \xrightarrow{g''} & E'' \end{array}$$

We are now in position to prove the main result of this section.

THEOREM 2.2. *The bundle maps g, g', and g'' map the connections Γ_P, $\Gamma_{P'}$, and $\Gamma_{P''}$ upon the canonical connections Γ_E, $\Gamma_{E'}$, and $\Gamma_{E''}$, respectively.*

Proof. Since f' (resp. f'') maps Γ_E upon $\Gamma_{E'}$ (resp. $\Gamma_{E''}$) and since h' (resp. h'') maps Γ_P upon $\Gamma_{P'}$ (resp. $\Gamma_{P''}$), it suffices to prove that g maps Γ_P upon Γ_E. The flat Riemannian connection of $\mathbf{R}^{n+p}$ is given by the form $\rho^*(\gamma)$ on $O(\mathbf{R}^{n+p})$ (cf. §9 of Chapter II), where $\rho \colon O(\mathbf{R}^{n+p}) = \mathbf{R}^{n+p} \times O(n + p) \to O(n + p)$ is the natural projection. The restriction of $\rho^*(\gamma)$ to P is given by $g^*(\gamma)$. The connection form ω on P is the $\mathfrak{o}(n) + \mathfrak{o}(p)$-component of $g^*(\gamma)$ and hence is equal to $g^*(\omega_E)$ since ω_E is the $\mathfrak{o}(n) + \mathfrak{o}(p)$-component of γ. QED.

We shall give a geometric interpretation of g, g', g'', and $\tilde{g}$. Let $y^1, \ldots, y^{n+p}$ be the coordinate system in $\mathbf{R}^{n+p}$ and let w_0 be the orthonormal frame at the origin given by $\partial/\partial y^1, \ldots, \partial/\partial y^{n+p}$. To each element $a \in O(n + p)$ we assign the frame aw_0 at the origin of $\mathbf{R}^{n+p}$. This gives a one-to-one correspondence between $O(n + p)$ and the set of orthonormal frames at the origin of $\mathbf{R}^{n+p}$. For each adapted frame $v \in P$ at $x \in M$, $g(v)$ is the frame at the origin of

$\mathbf{R}^{n+p}$ parallel to v; this follows from the fact that the mapping $\rho: O(\mathbf{R}^{n+p}) = \mathbf{R}^{n+p} \times O(n+p) \to O(n+p)$ is the parallel displacement of $O(\mathbf{R}^{n+p})$ into the frames at the origin. This interpretation of g implies that g' maps a frame $u \in O(M)$ upon the n-frame in $\mathbf{R}^{n+p}$ parallel to u. Similarly, g'' maps a normal frame u at $x \in M$ upon the p-frame in $\mathbf{R}^{n+p}$ parallel to u. Finally, $\tilde{g}$ maps $x \in M$ into the element of $G(n, p)$ represented by the n-dimensional subspace of $\mathbf{R}^{n+p}$ parallel to $T_x(M)$.

When we consider the orientation in $\mathbf{R}^{n+p}$, we take the following three principal fibre bundles over $\tilde{G}(n, p) = SO(n+p)/SO(n) \times SO(p)$:

$\tilde{E} = SO(n+p)$ over $\tilde{G}(n, p)$ with group $SO(n) \times SO(p)$,
$\tilde{E}' = SO(n+p)/SO(p)$ over $\tilde{G}(n, p)$ with group $SO(n)$,
$\tilde{E}'' = SO(n+p)/SO(n)$ over $\tilde{G}(n, p)$ with group $SO(p)$.

The base space $\tilde{G}(n, p)$ is the *Grassmann manifold of oriented n-planes* in $\mathbf{R}^{n+p}$. If M is orientable and oriented, taking those adapted frames which are compatible with the orientations of M and $\mathbf{R}^{n+p}$ we obtain a subbundle $\tilde{P}$ of P with group $SO(n) \times SO(p)$. We set $\tilde{P}' = \tilde{P}/SO(p)$ and $\tilde{P}'' = \tilde{P}/SO(n)$ and we define mappings $\tilde{h}', \tilde{h}'', \tilde{f}', \tilde{f}'', \tilde{g}, \tilde{g}',$ and $\tilde{g}''$ analogously. Then we obtain a theorem similar to Theorem 2.2. Moreover, if M is an oriented hypersurface in $\mathbf{R}^{n+1}$, then the bundle map $\tilde{g}: \tilde{P} \to \tilde{E}$ induces a map ϕ from the base space M of $\tilde{P}$ into the base space $\tilde{G}(n, 1) = S^n$ (n-sphere) of E, called the *spherical map of Gauss*. Geometrically speaking, ϕ assigns to each point $x \in M$ the unit vector at the origin of $\mathbf{R}^{n+1}$ which is parallel to the unit normal vector ξ_x of M at x chosen in such a way that ξ_x is compatible with the orientations of M and $\mathbf{R}^{n+1}$. We shall often identify $\phi(x)$ with ξ_x.

If M is not orientable, we can assign to each $x \in M$ the normal line, but we cannot obtain a continuous field of unit normal vectors, and thus we obtain the mapping $\tilde{g}$ of M into the real projective space $G(n, 1)$ induced by the bundle map $g: P \to E$.

Example 2.1. Let M be an oriented hypersurface in $\mathbf{R}^{n+1}$ as above. The bundle $\tilde{E} = \tilde{E}' = SO(n+1)$ over $\tilde{G}(n, 1) = S^n$ with group $SO(n)$ can be identified with the bundle of oriented orthonormal frames over S^n in a natural manner and the canonical connection in $\tilde{E}$ (or $\tilde{E}'$) can be identified with the Riemannian connection of S^n (cf. the proof of Theorem 3.1 of Chapter V). Then Theorem 2.2 means that the spherical map $\phi: M \to S^n$ together with the bundle map $g': \tilde{P}' \to \tilde{E}'$ sends the Riemannian connection of M into the Riemannian connection of S^n in the following sense. Let τ be a curve from $x \in M$ to $y \in M$. The parallel displacement $\tau: T_x(M) \to T_y(M)$ along τ corresponds to the parallel displacement $\phi(\tau): T_{\phi(x)}(S^n) \to T_{\phi(y)}(S^n)$ along the curve $\phi(\tau)$ as follows. Since both $T_x(M)$ and $T_{\phi(x)}(S^n)$ are perpendicular to ξ_x, they are parallel to each other in $\mathbf{R}^{n+1}$ and can be identified by a linear isomorphism $\psi_x: T_x(M) \to T_{\phi(x)}(S^n)$. Similarly, we have $\psi_y: T_y(M) \to T_{\phi(y)}(S^n)$. Then $\tau: T_x(M) \to T_y(M)$ coincides with $\psi_y^{-1} \circ \phi(\tau) \circ \psi_x$.

3. Covariant differentiation and second fundamental form

In this section we shall discuss the Riemannian connection of a submanifold by using the formalism of covariant differentiation $\nabla_X Y$. We shall also define the second fundamental form.

Let M be an n-dimensional manifold immersed in a Riemannian manifold N of dimension $n + p$. We denote by ∇' covariant differentiation in N. Since the discussion is local, we may assume, if we want, that M is imbedded in N and that we can choose p cross sections $\xi_1, \ldots, \xi_p$ of the normal bundle $T(M)^\perp$, namely, p differentiable fields of normal vectors, that are linearly independent at each point of M. They may be further assumed to be orthonormal at each point.

Let X and Y be vector fields on M. Since $(\nabla'_X Y)_x$ is defined for each $x \in M$, we shall denote by $(\nabla_X Y)_x$ its tangential component and by $\alpha_x(X, Y)$ its normal component so that

$$(\nabla'_X Y)_x = (\nabla_X Y)_x + \alpha_x(X, Y),$$

where
$$(\nabla_X Y)_x \in T_x(M) \quad \text{and} \quad \alpha_x(X, Y) \in T_x(M)^\perp.$$

Here $(\nabla_X Y)_x$ is introduced just as a symbol for the tangential component; the point is to show that it is in fact covariant differentiation for the Riemannian connection of M.

VII. SUBMANIFOLDS

It is easily verified that the vector field $\nabla_X Y$ which assigns the vector $(\nabla_X Y)_x$ to each point $x \in M$ is differentiable and that $\alpha(X, Y)$ is a differentiable field of normal vectors to M. We prove

PROPOSITION 3.1. $\nabla_X Y$ *is the covariant differentiation for the Riemannian connection of* M.

Proof. We verify the properties (1), (2), (3), and (4) in Proposition 2.8 of Chapter III. (1), (2), and (3) are obvious from the corresponding properties of ∇' on N and linearity of the projection $T_x(N) \to T_x(M)$. To verify (4), let f be a differentiable function on M. Then

$$\nabla'_X(fY) = (Xf)Y + f(\nabla'_X Y),$$

where $(Xf)Y$ is tangential to M. Thus taking the tangential components of both sides, we obtain

$$\nabla_X(fY) = (Xf)Y + f(\nabla_X Y),$$

proving property (4) for ∇. By Proposition 7.5 of Chapter III we see that there is a unique linear connection Γ on M for which $\nabla_X Y$ is the covariant differentiation. To show that Γ is the Riemannian connection for the induced metric on M, it is sufficient to show:

(a) the torsion tensor of Γ is 0, that is,

$$\nabla_X Y - \nabla_Y X = [X, Y],$$

(b) $\nabla g = 0$.

In order to prove (a), let us write

$$\nabla'_X Y = \nabla_X Y + \alpha(X, Y)$$

and

$$\nabla'_Y X = \nabla_Y X + \alpha(Y, X).$$

If we extend X and Y to vector fields X' and Y' on N (as we may do locally), then the restriction of $[X', Y']$ to M is tangent to M and coincides with $[X, Y]$. Thus

$$[X', Y']_x = [X, Y]_x, \quad \text{where} \quad x \in M.$$

Of course, we also have

$$\nabla'_{X'} Y' = \nabla'_X Y \quad \text{and} \quad \nabla'_{Y'} X' = \nabla'_Y X \quad \text{on} \quad M.$$

12   FOUNDATIONS OF DIFFERENTIAL GEOMETRY

From the equations above we obtain

$$\nabla'_{X'}Y' - \nabla'_{Y'}X' - [X', Y']$$
$$= \nabla_X Y - \nabla_Y X - [X, Y] + \alpha(X, Y) - \alpha(Y, X).$$

Since the left hand side is 0 (because the torsion tensor of the Riemannian connection ∇' of N is 0), we see that

$$\nabla_X Y - \nabla_Y X - [X, Y] = 0,$$

proving (a). Furthermore we have

$$\alpha(X, Y) = \alpha(Y, X).$$

To prove (b), we start from $\nabla' g = 0$, which implies

$$X \cdot g(Y, Z) = g(\nabla'_X Y, Z) + g(Y, \nabla'_X Z) \quad \text{on} \quad M$$

for any vector fields X, Y, and Z on M. We have, however,

$$g(\nabla'_X Y, Z) = g(\nabla_X Y + \alpha(X, Y), Z) = g(\nabla_X Y, Z),$$

because $\alpha(X, Y)$ is normal to M. Similarly, we have

$$g(Y, \nabla'_X Z) = g(Y, \nabla_X Z).$$

Thus we obtain

$$X \cdot g(Y, Z) = g(\nabla_X Y, Z) + g(Y, \nabla_X Z),$$

which means $\nabla g = 0$. We have thus proved Proposition 3.1.   QED.

We now prove the basic properties concerning the normal component $\alpha(X, Y)$. We denote by $\mathfrak{X}(M)^\perp$ the set of all differentiable fields of normal vectors to M; it is a real vector space and a module over the algebra $\mathfrak{F}(M)$ of differentiable functions on M.

PROPOSITION 3.2. *The mapping* $\alpha \colon \mathfrak{X}(M) \times \mathfrak{X}(M) \to \mathfrak{X}(M)^\perp$ *is symmetric (i.e., $\alpha(X, Y) = \alpha(Y, X)$) and bilinear over $\mathfrak{F}(M)$. Consequently, $\alpha_x(X, Y)$ depends only on X_x and Y_x, and there is induced a symmetric bilinear mapping α_x of $T_x(M) \times T_x(M)$ into $T_x(M)^\perp$.*

Proof. The symmetry of α has been proved in the proof of Proposition 3.1. Additivity in X or Y (when the other is fixed) is obvious. For any $f \in \mathfrak{F}(M)$, we have

$$\nabla_{fX} Y + \alpha(fX, Y) = \nabla'_{fX} Y = f \nabla'_X Y$$
$$= f(\nabla_X Y + \alpha(X, Y)),$$

which implies
$$\alpha(fX, Y) = f \cdot \alpha(X, Y).$$
By symmetry, we have $\alpha(X, fY) = f \cdot \alpha(X, Y)$, thus proving that α is bilinear over $\mathfrak{F}(M)$. The rest of Proposition 3.2 is similar to the situation in Proposition 3.1 of Chapter I.   QED.

We define $\alpha: \mathfrak{X}(M) \times \mathfrak{X}(M) \to \mathfrak{X}(M)^\perp$ as the *second fundamental form of M* (for the given immersion in N). For each $x \in M$, $\alpha_x: T_x(M) \times T_x(M) \to T_x(M)^\perp$ is called the *second fundamental form of M at x*.

In the case where M is a hypersurface immersed in N, choose a field of unit normal vectors ξ in a neighborhood U of a point $x_0 \in M$. For any vector fields X and Y on U, we may write
$$\alpha(X, Y) = h(X, Y)\xi,$$
where $h(X, Y)$ is a symmetric mapping of $\mathfrak{X}(U) \times \mathfrak{X}(U)$ into $\mathfrak{F}(U)$ which is bilinear over $\mathfrak{F}(U)$. At each $x \in U$, h_x is a symmetric bilinear function on $T_x(M) \times T_x(M)$. In classical literature, h is called the second fundamental form of M for the choice of ξ. If it is possible to choose a field of unit normal vectors ξ globally on M, then we can define h globally as a mapping of $\mathfrak{X}(M) \times \mathfrak{X}(M) \to \mathfrak{F}(M)$.

More generally, if M has codimension p, then we may locally choose p fields of unit normal vectors $\xi_1, \ldots, \xi_p$ that are orthogonal at each point. We may then express α by
$$\alpha(X, Y) = \sum_{i=1}^{p} h^i(X, Y)\xi_i,$$
thus getting p second fundamental forms in the classical sense.

Next, let $X \in \mathfrak{X}(M)$ and $\xi \in \mathfrak{X}(M)^\perp$ and write
$$(\nabla'_X \xi)_x = -(A_\xi(X))_x + (D_X \xi)_x,$$
where, for the moment, $-(A_\xi(X))$ and $D_X \xi$ are just symbols for the tangential and normal components that depend on X and ξ. It is easily verified that the vector field $x \to (A_\xi(X))_x$ and the field of normal vectors $x \to (D_X \xi)_x$ are differentiable on M. About A_ξ we prove

FOUNDATIONS OF DIFFERENTIAL GEOMETRY

PROPOSITION 3.3. (1) *The mapping* $(X, \xi) \in \mathfrak{X}(M) \times \mathfrak{X}(M)^\perp \to A_\xi(X) \in \mathfrak{X}(M)$ *is bilinear over* $\mathfrak{F}(M)$*; consequently,* $(A_\xi(X))_x$ *depends only on* X_x *and* ξ_x*, and there is induced a bilinear mapping of* $T_x(M) \times T_x(M)^\perp$ *into* $T_x(M)$*, where x is an arbitrary point of M.*
(2) *For each* $\xi \in T_x(M)^\perp$*, we have*

$$g(A_\xi(X), Y) = g(\alpha(X, Y), \xi)$$

for all $X, Y \in T_x(M)$*; consequently,* A_ξ *is a symmetric linear transformation of* $T_x(M)$ *with respect to* g_x.

Proof. (1): Additivity in X or ξ (when the other is fixed) is obvious. For any $f \in \mathfrak{F}(M)$, we have

$$\nabla'_X(f\xi) = f \cdot \nabla'_X \xi + (Xf)\xi = -f \cdot (A_\xi(X)) + f \cdot D_X \xi + (Xf)\xi,$$

Comparing this with

$$\nabla'_X(f\xi) = -A_{f\xi}(X) + D_X(f\xi),$$

we obtain $A_{f\xi}(X) = f A_\xi(X)$ for the tangential components and $D_X(f\xi) = (Xf)\xi + f \cdot D_X \xi$ for the normal components. (The second identity is used in the next proposition.) On the other hand, a similar argument for $\nabla'_{fX}(\xi)$ implies that $A_\xi(fX) = f \cdot A_\xi(X)$ This shows that $A_\xi(X)$ is bilinear in X and ξ over $\mathfrak{F}(M)$. (We have also $D_{fX}\xi = f \cdot D_X \xi$, which will be used for the next proposition.)

(2): For any $Y \in \mathfrak{X}(M)$, we have $g(Y, \xi) = 0$. Differentiating covariantly with respect to X (for the Riemannian connection ∇'), we have

$$g(\nabla'_X Y, \xi) + g(Y, \nabla'_X \xi) = 0$$

so that

$$g(\nabla_X Y + \alpha(X, Y), \xi) - g(Y, A_\xi(X) + D_X \xi) = 0.$$

Since $g(\nabla_X Y, \xi) = g(Y, D_X \xi) = 0$, we get

$$g(\alpha(X, Y), \xi) = g(A_\xi(X), Y).$$

This shows that A_ξ is the linear transformation of $T_x(M)$ which corresponds to the symmetric bilinear function α on $T_x(M) \times T_x(M)$. Thus A_ξ is symmetric: $g(A_\xi(X), Y) = g(X, A_\xi(Y))$, proving Proposition 3.3.   QED.

VII. SUBMANIFOLDS

As for D_X, we have

PROPOSITION 3.4. *The mapping* $(X, \xi) \in \mathfrak{X}(M) \times \mathfrak{X}(M)^\perp \to D_X \xi \in \mathfrak{X}(M)^\perp$ *coincides with covariant differentiation of the cross section ξ of the normal bundle $T(M)^\perp$ in the direction of X with respect to the connection in $T(M)^\perp$ defined in §1.*

Proof. In the preceding proof we have verified the same formal properties of D_X as those in Proposition 1.1 of Chapter III. Thus $D_X \xi$ is actually covariant differentiation of a certain linear connection in the normal bundle. Moreover, for $\xi, \eta \in \mathfrak{X}(M)^\perp$ we have

$$\nabla'_X \xi = -A_\xi(X) + D_X \xi, \quad \nabla'_X \eta = -A_\eta(X) + D_X \eta$$

so that

$$g(D_X \xi, \eta) + g(\xi, D_X \eta) = g(\nabla'_X \xi, \eta) + g(\xi, \nabla'_X \eta)$$
$$= X \cdot g(\xi, \eta),$$

which shows that our connection D_X is metric for the fibre metric in $T(M)^\perp$, namely, the restriction of g to the normal spaces.

It remains to show that the metric connection D_X coincides with the connection defined in §1. However, we shall omit the proof. QED.

We have thus developed the first set of basic formulas for submanifolds, namely,

(I) $$\nabla'_X Y = \nabla_X Y + \alpha(X, Y)$$
(II) $$\nabla'_X \xi = -A_\xi(X) + D_X \xi.$$

(I) is called *Gauss's formula*, and (II) is called *Weingarten's formula*.

In the case of a hypersurface M, (II) takes a simpler form. In fact, if we take a field of unit normal vector fields ξ, then differentiating $g(\xi, \xi) = 1$, we obtain

$$g(\nabla'_X \xi, \xi) = 0, \quad \text{and hence} \quad g(D_X \xi, \xi) = 0.$$

Since $D_X \xi$ is normal and is therefore a scalar multiple of ξ, we must have $D_X \xi = 0$ at each point. Thus $D_X \xi = 0$ (when $g(\xi, \xi) = 1$).

In the following examples, we shall discuss some special cases of (I) and (II) and also give some geometric consequences of Propositions 3.1–3.3.

Example 3.1. Let M_1 and M_2 be submanifolds, both of dimension n, in a Riemannian manifold N of dimension $n + p$. Let $\tau = x(t)$, $0 \leq t \leq 1$, be a differentiable curve in $M_1 \cap M_2$. We shall say that M_1 and M_2 are *tangent to each other along* τ if $T_{x(t)}(M_1) = T_{x(t)}(M_2)$ for each t, $0 \leq t \leq 1$. In this case, the parallel displacement along τ in M_1 coincides with the parallel displacement along τ in M_2. In fact, if $X = \dot{x}_t$, then for any vector field Y along τ, we have

$$\nabla'_X Y = \nabla^{(1)}_X Y + \alpha^{(1)}(X, Y) = \nabla^{(2)}_X Y + \alpha^{(2)}(X, Y),$$

where $\nabla^{(1)}$ (resp. $\nabla^{(2)}$) is the covariant differentiation for M_1 (resp. M_2) and $\alpha^{(1)}$ (resp. $\alpha^{(2)}$) is the second fundamental form for M_1 (resp. M_2). Now if we assume that Y is parallel along τ in M_1, then

$$\nabla^{(1)}_X Y = 0,$$

which implies that $\nabla'_X Y$ is normal to M_1 (and to M_2). This in turn means that $\nabla^{(2)}_X Y = 0$, that is, Y is parallel along τ in M_2. *In particular, if τ is a geodesic in M_1, τ is a geodesic in M_2.*

Example 3.2. Let M be a submanifold of N. Let $\tau = x_t$, $0 \leq t \leq 1$, be a curve in M. Then τ is a geodesic in M if and only if $\nabla'_X X$, where $X = \dot{x}_t$, is normal to M. In particular, if τ is a geodesic of N contained in M, it is a geodesic in M. (A geodesic in M is not in general a geodesic in N; we shall discuss this question in detail in §8.)

Example 3.3. Let M be a submanifold of dimension n in a Riemannian manifold N of dimension $n + p$. Let x_0 be a point of M. It is possible to take a system of normal coordinates $y^1, \ldots, y^{n+p}$ with origin x_0 such that $(\partial/\partial y^1)_{x_0}, \ldots, (\partial/\partial y^n)_{x_0}$ span $T_{x_0}(M)$. In fact, let $Y_1, \ldots, Y_n, Y_{n+1}, \ldots, Y_{n+p}$ be an orthonormal basis of $T_{x_0}(N)$ such that $Y_1, \ldots, Y_n$ form a basis of $T_{x_0}(M)$. We may choose a system of normal coordinates $y^1, \ldots, y^{n+p}$ such that $(\partial/\partial y^i)_{x_0} = Y_i$, $1 \leq i \leq n + p$ (cf. §3 of Chapter IV). Note that $Y_{n+1}, \ldots, Y_{n+p}$ form a basis of $T_{x_0}(M)^\perp$.

Let $x^1, \ldots, x^n$ be an arbitrary coordinate system in a neighborhood U of x_0 in M and let

$$y^i = y^i(x^1, \ldots, x^n), \qquad 1 \leq i \leq n + p,$$

be the system of equations that defines the imbedding of U into N. We shall show that

$$\alpha((\partial/\partial x^\lambda)_{x_0}, (\partial/\partial x^\mu)_{x_0}) = \sum_{k=n+1}^{n+p} (\partial^2 y^k/\partial x^\lambda \partial x^\mu)_{x_0} Y_k;$$

namely, the coefficients of α_{x_0} with respect to the basis $(\partial/\partial x^1)_{x_0}$, ..., $(\partial/\partial x^n)_{x_0}$ in $T_{x_0}(M)$ and the basis $Y_{n+1}, \ldots, Y_{n+p}$ in $T_{x_0}(M)^\perp$ are those of the Hessian $\partial^2 y^k/\partial x^\lambda \partial x^\mu$ at x_0. In order to prove this we compute:

$$\nabla'_{\partial/\partial x^\lambda}(\partial/\partial x^\mu) = \nabla'_{\partial/\partial x^\lambda}\left(\sum_{k=1}^{n+p}(\partial y^k/\partial x^\mu)(\partial/\partial y^k)\right)$$

$$= \sum_{k,l=1}^{n+p}(\partial y^k/\partial x^\mu)(\partial y^l/\partial x^\lambda)\nabla'_{\partial/\partial y^l}(\partial/\partial y^k)$$

$$+ \sum_{k=1}^{n+p}(\partial^2 y^k/\partial x^\lambda \partial x^\mu)(\partial/\partial y^k)$$

$$= \sum_{m,k,l=1}^{n+p}(\partial y^k/\partial x^\mu)(\partial y^l/\partial x^\lambda)\Gamma'^m_{kl}(\partial/\partial y^m)$$

$$+ \sum_{k=1}^{n+p}(\partial^2 y^k/\partial x^\lambda \partial x^\mu)(\partial/\partial y^k),$$

where Γ'^m_{kl} are the Cristoffel symbols for the Riemannian connection in N with respect to $y^1, \ldots, y^{n+p}$. Note that Γ'^m_{kl} are 0 at the origin x_0 of the normal coordinate system. Taking the normal components of both sides at x_0 of the equation above, we have

$$\alpha_{x_0}((\partial/\partial x^\lambda)_{x_0}, (\partial/\partial x^\mu)_{x_0}) = \sum_{k=n+1}^{n+p}(\partial^2 y^k/\partial x^\lambda \partial x^\mu)_{x_0} Y_k.$$

Example 3.4. Let $N = \mathbf{R}^{n+1}$, the Euclidean space with standard Euclidean metric, and let M be an n-dimensional submanifold of $\mathbf{R}^{n+1}$, that is, a hypersurface in $\mathbf{R}^{n+1}$. We may represent M locally by

$$y^i = y^i(x^1, \ldots, x^n), \qquad 1 \leq i \leq n+1,$$

where $y^1, \ldots, y^{n+1}$ is the standard rectangular coordinate system and $x^1, \ldots, x^n$ is an arbitrary local coordinate system of M. Or we may consider $y = (y^1, \ldots, y^{n+1})$ as the position vector of a point with coordinates $(y^1, \ldots, y^{n+1})$ and represent M locally by the equation for the vector-valued function:

$$y = y(x^1, \ldots, x^n).$$

For each i, $1 \leq i \leq n$, the vector field $\partial/\partial x^i$ can be expressed by the vector-valued function $e_i = \partial y/\partial x^i$. The induced metric g on M is given by
$$g_{ij} = g(\partial/\partial x^i, \partial/\partial x^j) = (e_i, e_j),$$
where $(\ ,\ )$ denotes the standard inner product in the vector space $\mathbf{R}^{n+1}$. Choose a field of unit normal vectors ξ on M which is represented by the vector-valued function $\xi(x^1, \ldots, x^n)$ on the coordinate neighborhood of M. Thus
$$(\xi, \xi) = 1 \quad \text{and} \quad (\xi, e_i) = 0, \qquad 1 \leq i \leq n.$$
Since the Riemannian connection of $\mathbf{R}^{n+1}$ is flat, we have
$$\nabla'_{e_i} e_j = \partial e_j/\partial x^i,$$
which is the partial derivative of the vector-valued function e_j. Thus the formula of Gauss can be written as

(I) $$\partial e_j/\partial x^i = \sum_{k=1}^n \Gamma^k_{ij} e_k + h_{ij} \xi,$$

where we know that Γ^k_{ij} are the Christoffel symbols for the Riemannian connection of the hypersurface M, that is,
$$\nabla_{\partial/\partial x^i}(\partial/\partial x^j) = \sum_{k=1}^n \Gamma^k_{ij}(\partial/\partial x^k)$$
and that h_{ij} are the coefficients of the second fundamental form. Similarly, the formula of Weingarten takes the form

(II) $$\partial \xi/\partial x^j = -\sum_{k=1}^n a^k_j e_k,$$

where (a^k_j) is the matrix representing $A = A_\xi$ with respect to $e_i = (\partial/\partial x^i)$, $1 \leq i \leq n$. As a special case of Proposition 3.3 (2) (or as can be directly verified), we have $g(Ae_i, e_j) = h(e_i, e_j)$, that is,
$$\sum_{k=1}^n g_{kj} a^k_i = h_{ij} \quad \text{and} \quad a^k_i = \sum_{j=1}^n g^{kj} h_{ji},$$
where (g^{kj}) is the inverse of the matrix (g_{kj}).

Example 3.5. Continuing Example 3.4, we shall reconsider the spherical map $\phi: x \in M \to \xi_x \in S^n$ defined in §2. For each point $x \in M$, the differential $(\phi_*)_x$ is a linear mapping of $T_x(M)$ into

$T_{\phi(x)}(S^n)$. Let us denote by p_x the natural linear isomorphism of $T_x(\mathbf{R}^{n+1})$ onto $T_{\phi(x)}(\mathbf{R}^{n+1})$; observe that p_x maps $T_x(M)$ onto $T_{\phi(x)}(S^n)$. We show that the linear transformation $p_x^{-1} \cdot (\phi^*)_x$ of $T_x(M)$ into itself coincides with $-A$ in Example 3.4. In fact, each $e_j = (\partial/\partial x^j)$ is mapped by $(\phi_*)_x$ upon the vector in $T_{\phi(x)}(S^n)$ represented by the vector $(\partial \xi/\partial x^j)_x$ in $\mathbf{R}^{n+1}$. Thus $p_x^{-1} \cdot (\phi_*)_x e_j$ is also represented by $(\partial \xi/\partial x^j)_x$. By (II) in Example 3.4, this means that $p_x^{-1} \cdot (\phi_*)_x$ maps e_j upon $-\Sigma_{k=1}^n a_j^k e_k$. Thus our mapping coincides with $-A$.

We may therefore say that, through the identification by p_x, $-A_x$ is nothing but the Jacobian of the spherical map. We may also relate the spherical map to parallel displacement on M (cf. Theorem 2.3). Let $x(t)$ be a curve on M and let $X_t = \dot{x}(t)$ be the field of tangent vectors of the curve. Suppose that Y_t is a field of vectors that are parallel along $x(t)$ on M. Then $p_{x(t)} \cdot Y_t$ is parallel along the curve $\phi(x(t))$ on S^n. To prove this, we write

$$dY_t/dt = \nabla_{X_t} Y_t + h(X_t, Y_t)\xi_{x(t)} = h(X_t, Y_t)\xi_{x(t)},$$

since $\nabla_{X_t} Y_t = 0$. This means that dY_t/dt is normal to M along $x(t)$. Since $p_{x(t)} \cdot Y_t$ is represented by the same $\mathbf{R}^{n+1}$-valued vector function of t as that for Y_t, and since $p_{x(t)} \cdot \xi_{x(t)}$ is equal to the unit normal vector to S^n at $\phi(x(t))$, it follows that $d(p_{x(t)} \cdot Y_t)/dt$ is normal to S^n along $\phi(x(t))$. This proves that $p_{x(t)} \cdot Y_t$ is parallel along $\phi(x(t))$ on S^n.

We shall conclude this section by expressing the second fundamental form of M in terms of the canonical form and the connection form of the bundle $O(N, M)$.

Let M be an n-dimensional manifold immersed in an $(n+p)$-dimensional Riemannian manifold M. Let φ and ψ be the canonical form and the Riemannian connection form on $O(N)$, respectively. We define an $\mathbf{R}^p$-valued quadratic form $\tilde{\alpha}$ on $O(N, M)$ as follows:

$$\tilde{\alpha}(X, Y) = \text{the } \mathbf{R}^p\text{-component of } \psi(X)\varphi(Y),$$

where X and Y are tangent vectors of $O(N, M)$ at a point $v \in O(N, M)$.

Take the natural basis for $\mathbf{R}^{n+p}$ so that the first n vectors of the basis span $\mathbf{R}^n$ and the last p vectors span $\mathbf{R}^p$. Then we can write φ and ψ in matrix forms (φ^A) and (ψ_B^A), respectively. We shall use the convention that indices $A, B, \ldots$ run from 1 to $n+p$, indices

$i, j, \ldots$, run from 1 to n, and indices $r, s, \ldots$, run from $n + 1$ to $n + p$. We know by Proposition 1.1 that, restricted to $O(N, M)$, the forms φ^r vanish identically. The first structure equation of $O(N)$ restricted to $O(N, M)$ yields

$$0 = d\varphi^r = -\sum_i \psi_i^r \wedge \varphi^i.$$

It follows that, restricted to $O(N, M)$, ψ_i^r can be written as follows:

$$\psi_i^r = \sum_j A_{ij}^r \varphi^j \quad \text{with} \quad A_{ij}^r = A_{ji}^r.$$

Hence $\tilde{\alpha} = (\tilde{\alpha}^r)$ can be written as follows:

$$\tilde{\alpha}^r = \sum_{i,j} A_{ij}^r \varphi^i \varphi^j.$$

PROPOSITION 3.5. *The second fundamental form α of an immersed manifold M in a Riemannian manifold N is related to the form $\tilde{\alpha}$ on $O(N, M)$ by*

$$\alpha(\pi X, \pi Y) = v(\tilde{\alpha}(X, Y)) \quad \text{for } X, Y \in T_v(O(N, M)),$$

where π denotes the projection $O(N, M) \to M$.

Proof. Since the statement to be proved is local, we may assume that $O(N, M)$ admits a cross section. We extend the given vector $Y \in T_v(O(N, M))$ to a vector field Y on $O(N) \mid M$ such that it is invariant by the action of the structure group $O(n + p)$ and that it is tangent to $O(N, M)$ at each point of $O(N, M)$. To construct such a vector field Y, take a cross section σ of $O(N, M)$ and extend first the given vector to a vector field on $\sigma(M)$ and then extend it further to a vector field on $O(N) \mid M$ by translating it with the action of $O(n + p)$. The restriction of Y to $O(N, M)$ will be denoted by the same letter Y. Then $\varphi(Y)$ may be considered as an $\mathbf{R}^{n+p}$-valued function defined on $O(N) \mid M$ or on $O(N, M)$ according as Y is considered as a vector field defined on $O(N) \mid M$ or on $O(N, M)$. From the definition of the canonical form φ, we have

$$\varphi(Y)_u = u^{-1}(\pi(Y_u)) \quad \text{for } u \in O(N) \mid M.$$

(We denote by π the projection of $O(N) \mid M$ onto M as well as the projection $O(N, M) \to M$). We are now in a position to apply the Lemma in the proof of Proposition 1.1 of Chapter III (p. 115, Volume I); $\varphi(Y)$ will play the role of f there.

Extend $\pi X \in T_{\pi(v)}(M)$ to a vector field on M and denote by X' its horizontal lift to $O(N) \mid M$ with respect to the connection defined by the restriction of ψ to $O(N) \mid M$. Then
$$(\nabla'_{\pi X'} \pi Y)_{\pi(v)} = v((X'(\varphi(Y)))_v).$$
Similarly, denote by X^* the horizontal lift of $\pi X'$ to $O(N, M)$ with respect to the connection defined by the $(\mathfrak{o}(n) + \mathfrak{o}(p))$-component of the restriction of ψ to $O(N, M)$. Then
$$(\nabla_{\pi X^*} \pi Y)_{\pi(v)} = v((X^*(\varphi(Y)))_v).$$
Since $\pi X = \pi(X'_v) = \pi(X^*_v)$, we have
$$\nabla'_{\pi X} \pi Y - \nabla_{\pi X} \pi Y = v(((X' - X^*)(\varphi(Y)))_v).$$
Since $\psi(X')_v = 0$, we have $\psi(X' - X^*)_v = -\psi(X^*)_v$. We set
$$A = \psi(X' - X^*)_v = -\psi(X^*)_v \in \mathfrak{o}(n + p).$$
The fundamental vector field A^* on $O(N) \mid M$ corresponding to A coincides with the vertical vector field $X' - X^*$ at v. To evaluate $(X' - X^*)(\varphi(Y))$ at v we consider therefore $A^*(\varphi(Y))$. By Proposition 3.11 of Chapter I, we have
$$A^*(\varphi(Y)) = Y(\varphi(A^*)) + \varphi([A^*, Y]) + 2d\varphi(A^*, Y).$$
Since A^* is vertical, we have $\varphi(A^*) = 0$. Since Y is invariant by $O(n + p)$, we have $[A^*, Y] = 0$. Hence
$$A^*(\varphi(Y)) = 2d\varphi(A^*, Y) = -\psi(A^*)\varphi(Y) + \psi(Y)\varphi(A^*)$$
$$= -\psi(A^*)\varphi(Y).$$
We evaluate the equality above at v. Since $A^*_v = (X' - X^*)_v$ and $\psi(A^*) = A = -\psi(X^*)_v$, we have
$$((X' - X^*)(\varphi(Y)))_v = (\psi(X^*)\varphi(Y))_v = (\tilde{\alpha}(X^*, Y))_v.$$
Hence
$$\nabla'_{\pi X} \pi Y - \nabla_{\pi X} \pi Y = v(((X' - X^*)(\varphi(Y)))_v) = v(\tilde{\alpha}(X^*, Y)_v).$$
Since $\tilde{\alpha}^r = \Sigma_{i,j} A^r_{ij} \varphi^i \varphi^j$, it follows that $(\tilde{\alpha}(X^*, Y))_v = (\tilde{\alpha}(X, Y))_v$. By definition,
$$\alpha(\pi X, \pi Y) = \nabla'_{\pi X} \pi Y - \nabla_{\pi X} \pi Y.$$
Hence,
$$\alpha(\pi X, \pi Y) = v(\tilde{\alpha}(X, Y)). \qquad \text{QED.}$$

4. Equations of Gauss and Codazzi

Let M be an n-dimensional Riemannian manifold which is isometrically immersed in an $(n + p)$-dimensional Riemannian manifold N. We shall first find a relationship between the curvature tensor fields of M and N.

Since the discussion is local, we choose p orthonormal fields of normal vectors $\xi_1, \ldots, \xi_p$ to M. Let h^i be the corresponding second fundamental forms and let $A_i = A_{\xi_i}$. Using the formulas of Gauss and Weingarten we obtain for any vector fields X, Y, and Z tangent to M

$$\begin{aligned}
\nabla'_X(\nabla'_Y Z) &= \nabla'_X(\nabla_Y Z + \sum h^i(Y, Z)\xi_i) \\
&= \nabla_X(\nabla_Y Z) + \sum h^i(X, \nabla_Y Z)\xi_i \\
&\quad + \sum X \cdot h^i(Y, Z)\xi_i \\
&\quad + \sum h^i(Y, Z)\{-A_i(X) + D_X \xi_i\}, \\
&= \nabla_X(\nabla_Y Z) - \sum h^i(Y, Z)A_i(X) \\
&\quad + \sum \{X \cdot h^i(Y, Z) + h^i(X, \nabla_Y Z)\}\xi_i \\
&\quad + \sum h^i(Y, Z)D_X \xi_i,
\end{aligned}$$

where the summation extends from 1 to p. In the last expression, the first two terms give the tangential component, and the last two terms the normal component.

For $\nabla'_Y(\nabla'_X Z)$ we may simply interchange X and Y in the equation above. We have also

$$\begin{aligned}
\nabla'_{[X,Y]}Z &= \nabla_{[X,Y]}Z + \sum h^i([X, Y], Z)\xi_i \\
&= \nabla_{[X,Y]}Z + \sum \{h^i(\nabla_X Y, Z) - h^i(\nabla_Y X, Z)\}\xi_i,
\end{aligned}$$

by virtue of $[X, Y] = \nabla_X Y - \nabla_Y X$ on M. Using these equations, we find that the tangential component of

$$R'(X, Y)Z = \nabla'_X(\nabla'_Y Z) - \nabla'_Y(\nabla'_X Z) - \nabla'_{[X,Y]}Z$$

is equal to

$$R(X, Y)Z + \sum \{h^i(X, Z)A_i(Y) - h^i(Y, Z)A_i(X)\}.$$

If W is tangent to M, then we get

$$\begin{aligned}
g(R'(X, Y)Z, W) &= g(R(X, Y)Z, W) + \sum \{h^i(X, Z)h^i(Y, W) \\
&\quad - h^i(Y, Z)h^i(X, W)\} \\
&= g(R(X, Y)Z, W) + g(\alpha(X, Z), \alpha(Y, W)) \\
&\quad - g(\alpha(Y, Z), \alpha(X, W)),
\end{aligned}$$

VII. SUBMANIFOLDS 23

since, for example,

$$\sum h^i(X, Z)g(A_i(Y), W) = \sum h^i(X, Z)h^i(Y, W)$$
$$= g(\alpha(X, Z), \alpha(Y, W))$$

because of orthonormality of $\xi_1, \ldots, \xi_p$. Thus the relationship between the Riemannian curvature tensors of N and M (cf. §2 of Chapter V) is given by

PROPOSITION 4.1 *(Equation of Gauss.)*

$$R'(W, Z, X, Y) = R(W, Z, X, Y) + g(\alpha(X, Z), \alpha(Y, W))$$
$$- g(\alpha(Y, Z), \alpha(X, W)),$$

where X, Y, Z, and W are arbitrary tangent vectors to M.

If we wish, we may state the equation of Gauss in terms of the curvature transformations as follows. For $X, Y, Z \in T_x(M)$, there is a unique element in $T_x(M)$, which we denote by $B(X, Y)Z$, such that

$$g(B(X, Y)Z, W) = g(\alpha(X, Z), \alpha(Y, W)) - g(\alpha(Y, Z), \alpha(X, W))$$

for every $W \in T_x(M)$. It is obvious that $B(X, Y)Z$ is trilinear in X, Y, and Z, and that $B(Y, X) = -B(X, Y)$. The equation of Gauss says that *the curvature transformation $R'(X, Y)$ followed by the projection: $T_x(N) \to T_x(M)$ is equal to $R(X, Y) + B(X, Y)$.*

COROLLARY 4.2. *If N is of constant sectional curvature k, then*

$$R(X, Y)Z = k\{g(Y, Z)X - g(X, Z)Y\} - B(X, Y)Z.$$

In particular, if $N = \mathbf{R}^{n+p}$ (with flat metric), then

$$R(X, Y) = -B(X, Y).$$

Proof. This follows from the expression of R' given in Corollary 2.3 of Chapter V. QED.

Example 4.1. If the codimension p is 1, that is, if M is a hypersurface of N, then we have

$$B(X, Y)Z = h(X, Z)AY - h(Y, Z)AX$$
$$= g(AX, Z)AY - g(AY, Z)AX.$$

In particular, if $N = \mathbf{R}^{n+1}$, then we have
$$R(X, Y)Z = g(AY, Z)AX - g(AX, Z)AY.$$
This is the classical equation of Gauss (for $n = 2$). Note that the right hand side is independent of the choice of a unit normal field.

Example 4.2. Let M be a hypersurface of $\mathbf{R}^{n+1}$. For a choice of a unit normal field ξ, $A = A_\xi$ is a symmetric transformation of $T_x(M)$. Thus there exists an orthonormal basis $X_1, \ldots, X_n$ in $T_x(M)$ such that $AX_i = \lambda_i X_i$, $1 \leq i \leq n$, where $\lambda_1, \ldots, \lambda_n$ are the eigenvalues of A. For any pair (i, j), where $i < j$, we have

$$\begin{aligned} R(X_i, X_j)X_k &= g(AX_j, X_k)AX_i - g(AX_i, X_k)AX_j \\ &= \lambda_i \lambda_j \delta_{jk} X_i - \lambda_i \lambda_j \delta_{ik} X_j \\ &= \begin{cases} 0 & \text{if } k \neq i, j \\ -\lambda_i \lambda_j X_j & \text{if } k = i \\ \lambda_i \lambda_j X_i & \text{if } k = j. \end{cases} \end{aligned}$$

Thus $R(X_i, X_j)$ is represented by the skew-symmetric matrix

$$\begin{array}{c} \;\;\;i\;\;\;\;\;j \\ \begin{array}{c} i \\ j \end{array}\!\!\left(\begin{array}{cc} & \lambda_i \lambda_j \\ -\lambda_i \lambda_j & \end{array} \right) \end{array}$$

with respect to $X_1, \ldots, X_n$. Let $M = S^n(r)$ be the hypersphere of radius $r > 0$ and with origin 0 in $\mathbf{R}^{n+1}$. If we choose the outward unit normal $\xi_x = x/\|x\|$ for $x \in M$, then $A_\xi = -(1/r)I$ by Example 3.5, where I denotes the identity transformation of $T_x(M)$. By the classical equation of Gauss, we have
$$R(X, Y)Z = (1/r^2)(g(Y, Z)X - g(X, Z)Y),$$
which shows that $S^n(r)$ has constant sectional curvature $1/r^2$. (This result was established in Theorem 3.2 of Chapter V by a different method; our present method is classical and more elementary.)

Let us now look at the normal component of $R'(X, Y)Z$ for an arbitrary submanifold M in N. Using the expression for $\nabla'_X(\nabla'_Y Z)$, $\nabla'_Y(\nabla'_X Z)$, and $\nabla'_{[X,Y]} Z$ and observing that
$$X \cdot h^i(Y, Z) - h^i(\nabla_X Y, Z) - h^i(Y, \nabla_X Z) = (\nabla_X h^i)(Y, Z),$$

we see that the normal component of $R'(X, Y)Z$ is equal to

$$\sum \{(\nabla_X h^i)(Y, Z) - (\nabla_Y h^i)(X, Z)\} \xi_i$$
$$+ \sum \{h^i(Y, Z) D_X \xi_i - h^i(X, Z) D_Y \xi_i\}.$$

Actually, we may give a simpler expression to the normal vector above. For the second fundamental form α, we define the covariant derivative, denoted by $\tilde{\nabla}_X \alpha$, to be

$$(\tilde{\nabla}_X \alpha)(Y, Z) = D_X(\alpha(Y, Z)) - \alpha(\nabla_X Y, Z) - \alpha(Y, \nabla_X Z).$$

(This is the covariant derivative of α with respect to the connection in $T(M) + T(M)^\perp$ obtained by combining the connections ∇_X in $T(M)$ and D_X in $T(M)^\perp$; see Proposition 6.3 of Chapter II and Proposition 1.2 of the present chapter, although we shall not elaborate on this matter.) Using $\xi_1, \ldots, \xi_p$, we have

$$(\tilde{\nabla}_X \alpha)(Y, Z) = D_X(\sum h^i(Y, Z) \xi_i)$$
$$- \sum \{h^i(\nabla_X Y, Z) + h^i(Y, \nabla_X Z)\} \xi_i$$
$$= \sum X \cdot h^i(Y, Z) \xi_i + \sum h^i(Y, Z) D_X \xi_i$$
$$- \sum \{h^i(\nabla_X Y, Z) + h^i(Y, \nabla_X Z)\} \xi_i$$
$$= \sum (\nabla_X h^i)(Y, Z) \xi_i + \sum h^i(Y, Z) D_X \xi_i.$$

It follows that the normal component of $R'(X, Y)Z$ is expressed by $(\tilde{\nabla}_X \alpha)(Y, Z) - (\tilde{\nabla}_Y \alpha)(X, Z)$. We have thus

PROPOSITION 4.3 (*Equation of Codazzi*). *The normal component of $R'(X, Y)Z$ is given by*

$$(\tilde{\nabla}_X \alpha)(Y, Z) - (\tilde{\nabla}_Y \alpha)(X, Z)$$
$$= \sum \{(\nabla_X h^i)(Y, Z) - (\nabla_Y h^i)(X, Z)\} \xi_i$$
$$+ \sum \{h^i(Y, Z) D_X \xi_i - h^i(X, Z) D_Y \xi_i\}.$$

COROLLARY 4.4. *If N is of constant sectional curvature, then we have*

$$(\tilde{\nabla}_X \alpha)(Y, Z) = (\tilde{\nabla}_Y \alpha)(X, Z).$$

Proof. By Corollary 2.3 of Chapter V, we know that $R'(X, Y)Z$ is tangent to M; hence its normal component is 0. QED.

Example 4.3. If $N = \mathbf{R}^{n+1}$ and if M is a hypersurface, then we have $D_X \xi = 0$, and hence

$$(\nabla_X h)(Y, Z) = (\nabla_Y h)(X, Z),$$

or, equivalently,

$$(\nabla_X A)(Y) = (\nabla_Y A)(X).$$

This is the classical equation of Codazzi (when $n = 2$).

Remark. At the end of §3, we showed how the second fundamental form α can be expressed in terms of the canonical form φ and the connection form ψ on the bundle $O(N, M)$. We shall indicate how the equations of Gauss and Codazzi can be derived from that viewpoint. Using the same convention as in §3 that $1 \leq A, B, \ldots \leq n + p$, while $1 \leq i, j, \ldots \leq n$ and $n + 1 \leq r, s, \ldots \leq n + p$, we express the second structure equation of N as follows:

$$\Psi^A_B = d\psi^A_B + \sum_k \psi^A_k \wedge \psi^k_B + \sum_r \psi^A_r \wedge \psi^r_B,$$

where (Ψ^A_B) is the curvature form of N. Setting $A = i$ and $B = j$ and restricting this second structure equation to $O(N, M)$, we obtain

$$\Psi^i_j = \Omega^i_j + \sum_r \psi^i_r \wedge \psi^r_j,$$

where (Ω^i_j) is the curvature form of M (lifted from $O(M)$ to $O(N, M)$). Since $\psi^i_r = -\psi^r_i = -\Sigma_k A^r_{ik} \varphi^k$ and $\psi^r_j = \Sigma_l A^r_{jl} \varphi^l$, we have

$$\Psi^i_j = \Omega^i_j - \tfrac{1}{2} \sum_{r,k,l} (A^r_{ik} A^r_{jl} - A^r_{il} A^r_{jk}) \varphi^k \wedge \varphi^l.$$

This is equivalent to Proposition 4.1 (Equation of Gauss). Similarly, setting $A = s$ and $B = j$ in the second structure equation of N and restricting it to $O(N, M)$, we obtain an equation equivalent to Proposition 4.3 (Equation of Codazzi).

In the rest of the section, we shall prove some theorems involving sectional curvature. First we prove preparatory results.

PROPOSITION 4.5. *Let M be an n-dimensional submanifold immersed in an $(n + p)$-dimensional Riemannian manifold N. Let X and Y be a*

VII. SUBMANIFOLDS

pair of orthonormal vectors in $T_x(M)$, *where* $x \in M$. *For the plane* $X \wedge Y$ *spanned by* X *and* Y, *we have*

$$k_N(X \wedge Y) = k_M(X \wedge Y) + g(\alpha(X, Y), \alpha(X, Y))$$
$$- g(\alpha(X, X), \alpha(Y, Y)),$$

where k_N (*resp.* k_M) *denotes the sectional curvature in* N (*resp.* M). *In particular, if* $N = \mathbf{R}^{n+p}$, *then*

$$k_M(X \wedge Y) = g(\alpha(X, X), \alpha(Y, Y)) - g(\alpha(X, Y), \alpha(X, Y)).$$

Proof. This is immediate from the equation of Gauss in Proposition 4.1, since

$$k_N(X \wedge Y) = R'(X, Y, X, Y) \quad \text{and} \quad k_M(X \wedge Y) = R(X, Y, X, Y).$$

Q.E.D.

PROPOSITION 4.6. *Let* M *be an n-dimensional compact manifold immersed in* $\mathbf{R}^{n+p}$. *Then there is a point* $x_0 \in M$ *such that*

$$\alpha(X, X) \neq 0 \quad \text{for any } X \in T_{x_0}(M), X \neq 0.$$

Proof. Let $y(x)$ denote the position vector for the point in $\mathbf{R}^{n+p}$ corresponding to $x \in M$. Let $\varphi(x) = (y(x), y(x))/2$, where (,) is the Euclidean inner product. The differentiable function φ on M takes a maximum, say, at x_0. Namely, x_0 is a point such that the point $y(x_0) = y_0$ is of maximum distance from the origin in $\mathbf{R}^{n+p}$. The following argument is valid if φ has a local maximum at x_0; hence we now assume that a neighborhood U of x_0 is imbedded in $\mathbf{R}^{n+p}$, and we identify $x \in U$ with $y(x)$, so that $\varphi(x) = (x, x)$. For any vector field X on U, Xx (that is, X applied to the vector-valued function x) is the vector-valued function which expresses X.

This being said, we have $X\varphi = (X, x)$ and this is 0 at x_0. Thus $(X_{x_0}, x_0) = 0$. Since X is arbitrary, it shows that the vector x_0 is normal to M at x_0. We have further

$$X^2\varphi = (\nabla'_X X, x) + (X, X) = (\alpha(X, X), x) + g(X, X) \quad \text{at } x_0,$$

since $\nabla'_X X = \nabla_X X + \alpha(X, X)$. Since φ has a local maximum at x_0, it follows that $X^2\varphi \leq 0$ at x_0. Thus we obtain

$$(\alpha(X_{x_0}, X_{x_0}), x_0) + g(X_{x_0}, X_{x_0}) \leq 0,$$

that is,

$$(\alpha(X_{x_0}, X_{x_0}), x_0) \leqq -g(X_{x_0}, X_{x_0}) < 0, \text{ and } \alpha(X_{x_0}, X_{x_0}) \neq 0$$

for $X_{x_0} \neq 0$. This proves that $\alpha(X, X) \neq 0$ for every non-zero $X \in T_{x_0}(M)$. QED.

We now state the main result.

THEOREM 4.7. *Let M be an n-dimensional compact Riemannian manifold isometrically immersed in $\mathbf{R}^{n+p}$. If, at every point x of M, the tangent space $T_x(M)$ contains an m-dimensional subspace T'_x such that the sectional curvature for any plane in T'_x is non-positive, then we have $p \geqq m$.*

Proof. By Proposition 4.6, there exists a point x of M such that $\alpha(X, X) \neq 0$ for any non-zero vector X in $T_x(M)$. Consider the restriction of α to $T'_x \times T'_x$. By assumption, the sectional curvature $k(X \wedge Y)$ is non-positive for any plane $X \wedge Y$ in $T'_x(M)$. By Proposition 4.5, we have

$$g(\alpha(X, X), \alpha(Y, Y)) - g(\alpha(X, Y), \alpha(X, Y)) \leqq 0$$

whenever X and Y are a pair of orthonormal vectors in $T'_x(M)$. Actually, the preceding inequality holds for all X and Y in $T'_x(M)$ by Proposition 1.3 of Chapter V (or as can be directly verified by orthonormalizing X and Y). Our conclusion will therefore follow from the following lemma.

LEMMA. *Let $\alpha: \mathbf{R}^m \times \mathbf{R}^m \to \mathbf{R}^p$ be a symmetric bilinear mapping and let g be a positive definite inner product in $\mathbf{R}^p$. If*

(1) $\qquad g(\alpha(x, x), \alpha(y, y)) - g(\alpha(x, y), \alpha(x, y)) \leqq 0$

for all $x, y \in \mathbf{R}^m$ and if

(2) $\qquad\qquad\qquad \alpha(x, x) \neq 0$

for all non-zero $x \in \mathbf{R}^m$, then we have $p \geqq m$.

Proof. We may extend α to a symmetric complex bilinear mapping of $\mathbf{C}^m \times \mathbf{C}^m \to \mathbf{C}^p$. Consider the equation

$$\alpha(z, z) = 0.$$

Since α is $\mathbf{C}^p$-valued, this equation is equivalent to a system of p quadratic equations:

$$\alpha^1(z, z) = 0, \ldots, \alpha^p(z, z) = 0.$$

Suppose $p < m$. Then the above system of equations has a non-zero solution z. By assumption (2), z is not contained in $\mathbf{R}^m$. Let $z = x + \sqrt{-1}\,y$, where $x, y \in \mathbf{R}^m$ and $y \neq 0$. Since

$$0 = \alpha(z, z) = \alpha(x, x) - \alpha(y, y) + 2\sqrt{-1}\,\alpha(x, y),$$

we have

$$\alpha(x, x) = \alpha(y, y) \neq 0 \quad \text{and} \quad \alpha(x, y) = 0.$$

This pair of vectors x and y contradict our assumption (1). QED.

Remark. Since we make use of the curvature, the immersion in Theorem 4.7 must be at least of class C^3. Chern and Kuiper [1] conjectured the lemma above and showed that it implied Theorem 4.7. The lemma was then proved by Otsuki [1]. The proof above is due to T. A. Springer.

From Theorem 4.7, we obtain

COROLLARY 4.8. *An n-dimensional compact Riemannian manifold with non-positive sectional curvature cannot be isometrically immersed into* $\mathbf{R}^{2n-1}$.

In particular, we have the following result of Tompkins [1]:

COROLLARY 4.9. *An n-dimensional compact flat Riemannian manifold cannot be isometrically immersed into* $\mathbf{R}^{2n-1}$.

Corollary 4.8 has been generalized by O'Neill [2] as follows. Let M be a compact n-dimensional Riemannian manifold and $\bar{M}$ a complete simply connected Riemannian manifold of dimension less than $2n$. If the sectional curvatures K and $\bar{K}$ of M and $\bar{M}$ satisfy $K \leq \bar{K} \leq 0$, then M cannot be isometrically immersed in $\bar{M}$. As O'Neill indicates, Theorem 4.7 itself can be generalized in a similar manner.

For other applications of the idea of Chern-Kuiper in Theorem 4.7, see Otsuki [2], [3].

5. *Hypersurfaces in a Euclidean space*

In this and the following sections we shall deal with hypersurfaces in a Euclidean space. We shall denote by M an n-dimensional manifold isometrically immersed in $\mathbf{R}^{n+1}$. When we want to emphasize a given immersion f, we shall write (M, f) instead of M alone.

We summarize the basic formulas for M. In the following, $X, Y, Z, \ldots$ will denote vector fields tangent to M, and ξ is a field of unit normal vectors defined locally on M (that is, in a neighborhood of a point in question) or globally on M if this is the case. The formulas of Gauss and Weingarten are

(I) $$\nabla'_X Y = \nabla_X Y + h(X, Y)\xi$$

and

(II) $$\nabla'_X \xi = -AX,$$

where ∇' denotes covariant differentiation in $\mathbf{R}^{n+1}$, and $A = A_\xi$ is the symmetric transformation of each tangent space $T_x(M)$ corresponding to the symmetric bilinear function h on $T_x(M) \times T_x(M)$. The equations of Gauss and Codazzi (Examples 4.1 and 4.3) are

(III) $$R(X, Y)Z = g(AY, Z)AX - g(AX, Z)AY$$

and

(IV) $$(\nabla_X A)(Y) = (\nabla_Y A)(X).$$

A point x of M is called an *umbilic* (or an *umbilical point*) if A_x is equal to λI, where λ is a scalar and I is the identity transformation. This property is, of course, independent of whether we use ξ or $-\xi$ in a neighborhood of x.

Example 5.1. If M is a hyperplane in $\mathbf{R}^{n+1}$, then A is identically equal to 0. If M is a hypersphere of radius r in $\mathbf{R}^{n+1}$, then $A = -I/r$ (cf. Example 4.2). In both cases, every point of M is umbilical.

We shall prove the converse of this Example in the following global form.

THEOREM 5.1. *Let (M, f) be a connected hypersurface in $\mathbf{R}^{n+1}$, where M is complete (as a Riemannian manifold). If every point x is umbilical, then either $f(M)$ is a hyperplane or $f(M)$ is a hypersphere of a certain radius. In both cases, f is an isometric imbedding.*

Proof. We first prove a local result. Let U be a coordinate neighborhood of a point x_0 on which f is one-to-one. We show that $f(U)$ is either part of a hyperplane or part of a hypersphere. Since every point of U is umbilical, there exists a real-valued function $\lambda(x)$ on U such that $A = \lambda I$ at each point of U. The tensor field A being differentiable, trace $A = n\lambda$ is differentiable, which shows that the function λ is differentiable. We shall prove that it is, in fact, a constant function. For any vector fields X and Y, we have

$$(\nabla_X A)(Y) = \nabla_X(AY) - A(\nabla_X Y) = \nabla_X(\lambda Y) - A(\nabla_X Y)$$
$$= (X\lambda)Y + \lambda(\nabla_X Y) - A(\nabla_X Y) = (X\lambda)Y.$$

Similarly,
$$(\nabla_Y A)(X) = (Y\lambda)X.$$

By the equation of Codazzi we obtain

$$(X\lambda)Y - (Y\lambda)X = 0.$$

For each $x \in U$, we may choose vector fields X and Y in U so that X_x and Y_x are linearly independent. Then the preceding equation implies that $X_x\lambda = Y_x\lambda = 0$. It follows that $Z\lambda = 0$ for every $Z \in T_x(M)$. This implies that λ is equal to a constant on U.

Identifying $x \in U$ with the position vector of the corresponding point in $\mathbf{R}^{n+1}$ and expressing the unit normal ξ by the vector in $\mathbf{R}^{n+1}$, we consider $x + \xi_x$ as an $\mathbf{R}^{n+1}$-valued vector function on U. If we further identify a tangent vector $X \in T_x(M)$ with the vector $f_*(X)$ in $\mathbf{R}^{n+1}$, we have

$$\nabla'_X(\lambda x + \xi) = \lambda X + (\nabla'_X \xi)$$
$$= \lambda X + (-\lambda X) = 0,$$

which shows that $\lambda x + \xi$ is a constant vector, say, a, in $\mathbf{R}^{n+1}$.

If $\lambda = 0$, then we have for any vector field X

$$X \cdot (x - x_0, a) = (X, a) = 0,$$

where $(\ ,\ )$ denotes the Euclidean inner product. This shows that the vector $x - x_0$ in $\mathbf{R}^{n+1}$ is perpendicular to the constant vector a so that $f(U)$ lies on the hyperplane through x_0 and perpendicular to a.

If $\lambda \neq 0$, then $\lambda x + \xi_x = a$ implies $x - a/\lambda = -\xi_x/\lambda$ and hence $\|x - a/\lambda\| = |1/\lambda|$, which shows that $f(U)$ lies on the hypersphere with center a/λ and of radius $1/|\lambda|$. This completes the proof of the local result.

We now complete the proof of Theorem 5.1. Let x_0 be a point of M and assume that a neighborhood U of x_0 exists such that $f(U)$ lies on a hyperplane Π in $\mathbf{R}^{n+1}$ (or on a hypersphere Σ in $\mathbf{R}^{n+1}$; the second case can be handled exactly in the same way as the first case). We show that $f(M)$ then lies in Π. If M_1 denotes the set of all points x_1 of M such that there is a neighborhood U_1 of x_1 for which $f(U_1)$ lies in Π, then M_1 is a non-empty open subset of M. If $x \in M$ is a limit point of M_1, let V be a neighborhood of x such that $f(V)$ lies either on a certain hyperplane or a certain hypersphere. Since V contains a point x_1 of M_1, there is a neighborhood U_1 of x_1, which may be assumed to be an open subset of V, such that $f(U_1)$ lies on Π. This means that $f(V)$ must lie on Π. This argument proves that $x_1 \in M_1$; that is, M_1 is closed. Since M is connected, we have $M_1 = M$; that is, $f(M) \subset \Pi$.

Finally, consider the isometric immersion f of M into the hyperplane Π (or the hypersphere Σ). By Theorem 4.6 of Chapter IV, we conclude that f is a covering map of M onto Π (or Σ). Since Π (or Σ) is simply connected, we conclude that f is one-to-one and $f(M) = \Pi$ (or Σ), proving Theorem 5.1.   QED.

We shall define a few notions for a hypersurface. At each $x \in M$, let $\lambda_1, \ldots, \lambda_n$ be the eigenvalues (some of which may coincide) of the symmetric transformation A_x of the tangent space $T_x(M)$. We take an orthonormal basis $X_1, \ldots, X_n$ of $T_x(M)$ such that $AX_i = \lambda_i X_i$, $1 \leq i \leq n$. In Example 4.2 we have seen that the curvature transformation $R(X_i, X_j)$ is represented by the skew-symmetric matrix

$$\begin{matrix} & i & j \\ i & \begin{pmatrix} & \lambda_i \lambda_j \\ -\lambda_i \lambda_j & \end{pmatrix} \\ j & \end{matrix}.$$

From Proposition 4.5, we see that *the sectional curvature* $k(X_i \wedge X_j)$ *of the plane* $X_i \wedge X_j$ *is given by*

$$k(X_i \wedge X_j) = \lambda_i \lambda_j.$$

The eigenvalues $\lambda_1, \ldots, \lambda_n$ are traditionally called the *principal curvatures* at x. If they are all distinct, then the corresponding eigenvectors of unit length are determined up to a sign, and they are called the *principal directions* at x.

Let $\sigma_1(t), \ldots, \sigma_n(t)$ be the elementary symmetric functions of $t = (t_1, \ldots, t_n)$, namely,

$$\sigma_1(t) = \sum_{i=1}^{n} t_i, \qquad \sigma_2(t) = \sum_{i<j} t_i t_j,$$

$$\sigma_3(t) = \sum_{i<j<k} t_i t_j t_k, \ldots$$

$$\sigma_n(t) = t_1 t_2 \cdots t_n.$$

They appear as the coefficients of the characteristic polynomial $\det(\lambda I - A)$ of a symmetric matrix A with eigenvalues $t_1, \ldots, t_n$ in the following fashion:

$$\det(\lambda I - A) = \lambda^n + (-1)\sigma_1 \lambda^{n-1} + \sigma_2 \lambda^{n-2} + (-1)\sigma_3 \lambda^{n-3}$$
$$+ \cdots + (-1)^n \sigma_n.$$

We define at each point x of M

$$K_i = \sigma_i(\lambda_1, \ldots, \lambda_n), \qquad 1 \leq i \leq n.$$

Observe that $\lambda_1, \ldots, \lambda_n$ are determined up to a sign common to all of them, because $A = A_\xi$ is determined up to a sign depending on whether we use ξ or $-\xi$ for a field of unit normals. This means that $K_1, K_3, K_5, \ldots$ are determined up to a sign whereas $K_2, K_4, K_6, \ldots$ are defined uniquely.

In particular,

$$K_1/n = \sum_{i=1}^{n} \lambda_i/n$$

is called the *mean curvature* of M at x (defined up to a sign). $K_n = \lambda_1 \lambda_2 \cdots \lambda_n$ is called the *Gaussian curvature* (or *Gauss-Kronecker curvature*) of M at x (again defined up to a sign if n is odd, and uniquely determined if n is even).

Example 5.2. For the case where $n = 2$, the Gaussian curvature $K_2 = \lambda_1 \lambda_2$ is the classical definition of the Gaussian curvature for a surface. It coincides with the sectional curvature at x (for the unique choice of the tangent plane); thus it is completely determined by the metric on the surface—this result is the so-called *Theorema egregium* of Gauss.

Example 5.3. For a surface $(n = 2)$, the mean curvature $(\lambda_1 + \lambda_2)/2$ is traditionally denoted by H. For a sphere of radius r, we have $H = 1/r$ (hence a constant), if we choose an inward normal at each point. In order to eliminate the indeterminacy of sign for the mean curvature, we may define the *mean curvature normal* at each point x of a hypersurface. Choose a field of unit normals ξ in a neighborhood of x, and define $\eta_x = (H)_x \xi_x$. If we change the sign of ξ, then H will change the sign and, consequently, η_x remains unchanged. Thus a unique normal vector η_x is defined at each point. We call η the *mean curvature normal*.

Remark. More generally, we may define the mean curvature normal for an n-dimensional Riemannian manifold M isometrically immersed in an $(n + p)$-dimensional Riemannian manifold N. Let $x \in M$. By Proposition 3.3, we have a mapping $\xi \in T_x(M)^\perp \to A_\xi$. Thus $1/n$ (trace A_ξ) is a linear function on $T_x(M)^\perp$. There is a unique element in $T_x(M)^\perp$, say, η, such that

$$\frac{1}{n} (\text{trace } A_\xi) = g(\xi, \eta) \qquad \text{for every } \xi \in T_x(M)^\perp.$$

We call η the *mean curvature normal at x*. If $\xi_1, \ldots, \xi_p$ is an orthonormal basis in $T_x(M)^\perp$, and if $A_i = A_{\xi_i}$, then

$$\text{trace } A_i = g(\xi_i, \eta), \qquad 1 \leq i \leq p,$$

so that

$$\eta = \frac{1}{n} \sum_{i=1}^{p} (\text{trace } A_i) \xi_i.$$

In particular, for $p = 1$ we have the definition given in Example 5.3 in the case of $N = \mathbf{R}^{n+1}$. M is said to be *minimal* in N (for the immersion) if the mean curvature normal vanishes at each point. We may prove the following: *There is no compact minimal submanifold in a Euclidean space*. From the proof of Proposition 4.6, we know that there is a point x_0 of the compact submanifold such that the position vector x_0 is normal and such that $(\alpha(X, X), x_0) < 0$ for every tangent vector $X \neq 0$ at x_0. If we let $\xi = x_0$, then from Proposition 3.3 (2), we have

$$g(A_\xi(X), X) = g(\alpha(X, X), \xi) < 0$$

for every tangent vector $X \neq 0$, which implies that A_ξ is negative-definite. Thus trace A_ξ cannot be 0.

Remark. Let $\lambda_1, \ldots, \lambda_n$ be the principal curvatures of a hypersurface M at $x \in M$, and let $X_1, \ldots, X_n$ be the corresponding orthonormal basis for $T_x(M)$ so that $AX_i = \lambda_i X_i$ for $i = 1, \ldots, n$.

VII. SUBMANIFOLDS

The Riemannian curvature tensor R may be considered as a symmetric bilinear mapping $\bigwedge^2 T_x(M) \times \bigwedge^2 T_x(M) \to \mathbf{R}$ at each point $x \in M$. Let $\hat{R}$ be the corresponding symmetric linear transformation $\bigwedge^2 T_x(M) \to \bigwedge^2 T_x(M)$. By the equation of Gauss (cf. Example 4.1) we have

$$\hat{R}(X \wedge Y) = AX \wedge AY \qquad \text{for } X, Y \in T_x(M).$$

Hence
$$\hat{R}(X_i \wedge X_j) = \lambda_i \lambda_j X_i \wedge X_j \qquad \text{for } 1 \leq i, j \leq n.$$

Since $\{X_i \wedge X_j;\ 1 \leq i < j \leq n\}$ is an orthonormal basis for $\bigwedge^2 T_x(M)$, the set $\{\lambda_i \lambda_j;\ 1 \leq i < j \leq n\}$ is the set of eigenvalues of $\hat{R}$ counting multiplicity. Since $\hat{R}$ is intrinsic, that is, depends only on the Riemannian metric of M and not on the immersion of M in $\mathbf{R}^{n+1}$, its eigenvalues $\{\lambda_i \lambda_j;\ 1 \leq i < j \leq n\}$ are also intrinsic although the eigenvalues $\lambda_1, \ldots, \lambda_n$ of A are not. In particular, every symmetric function of $\{\lambda_i \lambda_j;\ 1 \leq i < j \leq n\}$ such as $K_2 = \Sigma_{i<j} \lambda_i \lambda_j$ is intrinsic.

We shall find the expression for the Ricci tensor of a hypersurface.

PROPOSITION 5.2. *Let M be a hypersurface immersed in $\mathbf{R}^{n+1}$. At each point x of M, the Ricci tensor S is given by*

$$S(X, Y) = g(AX, Y) \operatorname{trace} A - g(A^2 X, Y), \quad X, Y \in T_x(M).$$

Proof. By definition (§5 of Chapter VI), we have

$$S(X, Y) = \text{trace of the map } Z \to R(Z, X)Y.$$

By the classical equation of Gauss (Example 4.1), we have

$$R(Z, X)Y = g(AX, Y)AZ - g(AZ, Y)AX.$$

The trace of the map $Z \to g(AX, Y)AZ$ is equal to $g(AX, Y) \operatorname{trace} A$. The trace of the map $Z \to g(AZ, Y)AX$ is seen to be $g(A(AX), Y)$ by virtue of the lemma below. Thus we get the desired formula.

LEMMA. *Let V be an n-dimensional vector space and V^* its dual space. For any $X \in V$ and $\omega \in V^*$, the trace of the linear map: $Z \to \omega(Z)X$ is equal to $\omega(X)$.*

The proof of the lemma is a simple exercise in linear algebra. In the proof of Proposition 5.2, we consider $Z \to g(AZ, Y)$ as a linear function (for the fixed Y). QED.

We recall that a Riemannian manifold is called an Einstein manifold if $S = \rho g$, where ρ is a constant. For dimensions 2 and 3, this condition implies that the space is of constant sectional curvature (p. 293, Volume I). We shall prove

THEOREM 5.3. *Let M be a hypersurface immersed in $\mathbf{R}^{n+1}$, where $n \geq 3$. If M is Einstein: $S = \rho g$, then*
(1) *ρ cannot be negative;*
(2) *if $\rho = 0$, then M is locally Euclidean;*
(3) *if $\rho > 0$, then every point of M is umbilical and M is locally a hypersphere.*

Proof. We first prove (2) and (3). If $S = \rho g$, then by Proposition 5.2 we have

$$(\text{trace } A)A - A^2 = \rho I,$$

where I is the identity transformation. At a point x of M we take an orthonormal basis $X_1, \ldots, X_n$ in $T_x(M)$ such that $AX_i = \lambda_i X_i$, $1 \leq i \leq n$. Then the preceding equation gives

$$\left(\sum_i \lambda_i\right)\lambda_i - \lambda_i^2 = \rho, \quad 1 \leq i \leq n.$$

Let $s = \Sigma_i \lambda_i = \text{trace } A$. Each of $\lambda_1, \ldots, \lambda_n$ is a root of the quadratic equation $\lambda^2 - s\lambda + \rho = 0$.

In the case where $\rho = 0$, we have $\lambda(s - \lambda) = 0$. Thus either all λ_i's are 0 (and $s = 0$) or we may assume, by rearranging the indices, that $\lambda_1 = \cdots = \lambda_p = s \neq 0$ and $\lambda_{p+1} = \cdots = \lambda_n = 0$, where $1 \leq p \leq n$. In the latter case, we have $s = ps$, which implies that $p = 1$. Thus the eigenvalues of A are either all 0 or all 0 except one. At any rate, we find that all the sectional curvatures at x are 0, that is, $R = 0$ at x. Since x is arbitrary, we conclude that M is locally Euclidean. This proves (2).

In the case where $\rho > 0$, we show that all λ_i's are equal. Suppose that $\lambda_1 \neq \lambda_2$. Then the other λ_i's are equal to λ_1 or λ_2. Thus we assume that λ_1 appears p times and λ_2 appears q times (so that $n = p + q$) among the eigenvalues of A. Then we have from the quadratic equation

$$\lambda_1 + \lambda_2 = s = p\lambda_1 + q\lambda_2, \text{ that is, } (p-1)\lambda_1 + (q-1)\lambda_2 = 0$$

and

$$\lambda_1 \lambda_2 = \rho > 0.$$

The second relation shows that λ_1 and λ_2 have the same sign. Then the first relation implies that $p = 1$, $q = 1$ and hence $n = 2$, contrary to the assumption. This proves that all λ_i's are equal (and not equal to 0, because $\rho \neq 0$).

We now prove (1). Consider the equation

(1) $\qquad \lambda^2 - s\lambda + \rho = 0, \qquad$ where $\rho < 0$.

If all λ_i's are equal, then $s = n\lambda$, and hence $\rho = (n - 1)\lambda^2$, which contradicts the assumption $\rho < 0$. Thus the equation above has two distinct roots. Assume that the eigenvalues of A are given by

$$\lambda_1 = \lambda_2 = \cdots = \lambda_p = \lambda \quad \text{and} \quad \lambda_{p+1} = \cdots = \lambda_n = \mu,$$

where

$$\lambda + \mu = s, \qquad p\lambda + (n - p)\mu = s,$$

so that

$$(p - 1)\lambda + (n - p - 1)\mu = 0$$

and

$$\lambda\mu = \rho.$$

From these relations we obtain

$$(p - 1)\lambda^2 + (n - p - 1)\rho = 0.$$

If $p = 1$, then $(n - p - 1)\rho = 0$, and hence $n = p + 1 = 2$, which is not the case. Thus $p \neq 1$ and we obtain

(2) $\qquad \lambda^2 = -\rho(n - p - 1)/(p - 1).$

The argument above applies to each point x of M. Since $s = \operatorname{trace} A$ is differentiable and since the equation (1) has two distinct roots at each point, it follows that the two roots λ and μ are differentiable functions. From the equation (2) we conclude that λ is a constant function (since M is connected), and so is μ. It also follows that p and $n - p$ are also constant.

We define two distributions Δ_1 and Δ_2 on M as follows:

$$\Delta_1(x) = \{X \in T_x(M); AX = \lambda X\}$$
$$\Delta_2(x) = \{X \in T_x(M); AX = \mu X\}.$$

We shall show that both Δ_1 and Δ_2 are differentiable and involutive, and that M is locally a direct product of the maximal integral manifolds M_1 and M_2 of Δ_1 and Δ_2 as a Riemannian manifold. This will finish the proof, because if X and Y are tangent to M_1 and M_2, respectively, we have $0 = \hat{R}(X \wedge Y) = AX \wedge AY = \lambda \mu X \wedge Y$, which implies $\lambda \mu = 0$, contrary to $\rho = \lambda \mu < 0$.

First, to prove that Δ_1 and Δ_2 are differentiable, let $X_1, \ldots, X_p, X_{p+1}, \ldots, X_n$ be differentiable vector fields such that X_i, $1 \leq i \leq p$, and X_j, $p + 1 \leq j \leq n$, form a basis of $\Delta_1(x_0)$ and $\Delta_2(x_0)$, respectively, at a point x_0. We define vector fields $Y_1, \ldots, Y_n$ by
$$Y_i = (A - \mu)X_i, \quad 1 \leq i \leq p,$$
and
$$Y_j = (A - \lambda)X_j, \quad p + 1 \leq j \leq n.$$
Since $(A - \lambda)Y_i = (A - \lambda)(A - \mu)X_i = 0$ for $1 \leq i \leq p$ and $(A - \mu)Y_j = (A - \mu)(A - \lambda)X_j = 0$ for $p + 1 \leq j \leq n$, we see that Y_i, $1 \leq i \leq p$, belong to Δ_1 and Y_j, $p + 1 \leq j \leq n$, belong to Δ_2. At x_0, we have $Y_i = (\lambda - \mu)X_i$, $1 \leq i \leq p$, and $Y_j = (\mu - \lambda)X_j$, $p + 1 \leq j \leq n$. Thus $Y_1, \ldots, Y_n$ are linearly independent at x_0 and therefore in a neighborhood of x_0. Thus we have shown that Δ_1 and Δ_2 have local bases $Y_1, \ldots, Y_p$ and $Y_{p+1}, \ldots, Y_n$, respectively.

Second, to prove that Δ_1 and Δ_2 are involutive, let X and Y be vector fields belonging to Δ_1. Since
$$(\nabla_X A)(Y) = \nabla_X(AY) - A(\nabla_X Y) = \lambda \nabla_X Y - A(\nabla_X Y)$$
and
$$(\nabla_Y A)(X) = \nabla_Y(AX) - A(\nabla_Y X) = \lambda \nabla_Y X - A(\nabla_Y X),$$
the Codazzi equation implies
$$A(\nabla_X Y - \nabla_Y X) = \lambda(\nabla_X Y - \nabla_Y X).$$
Since $\nabla_X Y - \nabla_Y X = [X, Y]$, we have
$$A([X, Y]) = \lambda[X, Y],$$
which shows that $[X, Y]$ belongs to Δ_1. Thus Δ_1 is involutive. The argument for Δ_2 is quite similar.

We now show that Δ_1 and Δ_2 are parallel with respect to the Riemannian connection in M. Let X be a vector field belonging to Δ_1 and let Y be a vector field belonging to Δ_2. The Codazzi equation in this case gives
$$A(\nabla_X Y - \nabla_Y X) = \mu \nabla_X Y - \lambda \nabla_Y X.$$

Denoting by $(\nabla_X Y)_1$ and $(\nabla_X Y)_2$ the Δ_1-component and the Δ_2-component of the vector field $\nabla_X Y$, we have

$$\lambda(\nabla_X Y)_1 + \mu(\nabla_X Y)_2 - \lambda(\nabla_Y X)_1 - \mu(\nabla_Y X)_2$$
$$= \mu(\nabla_X Y)_1 + \mu(\nabla_X Y)_2 - \lambda(\nabla_Y X)_1 - \lambda(\nabla_Y X)_2$$

so that comparing the Δ_1-components and the Δ_2-components we obtain

$$\lambda(\nabla_X Y)_1 = \mu(\nabla_X Y)_1 \quad \text{and} \quad \mu(\nabla_Y X)_2 = \lambda(\nabla_Y X)_2.$$

Since $\lambda \neq \mu$, we have

$$(\nabla_X Y)_1 = 0, \quad \text{that is,} \quad \nabla_X Y \in \Delta_2$$

and

$$(\nabla_Y X)_2 = 0, \quad \text{that is,} \quad \nabla_Y X \in \Delta_1.$$

Let Z be a vector field belonging to Δ_1. Differentiating $g(X, Y) = 0$ covariantly with respect to Z, we have

$$g(\nabla_Z X, Y) + g(X, \nabla_Z Y) = 0.$$

Since $\nabla_Z Y$ belongs to Δ_2 by what we have already shown, we get $g(X, \nabla_Z Y) = 0$ and hence $g(\nabla_Z X, Y) = 0$. This means that $\nabla_Z X \in \Delta_1$. Thus we have shown that if $X \in \Delta_1$, then $\nabla_Y X \in \Delta_1$ for any $Y \in \Delta_1$ or $Y \in \Delta_2$, and hence for any arbitrary vector field Y. This proves that Δ_1 is parallel. Similarly Δ_2 is parallel. By Proposition 5.2 of Chapter IV we see that for each point x of M there is a neighborhood U of x which is the Riemannian direct product of integral manifolds of Δ_1 and Δ_2. We have thus completed the proof of (1) of Theorem 5.3. Q E D.

The third part of Theorem 5.3 was proved by T. Y. Thomas [5] and the proof was simplified by E. Cartan (see T. Y. Thomas [3]). The first part of Theorem 5.3 is a special case of the results on Einstein hypersurfaces of spaces of constant sectional curvature by A. Fialkow [1]; the proof we presented here for (1) is due to B. Smyth.

COROLLARY 5.4. *Let M be a connected, complete Riemannian manifold of dimension $n \geq 3$ which is isometrically immersed in $\mathbf{R}^{n+1}$. If M has constant sectional curvature $k \neq 0$, then k is positive and M is a hypersphere.*

Proof. This follows from Theorems 5.3 and 5.1. (The case $k < 0$ can be excluded more easily by the following argument. If X_i, $1 \leq i \leq n$, is an orthonormal basis consisting of eigenvectors for the eigenvalues $\lambda_1, \ldots, \lambda_n$, then we know that $k(X_i \wedge X_j) = \lambda_i \lambda_j$. Thus we have $\lambda_i \lambda_j = k$ for $i \neq j$. If $n \geq 3$ and $k \neq 0$, then it follows that $\lambda_1 = \cdots = \lambda_n$. Thus $k > 0$.) QED.

Corollary 5.4 is valid for the case $n = 2$ as well, but the proof is more difficult (see Note 15).

Both the second fundamental form and the Gaussian curvature are closely related to the convexity of a hypersurface. A hypersurface M in $\mathbf{R}^{n+1}$ is said to be *convex at a point* $x \in M$ if the hyperplane H_x of $\mathbf{R}^{n+1}$ tangent to M at x does not separate a neighborhood of x in M into two parts. Moreover, if x is the only point of a neighborhood which lies on H_x, then M is said to be *strictly convex at* x. If, for every $x \in M$, H_x does not separate M into two parts, then M is said to be *convex*. Moreover, if, for every $x \in M$, x is the only point of M which lies on H_x, then M is said to be *strictly convex*. A convex hypersurface is always orientable. Choosing at each point x of M the unit normal vector pointed outward (i.e., to the opposite side of M with respect to H_x), we obtain a continuous field of unit normal vectors.

We say that the second fundamental form α of a hypersurface M is *definite* at $x \in M$ if $\alpha(X, X) \neq 0$ for all non-zero vectors $X \in T_x(M)$. If we choose a unit normal vector ξ at x and write $\alpha = h\xi$ (cf. §3), then α is definite if and only if the classical second fundamental form h is either positive or negative definite. Similarly, we say that α is *non-degenerate* at x if h is non-degenerate at x.

PROPOSITION 5.5. *A hypersurface M in $\mathbf{R}^{n+1}$ is strictly convex at a point x if its second fundamental form α is definite at x.*

Proof. Fixing the point x, we choose a Euclidean coordinate system $y^1, \ldots, y^{n+1}$ for $\mathbf{R}^{n+1}$ such that $dy^{n+1} = 0$ defines the tangent space $T_x(M)$. Then y^{n+1}, considered as a function on M, is critical at x. Its Hessian at x is nothing but the second fundamental form of M at x (cf. Examples 3.3 and 3.4). If α is definite at x, then y^{n+1}, as a function on M, attains an isolated local minimum or maximum. Hence a neighborhood of x in M lies strictly in one side of the tangent hyperplane at x. QED.

THEOREM 5.6. *For a connected compact hypersurface M in $\mathbf{R}^{n+1}$ ($n \geq 2$), the following conditions are equivalent:*
  (1) *The second fundamental form α of M is definite everywhere on M;*
  (2) *M is orientable and the spherical map of Gauss $M \to S^n$ is a diffeomorphism;*
  (3) *The Gaussian curvature K_n of M never vanishes on M.*

Moverover, any one of the conditions above implies that M is strictly convex.

Proof. (1) $\to$ (2). At each point $x \in M$ we choose a unit normal vector ξ_x in such a way that the classical second fundamental form h defined by $\alpha(X, Y) = h(X, Y)\xi_x$ for $X, Y \in T_x(M)$ is negative definite. Then ξ is continuous and hence M is orientable. At a point where α is non-degenerate, the Jacobian of the spherical map $M \to S^n$ is non-degenerate (cf. Example 3.5). Since M is compact, the spherical map $M \to S^n$ is a covering projection by Corollary 4.7 of Chapter IV. Since S^n is simply connected, the spherical map is a diffeomorphism.

(2) $\to$ (3). Since the Jacobian of the spherical map $M \to S^n$ is non-degenerate everywhere, so is the second fundamental form (cf. Example 3.5). Hence $K_n \neq 0$ everywhere.

(3) $\to$ (1). Since $K_n \neq 0$ everywhere, α is non-degenerate everywhere. Since M is compact, there is a point $x_0 \in M$ such that

$$\alpha(X, X) \neq 0 \qquad \text{for all } X \in T_{x_0}(M), X \neq 0$$

by Proposition 4.6. Since α is definite at x_0 and is non-degenerate everywhere, α is definite everywhere.

To prove the last assertion, let x be any point of M, and choose a Euclidean coordinate system $y^1, \ldots, y^{n+1}$ of $\mathbf{R}^{n+1}$ such that the tangent hyperplane H_x is given by $y^{n+1} = 0$ and a neighborhood of x in M lies in the region $y^{n+1} \leq 0$. Let x^* be a point of M where the function y^{n+1} attains its maximum on M. Then H_{x^*} is parallel to H_x, and the outward unit normal vector at x^* is parallel to the one at x (in the same direction). Since the spherical map is one-to-one, we have $x^* = x$. Hence $M - \{x\}$ lies in the region $y^{n+1} < 0$.
QED.

The last assertion of Theorem 5.6 is due to Hadamard [1]. Chern and Lashof [1] proved that a compact surface in $\mathbf{R}^3$ with $K_2 \geq 0$ is convex and constructed for $n \geq 3$ a non-convex compact hypersurface M in $\mathbf{R}^{n+1}$ with $K_n \geq 0$, For hypersurfaces with $K_n \geq 0$ or $K_n \leq 0$ which are not necessarily compact, see Hartman and Nirenberg [1] and Nirenberg [3]. (See Note 15 for these and other results.)

6. Type number and rigidity

Let M be a hypersurface immersed in $\mathbf{R}^{n+1}$. At each point x of M, the *type number* of M at x, denoted by $t(x)$, is defined to be the rank of the linear transformation A of $T_x(M)$. Of course, it is determined independently of the choice of the field of unit normals ξ because $A_{-\xi}\ (= -A_\xi)$ and A_ξ have the same rank. By Example 3.5 we know that $t(x)$ is the rank of the Jacobian of the spherical map $\phi\colon M \to S^n$, which is at least locally defined.

We shall prove

THEOREM 6.1. *For a hypersurface M immersed in $\mathbf{R}^{n+1}$,*
(1) $t(x)$ *is 0 or 1 if and only if $R = 0$ at x.*
(2) *If $t(x) \geq 2$, then $t(x) = n - \dim T_x^*$, where*

$$T_x^* = \{X \in T_x(M);\ R(X, Y) = 0 \text{ for all } Y \in T_x(M)\}.$$

Proof. (1) If $t(x) = 0$, that is, $A = 0$, then the equation of Gauss implies that $R = 0$ at x. Suppose $t(x) = 1$. Then there is a non-zero vector $W \in T_x(M)$ such that for each $X \in T_x(M)$ we have $AX = cW$, where c is a scalar depending on X. The equation of Gauss again implies that $R(X, Y) = 0$ for all X, Y. Conversely, suppose that $R = 0$ at x. If $t(x) \geq 2$, there exist a pair of non-zero vectors X and Y in $T_x(M)$ such that $AX = \lambda X$, $AY = \mu Y$, where $\lambda, \mu \neq 0$, and such that $g(X, Y) = 0$. We find that

$$g(R(X, Y)Y, X) = \lambda \mu \neq 0,$$

which is a contradiction to the assumption $R = 0$.

(2) Let T_x'' be the null space of A, and let T_x' be the orthogonal complement of T_x''. Since A is symmetric, we may easily see that A maps T_x' onto itself in a one-to-one manner. If $X \in T_x''$, then the equation of Gauss implies that $R(X, Y) = 0$ for all $Y \in T_x(M)$, that is, $X \in T_x^*$. Now assuming $\dim T_x' = t(x) \geq 2$, we prove that $T_x'' = T_x^*$. Suppose there is an $X \in T_x^*$ which is not in T_x''. Since $X = X_1 + X_2$, where $X_1 \in T_x'$ and $X_2 \in T_x'' \subset T_x^*$, we see that

$X_1 \in T_x^* \cap T_x'$. Since dim $T_x' \geq 2$, there exists a $Y \in T_x'$ such that X_1 and Y are linearly independent (and so are AX_1 and AY). By the equation of Gauss we have

$$R(X_1, Y)AY = g(AY, AY)AX_1 - g(AX_1, AY)AY.$$

Since $X_1 \in T_x^*$, we have $R(X_1, Y) = 0$, that is, the right hand side is also equal to 0. Since $g(AY, AY) \neq 0$, this relation means that AX_1 and AY are linearly dependent, which is a contradiction. We have thus shown that $T_x'' = T_x^*$. Hence $t(x) = \dim T_x' = n - \dim T_x'' = n - \dim T_x^*$, completing the proof of Theorem 6.1.
Q.E.D.

Theorem 6.1 shows that, unless M is locally Euclidean, the rank of the second fundamental form is independent of an immersion. We shall prove that under a stronger assumption the immersion itself is unique (up to an isometry of $\mathbf{R}^{n+1}$). First we show that the second fundamental form is uniquely determined (up to a sign) at each point.

THEOREM 6.2. *Let M be an n-dimensional Riemannian manifold and let f and $\bar{f}$ be isometric immersions of M into $\mathbf{R}^{n+1}$. If the type number $t(x)$ of the immersion f at x is ≥ 3, then the second fundamental form h for f and the second fundamental form $\bar{h}$ for $\bar{f}$ coincide at x up to a sign.*

Proof. Let us first remark that, by virtue of Theorem 6.1, R is not 0 at x, and hence the type number $\bar{t}(x)$ for $\bar{f}$ is equal to $t(x)$. As we see from the proof of Theorem 6.1, the null space of A and the null space of $\bar{A}$ coincide with $T_x^* = \{X; R(X, Y) = 0 \text{ for all } Y \in T_x\}$. This means that in the decomposition $T_x(M) = T_x' + T_x''$ in that proof, T_x'' is the null space for A as well as for $\bar{A}$, and both A and $\bar{A}$ map T_x' onto itself in a one-to-one manner.

This being said, consider the exterior product $T_x' \wedge T_x'$ and introduce an inner product s such that

$$s(X \wedge Y, Z \wedge W) = g(X, Z)g(Y, W) - g(X, W)g(Y, Z).$$

The equation of Gauss implies that

$$\begin{aligned}g(R(X, Y)Z, W) &= s(X \wedge Y, AZ \wedge AW) \\ &= s(X \wedge Y, \bar{A}Z \wedge \bar{A}W).\end{aligned}$$

Since s is positive definite on $T'_x \wedge T'_x$, we have

(1) $\quad AZ \wedge AW = \bar{A}Z \wedge \bar{A}W \qquad$ for all $Z, W \in T'_x$.

We shall now prove that, for each $X \in T'_x$, $\bar{A}X = cAX$ for some c (this constant c may depend on X, although we shall eventually show that $c = \pm 1$ independently of X). Since $AX \neq 0$ for $X \neq 0$, it is sufficient to show that AX and $\bar{A}X$ are linearly dependent. Suppose that they are linearly independent so that $AX \wedge \bar{A}X \neq 0$. Since $\dim T'_x = t(x) \geq 3$ by assumption, we may choose a $Y \in T'_x$ such that $AX, \bar{A}X, AY$ are linearly independent (and hence $AX \wedge AY \wedge \bar{A}X \neq 0$).

By property (1), we have $AX \wedge AY = \bar{A}X \wedge \bar{A}Y$ so that $AX \wedge AY \wedge \bar{A}X = \bar{A}X \wedge \bar{A}Y \wedge \bar{A}X = 0$, which is a contradiction. This proves our assertion that $\bar{A}X = cAX$ for some c.

Let $X_1, \ldots, X_r$ be a basis in T'_x. There exist $c_1, \ldots, c_r$ such that $\bar{A}X_i = c_i AX_i$ for $1 \leq i \leq r$. At the same time, $\bar{A}(X_i + X_j) = cA(X_i + X_j)$ for some c. We have then

$$\bar{A}(X_i + X_j) = c_i AX_i + c_j AX_j$$
$$= c(AX_i + AX_j).$$

For any $i \neq j$, $1 \leq i, j \leq r$, we know that AX_i and AX_j are linearly independent, since A is one-to-one. Thus $c = c_i = c_j$. This proves that $c_1 = \cdots = c_r$. Thus $\bar{A} = cA$ for some c. From property (1) we conclude that $c^2 = 1$, that is, $c = \pm 1$. This proves that $\bar{A} = \pm A$ and $\bar{h} = \pm h$ at x. QED.

COROLLARY 6.3. *Let M be an n-dimensional connected and orientable Riemannian manifold, and let f and $\bar{f}$ be isometric immersions of M into $\mathbf{R}^{n+1}$. If the type number $t(x)$ of the immersion f is ≥ 3 at every point x of M, then we have, by appropriate choice of the fields of unit normals for f and $\bar{f}$, the same second fundamental form for f and $\bar{f}$.*

Proof. Let ξ and $\bar{\xi}$ be fields of unit normal vectors globally defined on M for the immersions f and $\bar{f}$, respectively. Let M^+ (resp. M^-) be the set of points $x \in M$ at which $h = \bar{h}$ (resp. $h = -\bar{h}$). Then both M^+ and M^- are closed. By Theorem 6.2, M is a disjoint union of M^+ and M^-. Since M is connected, either $M = M^+$ or $M = M^-$. If $M = M^+$, we are done. If $M = M^-$, we change $\bar{\xi}$ into $-\bar{\xi}$. QED.

VII. SUBMANIFOLDS   45

We shall prove that a hypersurface M is *rigid* in $\mathbf{R}^{n+1}$ if the type number is greater than 2 at every point; rigidity means that an isometric immersion is unique up to an isometry of $\mathbf{R}^{n+1}$. In view of the results above, we shall first state the following classical result:

THEOREM 6.4. *Let M be a connected n-dimensional Riemannian manifold, and let f and $\bar{f}$ be isometric immersions of M into $\mathbf{R}^{n+1}$ with fields of unit normal vectors ξ and $\bar{\xi}$, respectively. If the second fundamental forms h and $\bar{h}$ of f and $\bar{f}$ (with respect to ξ and $\bar{\xi}$), respectively, coincide on M, then there is an isometry τ of $\mathbf{R}^{n+1}$ such that $\bar{f} = \tau \circ f$.*

Proof. We first prove the local version, that is, each point x_0 has a neighborhood U such that $\bar{f}(x) = \tau \circ f(x)$ for every $x \in U$, where τ is a certain isometry of $\mathbf{R}^{n+1}$ which depends only on U. By composing f with a suitable isometry of $\mathbf{R}^{n+1}$ we may assume that $\bar{f}(x_0) = f(x_0)$ and that $\bar{f}_*(T_{x_0}(M)) = f_*(T_{x_0}(M))$ (namely, the tangent hyperplane of $\bar{f}(M)$ at $\bar{f}(x_0)$ coincides with that of $f(M)$ at $f(x_0) = \bar{f}(x_0)$). We may also assume that ξ and $\bar{\xi}$ coincide at this point.

Let
$$y = y(x^1, \ldots, x^n) \quad \text{and} \quad \bar{y} = \bar{y}(x^1, \ldots, x^n)$$
be the equations for the immersions f and $\bar{f}$, respectively, where $x^1, \ldots, x^n$ are local coordinates in U and y and $\bar{y}$ are the position vectors in $\mathbf{R}^{n+1}$ with components $(y^1, \ldots, y^{n+1})$ and $(\bar{y}^1, \ldots, \bar{y}^{n+1})$. In the notation of Example 3.4, let $e_j = \partial y/\partial x^j$ and $\bar{e}_j = \partial \bar{y}/\partial x^j$. Then we have

(I) $\qquad \partial e_j/\partial x^i = \sum_k \Gamma_{ij}^k e_k + h_{ij}\xi$

(II) $\qquad \partial \xi/\partial x^j = -\sum_k a_j^k e_k$

for the immersion f and the corresponding equations for $\bar{f}$

($\bar{\text{I}}$) $\qquad \partial \bar{e}_j/\partial x^i = \sum \Gamma_{ij}^k \bar{e}_k + h_{ij}\bar{\xi}$

($\overline{\text{II}}$) $\qquad \partial \bar{\xi}/\partial x^j = -\sum a_j^k \bar{e}_k.$

Note that h_{ij} are the components of $h = \bar{h}$ with respect to $(x^1, \ldots, x^n)$, $a_j^k = \Sigma_m g^{km}h_{mj}$ are common to f and $\bar{f}$, and, of course, we use the same Christoffel symbols Γ_{ij}^k in (I) and ($\bar{\text{I}}$). We may also assume that e_j and $\bar{e}_j$ coincide at x_0 for every j, $1 \leq j \leq n$; in fact, since $(e_j, e_k) = g_{jk} = (\bar{e}_j, \bar{e}_k)$, we may choose an isometry of $\mathbf{R}^{n+1}$

which maps the frame $(e_1, \ldots, e_n, \xi)$ at $f(x_0)$ upon the frame $(\bar{e}_1, \ldots, \bar{e}_n, \bar{\xi})$ at the same point. By composing this isometry with $\bar{f}$, we may satisfy our requirement.

Thus we see that $(e^1, \ldots, e^n, \xi)$ for f and $(\bar{e}^1, \ldots, \bar{e}^n, \bar{\xi})$ for $\bar{f}$, both as a set of $\mathbf{R}^{n+1}$-valued vector functions of $(x^1, \ldots, x^n)$, satisfy the same set of equations I and II. Since the initial conditions at x_0 coincide, uniqueness of the solution set of the system I and II implies that $e_j = \bar{e}_j$, $1 \leq j \leq n$, and $\xi = \bar{\xi}$. Thus $\partial y/\partial x^j = \partial \bar{y}/\partial x^j$, $1 \leq j \leq n$. Since $y(x_0) = \bar{y}(x_0)$, we conclude that $y = \bar{y}$ on U. We have thus shown that there is an isometry, say τ_{x_0}, of $\mathbf{R}^{n+1}$ such that $\bar{f}(x) = \tau_{x_0} \circ f(x)$ and $\bar{\xi}(x) = \tau_{x_0} \circ \xi(x)$ for every $x \in U$.

The global version now follows easily. Suppose that U and V are two connected open sets in M such that $U \cap V \neq 0$ and that there are isometries τ and μ of $\mathbf{R}^{n+1}$ such that $\bar{f} = \tau \circ f$ and $\bar{\xi} = \tau \circ \xi$ on U, and $\bar{f} = \mu \circ f$ and $\bar{\xi} = \mu \circ \xi$ on V. For any $z \in U \cap V$, let $X_1, \ldots, X_n$ be a basis of $T_z(M)$. Then both τ and μ map the frame $(f(X_1), \ldots, f(X_n), \xi)$ at $f(z)$ upon the frame $(\bar{f}(X_1), \ldots, \bar{f}(X_n), \bar{\xi})$ at $\bar{f}(z)$. Hence τ and μ coincide. Since M is connected, it follows that there is an isometry τ of $\mathbf{R}^{n+1}$ such that $\bar{f} = \tau \circ f$ on M, proving Theorem 6.4.   QED.

COROLLARY 6.5. *Let M be a connected, orientable, n-dimensional Riemannian manifold, and let f and $\bar{f}$ be isometric immersions of M into $\mathbf{R}^{n+1}$. If the type number $t(x)$ of the immersion f is ≥ 3 at every point x of M, then there is an isometry τ of $\mathbf{R}^{n+1}$ such that $\bar{f} = \tau \circ f$.*

Proof. This follows from Corollary 6.3 and Theorem 6.4.

QED.

Corollary 6.5 is due to Beez [1] and Killing [1]. (See also Thomas [1].) A proof of Corollary 6.5 making use of differential forms can be found in E. Cartan [10]. Thomas [1] defines the type number to be 1 if the rank of A is either 0 or 1 so that with his definition the type number depends only on the curvature tensor R (cf. Theorem 6.1).

Sacksteder [1] obtained a number of rigidity theorems for hypersurfaces. Among other things he proved that two mutually isometric complete convex hypersurfaces in $\mathbf{R}^{n+1}$, $n \geq 3$, are congruent if their type numbers are at least 3 at one point. (See Allendoerfer [3] and Note 17 for rigidity questions related to this section.)

7. Fundamental theorem for hypersurfaces

When a Riemannian manifold M with metric g is isometrically immersed in a Euclidean space $\mathbf{R}^{n+1}$ as a hypersurface, we have the second fundamental form h and the corresponding operator A, which are defined locally for a given choice of the field of unit normal vectors. We know that A is related to the metric g by the equations of Gauss and Codazzi (Examples 4.1 and 4.3)

(III) $\qquad R(X, Y)Z = g(AY, Z)AX - g(AX, Z)AY$

and

(IV) $\qquad (\nabla_X A)(Y) = (\nabla_Y A)(X).$

Here one should, of course, note that covariant differentiation ∇ and the curvature tensor field R are uniquely determined by the metric g.

In this section we shall prove the converse, first in a local form and then in a global form.

THEOREM 7.1. *Let M be an n-dimensional Riemannian manifold with metric g. Let A be a tensor field of type $(1, 1)$ on M which defines a symmetric transformation of each tangent space $T_x(M)$ and satisfies the equations of Gauss and Codazzi. Let x_0 be a point M and let $\{X_1, \ldots, X_n\}$ be an orthonormal basis of $T_{x_0}(M)$. Also, let y_0 be a point of $\mathbf{R}^{n+1}$ and let $(e_1)_0, \ldots, (e_n)_0, \xi_0$ be an orthonormal frame at y_0. Then there exists an isometric imbedding f of a neighborhood U of x_0 into $\mathbf{R}^{n+1}$ such that $f(x_0) = y_0$, $f_*(X_i) = (e_i)_0$, $1 \leq i \leq n$, ξ_0 is normal to $f(U)$ at y_0, and such that A is the symmetric operator corresponding to the second fundamental form for f. Moreover, any two such isometric imbeddings coincide in a neighborhood of x_0.*

We prepare a few lemmas. Let $x^1, \ldots, x^n$ be a system of local coordinates with origin x_0 such that $X_i = (\partial/\partial x^i)_0$, and let $(y^1, \ldots, y^{n+1})$ be a system of rectangular coordinates with origin y_0 such that $(\partial/\partial y^i)_0 = (e_i)_0$ for $1 \leq i \leq n$ and $(\partial/\partial y^{n+1})_0 = \xi_0$. For an isometric imbedding f of a neighborhood U of x_0 into $\mathbf{R}^{n+1}$ satisfying the conditions stated in the theorem, let

$$y^p = f^p(x^1, \ldots, x^n), \quad 1 \leq p \leq n+1,$$

be the set of equations which defines f. As in Example 3.4 the vector field $(\partial/\partial x^i)$ is mapped by f_* upon $e_i = \partial y/\partial x^i$, where $y = (y^1, \ldots, y^{n+1})$. From the conditions $f_*(X_i) = (e_i)_0$ we obtain

(1) $(\partial f^p/\partial x^i)_0 = \delta_i^p$, $\quad 1 \leq p \leq n$,
(2) $(\partial f^{n+1}/\partial x^i)_0 = 0$.

The components of ξ_0 satisfy

(3) $\xi_0^p = 0$, $\quad 1 \leq p \leq n$,
(4) $\xi_0^{n+1} = 1$.

For each p, $1 \leq p \leq n$, we have the formula of Gauss:

(I$_p$) $\partial e_j^p/\partial x^i = \sum\limits_{k=1}^n \Gamma_{ij}^k e_k^p + h_{ij} \xi^p$,

where $e_j^p = \partial f^p/\partial x^j$. The formula of Weingarten gives

(II$_p$) $\partial \xi^p/\partial x^j = -\sum\limits_{k=1}^n a_j^k e_k^p$,

where (a_j^k) are the components of A with respect to $(x^1, \ldots, x^n)$. Let Z^p be the vector field which corresponds to the differential df^p of the function f^p (so that $g(Z^p, X) = Xf^p$ for any vector field X on U).

LEMMA 1. *For each p, $1 \leq p \leq n+1$, the pair (Z^p, ξ^p) is a solution of the system of differential equations*

(5) $\nabla_Y Z^p - \xi^p A Y = 0$
(6) $Y\xi^p + g(AY, Z^p) = 0$,

where Y is an arbitrary vector field, with initial conditions

(7) $(Z^p)_{x_0} = (\partial/\partial x^p)_{x_0}$, $\quad \xi^p(x_0) = 0$, $\quad$ for $1 \leq p \leq n$;
(8) $(Z^{n+1})_{x_0} = 0$, $\quad \xi^{n+1}(x_0) = 1$.

Proof. (7) is equivalent to (1) and (3). (8) is equivalent to (2) and (4). The formula of Gauss (I$_p$) is equivalent to

$$\nabla_{\partial/\partial x^i}(df^p) = \sum_{i=1}^n \xi^p h_{ij}\, dx^i,$$

that is,

$$(\nabla_Y(df^p))(X) = \xi^p h(X, Y) = \xi^p g(X, AY).$$

Since Z^p corresponds to df^p under the duality defined by g, we have $g(\nabla_Y Z^p, X) = (\nabla_Y(df^p))(X)$. Hence (I_p) is equivalent to (5). Similarly, (II_p) is equivalent to (6).

LEMMA 2. *Suppose that, for each p, $1 \le p \le n+1$, we have a vector field Z^p and a function ξ^p which form a solution of the system (5) and (6) with initial conditions (7) and (8). Then there exists an isometric imbedding f of a neighborhood of x_0 into $\mathbf{R}^{n+1}$: $y^p = f^p(x^1, \ldots, x^n)$, $1 \le p \le n+1$, such that df^p corresponds to Z^p, $1 \le p \le n+1$, and such that $\xi = (\xi^1, \ldots, \xi^{p+1})$ is a field of unit normals.*

Proof. If we know the existence of functions f^p, $1 \le p \le n+1$, such that df^p correspond to Z^p, then (5) and (6) imply (I_p) and (II_p), $1 \le p \le n+1$, namely, the formulas of Gauss and Weingarten. Differentiating the functions $\varphi_{jk} = (e_j, e_k)$, $\varphi_k = (\xi, e_k)$ and $\varphi = (\xi, \xi)$ with respect to x^i and using (I_p) and (II_p), we find

$$\partial \varphi_{jk}/\partial x^i = \sum_l (\Gamma^l_{ij}\varphi_{lk} + \Gamma^l_{ik}\varphi_{lj}) + h_{ij}\varphi_k + h_{ik}\varphi_j$$

$$\partial \varphi_k/\partial x^i = -\sum_l a^l_i \varphi_{lk} + \sum_l \Gamma^l_{ik}\varphi_l + h_{ik}\varphi$$

$$\partial \varphi/\partial x^i = -2\sum_l a^l_i \varphi_l.$$

We observe that the functions $\varphi_{jk} = g_{jk}$, $\varphi_k = 0$ and $\varphi = 1$ also satisfy the above system of equations (cf. Corollary 2.4 of Chapter IV). Because of the same initial conditions at x_0 we conclude that $(e_j, e_k) = g_{jk}$, $(\xi, e_k) = 0$ and $(\xi, \xi) = 1$, which mean that f is isometric and that ξ is a field of unit normals.

It remains therefore to show the existence of functions f^p such that df^p correspond to Z^p. Let ω^p be the 1-form corresponding to Z^p. To prove that ω^p is exact (i.e., $\omega^p = df^p$ for some function f^p), it is sufficient to show that $d\omega^p = 0$. This property of ω^p is equivalent to the condition

$$X \cdot \omega^p(Y) - Y \cdot \omega^p(X) - \omega^p([X, Y]) = 0$$

for all vector fields X, Y. The left hand side is equal to

$$X \cdot g(Z^p, Y) - Y \cdot g(Z^p, X) - g(Z^p, [X, Y])$$
$$= g(\nabla_X Z^p, Y) - g(\nabla_Y Z^p, X).$$

Hence $d\omega^p = 0$ if and only if

$$g(\nabla_X Z^p, Y) = g(X, \nabla_Y Z^p)$$

for all vector fields X, Y. This last condition is indeed satisfied, since

$$g(\nabla_X Z^p, Y) = g(\xi^p AX, Y) = g(X, \xi^p AY) = g(X, \nabla_Y Z^p)$$

by virtue of (5) and of the symmetry of A.

In Lemma 2 it is clear that, given any point, say, y_0 in $\mathbf{R}^{n+1}$, we may choose an isometry f which maps x_0 upon y_0. Thus it follows that in order to prove the existence of an isometric imbedding in Theorem 7.1 it is sufficient to prove that the system of equations (5) and (6) have a solution with any arbitrary initial conditions. The uniqueness follows from the uniqueness of the solution of (5) and (6) with a preassigned initial condition, which is almost obvious. We shall now prove

LEMMA 3. *Let x_0 be a point of M and let U be a neighborhood with normal coordinates $x^1, \ldots, x^n$, $|x^i| < d$. For any $Z_0 \in T_{x_0}(M)$ and for any real number c, there exists a unique vector field Z and a unique function ξ which satisfy*

(9) $\nabla_Y Z - \xi AY = 0$

and

(10) $Y\xi + g(AY, Z) = 0$

with initial conditions: $Z_{x_0} = Z_0$ *and* $\xi(x_0) = c$.

Proof. For any point $a = (a^1, \ldots, a^n)$ of U, we take the geodesic $a_t = (a^1 t, \ldots, a^n t)$ and define $Z(t)$ and $\xi(t)$ as a solution of the system of ordinary differential equations

(11) $\nabla_X Z - \xi AX = 0$

(12) $X\xi + g(AX, Z) = 0$,

with initial conditions $Z(0) = Z_0$ and $\xi(0) = c$, where X is the tangent vector of the geodesic: $X = \sum_{i=1}^{n} a^i (\partial/\partial x^i)$. We then define $Z_a = Z(1)$ and $\xi(a) = \xi(1)$, thus getting a vector field Z and a function ξ on U. We have to show that Z and ξ satisfy (9) and (10) for any tangent vector Y at an arbitrary point a.

We extend the vector Y to a vector field $Y = \sum_{j=1}^{n} b^j (\partial/\partial x^j)$ with constant components b^j on U. Similarly, we extend the family of tangent vectors X of the geodesic a_t to the vector field

$X = \Sigma_{i=1}^n a^i(\partial/\partial x^i)$ with constant components a^i on U. We shall show that along a_t

(13) $\nabla_X(\nabla_Y Z - \xi AY) = (Y\xi + g(AY, Z))AX$

(14) $X(Y\xi + g(AY, Z)) = g(AX, \nabla_Y Z - \xi AY)$.

If this is done, then we see that $\nabla_Y Z - \xi AY$ and $Y\xi + g(AY, Z)$, which obviously have initial conditions

$\nabla_Y Z - \xi AY = 0$ and $Y\xi + g(AY, Z) = 0$ at x_0,

must be equal to the zero vector field and the constant function 0, which satisfy (13) and (14) with the same initial conditions; this is by virtue of the uniqueness of the solution of the system (13) and (14) along a_t.

The verification of (13) and (14) is based on the equations of Gauss and Codazzi for A. Since $[X, Y] = 0$, we have

$\nabla_X(\nabla_Y Z - \xi AY) = \nabla_Y(\nabla_X Z) + R(X, Y)Z$
$\qquad - \xi(\nabla_X A)(Y) - \xi A(\nabla_X Y) - (X\xi)AY,$

using

$R(X, Y) = [\nabla_X, \nabla_Y].$

Since the computation is being done along a_t, we have (11) and (12). Thus we see, using $\nabla_X Y = \nabla_Y X$ as well, that along a_t

$\nabla_X(\nabla_Y Z - \xi AY) = \nabla_Y(\xi AX) + R(X, Y)Z - \xi(\nabla_X A)(Y)$
$\qquad - \xi A(\nabla_Y X) + g(AX, Z)AY$
$\qquad = (Y\xi)AX + \xi(\nabla_Y A)(X) + \xi A(\nabla_Y X)$
$\qquad + R(X, Y)Z - \xi(\nabla_X A)(Y) - \xi A(\nabla_Y X)$
$\qquad + g(AX, Z)AY$
$\qquad = (Y\xi)AX + R(X, Y)Z + g(AX, Z)AY$
$\qquad + \xi(\nabla_Y A)(X) - \xi(\nabla_X A)(Y).$

The equation of Codazzi says that the last two terms cancel each other. The equation of Gauss implies

$R(X, Y)Z + g(AX, Z)AY = g(AY, Z)AX.$

Thus we get (13). The verification of (14) is quite similar except that we need only the equation of Codazzi.

With the proof of Lemma 3 we have concluded the proof of Theorem 7.1. QED.

Remark. The equations of Gauss and Codazzi are exactly the integrability conditions for the system (9) and (10) and hence are also necessary conditions for the system (9) and (10) to have a solution for any preassigned initial conditions.

We shall say that an imbedding or immersion f of an open subset W of M into $\mathbf{R}^{n+1}$ is *allowable* (for the given metric g and A on M) if it is isometric and if there is a field of unit normals ξ on W (that is, ξ_x is normal to $f(W)$ at $f(x)$) so that A is the symmetric operator corresponding to the second fundamental form for f with respect to ξ. We have proved that given any point x of M there is an allowable imbedding of a normal neighborhood of x, and that, furthermore, an allowable imbedding of U and a field of unit normals ξ are uniquely determined by the differential of f at x and by ξ_x.

We shall now prove the following global version of Theorem 7.1.

THEOREM 7.2. *Let M, g, and A be as in the assumptions of Theorem 7.1. If M is simply connected and connected, there is an allowable immersion of M into $\mathbf{R}^{n+1}$; any two allowable immersions differ by an isometry of $\mathbf{R}^{n+1}$.*

Proof. Let x_0 be an arbitrary point of M and let f_U be an allowable imbedding of a neighborhood U of x_0. For any curve $x(t)$, $0 \leq t \leq 1$, in M such that $x(0) = x_0$, we define a *continuation* of f_U along $x(t)$ almost in the same way as p. 254 of Volume I; more precisely, a continuation f_t is a family such that

(1) for each t, f_t is an allowable imbedding of a neighborhood U_t of $x(t)$ into $\mathbf{R}^{n+1}$;

(2) for each t, there is a positive number $\delta > 0$ such that if $|s - t| < \delta$, then $x(s) \in U_t$, and f_s and its associated field of unit normals coincide with f_t and its associated field of unit normals in a neighborhood of $x(s)$;

(3) f_0 coincides with f_U in a neighborhood of x_0, and so do their associated fields of unit normals.

If a continuation exists along a curve, it is unique. More precisely, if f_t and f_t' are continuations of f_U along $x(t)$ with fields of unit normals ξ and ξ', respectively, then each point $x(t)$ has a neighborhood on which $f_t = f_t'$ and $\xi = \xi'$. This assertion follows from the uniqueness part of Theorem 7.1.

Now a continuation exists along any curve $x(t)$. Let t_0 be the supremum of $t_1 > 0$ such that a continuation of f_U exists for $0 \leq t < t_1$. Let W be a convex neighborhood of $x(t_0)$ such that every point of W has a normal neighborhood containing W (cf. Theorem 8.7 of Chapter III). Take $t_1 < t_0$ such that $x(t_1) \in W$ and let V be a normal neighborhood of $x(t_1)$ which contains W. By the existence part of Theorem 7.1 we may extend the imbedding f_{t_1} to an allowable imbedding f' of V. For a $\delta > 0$ such that $x(t) \in W$ for $t_0 \leq t < t + \delta$, we define $f_t = f'$. Thus we obtain a continuation beyond t_0. (The arguments here are almost identical with those for the proof of Theorem 6.1 of Chapter VI, and indeed we may extend f_U to an allowable immersion of the whole manifold M in the same fashion as Theorem 6.1, namely, by using the factorization lemma (cf. Appendix 7); thus we omit the details here.)

Finally, we prove that any two allowable immersions f and f' of M differ only by an isometry of $\mathbf{R}^{n+1}$. Let x_0 be an arbitrary point and let U be a normal neighborhood of x_0. Then f'_U and f_U (restrictions of f and f' to U) differ by an isometry τ of $\mathbf{R}^{n+1}$ by virtue of the uniqueness part of Theorem 7.1. If $f'_U = \tau \circ f_U$, $\tau \circ f'$ is obtained by continuing $\tau \circ f'_U$; it follows that $\tau \circ f'$ coincides with f. QED.

Theorem 7.1 or 7.2 is called the fundamental theorem for hypersurfaces. For $n = 2$, it is due to Bonnet [2]. Sasaki [2] gave a different proof based on the construction of a certain involutive distribution on a suitable bundle over M.

8. Auto-parallel submanifolds and totally geodesic submanifolds

In this section, we shall consider not only Riemannian manifolds but, more generally, manifolds with affine connection. Let N be an $(n + p)$-dimensional manifold with an affine connection. A submanifold M of N is called *auto-parallel* if, for each vector $X \in T_x(M)$ and for each curve τ in M starting from x, the parallel displacement of X along τ (with respect to the affine connection of the ambient space N) yields a vector tangent to M. Thus, if M is auto-parallel, the affine connection of N induces an affine connection of M in a natural manner; this is intuitively clear but will be proved rigorously later. We recall (cf. §5 of Chapter IV) that a submanifold M of N is *totally geodesic at a point* $x \in M$ if every geodesic $\tau = x_t$ with $x = x_0$ which is tangent to M at x is contained in M for small values of t. If M is totally geodesic at every point of M, then M is called a *totally geodesic submanifold* of N.

Our immediate objective is to prove that every auto-parallel submanifold is totally geodesic and that the converse is true when the affine connection of N has no torsion.

A linear frame v of N at a point x of M is said to be *adapted* if it is of the form $(Y_1, \ldots, Y_n, Y_{n+1}, \ldots, Y_{n+p})$ with $Y_1, \ldots, Y_n$ tangent to the submanifold M. Let $\mathbf{R}^n$ be the subspace of $\mathbf{R}^{n+p}$ consisting of elements whose last p components are zero. Then, considered as a linear transformation $\mathbf{R}^{n+p} \to T_x(N)$, a linear frame v is adapted if and only if it maps $\mathbf{R}^n$ onto $T_x(M)$. Let $GL(n + p, n; \mathbf{R})$ be the subgroup of $GL(n + p; \mathbf{R})$ consisting of elements of the form

$$\begin{pmatrix} A & B \\ 0 & C \end{pmatrix},$$

where A and C are nonsingular matrices of degree n and p respectively and B is a matrix with n rows and p columns. In other words, it is the group of linear transformations of $\mathbf{R}^{n+p}$ which send $\mathbf{R}^n$ onto itself. Let $L(N, M)$ be the set of all adapted linear frames. It is an easy matter to verify that $L(N, M)$ is a principal fibre bundle over M with group $GL(n + p, n; \mathbf{R})$. As in §1 we have the following diagram

$$\begin{array}{ccccccc} L(M) & \xleftarrow{h} & L(N, M) & \xrightarrow{i} & L(N) \mid M & \xrightarrow{j} & L(N) \\ \downarrow{\pi'} & & \downarrow{\pi} & & \downarrow{\pi} & & \downarrow{\pi} \\ M & \longleftarrow & M & \longrightarrow & M & \longrightarrow & N, \end{array}$$

where both i and j are the injections and h is the bundle homomorphism defined by

$$h(v) = (Y_1, \ldots, Y_n) \quad \text{for } v = (Y_1, \ldots, Y_{n+p}) \in L(N, M).$$

Let θ and φ be the canonical forms on $L(M)$ and $L(N)$, respectively (cf. §2 of Chapter III). Then we have

PROPOSITION 8.1. *$h^*\theta$ coincides with the restriction $i^* \circ j^*\varphi$ of φ to $L(N, M)$. In particular, $i^* \circ j^*\varphi$ is $\mathbf{R}^n$-valued.*

VII. SUBMANIFOLDS

Proof. The proof is the same as that of Proposition 1.1.
QED.

As in §3 and §4, we shall use the following convention for the ranges of indices: $1 \leq A, B, \ldots \leq n + p$; $1 \leq i, j \ldots \leq n$ and $n + 1 \leq r, s, \ldots \leq n + p$.

Let ψ be the connection form on $L(N)$ defining the affine connection of N and let $\psi = (\psi_B^A)$. We denote by $\bar{\nabla}$ covariant differentiation on N.

PROPOSITION 8.2. *Let M be a submanifold of N. The following three conditions are equivalent:*

(1) *M is an auto-parallel submanifold of N;*
(2) *If X and Y are vector fields on M, then $\bar{\nabla}_X Y$ is tangent to M at every point of M;*
(3) *The restriction of ψ to $L(N, M)$ is $\mathfrak{gl}(n + p, n; \mathbf{R})$-valued, that is, the restriction of ψ_i^r to $L(N, M)$ vanishes for $1 \leq i \leq n$ and $n + 1 \leq r \leq n + p$.*

Proof. (1) → (2) follows from the definition of the covariant derivative in terms of parallel displacement (cf. p. 114 of Volume I). To prove (2) → (3), let o be an arbitrary point of M and let V be a neighborhood of o in M with a system of coordinates $x^1, \ldots, x^{n+p}$ such that

$$U = \{x \in V; x^{n+1}(x) = \cdots = x^{n+p}(x) = 0\}$$

is a neighborhood of o in M on which $x^1, \ldots, x^n$ form a system of coordinates. (The existence of U with such a coordinate system can be easily shown by modifying Proposition 1.1 of Chapter I.) If $\bar{\Gamma}_{BC}^A$ are the Christoffel symbols of $\bar{\nabla}$ with respect to $x^1, \ldots, x^{n+p}$, then $\bar{\Gamma}_{ij}^r(x) = 0$ for every $x \in U$ and for $1 \leq i, j \leq n$ and $n + 1 \leq r \leq n + p$ by assumption (2) (cf. p. 143 of Volume 1). For the natural coordinate system (x^A, X_C^B) in $\pi^{-1}(V) \subset L(N)$ (cf. §7 of Chapter III), the points of $L(N, M)$ over M satisfy $x^r = 0$, $X_i^r = 0$ and $Y_i^r = 0$ for $1 \leq i \leq n$ and $n + 1 \leq r \leq n + p$. Recalling the expression for the connection form in Proposition 7.3 of Chapter III:

$$\psi_B^A = \sum_C Y_C^A \left(dX_B^C + \sum_{D,E} \bar{\Gamma}_{DE}^C X_B^E \, dx^D \right)$$

and making use of the fact that

$$dx^r = 0,\ X_i^r = 0,\ Y_i^r = 0,\ dX_i^r = 0,\ \Gamma_{ij}^r = 0 \quad \text{on } L(N, M)$$

we see that the restriction of ψ_i^r to $L(N, M)$ vanishes for $1 \leq i \leq n$ and $n + 1 \leq r \leq n + p$.

We shall finally prove $(3) \to (1)$. Let X be any horizontal vector of $L(N)$ at a point v of $L(N, M)$. To prove that X is actually tangent to $L(N, M)$, let X' be a tangent vector of $L(N, M)$ at v such that $X' - X$ is vertical. We set $A = \psi(X') - \psi(X)$. Since $\psi(X) = 0$ and since the restriction of ψ to $L(N, M)$ takes its values in $\mathfrak{gl}(n + p, n; \mathbf{R})$, A is an element of $\mathfrak{gl}(n + p, n; \mathbf{R})$. Since the fundamental vector field A^* corresponding to A coincides with $X' - X$ at v and since $GL(n + p, n; \mathbf{R})$ is the structure group of $L(N, M)$, the vector $X' - X$ is tangent to $L(N, M)$. Since X' is also tangent to $L(N, M)$, so is X. Hence every horizontal curve of $L(N)$ starting from a point of $L(N, M)$ lies in $L(N, M)$. In other words, every adapted frame $v \in L(N, M)$ remains in $L(N, M)$ under the parallel displacement along any curve of M. QED.

When M is an auto-parallel submanifold of N, the restriction of ψ to $L(N, M)$ defines a connection in $L(N, M)$ by virtue of Proposition 8.2. From this connection, the homomorphism $h: L(N, M) \to L(M)$ induces a connection in $L(M)$ by Proposition 6.1 of Chapter II. The connection form ω on $L(M)$ so determined is related to the connection form ψ on $L(N)$ by $h^*\omega = h(i^* \circ j^*\psi)$, where $j \circ i: L(N, M) \to L(N)$ is the natural imbedding, and h on the right hand side is the natural homomorphism of $\mathfrak{gl}(n + p, n; \mathbf{R})$ onto $\mathfrak{gl}(n, \mathbf{R})$.

PROPOSITION 8.3. *Every auto-parallel submanifold M of N is totally geodesic.*

Proof. Let τ be a geodesic of N starting from a point x of M and tangent to M at x. Let τ' be the geodesic of M (with respect to the induced affine connection of M) starting from x and tangent to τ at x. Then τ' is a geodesic of N. Indeed, if $\tau' = x_t$, then the vector field $\dot{x}_t$ along τ' is parallel with respect to the connection of M by definition of a geodesic. It follows that this vector field is parallel with respect to the affine connection of N.

VII. SUBMANIFOLDS

Thus τ' is a geodesic of N. Since any two geodesics tangent to each other at a point must coincide, we have $\tau = \tau'$, that is, τ is in M.
QED.

The following result is in E. Cartan [8; Chapter V].

THEOREM 8.4. *Let N be a manifold with an affine connection with zero torsion. Then every totally geodesic submanifold M of N is auto-parallel.*

Proof. For a given point o of M we choose V and $x^1, \ldots, x^{n+p}$ and define U as in the proof of $(2) \to (3)$ in Proposition 8.2. For any $x \in U$, we take a geodesic x_t such that $x_0 = x$ and $\dot{x}_0 = \sum_{i=0}^{n} a^i (\partial/\partial x^i)_x$ (that is, tangent to M at the initial point x), where $a^1, \ldots, a^n$ are arbitrarily fixed. Since M is totally geodesic by assumption, the geodesic x_t stays in M for small values of t. The equation (cf. Proposition 7.8 of Chapter III)

$$\frac{d^2 x^A}{dt^2} + \sum_{B,C} \Gamma^A_{BC}(x_t) \frac{dx^B}{dt} \frac{dx^C}{dt} = 0$$

reduces, for $n + 1 \leq r \leq n + p$, to

$$\sum_{j,k=1}^{n} \Gamma^r_{jk}(x_t) \frac{dx^j}{dt} \frac{dx^k}{dt} = 0.$$

In particular, we have for $t = 0$

$$\sum_{j,k=1}^{n} \Gamma^r_{jk}(x) a^j a^k = 0.$$

Since the torsion is 0, we have $\Gamma^r_{jk} = \Gamma^r_{kj}$. Since $a^1, \ldots, a^n$ are arbitrary, we may conclude $\Gamma^r_{jk}(x) = 0$ for $1 \leq j, k \leq n$ and $n + 1 \leq r \leq n + p$. This is valid at every point x of U. Thus $\overline{\nabla}_{\partial/\partial x^i}(\partial/\partial x^j)$, $1 \leq i, j \leq n$, are tangent to U at every point of U. It follows that if X and Y are vector fields on U (or on M), then $\overline{\nabla}_X Y$ is tangent to U (or M) everywhere. By Proposition 8.2 we conclude that M is an auto-parallel submanifold of M. QED.

Let M be a submanifold of N. For any covariant tensor field $\bar{K}$ on N we speak of its *restriction* to M which is defined in a natural way. For a tensor field $\bar{K}$ of type $(1, s)$ on N, we shall say that $\bar{K}$ *can be restricted* to M if, by considering $\bar{K}$ as an s-linear mapping $\mathfrak{X}(N) \times \cdots \times \mathfrak{X}(N) \to \mathfrak{X}(N)$ in the manner of Proposition 3.1 of Chapter I, $\bar{K}(\bar{X}, \ldots, \bar{X}_s)$ is tangent to M at every point of M whenever the vector fields $\bar{X}_1, \ldots, \bar{X}_s$ are tangent to M at every point of M. Under this condition it is obvious that $\bar{K}$ induces a tensor field of type $(1, s)$, called the *restriction* of $\bar{K}$ to M, in the natural fashion; at each point x of M, K_x is an s-linear mapping of $T_x(M) \times \cdots \times T_x(M)$ into $T_x(M)$ which is the restriction of the s-linear mapping $\bar{K}$ of $T_x(N) \times \cdots \times T_x(N)$ into $T_x(N)$.

Now we have

PROPOSITION 8.5. *Let N be a manifold with an affine connection $\bar{\nabla}$ and let M be an auto-parallel submanifold of N with induced affine connection ∇.*

(1) *If X and Y are vector fields on M, then $\bar{\nabla}_X Y$ is tangent to M at every point of M and $\bar{\nabla}_X Y = \nabla_X Y$ on M.*

(2) *If K is the restriction to M of a covariant tensor field $\bar{K}$ on N, then ∇K is the restriction to M of $\bar{\nabla} \bar{K}$.*

(3) *Let $\bar{K}$ be a tensor field of type $(1, s)$ on N which can be restricted to M and let K be its restriction. Then $\bar{\nabla} \bar{K}$ can be restricted to M and its restriction coincides with ∇K.*

Proof. (1) is contained in Proposition 8.2. To prove (2) for K of type $(0, s)$, let $X, X_1, \ldots, X_s$ be vector fields on M and extend them to vector fields $\bar{X}, \bar{X}_1, \ldots, \bar{X}_s$ on N; actually, we do this locally and that is sufficient for our purpose. Then

$$(\bar{\nabla}\bar{K})(\bar{X}_1, \ldots, \bar{X}_s; \bar{X})$$
$$= \bar{\nabla}_{\bar{X}}(\bar{K}(\bar{X}_1, \ldots, \bar{X}_s)) - \sum_{i=1}^{s} \bar{K}(\bar{X}_1, \ldots, \bar{\nabla}_{\bar{X}} \bar{X}_i, \ldots, \bar{X}_s)$$

and its value at $x \in M$ is equal to the value of

$$(\nabla K)(X_1, \ldots, X_s; X)$$
$$= \nabla_X (K(X_1, \ldots, X_s)) - \sum_{j=1}^{s} K(X_1, \ldots, \nabla_X X_i, \ldots, X_s)$$

at x by virtue of (1). The proof of (3) is quite similar.   QED.

PROPOSITION 8.6. *Let N and M be as in Proposition 8.5.*

(1) *The torsion tensor field $\bar{T}$ and the curvature tensor field $\bar{R}$ of N can be restricted to M and their restrictions are equal to the torsion tensor field T and the curvature tensor field R of M, respectively.*

(2) *The m-th covariant differentials $\bar{\nabla}^m \bar{T}$ and $\bar{\nabla}^m \bar{R}$ can be restricted to M and their restrictions are equal to $\nabla^m T$ and $\nabla^m R$, respectively, where m is an arbitrary positive integer.*

VII. SUBMANIFOLDS 59

Proof. We first prove (1). If $\bar{X}$ and $\bar{Y}$ are vector fields on N which are tangent to M at every point of M, then $\overline{\nabla}_{\bar{X}} \bar{Y}$, $\overline{\nabla}_{\bar{Y}} \bar{X}$ and $[\bar{X}, \bar{Y}]$ have the same property. Thus $\bar{T}(\bar{X}, \bar{Y})$ is also tangent to M at every point of M and is equal to $T(X, Y)$ on M. The proof for $\bar{R}$ is similar. (2) follows from (1) and (3) of Proposition 8.5.
QED.

COROLLARY 8.7. *Let M be an auto-parallel submanifold of a manifold N with an affine connection. Each of the following properties for N is inherited by M (with induced affine connection):*
(1) *the torsion tensor is zero;*
(2) *the curvature tensor is zero;*
(3) *the torsion tensor field is parallel;*
(4) *the curvature tensor field is parallel.*

In the Riemannian case we have

PROPOSITION 8.8. *Let N be a Riemannian manifold. For a submanifold M of N, the following conditions are equivalent.*
(1) *M is auto-parallel;*
(2) *M is totally geodesic;*
(3) *the second fundamental form of M is identically zero.*
When M is auto-parallel, the Riemannian connection of M with respect to the induced Riemannian metric coincides with the induced affine connection.

Proof. By definition of the second fundamental form α in §3:

$$\overline{\nabla}_X Y = \nabla_X Y + \alpha(X, Y),$$

we see that α is identically 0 if and only if $\overline{\nabla}_X Y$ is tangent to M whenever X and Y are tangent vector fields on M. Thus the equivalence of (1), (2), and (3) follows immediately from Proposition 8.2, Proposition 8.3, and Theorem 8.4. Also when α is identically 0, we know that the affine connection $\overline{\nabla}$ induces the affine connection ∇ on M, which is the Riemannian connection for the induced Riemannian metric on M by Proposition 3.1.
QED.

THEOREM 8.9. *Let M be an auto-parallel submanifold of a Riemannian manifold N. Let X be an infinitesimal isometry of N. At each point of M, decompose X into a vector tangent to M and a vector normal to M. Then the tangential component of X is an infinitesimal isometry of M.*

Proof. We first prove the following

60   FOUNDATIONS OF DIFFERENTIAL GEOMETRY

LEMMA. *Every normal vector to M remains normal under the parallel displacement along any curve contained in M.*

Proof. Let τ be a curve in M. Let X and Y be vector fields of N along τ parallel with respect to the Riemannian connection of N. If g is the Riemannian metric tensor of N, then $g(X, Y)$ is constant along τ. If Y is tangent to M at a point, then Y is tangent to M at every point of τ as M is auto-parallel. If X is furthermore normal to M at a point, then the constant function $g(X, Y)$ vanishes everywhere along τ. Hence, X is normal to M at every point of τ. This proves our lemma.

At each point of M, we decompose an infinitesimal isometry X of N as follows:
$$X = X' + X'',$$
where X' is tangent to M and X'' is normal to M. Let g denote also the induced Riemannian metric tensor on M. By Proposition 2.5 and Proposition 3.2 of Chapter VI, it suffices to prove that $g(\nabla_Y X', Y) = 0$ for all vector fields Y of M. By the same propositions, we have $g(\bar{\nabla}_Y X, Y) = 0$. Hence, we have
$$0 = g(\bar{\nabla}_Y X, Y) = g(\nabla_Y X', Y) + g(\bar{\nabla}_Y X'', Y).$$
By the lemma above, $\bar{\nabla}_Y X''$ is normal to M so that $g(\bar{\nabla}_Y X'', Y) = 0$. Hence, $g(\nabla_Y X', Y) = 0$.   QED.

COROLLARY 8.10. *Every closed auto-parallel submanifold of a homogeneous Riemannian manifold is homogeneous.*

Proof. Assume N is homogeneous and M is auto-parallel in N. Given a tangent vector of M, there exists an infinitesimal isometry X of N which extends it as N is homogeneous. The tangential component X' of X in Theorem 8.9 is an infinitesimal isometry of M which extends the given vector of M. On the other hand, every Cauchy sequence of M is a Cauchy sequence of N and, being homogeneous, N is complete (cf. Theorem 4.5 of Chapter IV). Hence, M is also complete. By Theorem 2.4 of Chapter VI, X' generates a *global* 1-parameter group of isometries of M. This shows that the set of all X' thus obtained generates a transitive group of isometries of M.   QED.

Remark. Evidently, the above argument shows that the assumption that M is closed can be replaced by the one that M is complete.

VII. SUBMANIFOLDS

We shall conclude this section by an important example of auto-parallel submanifold.

Example 8.1. Let N be a manifold with an affine connection and let G be any set of affine transformations of N. Let F be the set of points of N which are left fixed by G. Then each connected component M of F is a closed auto-parallel submanifold of N (provided that F is non-empty). Indeed, let $x \in M$. Then every element of G induces an endomorphism of $T_x(N)$. Let V be the subspace of $T_x(N)$ consisting of vectors which are left fixed by the endomorphisms of $T_x(N)$ induced by G. Let U^* be a neighborhood of the origin in $T_x(N)$ such that the exponential mapping $\exp_x \colon U^* \to N$ is one-to-one. Let $U = \exp_x(U^*)$. It is easy to see that $U \cap F = \exp_x(U^* \cap V)$. Since $U \cap F$ is a neighborhood of x in F and $\exp_x(U^* \cap V)$ is a submanifold of N, M is also a submanifold of N. Evidently, M is closed in N. If X is a parallel vector field along a curve τ in N, then every affine transformation f of N maps X into a parallel vector field along $f(\tau)$. Assuming that τ is in M and X is tangent to M at a single point of τ, we shall show that X is tangent to M at every point of τ. Since τ is in M, $f(\tau) = \tau$, and hence $f(X)$ is a parallel vector field along τ for every $f \in G$. Since f leaves M fixed, X and $f(X)$ coincide at the point where X is tangent to M. The uniqueness of the parallel displacement implies that X and $f(X)$ coincide at every point of τ. Thus, X is invariant by every element f of G. This implies that X is tangent to M at every point of τ, thus completing the proof of our assertion. Similarly, if F is the set of common zeros of any set of infinitesimal affine transformations of N, then each connected component M of F is a closed, auto-parallel submanifold of N. We have only to apply the same argument to the set of local 1-parameter group of local affine transformations of N generated by the given set of infinitesimal affine transformations.

CHAPTER VIII

Variations of the Length Integral

1. Jacobi fields

Let M be an n-dimensional manifold with an affine connection and let T and R be the torsion and the curvature tensor fields on M. A vector field X along a geodesic $\tau = x_t$ of M is called a *Jacobi field* if it satisfies the following second order ordinary linear differential equation, called the *Jacobi equation*:

$$\nabla^2_{\dot{x}_t} X + \nabla_{\dot{x}_t}(T(X, \dot{x}_t)) + R(X, \dot{x}_t)\dot{x}_t = 0,$$

where $\dot{x}_t$ is the vector tangent to τ at the point x_t. We denote by J_τ the set of Jacobi fields along τ; obviously, it forms a vector space over **R**.

PROPOSITION 1.1. *A Jacobi field X along $\tau = x_t$ is uniquely determined by the values of X and $\nabla_{\dot{x}_t} X$ at one point x_a of τ. In particular, we have*
$$\dim J_\tau = 2n.$$

Proof. This is a consequence of the fact that the Jacobi equation is a second order ordinary differential equation. QED.

We shall now give a geometric interpretation of a Jacobi field. By a *variation* of a geodesic $\tau = x_t$, $0 \leq t \leq 1$, we shall mean a 1-parameter family of geodesics τ^s, $-\varepsilon < s < \varepsilon$, such that $\tau = \tau^0$. More precisely, it is a differentiable mapping of class C^∞ of $[0, 1] \times (-\varepsilon, \varepsilon)$ into M, $(t, s) \to x_t^s$, such that

(1) For each fixed $s \in (-\varepsilon, \varepsilon)$, $\tau^s = x_t^s$ is a geodesic;

(2) $x_t^0 = x_t$ for $0 \leq t \leq 1$.

An *infinitesimal variation* X of a geodesic τ is a vector field along τ induced by some variation $\tau^s = x_t^s$ of τ in the following manner.

For each fixed t, we denote by $\tau_t = x^s_{(t)}$ the curve described by x^s_t, $-\varepsilon < s < \varepsilon$. The tangent vector of τ_t at $x^s_{(t)}$ is denoted by $\dot{x}^s_{(t)}$. Then X is defined by

$$X_x = \dot{x}^0_{(t)} \quad \text{for } 0 \leq t \leq 1.$$

THEOREM 1.2. *A vector field X along a geodesic τ is a Jacobi field if and only if it is an infinitesimal variation of τ.*

Proof. Supposing that X is an infinitesimal variation of τ, let $\tau^s = x^s_t$ be a variation of τ which induces X. Let γ be a mapping of $[0, 1] \times (-\varepsilon, \varepsilon)$ into the bundle $L(M)$ of linear frames such that, if we set $u^s_t = \gamma(s, t)$, then

(1) $\pi(u^s_t) = x^s_t \quad \text{for } 0 \leq t \leq 1, \; -\varepsilon < s < \varepsilon,$

where π is the projection of $L(M)$ onto M;

(2) For each fixed s, u^s_t, $0 \leq t \leq 1$, is a horizontal curve.

Let U and V be vector fields on $[0, 1] \times (-\varepsilon, \varepsilon)$ defined by

$$U = \partial/\partial s, \quad V = \partial/\partial t.$$

Let θ, ω, Θ, and Ω be the canonical form, the connection form, the torsion form, and the curvature form on $L(M)$ respectively. We set

$$\theta^* = \gamma^*\theta, \; \omega^* = \gamma^*\omega, \; \Theta^* = \gamma^*\Theta, \; \Omega^* = \gamma^*\Omega.$$

Then we have

(3) $[U, V] = 0$;
(4) $\omega^*(V) = 0$;
(5) $V(\theta^*(V)) = 0$.

The assertion (3) is evident. (4) follows from (2). (1) and (2) imply that $\theta^*(V)$ is constant (cf. Proposition 6.3 of Chapter III) and hence (5).

Proposition 3.11 of Chapter I and the first structure equation imply

(6) $U(\theta^*(V)) - V(\theta^*(U)) - \theta^*([U, V]) = 2\, d\theta^*(U, V)$
$= -\omega^*(U) \cdot \theta^*(V) + \omega^*(V) \cdot \theta^*(U) + 2\Theta^*(U, V).$

By (3), (4), and (6), we have

(7) $U(\theta^*(V)) - V(\theta^*(U)) = -\omega^*(U) \cdot \theta^*(V) + 2\Theta^*(U, V).$

VIII. VARIATIONS OF THE LENGTH INTEGRAL

If we apply V to (7), then by (5) we have

(8) $\quad VU(\theta^*(V)) - V^2(\theta^*(U))$
$$= -V(\omega^*(U)) \cdot \theta^*(V) + 2V(\Theta^*(U, V)).$$

On the other hand, (3) and (4) imply

(9) $\quad VU(\theta^*(V)) = UV(\theta^*(V)) - [U, V](\theta^*(V)) = 0.$

Combining (8) and (9), we obtain

(10) $\quad V^2(\theta^*(U)) = V(\omega^*(U)) \cdot \theta^*(V) - 2V(\Theta^*(U, V)).$

Similarly, Proposition 3.11 of Chapter I and the second structure equation imply

(11) $\quad U(\omega^*(V)) - V(\omega^*(U)) - \omega^*([U, V]) = 2\, d\omega^*(U, V)$
$$= -\omega^*(U)\omega^*(V) + \omega^*(V)\omega^*(U) + 2\Omega^*(U, V).$$

(3), (4), and (11) imply

(12) $\quad V(\omega^*(U)) = -2\Omega^*(U, V).$

Combining (10) and (12), we obtain finally

(13) $\quad V^2(\theta^*(U)) + 2V(\Theta^*(U, V)) + 2\Omega^*(U, V) \cdot \theta^*(V) = 0.$

We shall show that this is nothing but the Jacobi equation. Indeed, we have

(14) $\quad \theta^*(U)_{(t,0)} = \theta(\dot{u}^0_{(t)}) = u_t^{-1}(X_{x_t}),$ where $u_t = u_t^0.$

The second equality in (14) follows from the definition of θ and from $\pi(\dot{u}^0_{(t)}) = X_{x_t}$. In general, if Z is a vector field along x_t, then (cf. Lemma in the proof of Proposition 1.1 of Chapter III)

$$\nabla_{\dot{x}_t} Z = u_t \left(\frac{d}{dt} (u_t^{-1} Z_{x_t}) \right).$$

Since $V = \partial/\partial t$, from (14) we obtain

(15) $\quad u_t((V^2(\theta^*(U)))_{(t,0)}) = \nabla^2_{\dot{x}_t} X.$

Similarly, from $\pi(\dot{u}^0_{(t)}) = X_{x_t}$ and $\pi(\dot{u}^{(0)}_t) = \dot{x}_t$, we obtain (cf. the definition of the torsion tensor T in §5 of Chapter III)

(16) $\quad 2\Theta^*(U, V)_{(t,0)} = 2\Theta(\dot{u}^0_{(t)}, \dot{u}^{(0)}_t) = u_t^{-1}(T(X_{x_t}, \dot{x}_t)).$

Hence,

(17) $u_t((V(2\Theta^*(U, V)))_{(t,0)}) = \nabla_{\dot{x}_t}(T(X_{x_t}, \dot{x}_t))$.

From the definition of the curvature tensor R, we obtain

(18) $u_t((2\Omega^*(U, V)\theta^*(V))_{(t,0)}) = R(X_{x_t}, \dot{x}_t)\dot{x}_t$.

Restricting (13) to the line $s = 0$ and applying (15), (17), and (18), we obtain

(19) $\nabla^2_{\dot{x}_t}X + \nabla_{\dot{x}_t}(T(X, \dot{x}_t)) + R(X, \dot{x}_t)\dot{x}_t = 0$.

Conversely, let X be a Jacobi field along $\tau = x_t$, $0 \leq t \leq 1$. Choose two parameter values a and b close to each other so that x_a and x_b are in a convex neighborhood (cf. Theorem 8.7 of Chapter III) and so that every Jacobi field along τ is uniquely determined by its values at x_a and x_b. This is possible because the Jacobi equation is a second order ordinary linear differential equation.

Let x_a^s (resp. x_b^s), $-\varepsilon < s < \varepsilon$, be a curve such that $x_a = x_a^0$ (resp. $x_b = x_b^0$) and such that X_{x_a} (resp. X_{x_b}) is its tangent vector at x_a (resp. x_b). If we take ε sufficiently small, then for each s, we have a unique geodesic x_t^s, $0 \leq t \leq 1$, through x_a^s and x_b^s such that x_t^s, $a \leq t \leq b$, lies in the convex neighborhood chosen above. Then the family of geodesics $\tau^s = x_t^s$ is a variation of τ. Let Y be the Jacobi field along τ induced by this variation. Since X and Y coincide at x_a and x_b, they are identically equal, thus proving that X is the infinitesimal variation induced by the variation τ^s.
QED.

PROPOSITION 1.3. *Let X be an infinitesimal affine transformation of M. Then X is a Jacobi field along every geodesic $\tau = x_t$, $0 \leq t \leq 1$, of M.*

Proof. We shall give two proofs. By Propositions 2.5 and 2.6 of Chapter VI, a vector field X is an infinitesimal affine transformation if and only if

$$\nabla^2_Y X + \nabla_Y(T(X, Y)) + R(X, Y)Y + A_X(\nabla_Y Y) = 0$$

for every vector field Y, where A_X is the tensor field of type $(1, 1)$ induced by the derivation $L_X - \nabla_X$ (cf. §2 of Chapter VI). Applying the above formula to the vector field $\dot{x}_t$ along τ and noting the fact that τ is a geodesic, we see that X satisfies the Jacobi equation.

Another proof can be given by means of Theorem 1.2. Let φ_s, $-\varepsilon < s < \varepsilon$, be a local 1-parameter group of local affine transformations generated by X. Taking ε sufficiently small we may assume that $\varphi_s(x_t)$ is defined for $0 \leq t \leq 1$ and $-\varepsilon < s < \varepsilon$. Then it is clear that X restricted to τ is the infinitesimal variation induced by the variation $\tau^s = \varphi_s(\tau)$. QED.

Let τ be a geodesic in M. Two points x and y on τ are said to be *conjugate* to each other along τ if there is a non-zero Jacobi field X along τ which vanishes both at x and at y.

Let $\tau^s = x_t^s$, $0 \leq t \leq 1$ and $-\varepsilon < s < \varepsilon$, be a variation of a geodesic τ such that $x = x_0^s$ and $y = x_1^s$ for all s. Then x and y are conjugate along τ provided that the Jacobi field induced from τ^s is non-zero. It should be noted that not every Jacobi field vanishing at x and y can be induced by such a variation. The proof of Theorem 1.2 shows that we can always construct a variation τ^s of τ such that the given Jacobi field is induced by τ^s and that $x = x_0^s$ for all s. Although the tangent vector of x_1^s at $y = x_1^0$ is zero, x_1^s may not coincide with y.

We shall now interpret conjugate points in terms of the exponential mapping. Let $\exp_x \colon T_x(M) \to M$ be the exponential mapping defined in §6 of Chapter III. For the sake of simplicity, we shall assume that the affine connection is complete. A point X of $T_x(M)$ is called a *conjugate point of x in* $T_x(M)$ if $\exp_x$ is singular at X (i.e., the Jacobian matrix of $\exp_x$ is singular at X).

THEOREM 1.4. *If $X \in T_x(M)$ is a conjugate point of x in $T_x(M)$, then $\exp_x(X) \in M$ is a conjugate point of x along the geodesic $\exp_x(tX)$, $0 \leq t \leq 1$. Conversely, every conjugate point of x can be thus obtained.*

Proof. Set $T_x = T_x(M)$. In order to avoid any notational confusion, we denote by f the differential of $\exp_x$ at X. Then f is a linear mapping of $T_X(T_x)$ into $T_y(M)$, where $y = \exp_x(X)$. If X is conjugate to x in $T_x(M)$, then f is singular. Consider a line through the point X in T_x whose tangent vector at X is annihilated by f. Let X^s, $-\varepsilon < s < \varepsilon$, be such a line, where $X = X^0$. Let $\tau^s = x_t^s$ be the variation of $\tau = \exp_x(tX)$ defined by $x_t^s = \exp_x(tX^s)$. It is clear that the Jacobi field induced by τ^s vanishes at x and y.

Conversely, every Jacobi field Y along $\tau = \exp_x(tX)$ vanishing at x and $y = \exp_x(X)$ is induced by a variation $\tau^s = x_t^s$ of τ such that $x = x_0^s$. Then there is a curve X^s through X in T_x such that $X = X^0$ and such that $x_t^s = \exp_x(tX^s)$. Since the tangent vector of the curve X^s at X^0 is mapped into $Y_y = 0$ by f, f is singular. This means that X is a conjugate point of x in $T_x(M)$. QED.

Remark. Let $\tau = x_t$, $0 \leq t < \infty$, be a geodesic such that $x_t = \exp_x(tX)$, where $X \in T_x(M)$. Since $\exp_x$ is nonsingular at the origin of $T_x(M)$, there is a positive number, say, a, such that there is no conjugate point of x on x_t, $0 \leq t \leq a$. If there are conjugate points of x on τ, let $S = \{s > 0; x_s \text{ is a conjugate point of } x \text{ along } x_t, 0 \leq t \leq s\}$, and let $b = \inf S$. Since $\exp_x$ is singular at sX, it is singular at bX, that is, x_b is a conjugate point of x. Thus we may speak of the *first conjugate point* of x along τ.

Jacobi equations of non-symmetrical invariant connections on homogeneous spaces have been considered by Chavel [1], [2] (see also Rauch [5]).

2. Jacobi fields in a Riemannian manifold

From now on we shall assume that M is a Riemannian manifold and that every geodesic $\tau = x_t$ is parametrized by its arc length unless otherwise stated. For any vector field X along τ, the vector fields $\nabla_{\dot{x}_t} X$ and $\nabla^2_{\dot{x}_t} X$ along τ will be denoted by X' and X'', respectively.

Every geodesic $\tau = x_t$ admits two Jacobi fields in a natural way. One is given by $\dot{x}_t$ and will be denoted by $\dot{\tau}$. The other is given by $t\dot{x}_t$ and will be denoted by $\hat{\tau}$. It is a trivial matter to verify that $\dot{\tau}$ and $\hat{\tau}$ satisfy the Jacobi equation.

PROPOSITION 2.1. *Every Jacobi field X along a geodesic $\tau = x_t$ of a Riemannian manifold M can be uniquely decomposed in the following form:*
$$X = a\dot{\tau} + b\hat{\tau} + Y,$$
where a and b are real numbers and Y is a Jacobi field along τ which is everywhere perpendicular to τ.

Proof. Let g be the Riemannian metric on M and set
$$a = g(\dot{x}_0, X_{x_0}), \quad b = g(\dot{x}_0, X'_{x_0}), \quad Y = X - a\dot{\tau} - b\hat{\tau}.$$

Since X, $\dot\tau$, and $\hat\tau$ satisfy the Jacobi equation, so does Y. Taking the inner product of $\dot\tau$ with the Jacobi equation satisfied by Y, we obtain
$$g(Y'', \dot\tau) + g(R(Y, \dot\tau)\dot\tau, \dot\tau) = 0.$$
Since $R(Y, \dot\tau)$ is a skew-symmetric linear transformation of the tangent space at each point (cf. Proposition 2.1 of Chapter V), the second term in the above equation vanishes identically. Hence, the first term also vanishes. From $\nabla_{\dot\tau}\dot\tau = 0$ and $\nabla g = 0$, we obtain
$$\frac{d^2}{dt^2} g(Y, \dot\tau) = \nabla_{\dot\tau}^2 g(Y, \dot\tau) = g(Y'', \dot\tau) = 0.$$
Thus, $g(Y, \dot\tau) = At + B$, where A and B are constants. Since $\hat\tau_{x_0} = 0$, we have
$$B = g(Y_{x_0}, \dot\tau_{x_0}) = g(X_{x_0} - a\dot\tau_{x_0}, \dot\tau_{x_0}) = g(X_{x_0}, \dot x_0) - g(a\dot x_0, \dot x_0)$$
$$= a - a = 0.$$
As τ is a geodesic, we have $Y' = X' - b\dot\tau$. Hence,
$$A = \frac{d}{dt} g(Y, \dot\tau)_{t=0} = g(X'_{x_0}, \dot x_0) - g(b\dot x_0, \dot x_0) = b - b = 0.$$
Thus, $g(Y, \dot\tau) = 0$.

In order to prove the uniqueness of the decomposition, let
$$X = a'\dot\tau + b'\hat\tau + Z$$
be another decomposition of X such that Z is perpendicular to τ. For each t, we have
$$X_{x_t} = (a + bt)\dot x_t + Y_{x_t} = (a' + b't)\dot x_t + Z_{x_t}.$$
Since both Y_{x_t} and Z_{x_t} are perpendicular to $\dot x_t$, we have
$$a + bt = a' + b't, \qquad Y_{x_t} = Z_{x_t},$$
and hence
$$a = a', \qquad b = b', \qquad Y = Z.$$
QED.

COROLLARY 2.2. *If a Jacobi field along a geodesic τ is perpendicular to τ at two points, it is perpendicular to τ at every point of τ.*

Proof. Suppose a Jacobi field X is perpendicular to $\tau = x_t$ at x_r and x_s. Let $X = a\dot{\tau} + b\hat{\tau} + Y$ as in Proposition 2.1. Then we have
$$0 = (a + br)\dot{x}_r, \qquad 0 = (a + bs)\dot{x}_s.$$
Since $r \neq s$, we have $a = b = 0$. QED.

We shall establish a formula which will be needed in subsequent sections.

PROPOSITION 2.3. *Let X be a Jacobi field and Y a piecewise differentiable vector field along a geodesic $\tau = x_t$. Then, for any parameter values a and b of t, we have*
$$\int_a^b [g(X', Y') - g(R(X, \dot{\tau})\dot{\tau}, Y)] \, dt + g(X', Y)_{t=a} - g(X', Y)_{t=b} = 0.$$

Proof. We obtain the desired formula by integrating the following equality:
$$\frac{d}{dt} g(X', Y) = g(X', Y') + g(X'', Y)$$
$$= g(X', Y') - g(R(X, \dot{\tau})\dot{\tau}, Y).$$
QED.

COROLLARY 2.4. *If $g(R(Y,\dot{\tau})\dot{\tau}, Y) \leq 0$ for every vector field Y along τ, no two points of τ are conjugate along τ. In particular, a Riemannian manifold with non-positive sectional curvature has no conjugate points.*

Proof. Let X be a Jacobi field along τ which vanishes at x_a and x_b. In the formula of Proposition 2.3, set $Y = X$. Then the unintegrated terms vanish. On the other hand, the integrand in the integrated term is non-negative and hence must be zero. This implies $g(X', X') = 0$. Since X vanishes at x_a, $X' = 0$ implies $X = 0$. QED.

The following proposition is also for later use.

PROPOSITION 2.5. *If X and Y are Jacobi fields along a geodesic τ, then*
$$g(X, Y') - g(X', Y) = constant.$$
Moreover, if X and Y vanish at some point, say x_a, of τ, then
$$g(X, Y') - g(X', Y) = 0.$$

Proof.
$$\frac{d}{dt} g(X, Y') = g(X', Y') + g(X, Y'')$$
$$= g(X', Y') - g(X, R(Y, \dot\tau)\dot\tau).$$
Similarly,
$$\frac{d}{dt} g(X', Y) = g(X', Y') - g(R(X, \dot\tau)\dot\tau, Y).$$

Since $g(X, R(Y, \dot\tau)\dot\tau) = g(R(X, \dot\tau)\dot\tau, Y)$ (cf. Proposition 2.1 of Chapter V), we have
$$\frac{d}{dt} (g(X, Y') - g(X', Y)) = 0.$$
QED.

Example 2.1. Let $\tau = x_t$ be a geodesic in a Riemannian manifold M with positive constant curvature k. Let $\dot x_0, Y_1, \ldots, Y_{n-1}$ be an orthonormal basis for $T_{x_0}(M)$. By the parallel displacement along τ, we extend $Y_1, \ldots, Y_{n-1}$ to parallel vector fields along τ so that $\dot\tau, Y_1, \ldots, Y_{n-1}$ form an orthonormal basis for the tangent space at every point of τ. Set
$$(U_i)_{x_t} = \sin(\sqrt{k}t)(Y_i)_{x_t}, \qquad (V_i)_{x_t} = \cos(\sqrt{k}t)(Y_i)_{x_t}.$$

By Corollary 2.3 of Chapter V we can verify that $U_1, \ldots, U_{n-1}$, $V_1, \ldots, V_{n-1}$ are Jacobi fields along τ. It follows (cf. Proposition 1.1) that these $2n-2$ Jacobi fields, $\dot\tau$ and $\hat\tau$ form a basis for the space J_τ of Jacobi fields along τ. It follows also that the conjugate points of x_0 along τ are given by the points with parameter values $t = m\pi/\sqrt{k}$, $m = \pm 1, \pm 2, \ldots$.

3. Conjugate points

As in §2, let M be a Riemannian manifold and $\tau = x_t$ a geodesic in M parametrized by its arc length. For each piecewise differentiable vector field X along τ, we set
$$I^b(X) = \int_a^b [g(X', X') - g(R(X, \dot\tau)\dot\tau, X)]\, dt,$$

where X' denotes the covariant derivative of X in the direction of $\dot{\tau}$. In this section, we shall investigate properties of $I_a^b(X)$ and shall derive some consequences on the distance between two consecutive conjugate points. The integral $I_a^b(X)$ is also closely related with the index form of Morse, which will be defined and studied later in §6.

PROPOSITION 3.1. *Let $\tau = x_t$, $a \geq t \geq b$, be a geodesic in M such that x_a has no conjugate point along $\tau = x_t$ for $a \leq t \leq b$. Let Y be a Jacobi field along τ which vanishes at x_a and is perpendicular to τ. Let X be a piecewise differentiable vector field along τ which vanishes at x_a and is perpendicular to τ. If $X_{x_b} = Y_{x_b}$, then we have*

$$I_a^b(X) \geq I_a^b(Y),$$

and the equality holds only when $X = Y$.

Proof. Let $J_{\tau,a}$ be the space of all Jacobi fields along τ which vanish at x_a and is perpendicular to τ. By Proposition 1.1, the space of all Jacobi fields along τ which vanish at x_a is of dimension n. It follows (cf. Proposition 2.1) that $\dim J_{\tau,a} = n - 1$. Let Y_1, ..., Y_{n-1} be a basis for $J_{\tau,a}$. Since $Y \in J_{\tau,a}$, we may write

$$Y = a_1 Y_1 + \cdots + a_{n-1} Y_{n-1},$$

where $a_1, \ldots, a_{n-1}$ are constants. Since there is no conjugate point of x_a on τ for $a \leq t \leq b$, $Y_1, \ldots, Y_{n-1}$ are linearly independent at every point x_t for $a < t \leq b$. There exist therefore functions $f_1(t), \ldots, f_{n-1}(t)$ such that

$$X = f_1 Y_1 + \cdots + f_{n-1} Y_{n-1}.$$

We have (Σ standing for $\Sigma_{i=1}^{n-1}$)

$$g(X', X') = g(\sum f_i' Y_i, \sum f_i' Y_i) + 2g(\sum f_i' Y_i, \sum f_i Y_i')$$
$$+ g(\sum f_i Y_i', \sum f_i Y_i'),$$

where each f_i' denotes df_i/dt. We also have

$$-g(R(X, \dot{\tau})\dot{\tau}, X) = -\sum f_i g(R(Y_i, \dot{\tau})\dot{\tau}, X) = \sum f_i g(Y_i'', X)$$
$$= g(\sum f_i Y_i'', \sum f_i Y_i),$$

VIII. VARIATIONS OF THE LENGTH INTEGRAL

where the second equality is a consequence of the assumption that each Y_i is a Jacobi field. On the other hand, we have

$$\frac{d}{dt} g(\sum f_i Y_i, \sum f_i Y_i')$$
$$= g(\sum f_i' Y_i, \sum f_i Y_i') + g(\sum f_i Y_i', \sum f_i Y_i')$$
$$+ g(\sum f_i Y_i, \sum f_i' Y_i') + g(\sum f_i Y_i, \sum f_i Y_i'').$$

Combining these three equalities we obtain

$$g(X', X') - g(R(X, \dot{\tau})\dot{\tau}, X)$$
$$= g(\sum f_i' Y_i, \sum f_i' Y_i) + \frac{d}{dt} g(\sum f_i Y_i, \sum f_i Y_i')$$
$$+ g(\sum f_i' Y_i, \sum f_i Y_i') - g(\sum f_i Y_i, \sum f_i' Y_i').$$

Since, by Proposition 2.5, we have

$$g(\sum f_i' Y_i, \sum f_i Y_i') - g(\sum f_i Y_i, \sum f_i' Y_i')$$
$$= \sum_{i,j} f_i' f_j (g(Y_i, Y_j') - g(Y_j, Y_i')) = 0,$$

we obtain

$$I_a^b(X) = \int_a^b g(\sum f_i' Y_i, \sum f_i' Y_i) \, dt + g(\sum f_i Y_i, \sum f_i Y_i')_{t=b}.$$

Similarly, we obtain

$$I_a^b(Y) = \int_a^b g(\sum a_i' Y_i, \sum a_i' Y_i) \, dt + g(\sum a_i Y_i, \sum a_i Y_i')_{t=b}.$$

Since $a_i' = da_i/dt = 0$, we have

$$I_a^b(Y) = g(\sum a_i Y_i, \sum a_i Y_i')_{t=b}.$$

By our assumption that $X_{x_b} = Y_{x_b}$, we have $a_i = f_i(b)$ for $i = 1, \ldots, n-1$. Hence,

$$I_a^b(X) - I_a^b(Y) = \int_a^b g(\sum f_i' Y_i, \sum f_i' Y_i) \geq 0.$$

Obviously, the equality holds if and only if $f_i' = 0$ for $i = 1, \ldots, n-1$. Since $a_i = f(b), f_i' = 0$ implies $a_i = f(t)$ for all t and hence $X = Y$. QED.

COROLLARY 3.2. *Let $\tau = x_t$, $a \leq t \leq b$, be a geodesic in M such that x_a has no conjugate point along $\tau = x_t$ for $a \leq t \leq b$. If X is a piecewise differentiable vector field along τ vanishing at x_a and x_b and perpendicular to τ, then*
$$I_a^b(X) \geq 0,$$
and the equality holds only when $X = 0$.

Proof. Set $Y = 0$ in Proposition 3.1. QED.

As an application of this corollary we prove

THEOREM 3.3. *Let M be a Riemannian manifold with sectional curvature $\geq k_0 > 0$. Then, for every geodesic τ of M, the distance between two consecutive conjugate points of τ is at most $\pi/\sqrt{k_0}$.*

Proof. Let $\tau = x_t$, $a \leq t \leq c$, be a geodesic such that x_c is the first conjugate point of x_a on τ. Let b be an arbitrary number such that $a < b < c$. Let Y be a parallel unit vector field along τ which is perpendicular to τ, and let $f(t)$ be a non-zero function such that $f(a) = f(b) = 0$. By Corollary 3.2 we have $I_a^b(fY) \geq 0$. On the other hand, since Y is parallel along τ, we have

$$I_a^b(fY) = \int_a^b [g(f'Y, f'Y) - f^2 g(R(Y, \dot{\tau})\dot{\tau}, Y)] \, dt$$
$$\leq \int_a^b (f'^2 - k_0 f^2) \, dt.$$

If we take
$$f(t) = \sin \pi \frac{(t-a)}{(t-b)}$$
and make use of
$$\int_0^\pi \cos^2 x \, dx = \int_0^\pi \sin^2 x \, dx = \pi/2,$$
then the above inequality implies $b - a \leq \pi/\sqrt{k_0}$. Since b is arbitrarily close to c, we have $c - a \leq \pi/\sqrt{k_0}$. QED.

By a similar method, we prove the following stronger result:

THEOREM 3.4. *Let M be a Riemannian manifold such that the Ricci tensor S is positive-definite with all eigenvalues $\geq (n-1)h_0 > 0$. Then, for every geodesic τ of M, the distance between any two consecutive conjugate points of τ is at most $\pi/\sqrt{h_0}$.*

VIII. VARIATIONS OF THE LENGTH INTEGRAL

Proof. Let $\tau = x_t$, a, c, and b be exactly as in the proof of Theorem 3.3. Let $Y_1, \ldots, Y_{n-1}$ be parallel unit vector fields along τ such that $\dot\tau, Y_1, \ldots, Y_{n-1}$ are orthonormal at each point of τ. Let $f(t)$ be a non-zero function with $f(a) = f(b) = 0$. By Corollary 3.2, we have $I_a^b(fY_i) \geq 0$ for $i = 1, \ldots, n-1$. On the other hand, we have

$$\sum_{i=1}^{n-1} I_a^b(fY_i) = \int_a^b [\sum g(f'Y_i, f'Y_i) - f^2 \sum g(R(Y_i, \dot\tau)\dot\tau, Y_i)] \, dt$$

$$\leq \int_a^b (n-1)(f'^2 - h_0 f^2) \, dt.$$

The rest of the proof is similar to that of Theorem 3.3.   QED.

PROPOSITION 3.5. *Let $\tau = x_t$, $a \leq t \leq b$, be a geodesic in a Riemannian manifold M. Then there is a conjugate point x_c, $a < c < b$, of x_a along τ if and only if there is a piecewise differentiable vector field X along τ such that (1) X is perpendicular to τ, (2) X vanishes at x_a and x_b, and (3) $I_a^b(X) < 0$.*

Proof. By Corollary 3.2, if there is a vector field X with properties (1), (2), and (3), there is a conjugate point x_c, $a < c < b$, of x_a. Conversely, let x_c, $a < c < b$, be a conjugate point of x_a and let Y be a Jacobi field along τ vanishing at x_a and x_c. By Corollary 2.2, Y is perpendicular to τ. We take a convex neighborhood U of x_c such that every point of U has a normal coordinate neighborhood containing U (cf. Theorem 8.7 of Chapter III), and let $\delta > 0$ be such that $x_{c-\delta}$ and $x_{c+\delta}$ are in U. Then since $x_{c+\delta}$ is not a conjugate point of $x_{c-\delta}$ along the segment τ' of τ from $x_{c-\delta}$ to $x_{c+\delta}$, the linear mapping of the set of Jacobi fields on τ' into $T_{x_{c-\delta}}(M) + T_{x_{c+\delta}}(M)$ given by $Z \to (Z_{x_{c-\delta}}, Z_{x_{c+\delta}})$ is one-to-one and therefore onto, because both vector spaces have dimension $2n$. Hence there is a Jacobi field on τ' with prescribed values at both ends. We now choose a Jacobi field Z on τ' such that $Z_{x_{c-\delta}} = Y_{x_{c-\delta}}$ and $Z_{x_{c+\delta}} = 0$.

We now define a vector field X along τ as follows:

$$\begin{aligned} X &= Y \quad \text{from} \quad x_a \quad \text{to} \quad x_{c-\delta}, \\ &= Z \quad \text{from} \quad x_{c-\delta} \quad \text{to} \quad x_{c+\delta}, \\ &= 0 \quad \text{from} \quad x_{c+\delta} \quad \text{to} \quad x_b. \end{aligned}$$

Since $0 = I_a^c(Y) = I_a^{c-\delta}(Y) + I_{c-\delta}^c(Y)$ by Proposition 2.3, we have

$$I_a^b(X) = I_a^b(X) - I_a^c(Y)$$
$$= I_a^{c-\delta}(Y) + I_{c-\delta}^{c+\delta}(Z) - I_a^{c-\delta}(Y) - I_{c-\delta}^c(Y)$$
$$= I_{c-\delta}^{c+\delta}(Z) - I_{c-\delta}^c(Y).$$

Let $\bar{Y}$ be a vector field along τ from $x_{c-\delta}$ to $x_{c+\delta}$ defined as follows:

$$\bar{Y} = Y \quad \text{from} \quad x_{c-\delta} \quad \text{to} \quad x_c,$$
$$= 0 \quad \text{from} \quad x_c \quad \text{to} \quad x_{c+\delta}.$$

Applying Proposition 3.1 to vector fields $\bar{Y}$ and Z, we have

$$I_{c-\delta}^{c+\delta}(Z) < I_{c-\delta}^{c+\delta}(\bar{Y}) = I_{c-\delta}^c(Y).$$

Hence, we have $I_a^b(X) < 0$. QED.

4. Comparison theorem

The main purpose of this section is to prove the following comparison theorem of Rauch [1]:

THEOREM 4.1. *Let M and N be Riemannian manifolds of dimension n with metric tensor g and h respectively. Let $\sigma = x_t$, $a \leq t \geq b$, be a geodesic in M and X a non-zero Jacobi field along σ perpendicular to σ. Let $\tau = y_t$, $a \leq t \leq b$, be a geodesic in N and Y a non-zero Jacobi field along τ perpendicular to τ. Assume:*

(1) *Both X and Y vanish at $t = a$;*
(2) *X' and Y' have the same length at $t = a$;*
(3) *x_a has no conjugate point on $\sigma = x_t$, $a \leq t \leq b$, and y_a has no conjugate point on $\tau = y_t$, $a \leq t \leq b$;*
(4) *For each t, $a \leq t \leq b$, if p is a plane in the tangent space at x_t and if q is a plane in the tangent space at y_t, then*

$$K(p) \geq K(q),$$

where $K(p)$ and $K(q)$ are the sectional curvatures for p and q respectively. Under these four assumptions, we have

$$g(X_{x_t}, X_{x_t}) \leq h(Y_{y_t}, Y_{y_t}) \quad \text{for every } t, \ a \leq t \leq b.$$

Proof. We set

$$u(t) = g(X_{x_t}, X_{x_t}), \quad v(t) = h(Y_{y_t}, Y_{y_t}) \quad \text{for } a \leq t \leq b.$$

Since $u(t) \neq 0$ and $v(t) \neq 0$ for $a < t \leq b$ by (1) and (3), we define functions $\mu(t)$ and $\nu(t)$ by

$$\mu(t) = I_a^t(X)/u(t), \quad \nu(t) = I_a^t(Y)/v(t) \quad \text{for } a < t \leq b.$$

Since $du/dt = 2g(X, X')$ and $dv/dt = 2h(Y, Y')$, by Proposition 2.3 we have

$$du/dt = 2\mu u, \quad dv/dt = 2\nu v.$$

Solving these differential equations, we have, for every ε, $a < \varepsilon < b$,

$$u(t) = u(\varepsilon) e^{2\int_\varepsilon^t \mu dt}, \quad v(t) = v(\varepsilon) e^{2\int_\varepsilon^t \nu dt} \quad \text{for } a < t \leq b.$$

By assumptions (1) and (2) and by l'Hospital's rule, we have

$$\lim_{\varepsilon \to a} u(\varepsilon)/v(\varepsilon) = \lim_{\varepsilon \to a} [g(X, X')/h(Y, Y')]_{t=\varepsilon}$$
$$= \lim_{\varepsilon \to a} [(g(X', X') + g(X, X''))/(h(Y', Y') + h(Y, Y''))]_{t=\varepsilon}$$
$$= 1.$$

Hence, we have

$$u(t)/v(t) = \lim_{\varepsilon \to a} e^{2\int_\varepsilon^t (\mu - \nu) dt}.$$

To complete the proof, it suffices to show that $\mu(t) \leq \nu(t)$ for $a < t \leq b$. We take an arbitrary number c, $a < c < b$, and fix it in the rest of the proof. We set

$$\bar{X} = X/u(c)^{1/2}, \quad \bar{Y} = Y/v(c)^{1/2},$$

so that $\bar{X}$ and $\bar{Y}$ are Jacobi fields along σ and τ respectively and $\bar{X}_{x_c}$ and $\bar{Y}_{y_c}$ are unit vectors. We shall construct a vector field Z along σ such that $g(Z, Z)_{x_t} = h(\bar{Y}, \bar{Y})_{y_t}$ and $g(Z', Z')_{x_t} = h(\bar{Y}', \bar{Y}')_{y_t}$ for $a \leq t \leq b$. To this end, let U_t (resp. V_t) be the normal space to σ at x_t (resp. the normal space to τ at y_t). Let f_c be a metric-preserving linear isomorphism of V_c onto U_c such that $f_c(\bar{Y}_{y_c}) = \bar{X}_{x_c}$. Let σ_t^r (resp. τ_t^r) be the parallel displacement along σ (resp. along τ) from x_r to x_t (resp. y_r to y_t). Let f_t be the linear isomorphism of V_t onto U_t defined by

$$f_t \circ \tau_t^c = \sigma_t^c \circ f_c.$$

If we set $Z_{x_t} = f_t(\bar{Y}_{y_t})$ for $a \leq t \leq b$, then we obtain the desired vector field Z along σ. We have
$$I_a^c(\bar{X}) \leq I_a^c(Z) \leq I_a^c(\bar{Y}).$$
The first inequality is a consequence of Proposition 3.1 and the second inequality follows from our assumption (4) and the definition of I_a^c. On the other hand, we have
$$\mu(c) = I_a^c(X)/u(c) = I_a^c(\bar{X}), \quad \nu(c) = I_a^c(Y)/v(c) = I_a^c(\bar{Y}).$$
Hence, we have
$$\mu(c) \leq \nu(c).$$
Since c is arbitrary, this completes the proof. QED.

COROLLARY 4.2. *Let M and N be Riemannian manifolds of dimension n. Let $\sigma = x_t$, $a \leq t \leq b$, and $\tau = y_t$, $a \leq t \leq b$, be geodesics in M and N respectively. Assume that, for each plane p in the tangent space at x_t and for each plane q in the tangent space at y_t, we have $K(p) \geq K(q)$. If x_a has no conjugate point on $\sigma = x_t$, $a \leq t \leq b$, then y_a has no conjugate point on $\tau = y_t$, $a \leq t \leq b$.*

Proof. Assume that y_c, $a < c < b$, is the first conjugate point of y_a along τ. Let Y be a non-zero Jacobi field along τ vanishing at y_a and y_c. Let X be a Jacobi field along σ vanishing at x_a such that the length of X' at x_a is the same as that of Y' at y_a (cf. Proposition 1.1). Then, by Theorem 4.1, we have
$$g(X_{x_t}, X_{x_t}) \leq h(Y_{y_t}, Y_{y_t}) \quad \text{for } a \leq t < c.$$
Hence, we have
$$g(X_{x_c}, X_{x_c}) = \lim_{t \to c} g(X_{x_t}, X_{x_t}) \leq \lim_{t \to c} h(Y_{y_t}, Y_{y_t}) = h(Y_{y_c}, Y_{y_c}) = 0.$$
This means that X is a Jacobi field vanishing at x_a and x_c and hence that x_c is conjugate to x_a along σ, thus contradicting our assumption. QED.

The following result is originally due to Bonnet [1]:

COROLLARY 4.3. *Let M be a Riemannian manifold whose sectional curvature K is bounded as follows: $0 < k_0 \leq K(p) \leq k_1$, where k_0 and k_1 are positive constants. If $\tau = x_t$, $a \leq t \leq b$, is a geodesic such that x_b is the first conjugate point of x_a along τ, then*
$$\pi/\sqrt{k_1} \leq b - a \leq \pi/\sqrt{k_0}.$$
Proof. The second inequality has been already proved in Theorem 3.4. However, it follows also from Corollary 4.2 and

Example 2.1 if we let N be a complete Riemannian manifold of constant curvature k_0 in Corollary 4.2. To obtain the first inequality, let N be a complete Riemannian manifold of constant curvature k_1 and interchange the roles of M and N in Corollary 4.2. QED.

Remark. In Corollary 4.2, let M be a Euclidean space and N a Riemannian manifold with non-positive curvature. Then we obtain an alternative proof for Corollary 2.4.

For extensions of the comparison theorem, see Toponogov [1], Tsukamoto [2], Berger [13], and Warner [2].

5. *The first and second variations of the length integral*

Let M be a Riemannian manifold with metric tensor g. Throughout this section we fix two points y and z of M and denote by Γ the set of all piecewise differentiable curves $\tau = x_t$, $a \leq t \leq b$, from y to z parametrized proportionally to arc length. The *tangent space of* Γ *at* τ, denoted by $T_\tau(\Gamma)$, is the vector space of all piecewise differentiable vector fields X along τ vanishing at the end points y and z. As we shall soon see, Γ is similar to a manifold and $T_\tau(\Gamma)$ plays the role of the tangent space. We shall use this analogy to motivate certain definitions.

The *length function* L on Γ assigns to each τ the length of τ. The main purpose of this section lies in the study of the Hessian of L at each critical point of L. The *total differential* dL of L assigns at each τ a linear functional on $T_\tau(\Gamma)$ in the following way. Given X in $T_\tau(\Gamma)$, consider a 1-parameter family of curves $\tau^s = x_t^s$, $a \leq t \leq b$ and $-\varepsilon < s < \varepsilon$ such that

(1) Each τ^s is an element of Γ;

(2) $\tau^0 = \tau$;

(3) There exists a finite set of numbers $t_i \in [a, b]$ with $a = t_0 < t_1 < \cdots < t_k = b$ such that $(t, s) \to x_t^s$ is differentiable on each rectangle $[t_i, t_{i+1}] \times (-\varepsilon, \varepsilon)$;

(4) For each fixed $t \in [a, b]$, the vector $\dot{x}_{(t)}^0$ tangent to the curve x_t^s, $-\varepsilon < s < \varepsilon$, at the point $x_t = x_t^0$ coincides with X_{x_t}.

We then set

$$dL(X) = \left(\frac{d}{ds} L(\tau^s)\right)_{s=0}.$$

We shall express $dL(X)$ explicitly by means of X and its covariant derivatives; in particular, we shall see that the above definition of $dL(X)$ is independent of the choice of the family τ^s.

THEOREM 5.1. *Let* $\tau^s = x_t^s$, $a \leq t \leq b$ *and* $-\varepsilon < s < \varepsilon$, *be a 1-parameter family of curves such that* $(t, s) \to x_t^s$ *is a differentiable mapping of* $[a, b] \times (-\varepsilon, \varepsilon)$ *into M and that each curve τ^s is parametrized proportionally to arc length. Set* $x_t = x_t^0$ *and* $\tau = \tau^0$. *Then we have*

$$\left(\frac{d}{ds} L(\tau^s)\right)_{s=0} = \frac{1}{r}\left\{g(X, \dot{\tau})_{x_b} - g(X, \dot{\tau})_{x_a} - \int_a^b g(X, \nabla_{\dot{\tau}} \dot{\tau})\, dt\right\},$$

where X is the vector field along τ defined by $X_{x_t} = \dot{x}_{(t)}^0$ *and r is the common length of tangent vectors to* τ.

The proof will be given later. As a direct consequence of this theorem, we have

THEOREM 5.2. *Let* $\tau \in \Gamma$ *and* $X \in T_\tau(\Gamma)$. *Let* $a = c_0 < c_1 < \cdots < c_h < c_{h+1} = b$ *be a partition of $[a, b]$ such that the restriction of τ to each $[c_j, c_{j+1}]$ is differentiable. Then*

$$dL(X) = \frac{1}{r}\left\{\sum_{j=1}^h g(X, \dot{\tau}^- - \dot{\tau}^+)_{x_{c_j}} - \int_a^b g(X, \nabla_{\dot{\tau}}\dot{\tau})\, dt\right\},$$

where $\dot{\tau}^-$ and $\dot{\tau}^+$ denote the left and right limits of the tangent vector field $\dot{\tau}$ at the points x_{c_j}.

As we shall soon see, this implies

THEOREM 5.3. *A curve $\tau \in \Gamma$ is a geodesic if and only if $dL(X) = 0$ for all $X \in T_\tau(\Gamma)$.*

This means that the geodesics belonging to Γ are precisely the critical points of L on Γ. We shall define the Hessian of L at a geodesic $\tau \in \Gamma$. Following Morse we shall call it the *index form* and denote it by I. It will be a real symmetric bilinear form on $T_\tau(\Gamma)$. As in the definition of dL, given $X \in T_\tau(\Gamma)$, we consider a 1-parameter family of curves τ^s with properties (1), (2), (3), and (4).

We set
$$I(X, X) = \left(\frac{d^2}{ds^2} L(\tau^s)\right)_{s=0}.$$
By polarization, we define the *index form* $I(X, Y)$ by
$$I(X, Y) = \tfrac{1}{2}[I(X + Y, X + Y) - I(X, X) - I(Y, Y)].$$
Then we have

THEOREM 5.4. *If $\tau \in \Gamma$ is a geodesic and if $X, Y \in T_\tau(\Gamma)$, then*
$$I(X, Y) = \frac{1}{r} \int_a^b [g(X^{\perp\prime}, Y^{\perp\prime}) - g(R(X^\perp, \dot\tau)\dot\tau, Y^\perp)]\, dt,$$
where $X^\perp = X - (1/r)g(X, \dot\tau)\dot\tau$ is the component of X perpendicular to τ and $X^{\perp\prime}$ denotes the covariant derivative $\nabla_{\dot\tau} X^\perp$ of $X^\perp$ along τ, and similarly for $Y^{\perp\prime}$.

This may be reformulated as follows:

THEOREM 5.5. *Let τ, X, and Y be as in Theorem 5.4. Then*
$$I(X, Y) = -\frac{1}{r} \int_a^b g(R(X^\perp, \dot\tau)\dot\tau + X^{\perp\prime\prime}, Y^\perp)\, dt$$
$$+ \frac{1}{r}\left[\sum_{j=1}^k g(X^{\perp\prime -} - X^{\perp\prime +}, Y^\perp)_{x_{t_j}}\right],$$
where $a = t_0 < t_1 < \cdots < t_h < t_{h+1} = b$ is a partition of $[a, b]$ such that X is differentiable in each interval $[t_j, t_{j+1}]$, $j = 0, 1, \ldots, h$, and $X^{\perp\prime -}$ (resp. $X^{\perp\prime +}$) denotes the left (resp. right) limit of the covariant derivatives of $X^\perp$ with respect to $\dot\tau$ at the points x_{t_j}.

From this we shall obtain

THEOREM 5.6. *Let $\tau \in \Gamma$ be a geodesic and $X \in T_\tau(\Gamma)$. Then $X^\perp$ is a Jacobi field if and only if $I(X, Y) = 0$ for all $Y \in T_\tau(\Gamma)$.*

The remainder of this section is devoted to the proofs of the above six theorems and to a few applications.

Proof of Theorem 5.1. We lift the mapping $(t, s) \to x_t^s$ to a mapping $\gamma\colon [a, b] \times (-\varepsilon, \varepsilon) \to O(M)$ (where $O(M)$ is the bundle of orthonormal frames over M) such that $\pi \circ \gamma(t, s) = x_t^s$ and that the curve γ^0 defined by $\gamma^0(t) = \gamma(t, 0)$ is horizontal.

82   FOUNDATIONS OF DIFFERENTIAL GEOMETRY

Let S and T be the vector fields in the rectangle $[a, b] \times (-\varepsilon, \varepsilon)$ defined by

$$S = \frac{\partial}{\partial s}, \qquad T = \frac{\partial}{\partial t}.$$

Let θ, ω, and Ω be the canonical form, the connection form, and the curvature form on $O(M)$ respectively. We define forms θ^*, ω^*, and Ω^* on the rectangle by

$$\theta^* = \gamma^*(\theta), \qquad \omega^* = \gamma^*(\omega), \qquad \Omega^* = \gamma^*(\Omega).$$

Then we have the following formulas (cf. §1):

(A) $[S, T] = 0$;
(B) $\omega^*(T) = 0$ at the points $(t, 0)$;
(C) $S(\theta^*(T)) = T(\theta^*(S)) + \omega^*(T)\theta^*(S) - \omega^*(S)\theta^*(T)$;
(D) $S(\omega^*(T)) = T(\omega^*(S)) + \omega^*(T)\omega^*(S)$
$\qquad\qquad\qquad - \omega^*(S)\omega^*(T) + 2\Omega^*(S, T).$

Observe that (B) is a consequence of the assumption that γ^0 is horizontal and that (C) and (D) follow from (A), the first and second structure equations and Proposition 3.11 of Chapter I.

We define a function F on the rectangle by setting

$$F = (\theta^*(T), \theta^*(T))^{1/2},$$

so that, at each (t, s), $F(t, s)$ is the length of the vector $\dot{x}_t^{(s)}$. The length $L(\tau^s)$ of each curve τ^s is then given by the integral:

$$L(\tau^s) = \int_a^b F(t, s) \, dt.$$

Since each τ^s is parametrized proportionally to arc length, the function $F(t, s)$ actually depends only on s. In particular, we have

(E) $r = F(t, 0)$.

Now we prove the following formula:

(F) $S(F) = \dfrac{1}{r} (T(\theta^*(S)), \theta^*(T))$ at the points $(t, 0)$.

VIII. VARIATIONS OF THE LENGTH INTEGRAL

In fact, by (C) we have
$$2F \cdot S(F) = S(F^2) = 2(S(\theta^*(T)), \theta^*(T))$$
$$= 2(T(\theta^*(S)), \theta^*(T)) + 2(\omega^*(T)\theta^*(S), \theta^*(T))$$
$$- 2(\omega^*(S)\theta^*(T), \theta^*(T)).$$

Since $\omega^*(S)$ is in $\mathfrak{o}(n)$ (i.e., skew-symmetric), the last term vanishes. Hence, (F) follows from (B) and (E).

We are now in position to compute the first variation
$$(dL(\tau^s)/ds)_{s=0}.$$

(G) $\quad \dfrac{d}{ds} L(\tau^s)_{s=0} = \displaystyle\int_a^b S(F)_{(t,0)} \, dt = \dfrac{1}{r} \int_a^b (T(\theta^*(S)), \theta^*(T))_{(t,0)} \, dt$

$$= \dfrac{1}{r} \int_a^b (T(\theta^*(S), \theta^*(T)))_{(t,0)} \, dt$$

$$- \dfrac{1}{r} \int_a^b (\theta^*(S), T(\theta^*(T)))_{(t,0)} \, dt$$

$$= \dfrac{1}{r} (\theta^*(S), \theta^*(T))_{(b,0)} - \dfrac{1}{r} (\theta^*(S), \theta^*(T))_{(a,0)}$$

$$- \dfrac{1}{r} \int_a^b (\theta^*(S), T(\theta^*(T)))_{(t,0)} \, dt.$$

On the other hand, we have

(H) $\theta^*(S)_{(t,0)} = \theta(\gamma(S_{(t,0)})) = \gamma^0(t)^{-1}(\pi \circ \gamma(S_{(t,0)}))$
$$= \gamma^0(t)^{-1}(X_{x_t});$$

(I) $\theta^*(T)_{(t,0)} = \theta(\gamma(T_{(t,0)})) = \gamma^0(t)^{-1}(\pi \circ \gamma(T_{(t,0)}))$
$$= \gamma^0(t)^{-1}(\dot{x}_t).$$

Since γ^0 is horizontal, we have
$$\gamma^0(t+h)^{-1}(\dot{x}_{t+h}) = \gamma^0(t)^{-1}[\tau_t^{t+h}(\dot{x}_{t+h})],$$
where τ_t^{t+h} denotes the parallel displacement along τ from x_{t+h} to x_t. This, together with (I), implies

(J) $(T(\theta^*(T)))_{(t,0)} = \displaystyle\lim_{h \to 0} \dfrac{1}{h} [\gamma^0(t+h)^{-1}(\dot{x}_{t+h}) - \gamma^0(t)^{-1}(\dot{x}_t)]$

$$= \gamma^0(t)^{-1} \left[\lim_{h \to 0} \dfrac{1}{h} (\tau_t^{t+h}(\dot{x}_{t+h}) - \dot{x}_t) \right]$$

$$= \gamma^0(t)^{-1}[(\nabla_{\dot{\tau}} \dot{\tau})_{x_t}].$$

Now, Theorem 5.1 follows from (G), (H), (I), and (J). QED.

Proof of Theorem 5.2. Subdivide each $[c_j, c_{j+1}]$ into a finite number of subintervals on each of which X is differentiable and apply Theorem 5.1 to each subinterval. QED.

Proof of Theorem 5.3. If $\tau \in \Gamma$ is a geodesic, then $\nabla_{\dot\tau}\dot\tau = 0$ and Theorem 5.2 implies $dL(X) = 0$ for all $X \in T_\tau(\Gamma)$. Conversely, let $\tau \in \Gamma$ be a curve such that $dL(X) = 0$ for all $X \in T_\tau(\Gamma)$. As in Theorem 5.2, assume that τ is differentiable in each interval $[c_j, c_{j+1}]$. Let f be a function along τ which vanishes at $c_1, \ldots, c_h$ and is positive everywhere else. If we set

$$X = f\nabla_{\dot\tau}\dot\tau,$$

then the formula in Theorem 5.2 reduces to

$$dL(X) = -\frac{1}{r}\int_a^b f \cdot g(\nabla_{\dot\tau}\dot\tau, \nabla_{\dot\tau}\dot\tau)\, dt \leq 0.$$

Thus, $dL(X) = 0$ implies that $\nabla_{\dot\tau}\dot\tau = 0$ wherever $\nabla_{\dot\tau}\dot\tau$ exists. In other words, τ is a broken geodesic. To prove that τ is differentiable even at $t = c_j, j = 1, \ldots, h$, we choose, for each fixed j, a vector field $X \in T_\tau(\Gamma)$ such that $X = \dot\tau^- - \dot\tau^+$ at $t = c_j$ and $X = 0$ at $t = c_k$ for $k \neq j$. Then Theorem 5.2 implies that $\dot\tau^- - \dot\tau^+ = 0$ at $t = c_j$. QED.

Proof of Theorem 5.4. We shall make use of notations and formulas in the proof of Theorem 5.1. In proving (F) we established the following formula:

(K) $S(F^2) = 2(T(\theta^*(S)), \theta^*(T)) + 2(\omega^*(T)\theta^*(S), \theta^*(T))$.

Hence, we have

(L) $\tfrac{1}{2}S^2(F^2) = (ST(\theta^*(S)), \theta^*(T)) + (T(\theta^*(S)), S(\theta^*(T)))$
$+ (S(\omega^*(T))\theta^*(S), \theta^*(T))$
$+ (\omega^*(T)S(\theta^*(S)), \theta^*(T))$
$+ (\omega^*(T)\theta^*(S), S(\theta^*(T)))$.

Of these five terms on the right hand side, the last two vanish at the points $(t, 0)$ by (B) in the proof of Theorem 5.1. We shall now compute the remaining three terms.

1st term $= (ST(\theta^*(S)), \theta^*(T)) = (TS(\theta^*(S)), \theta^*(T))$
$= T(S(\theta^*(S)), \theta^*(T)) - (S(\theta^*(S)), T(\theta^*(T)))$.

As τ is a geodesic and γ^0 is horizontal, $\theta^*(T)$ is constant along the line $s = 0$ and hence $T(\theta^*(T)) = 0$ at the points $(t, 0)$. Thus,

$$\text{1st term} = T(S(\theta^*(S)), \theta^*(T)) \quad \text{at } (t, 0).$$

By (C), we have

$$\begin{aligned}\text{2nd term} &= (T(\theta^*(S)), S(\theta^*(T))) \\ &= (T(\theta^*(S)), T(\theta^*(S))) + (T(\theta^*(S)), \omega^*(T)\theta^*(S)) \\ &\quad - (T(\theta^*(S)), \omega^*(S)\theta^*(T)).\end{aligned}$$

Since $\omega^*(S)$ is skew-symmetric and since $\omega^*(T) = 0$ at $(t, 0)$, by (B) we have

$$\begin{aligned}\text{2nd term} &= (T(\theta^*(S)), T(\theta^*(S))) \\ &\quad + (\omega^*(S)T(\theta^*(S)), \theta^*(T)) \quad \text{at } (t, 0).\end{aligned}$$

Finally, by (D) and (B) we have

$$\begin{aligned}\text{3rd term} &= (S(\omega^*(T))\theta^*(S), \theta^*(T)) \\ &= (T(\omega^*(S))\theta^*(S), \theta^*(T)) \\ &\quad + (2\Omega^*(S, T)\theta^*(S), \theta^*(T)) \quad \text{at } (t, 0).\end{aligned}$$

By adding these together we obtain

(M) $\tfrac{1}{2}S^2(F^2)$
$$\begin{aligned}&= (T(\theta^*(S)), T(\theta^*(S))) + (2\Omega^*(S, T)\theta^*(S), \theta^*(T)) \\ &\quad + T(S(\theta^*(S)) + \omega^*(S)\theta^*(S), \theta^*(T)) \quad \text{at } (t, 0).\end{aligned}$$

Since

$$F \cdot S^2(F) = \tfrac{1}{2}S^2(F^2) - S(F)S(F),$$

(F) and (M) imply

(N) $S^2(F)$
$$\begin{aligned}&= \frac{1}{r}\Big\{(T(\theta^*(S)), T(\theta^*(S))) + (2\Omega^*(S, T)\theta^*(S), \theta^*(T)) \\ &\quad + T(S(\theta^*(S)) + \omega^*(S)\theta^*(S), \theta^*(T)) \\ &\quad - \frac{1}{r^2}(T(\theta^*(S)), \theta^*(T))^2\Big\} \quad \text{at } (t, 0).\end{aligned}$$

The following formula follows from (H) in the same way as (J).

(O) $(T(\theta^*(S)))_{(t,0)} = \gamma^0(t)^{-1}(X'_{x_t})$.

86   FOUNDATIONS OF DIFFERENTIAL GEOMETRY

By (I) and (O) we may rewrite the first and last terms of the right hand side of (N) as follows:

first and last terms

$$= \frac{1}{r} \left\{ (T(\theta^*(S)), T(\theta^*(S))) - \frac{1}{r^2} (T(\theta^*(S)), \theta^*(T))^2 \right\}_{(t,0)}$$

$$= \frac{1}{r} \left\{ g(X', X') - \frac{1}{r^2} g(X', \dot{\tau})^2 \right\}_{x_t} = \frac{1}{r} g(X^{\perp\prime}, X^{\perp\prime})_{x_t}.$$

From (H) and (I) and from the definition of curvature tensor (cf. §5 of Chapter III), we obtain

$$\text{2nd term} = \frac{1}{r} (2\Omega^*(S, T)\theta^*(S), \theta^*(T))_{(t,0)}$$

$$= \frac{1}{r} g(R(X, \dot{\tau})X, \dot{\tau})_{x_t} = -\frac{1}{r} g(R(X, \dot{\tau})\dot{\tau}, X)_{x_t}$$

$$= -\frac{1}{r} g(R(X^\perp, \dot{\tau})\dot{\tau}, X^\perp)_{x_t}.$$

To prove that the integral of the third term from $t = a$ to $t = b$ is zero, we observe that $\pi \circ \gamma(a, s) = y$ and $\pi \circ \gamma(b, s) = z$ for all s so that $\gamma(S_{(a,s)})$ and $\gamma(S_{(b,s)})$ are vertical. This means $(\theta^*(S))_{(a,s)} = (\theta^*(S))_{(b,s)} = 0$. Hence, we have

$$\int_a^b [T(S(\theta^*(S)) + \omega^*(S)\theta^*(S), \theta^*(T))]_{(t,0)} \, dt$$
$$= (S(\theta^*(S)) + \omega^*(S)\theta^*(S), \theta^*(T))_{(b,0)}$$
$$- (S(\theta^*(S)) + \omega^*(S)\theta^*(S), \theta^*(T))_{(a,0)} = 0.$$

Finally, we have

$$I(X, X) = \int_a^b [S^2(F)]_{(t,0)} \, dt$$

$$= \frac{1}{r} \int_a^b [g(X^{\perp\prime}, X^{\perp\prime}) - g(R(X^\perp, \dot{\tau})\dot{\tau}, X^\perp)] \, dt.$$

Since $g(X^{\perp\prime}, Y^{\perp\prime}) - g(R(X^\perp, \dot{\tau})\dot{\tau}, Y^\perp)$ is symmetric in X and Y, we obtain the formula in Theorem 5.4.   QED.

Proof of Theorem 5.5. Integrating $g(X^{\perp\prime}, Y^{\perp\prime})$ by parts, we obtain

$$\int_{t_j}^{t_{j+1}} g(X^{\perp\prime}, Y^{\perp\prime})\, dt = \int_{t_j}^{t_{j+1}} \left[\frac{d}{dt} g(X^{\perp\prime}, Y^{\perp}) - g(X^{\perp\prime\prime}, Y^{\perp})\right] dt$$
$$= g(X^{\perp\prime -}, Y^{\perp})_{x_{t_{j+1}}} - g(X^{\perp\prime +}, Y^{\perp})_{x_{t_j}} - \int_{t_j}^{t_{j+1}} g(X^{\perp\prime\prime}, Y^{\perp})\, dt.$$

This combined with Theorem 5.4 yields Theorem 5.5.   QED.

Proof of Theorem 5.6. If $X^{\perp}$ is a Jacobi field, $X^{\perp}$ is differentiable and hence we have only to consider the integral term in the formula of Theorem 5.5. Since $X^{\perp}$ satisfies the Jacobi equation, the integrand in $I(X, Y)$ of Theorem 5.5 is identically zero. Hence, $I(X, Y) = 0$ for all $Y \in T_\tau(\Gamma)$. Conversely, assume that $I(X, Y) = 0$ for all $Y \in T_\tau(\Gamma)$. In Theorem 5.5, we choose Y as follows. Let f be a function along τ which vanishes at $t = t_0, t_1, \ldots, t_h$ and is positive everywhere else. If we set $Y = f \cdot R(X^{\perp}, \dot\tau)\dot\tau$, then the assumption $I(X, Y)$ implies that $X^{\perp}$ is a Jacobi field in each subinterval $[t_j, t_{j+1}]$. To prove that $X^{\perp}$ is a Jacobi field in the whole interval $[a, b]$, it suffices to prove that $X^{\perp}$ is differentiable of class C^1 at $t = t_j$ (cf. Proposition 1.1). To this end, we choose, for each fixed j, a vector field $Y \in T_\tau(\Gamma)$ such that $Y = X^{\perp\prime -} - X^{\perp\prime +}$ at $t = t_j$ and $Y = 0$ at $t = t_k$ for $k \neq j$. Then Theorem 5.5 implies that $X^{\perp\prime -} - X^{\perp\prime +} = 0$ at $t = t_j$.   QED.

The integral formula for $I(X, X)$ we obtained in the proof of Theorem 5.4 is known as Synge's formula (cf. Synge [1]). The exposition in this section is largely based on the paper of Ambrose [2].

As a consequence of Proposition 3.5 and Theorem 5.4, we have

THEOREM 5.7. *Let $\tau = x_t$, $a \leq t \leq b$, be a geodesic. If there is a conjugate point x_c, where $a < c < b$, of x_a, then τ is not a minimizing geodesic joining x_a to x_b, i.e., the length of τ is greater than the distance between x_a and x_b.*

Proof. By Proposition 3.5 and Theorem 5.4, there is a vector field X along τ with the following three properties:

(1) X is perpendicular to τ;
(2) X vanishes at x_a and x_b;
(3) $I(X, X) < 0$.

Let τ^s, $-\varepsilon < s < \varepsilon$, be a 1-parameter family of curves from x_a to x_b constructed in the definition of $dL(X)$ and $I(X, X)$. By Theorem 5.3, we have

$$\left(\frac{dL(\tau^s)}{ds}\right)_{s=0} = 0.$$

Since $I(X, X) = [d^2L(\tau^s)/ds^2]_{s=0}$ by definition, we have

$$\left(\frac{d^2L(\tau^s)}{ds^2}\right)_{s=0} < 0.$$

Hence, $L(\tau^s) < L(\tau)$ if $s \neq 0$ is sufficiently small.   QED.

Combining Theorems 3.3 and 3.4 with Theorem 5.7, we obtain

THEOREM 5.8. *Let M be a connected complete Riemannian manifold with sectional curvature $\geq k_0 > 0$ (or more generally, with Ricci tensor S whose eigenvalues are all $\geq (n-1)k_0 > 0$). Then*

(1) *The diameter of M is at most $\pi/\sqrt{k_0}$;*

(2) *M is compact;*

(3) *The fundamental group $\pi_1(M)$ is finite.*

Proof. Let x and y be arbitrary points of M and let τ be a minimizing geodesic joining x to y. By Theorems 3.3, 3.4, and 5.7, the length of τ is at most $\pi/\sqrt{k_0}$, which proves (1). Since M is bounded and complete, it is compact (cf. Theorem 4.1 of Chapter IV). The universal covering space $\tilde{M}$ of M with the naturally induced Riemannian metric satisfies the assumption of Theorem 5.8. Hence, $\tilde{M}$ is also compact. This means that $\pi_1(M)$ is finite.   QED.

Theorem 5.8 is due to Myers [1].

6. *Index theorem of Morse*

Let $\tau = x_t$, $a \leq t \leq b$, be a geodesic from $y = x_a$ to $z = x_b$ in a Riemannian manifold M. Given a conjugate point x_c, $a < c \leq b$, of x_a along τ, its *multiplicity* μ is defined to be the dimension of the space of all Jacobi fields along τ which vanish at x_a and x_c. By Proposition 1.1 and Corollary 2.2, $\mu \leq n - 1$ where $n = \dim M$.

VIII. VARIATIONS OF THE LENGTH INTEGRAL

In general, given a symmetric bilinear form B on a vector space V, the *index* $i(B)$, the *augmented index* $a(B)$ and the *nullity* $n(B)$ of B are defined by

$i(B) = $ the maximum dimension of those subspaces of V on which B is negative-definite;

$a(B) = $ the maximum dimension of those subspaces of V on which B is negative semi-definite;

$n(B) = \dim \{v \in V; B(v, v') = 0 \text{ for all } v' \in V\}$.

Let $T_\tau^\perp$ be the space of all piecewise differentiable vector fields along τ which are perpendicular to τ and vanish at the end points x_a and x_b. It is a subspace of the tangent space $T_\tau(\Gamma)$ defined in §5. We consider the index, the augmented index and the nullity of the index form I restricted to $T_\tau^\perp$. The object of this section is to give a proof of the Index Theorem of Morse:

THEOREM 6.1. *Let $\tau = x_t, a \leq t \leq b$, be a geodesic in a Riemannian manifold M. Then there are only finitely many points $x_{t_1}, \ldots, x_{t_k}$ $(a < t_1 < t_2 < \cdots < t_k < b)$ other than x_b which are conjugate to x_a along τ. Let μ_i be the multiplicity of x_{t_i}, $i = 1, \ldots, k$. Then*

(1) $i(I \mid T_\tau^\perp) = \mu_1 + \cdots + \mu_k$,

(2) $a(I \mid T_\tau^\perp) = i(I \mid T_\tau^\perp) + n(I \mid T_\tau^\perp)$,

(3) $n(I \mid T_\tau^\perp) = 0$ *if x_b is not conjugate to x_a,*

(4) $n(I \mid T_\tau^\perp) = $ *the multiplicity of x_b if x_b is conjugate to x_a.*

Proof. The essential part of the theorem is the finiteness of conjugate points and formula (1). We shall first prove (3) and (4). By Theorem 5.5,

$$I(X, Y) = I(X^\perp, Y^\perp) \qquad \text{for } X, Y \in T_\tau(\Gamma).$$

By Theorem 5.6, we have

$$\begin{aligned}
n(I \mid T_\tau^\perp) &= \dim \{X \in T_\tau^\perp; I(X, Y) = 0 \quad \text{for all } Y \in T_\tau^\perp\} \\
&= \dim \{X \in T_\tau^\perp; I(X, Y) = 0 \quad \text{for all } Y \in T_\tau(\Gamma)\} \\
&= \dim \{X \in T_\tau^\perp; X \text{ is a Jacobi field}\}.
\end{aligned}$$

This proves (3) and (4).

Our next step is to construct a finite-dimensional subspace J of $T_\tau^\perp$ such that $i(I \mid T_\tau^\perp) = i(I \mid J)$, $a(I \mid T_\tau^\perp) = a(I \mid J)$, and $n(I \mid T_\tau^\perp) = n(I \mid J)$. Choose a positive number δ such that the δ-neighborhood of every point of τ is a convex normal coordinate neighborhood. Let $a = a_0 < a_1 < \cdots < a_h = b$ be a subdivision of the interval $[a, b]$ such that $a_{i+1} - a_i < \delta$ for $i = 0, \ldots, h-1$. We set

$$J = J(a_0, \ldots, a_h)$$
$$= \{X \in T_\tau^\perp \, ; \, X \text{ is a Jacobi field along } \tau \mid [a_i, a_{i+1}] \text{ for all } i\}.$$

Let $N(a_i)$ be the normal space to τ at x_{a_i} and define a linear mapping

$$\alpha \colon J \to N = N(a_1) + \cdots + N(a_{h-1})$$

by

$$\alpha(X) = (X_1, \ldots, X_{h-1}),$$

where each X_i denotes the value of X at x_{a_i}.

LEMMA 1. (1) α is a linear isomorphism of J onto N;

(2) Define a retraction $\rho \colon T_\tau^\perp \to J$ by setting

$$\rho(X) = \alpha^{-1}(X_1, \ldots, X_{h-1}) \qquad \text{for } X \in T_\tau^\perp.$$

Then

$$I(X, X) \geq I(\rho X, \rho X) \qquad \text{for } X \in T_\tau^\perp,$$

and the equality holds if and only if $X \in J$.

(3) $i(I \mid T_\tau^\perp) = i(I \mid J)$, $a(I \mid T_\tau^\perp) = a(I \mid J)$, and $n(I \mid T_\tau^\perp) = a(I \mid J)$.

Proof. (1) Suppose $X \in J$ and $\alpha(X) = 0$ so that $X_0 = X_1 = \cdots = X_h = 0$. By our choice of the positive number δ, $x_{a_{i+1}}$ is not conjugate to x_{a_i} along τ. Hence, $X \equiv 0$ along τ, proving that α is injective. To show that α is surjective, it suffices to prove that, given vectors X_i and x_{a_i} and X_{i+1} at $x_{a_{i+1}}$, there is a Jacobi field X along $\tau \mid [a_i, a_{i+1}]$ which extends X_i and X_{i+1} (cf. Corollary 2.2). Since $x_{a_{i+1}}$ and x_{a_i} are not conjugate along τ, $X \to (X_i, X_{i+1})$ defines a linear isomorphism of the space of Jacobi fields along $\tau \mid [a_i, a_{i+1}]$ into the direct sum of the tangent spaces at x_{a_i} and $x_{a_{i+1}}$. Since it is a linear isomorphism of a vector space of dimension

VIII. VARIATIONS OF THE LENGTH INTEGRAL

$2n$ into a vector space of the same dimension (cf. Proposition 1.1), it must be surjective. This completes the proof of (1).

(2) With the notations in §3, we have

$$I(X, X) = \sum_{i=0}^{h-1} I_{a_i}^{a_{i+1}}(X), \qquad I(\rho X, \rho X) = \sum_{i=0}^{h-1} I_{a_i}^{a_{i+1}}(\rho X).$$

By Proposition 3.1, we have

$$I_{a_i}^{a_{i+1}}(X) \geq I_{a_i}^{a_{i+1}}(\rho X)$$

and the equality holds if and only if X is a Jacobi field along $\tau \mid [a_i, a_{i+1}]$.

(3) If U is a subspace of $T_\tau^\perp$ on which I is negative semi-definite, then I is negative semi-definite on ρU by (2). Moreover, $\rho: U \to \rho U$ is a linear isomorphism. In fact, if $X \in U$ and $\rho(X) = 0$, then (2) implies

$$0 \geq I(X, X) \geq I(\rho X, \rho X) = 0,$$

and hence $I(X, X) = I(\rho X, \rho X)$. Again by (2), we have $X = \rho X = 0$. Thus we have

$$a(I \mid T_\tau^\perp) \leq a(I \mid J).$$

The reverse inequality is obvious. The proof for the index $i(I \mid T_\tau^\perp)$ is similar. Finally, to prove $n(I \mid T_\tau^\perp) = n(I \mid J)$, let X be an element of J such that $I(X, Y) = 0$ for all $Y \in J$. Since X is a Jacobi field along $\tau \mid [a_i, a_{i+1}]$ for all i, the formula in Theorem 5.5 reduces to the following

$$I(X, Y) = \frac{1}{r} \left[\sum_{i=1}^{h-1} g(X'^- - X'^+, Y)_{x_{a_i}} \right].$$

In the same way as we proved Theorem 5.6, we conclude that $X'^- = X'^+$ at x_{a_i} for all i so that X is a Jacobi field along τ. This means that $n(I \mid T_\tau^\perp) \geq n(I \mid J)$. The reverse inequality is obvious.

Lemma 1 and the next lemma imply (2) of Theorem 6.1.

LEMMA 2. *If B is a symmetric bilinear form on a finite-dimensional vector space V, then*

$$a(B) = i(B) + n(B).$$

Proof. Let $v_1, \ldots, v_r$ be a basis for V with respect to which B is a diagonal matrix with diagonal entries $d_1, \ldots, d_r$. Set

$V_+ = $ the space spanned by $\{v_i; d_i > 0\}$;
$V_- = $ the space spanned by $\{v_i; d_i < 0\}$;
$V_0 = $ the space spanned by $\{v_i; d_i = 0\}$,

so that $V = V_+ + V_- + V_0$. We shall show that $n(B) = \dim V_0$, $i(B) = \dim V_-$ and $a(B) = \dim (V_0 + V_-)$. Clearly,

$$V_0 = \{X \in V; B(X, Y) = 0 \quad \text{for all } Y \in V\}$$

and hence $\dim V_0 = n(B)$. Let U be any subspace of V on which B is negative semi-definite. We claim that the projection $p: U \to V_0 + V_-$ along V_+ is injective. In fact, if $X \in U$ and $p(X) = 0$, then $X \in V_+$. Since B is negative semi-definite on U and positive definite on V_+, X must be zero. Thus, $\dim U \leq \dim (V_0 + V_-)$ and hence $a(B) = \dim (V_0 + V_-)$. Similarly, if U' is any subspace of V on which B is negative-definite, then the projection $p': U' \to V_-$ is injective and hence $i(B) = \dim V_-$. This completes the proof of Lemma 2.

Since $\dim J = (h-1)(n-1) < \infty$, Lemma 1 implies that both $a(I|T_\tau^\perp)$ and $i(I|T_\tau^\perp)$ are finite. The finiteness of conjugate points follows from the next lemma.

LEMMA 3. *For any finite number of conjugate points* $x_{t_1}, \ldots, x_{t_s}$ ($a < t_1 < \cdots < t_s \leq b$) *of* x_a *along* τ *with multiplicity* $\mu_1, \ldots, \mu_s$, *we have*

$$a(I|T_\tau^\perp) \geq \mu_1 + \cdots + \mu_s.$$

Proof. For each i, let $X_1^i, \ldots, X_{\mu_i}^i$ be a basis for the Jacobi fields along $\tau | [a, t_i]$ which vanish at $t = a$ and $t = t_i$ and extend them to be zero beyond $t = t_i$. It suffices to prove that $\mu_1 + \cdots + \mu_s$ vector fields $X_1^i, \ldots, X_{\mu_i}^i$, $i = 1, \ldots, s$, along τ are linearly independent and that I is negative semi-definite on the space spanned by them. Suppose

$$\sum_{i=1}^s X^i = 0,$$

where

$$X^i = c_1^i X_1^i + \cdots + c_{\mu_i}^i X_{\mu_i}^i.$$

VIII. VARIATIONS OF THE LENGTH INTEGRAL

Since $X^1, \ldots, X^{s-1}$ vanish on $\tau \mid [t_{s-1}, b]$, X^s must vanish along $\tau \mid [t_{s-1}, t_s]$. Being a Jacobi field along $\tau \mid [a, t_s]$, X^s must vanish identically along τ. Thus, $c_1^s = \cdots = c_{\mu_s}^s = 0$. Continuing this argument, we obtain $c_1^{s-1} = \cdots = c_{\mu_{s-1}}^{s-1} = 0$, and so on. To prove that I is negative semi-definite on the space spanned by $X_1^i, \ldots, X_{\mu_i}^i$, $i = 1, \ldots, s$, let

$$X = X^1 + \cdots + X^s,$$

where each X^i is a linear combination of $X_1^i, \ldots, X_{\mu_i}^i$ as above. Then

$$I(X, X) = \sum_{i=1}^{s} I(X^i, X^i) + 2 \sum_{1 \leq j < i \leq s} I(X^i, X^j).$$

For each pair (i, j) with $j \leq i$, we shall show that $I(X^i, X^j) = 0$. Let $\tau' = \tau \mid [a, t_i]$ and let I' be the index form for the geodesic τ'. Since X^i and X^j vanish beyond $t = t_i$, we have $I(X^i, X^j) = I'(X^i, X^j)$. As X^i is a Jacobi field along τ', $I'(X^i, X^j) = 0$ by Theorem 5.6. Thus, $I(X, X) = 0$, proving our assertion.

By proving (1) of Theorem 6.1, we shall complete the proof of Theorem 6.1. To this end, we define a 1-parameter family of symmetric bilinear forms B_r, $0 < r \leq 1$, on $J(a_0, \ldots, a_h)$ as follows. By translating the parameter t, we may assume that $a = 0$ so that $\tau = x_t$ is defined for $0 \leq t \leq b$. For each $r \in (0, 1]$, we define

$$N_r = N(ra_1) + \cdots + N(ra_{h-1}),$$

where $N(ra_i)$ is the normal space to τ at x_{ra_i}. We define also $J_r = J(ra_0, \ldots, ra_h)$ for the geodesic $\tau = x_t$, $0 \leq t \leq rb$, and a linear isomorphism

$$\alpha_r : J_r \to N_r$$

by setting

$$\alpha_r(X) = (X_{ra_1}, \ldots, X_{ra_{h-1}}),$$

where X_{ra_i} denotes the value of X at x_{ra_i}. Let $p_r : N \to N_r$ be the linear isomorphism induced by parallel displacement of $N(a_i)$ to $N(ra_i)$ along τ. Let I_r be the index form for the geodesic $\tau = x_t$, $0 \leq t \leq rb$. We then set

$$B_r(X, Y) = I_r(\alpha_r^{-1} \circ p_r \circ \alpha(X), \alpha_r^{-1} \circ p_r \circ \alpha(Y))$$

for $X, Y \in J(a_0, \ldots, a_h)$.

LEMMA 4. (1) $B_1(X, Y) = I(X, Y)$ for $X, Y \in J(a_0, \ldots, a_h)$;
(2) B_r is positive definite for small values of r;
(3) The nullity $n(B_r)$ is equal to the order of the point x_{rb} as a conjugate point of x_0 along τ. In particular, $n(B_r) = 0$ if x_{rb} is not conjugate to x_0 along τ;
(4) The family B_r is continuous in r.

Proof. (1) Since $\alpha_1 = \alpha$, $I_1 = I$, and $p_1 \colon N \to N_1 = N$ is the identity mapping, we have $B_1 = I$.

(2) Since $\alpha_r^{-1} \circ p_r \circ \alpha$ is an isomorphism of $J(a_0, \ldots, a_h)$ onto J_r, it suffices to show that I_r is positive-definite on J_r for small values of r. But Corollary 3.2 implies that if r is small enough so that $\tau \mid [0, rb]$ has no conjugate point of x_0, then I_r is positive-definite on J_r.

(3) Since $\alpha_r^{-1} \circ p_r \circ \alpha \colon J(a_0, \ldots, a_h) \to J_r$ is an isomorphism, we have $n(B_r) = n(I_r \mid J_r)$. Applying (3) of Lemma 1 and (3) and (4) of Theorem 6.1 to I_r, we see that $n(I_r \mid J_r)$ is equal to the order of x_{rb} as a conjugate point of x_0 along τ.

(4) For fixed $X, Y \in J(a_0, \ldots, a_h)$, we shall show that $B_r(X, Y)$ is continuous in r for $0 < r \leq 1$. Set

$$X(r) = \alpha_r^{-1} \circ p_r \circ \alpha(X), \quad Y(r) = \alpha_r^{-1} \circ p_r \circ \alpha(Y).$$

Since $X(r)$ is a Jacobi field along each piece $\tau \mid [ra_i, ra_{i+1}]$ of the geodesic τ, Theorem 5.5 gives the following formula:

$$B_r(X, Y) = A \sum_{i=1}^{h-1} g(X(r)'^{-} - X(r)'^{+}, Y(r))_{t=ra_i},$$

where A is the reciprocal of the length of $\tau \mid [0, rb]$. It is therefore sufficient to prove that, for each i, the mapping $r \to (X(r), X(r)'^{-}, X(r)'^{+})_{t=ra_i}$ is continuous. Define vectors $X_{0,r}, \ldots, X_{h,r}$ by

$$(X_{1,r}, \ldots, X_{h-1,r}) = p_r \circ \alpha(X), \quad X_{0,r} = 0, \quad X_{h,r} = 0.$$

We fix an integer i, $0 \leq i \leq h - 1$, and to simplify notations, we set

$$c = a_i, \quad d = a_{i+1}, \quad V_r = X_{i,r}, \quad W_r = X_{i+1,r}.$$

Then for each r, V_r is a vector at the point x_{rc}, and W_r is a vector at the point x_{rd}. Evidently, both V_r and W_r depend differentiably on r. We recall that $\tau \mid [c, d]$ is contained in a convex normal

VIII. VARIATIONS OF THE LENGTH INTEGRAL 95

coordinate neighborhood. Both $\exp uV_r$ and $\exp uW_r$ are also in the same neighborhood for $|u| < \varepsilon$ and $|s - 1| < \varepsilon$ for a small positive number ε. For fixed u and r, let $\tau^u(r) = x_t^u(r)$, $c \leq t \leq d$, be the geodesic from the point $\exp uV_r$ to the point $\exp uW_r$ in the neighborhood. For each fixed r, the family of geodesics $\tau^u(r)$, $-\varepsilon < u < \varepsilon$, is a variation of $\tau^0(r)$ and induces the unique Jacobi field along $\tau \mid [rc, rd]$ which coincides with V_r and W_r at the end points x_{rc} and x_{rd} of $\tau \mid [rc, rd]$. In other words, the induced Jacobi field coincides with the restriction of $X(r)$ to $\tau \mid [rc, rd]$. The continuity of $(X(r), X(r)'^+)_{t=rc}$ and $(X(r), X(r)'^-)_{t=rd}$ in r follows therefore from the differentiability of $x_t^u(r)$ in (r, u, t). But, the latter is evident from the following three facts:

(i) $\exp uV_r$ and $\exp uW_r$ are differentiable in (r, u);

(ii) $\exp uV_r$ and $\exp uW_r$ are in a convex normal coordinate neighborhood;

(iii) The mapping which sends $Z \in T_x(M)$ into $(x, \exp Z)$ is a diffeomorphism of a neighborhood of M in $T(M)$ onto a neighborhood of the diagonal of $M \times M$.

Note that (iii) is a trivial consequence of Propositions 8.1 and 8.2 of Chapter III and that (iii) implies that, given two points (e.g., $\exp uV_r$ and $\exp uW_r$) in a convex normal coordinate neighborhood depending differentiably on some parameters, the unique minimizing geodesic joining the two points depends also differentiably on the parameters. This completes the proof of Lemma 4.

We are now in position to complete the proof of (1) of Theorem 6.1. Let $x_{t_1}, \ldots, x_{t_k}$ ($0 < t_1 < t_2 < \cdots < t_k < b$), be the conjugate points of x_0 along τ with multiplicities $\mu_1, \ldots, \mu_k$ respectively. (The point x_b is not considered whether it is a conjugate point or not). Taking a basis for $J = J(a_0, \ldots, a_h)$, consider B_r, $0 < r \leq 1$, as a 1-parameter family of matrices. For each r, let $\beta_1(r) \leq \cdots \leq \beta_N(r)$ be the eigenvalues of B_r; they depend continuously on r. Applying (3) of Lemma 1 and (4) of Theorem 6.1 to the geodesic $\tau \mid [0, rb]$ and its index form I_r, we see that

$$n(B_r) = n(I_r \mid J_r) = \mu_i \quad \text{if} \quad rb = t_i,$$
$$n(B_r) = n(I_r \mid J_r) = 0 \quad \text{otherwise.}$$

Let
$$\beta_1(1) \leq \cdots \leq \beta_p(1) < 0 \leq \beta_{p+1}(1) \leq \cdots \leq \beta_N(1).$$

Since $\beta_1(r), \ldots, \beta_N(r)$ are all positive for small values of r by (2) of Lemma 4, each $\beta_j(r)$ for $1 \leq j \leq p$ vanishes for some r, $0 < r < 1$. Hence,

$$p \leq \sum_{0 < r < 1} n(B_r) = \mu_1 + \cdots + \mu_k.$$

On the other hand, $B_1 = (I \mid J)$ and hence $p = i(I \mid J)$. We have therefore the following inequality:

$$i(I \mid J) \leq \mu_1 + \cdots + \mu_k,$$

which, combined with Lemma 3 and (2) of Theorem 6.1, implies (1) of Theorem 6.1. QED.

For a proof which avoids the ad hoc subdivisions, see Osborn [1]. For generalizations of the index theorem, see Ambrose [4], Takahashi [1], Patterson [1] and Edwards [1].

7. Cut loci

Let M be a complete Riemannian manifold. Let us recall that a geodesic τ from a point x to a point y is said to be minimizing if its length $L(\tau)$ is equal to the distance $d(x, y)$ and that any two points of M can be joined by a minimizing geodesic (Theorem 4.2 of Chapter IV). A curve from x to y, parametrized proportionally to arc length, is a minimizing geodesic if and only if it is a shortest curve from x to y.

We fix a point x_0 of M and a geodesic $\tau = x_t$, $0 \leq t < \infty$, starting from x_0 and parametrized by arc length. Let A be the set of positive numbers s such that the segment of τ from x_0 to x_s is minimizing. Then A possesses the following two properties: (1) If $s \in A$ and $t < s$, then $t \in A$; (2) If r is a positive number such that every positive number s less than r lies in A, then r is in A. (1) is trivial. (2) may be proved as follows.

$$d(x_0, x_r) = \lim_{s \to r} d(x_0, x_s) = \lim_{s \to r} s = r.$$

These two properties of A imply that either $A = (0, \infty)$ or $A = (0, r]$ where r is some positive number. If $A = (0, r]$, then the point x_r is called the *cut point* (or the *minimum point*) of x_0 along τ. If $A = (0, \infty)$, we say that there is no cut point of x_0 along τ.

VIII. VARIATIONS OF THE LENGTH INTEGRAL

THEOREM 7.1. *Let x_r be the cut point of x_0 along a geodesic $\tau = x_t$, $0 \leq t < \infty$. Then, at least one (possibly both) of the following statements holds:*
  (1) *x_r is the first conjugate point of x_0 along τ;*
  (2) *There exist, at least, two minimizing geodesics from x_0 to x_r.*

Proof. Let $a_1, a_2, \ldots$ be a monotone decreasing sequence of real numbers converging to r. For each natural number k, let $\exp tX_k$, $0 \leq t \leq b_k$, be a minimizing geodesic from x_0 to x_{a_k}, where X_k is a unit tangent vector at x_0, and b_k is the distance from x_0 to x_{a_k}. Let X be the unit tangent vector at x_0 determined by $x_t = \exp tX$ for all t. Since x_r is the cut point of x_0 along τ and $a_k > r$, we have

$$X \neq X_k, \qquad a_k > b_k.$$

Since b_k is the distance between x_0 and x_{a_k}, we have

$$r = \lim_{k \to \infty} b_k.$$

Hence, the set of vectors $b_k X_k$ is contained in some compact subset of $T_{x_0}(M)$. By taking a subsequence if necessary, we may assume that the sequence $b_1 X_1, b_2 X_2, \ldots$ converges to some vector of length r, say rY, where Y is a unit tangent vector. Then, $\exp tY$, $0 \leq t \leq r$, is a geodesic from x_0 to x_r since $\exp rY = \lim_{k \to \infty} \exp b_k X_k = \lim_{k \to \infty} x_{a_k} = x_r$. It is of length r and hence minimizing. If $X \neq Y$, then $\exp tX$ and $\exp tY$, $0 \leq t \leq r$, are two distinct minimizing geodesics joining x_0 and x_r and (2) holds. Assume $X = Y$. Assuming also that x_r is not conjugate to x_0 along τ, we shall obtain a contradiction. Since the differential of $\exp: T_{x_0}(M) \to M$ is nonsingular at rX (cf. §1), $\exp$ is a diffeomorphism of a neighborhood U of rX in $T_{x_0}(M)$ onto an open neighborhood of x_r in M. Let k be a large integer such that both $a_k X$ and $b_k X_k$ are in U. Since $\exp a_k X = x_{a_k} = \exp b_k X_k$, we can conclude that $a_k X = b_k X_k$, thus contradicting the fact that $X \neq X_k$. We have shown that if $X = Y$, then x_r is conjugate to x_0 along τ. On the other hand, there is no conjugate point of x_0 before x_r along τ. Indeed, if x_s, where $0 \leq s \leq r$, were conjugate to x_0 along τ, then τ would not be minimizing beyond x_s by Theorem 5.7, which contradicts the definition of r. Hence, x_s is the first conjugate point of x_0 along τ.
QED.

COROLLARY 7.2. *If x_r is the cut point of x_0 along $\tau = x_t$, $0 \leq t < \infty$, then x_0 is the cut point of x_r along τ (in the reversed direction).*

Proof. Extending the geodesic τ in the negative direction, we may assume that x_t is defined for $-\infty < t < \infty$. Let $-a$ be any negative number. We claim that $\tau \mid [-a, r]$ is not minimizing. Assume (1) of Theorem 7.1. Then x_0 is a conjugate point of x_r along τ in the reversed direction and, by Theorem 5.7, $\tau \mid [-a, r]$ cannot be minimizing. Assume (2) of Theorem 7.1. Then the join of $\tau \mid [-a, 0]$ and a minimizing geodesic from x_0 to x_r other than $\tau \mid [0, r]$ gives us a non-geodesic curve of length $a + r$ joining x_{-a} and x_r. Hence, $\tau \mid [-a, r]$ whose length is also $a + r$, cannot be minimizing. This proves our claim. There exists therefore a non-negative number $b < r$ such that x_b is the cut point of x_r along τ in the reversed direction. Assuming that b is positive, we shall get a contradiction. We apply Theorem 7.1 to the geodesic $\tau \mid [b, r]$ in the reversed direction. Then x_b and x_r are conjugate to each other along τ or else there is another minimizing geodesic from x_r to x_b. By the reasoning we just used, we can prove that if $c < b$, then $\tau \mid [c, r]$ cannot be minimizing. In particular, $\tau \mid [0, r]$ cannot be minimizing, which contradicts the definition of r. Hence, $b = 0$. QED.

Let S_x be the set of unit tangent vectors at a point x of M; it is a unit sphere in $T_x(M)$. We define a function $\mu: S_x \to \mathbf{R}^+ \cup \infty$, where $\mathbf{R}^+$ denotes the set of positive real numbers, as follows. For each unit vector $X \in S_x$, consider the geodesic $\tau = \exp tX$, $0 \leq t < \infty$. If $\exp rX$ is the cut point of x along τ, then we set $\mu(X) = r$. If there is no cut point of x along τ, then we set $\mu(X) = \infty$. We introduce a topology in $\mathbf{R}^+ \cup \infty$ by taking intervals (a, b) and $(a, \infty] = (a, \infty) \cup \infty$ as a base for the open sets. Then we have

THEOREM 7.3. *The mapping $\mu: S_x \to \mathbf{R}^+ \cup \infty$ is continuous.*

Proof. Suppose that μ is not continuous at a point X of S_x and let $X_1, X_2, \ldots$ be a sequence of points of S_x converging to X such that

$$\mu(X) \neq \lim_{k \to \infty} \mu(X_k).$$

In general, $\lim_{k \to \infty} \mu(X_k)$ may not even exist. However, by taking a subsequence if necessary, we may assume that $\lim_{k \to \infty} \mu(X_k)$ exists in $R^+ \cup \infty$.

We consider first the case $\mu(X) > \lim_{k\to\infty} \mu(X_k)$. Set
$$a_k = \mu(X_k), \qquad a = \lim_{k\to\infty} \mu(X_k).$$

Since $\mu(X) > a$, $\exp aX$ cannot be conjugate to x along the geodesic $\exp tX$. By Theorem 1.4, $\exp: T_{x_0}(M) \to M$ maps a neighborhood, say U, of aX diffeomorphically onto a neighborhood of $\exp aX$. We may assume, by omitting a finite number of $a_k X_k$ if necessary, that all of $a_k X_k$ are in U. Since $\exp$ maps U diffeomorphically onto $\exp(U)$, $\exp a_k X_k$ cannot be conjugate to x_0 along the geodesic $\exp tX_k$. By Theorem 7.1, there is another minimizing geodesic from x_0 to $\exp a_k X_k$. In other words, there exists, for each k, a unit vector $Y_k \neq X_k$ at x_0 such that
$$\exp a_k X_k = \exp a_k Y_k.$$

Since $\exp$ maps U one-to-one into M, each $a_k Y_k$ does not lie in U. By taking a subsequence if necessary, we may assume that Y_1, $Y_2, \ldots$ converges to some unit vector, say Y. Then aY, which is the limit vector of $a_k Y_k$, does not lie in U. We have
$$\exp aY = \exp(\lim_{k\to\infty} a_k Y_k) = \lim_{k\to\infty}(\exp a_k Y_k) = \lim_{k\to\infty}(\exp a_k X_k)$$
$$= \exp(\lim_{k\to\infty} a_k X_k) = \exp aX.$$

Hence, both $\exp tX$ and $\exp tY$, $0 \leq t \leq a$, are minimizing geodesics from x_0 to $\exp aX = \exp aY$. This implies that if b is any number greater than a, then the geodesic $\exp tX$, $0 \leq t \leq b$, is no longer minimizing, contradicting the assumption $\mu(X) > a$.

We consider next the case $\mu(X) < \lim_{k\to\infty} \mu(X_k)$. Let b be a positive number such that $\lim_{k\to\infty} \mu(X_k) > \mu(X) + b$. By omitting a finite number of X_k if necessary, we may assume that
$$\mu(X_k) > \mu(X) + b \qquad \text{for all } k.$$

By the very definition of $\mu(X)$, there exists a unit vector $X' \neq X$ at x_0 such that $\exp tX'$, $0 \leq t \leq \mu(X) + b'$, is a minimizing geodesic from x_0 to $\exp(\mu(X) + b)X$, where $b' < b$. (Note that b' may be negative.) In particular,
$$\exp(\mu(X) + b)X = \exp(\mu(X) + b')X'.$$
We set
$$2c = b - b'.$$

Since the sequence of points $\exp (\mu(X) + b)X_k$ converges to $\exp (\mu(X) + b)X$, we may assume, by omitting a finite number of X_k if necessary, that the distances between $\exp (\mu(X) + b)X$ and $\exp (\mu(X) + b)X_k$ are all less than c. For each fixed k, consider the curve from x_0 to $\exp (\mu(X) + b)X_k$ defined as follows. It consists of the geodesic $\exp tX'$, $0 \leq t \leq \mu(X) + b'$, from x_0 to $\exp (\mu(X) + b')X' = \exp (\mu(X) + b)X$ and a minimizing geodesic from $\exp (\mu(X) + b)X$ to $\exp (\mu(X) + b)X_k$. Then the length of this curve is less than $\mu(X) + b' + c = \mu(X) + b - c$. This means that the geodesic $\exp tX_k$, $0 \leq t \leq \mu(X) + b$, is not minimizing, which contradicts the inequality $\mu(X_k) > \mu(X) + b$.
QED.

Let $\tilde{C}(x_0)$ denote the set of all $\mu(X)X$, where X are unit vectors at x_0 such that $\mu(X)$ are finite. Let $C(x_0)$ be the image of $\tilde{C}(x_0)$ under $\exp$. Obviously, $C(x_0)$ consists of all cut points of x_0 along all geodesics starting from x_0. We shall call $C(x_0)$ the *cut locus* of x_0 and $\tilde{C}(x_0)$ the *cut locus of x_0 in $T_{x_0}(M)$*.

THEOREM 7.4. *Let $E = \{tX; 0 \leq t < \mu(X)$ and X unit vectors at $x_0\}$. Then*
(1) *E is an open cell in $T_{x_0}(M)$;*
(2) $\exp$ *maps E diffeomorphically onto an open subset of M;*
(3) *M is a disjoint union of $\exp(E)$ and the cut locus $C(x_0)$ of x_0.*

Proof. (1) follows from Theorem 7.3. Obviously, $\exp$ maps E one-to-one into M. By Theorem 5.7, there cannot be any conjugate point of x_0 in E. Hence, the differential of $\exp: E \to M$ is non-singular at every point of E. This implies (2). Since E and $\tilde{C}(x_0)$ are disjoint, $\exp(E)$ and $C(x_0)$ are also disjoint. To show that M is a union of $\exp(E)$ and $C(x_0)$, let y be an arbitrary point of M. Let $\exp tX$, $0 \leq t \leq a$, be a minimizing geodesic from x_0 to y, where X is a unit tangent vector at x_0 and a is the distance from x_0 to y. From the very definition of $\mu(X)$, it follows that $a \leq \mu(X)$. Hence, aX is either in E or in $\tilde{C}(x_0)$. The point $y = \exp aX$ is therefore either in $\exp(E)$ or in $C(x_0)$. QED.

Remark. The open subset $\exp(E)$ of M is the largest open subset of M in which a normal coordinate system around x_0 can be defined.

VIII. VARIATIONS OF THE LENGTH INTEGRAL 101

THEOREM 7.5. *Let M be a complete Riemannian manifold and x_0 a point of M. Then M is compact if and only if, for every unit tangent vector X at x_0, $\mu(X)$ is finite.*

Proof. Suppose M is compact and let d be the diameter of M. If $a > d$, then $\exp tX$, $0 \leq t \leq a$, cannot be a minimizing geodesic from x_0 to $\exp aX$. Hence, $\mu(X) \leq d$. Conversely, assume that, for every unit vector X at x_0, $\mu(X)$ is finite. Since μ is a continuous function defined on the unit sphere in $T_{x_0}(M)$ (cf. Theorem 7.3), μ is bounded by a positive number, say b. Let B be the set of tangent vectors at x_0 whose length are less than or equal to b. Then B is a compact set containing E and $\tilde{C}(x_0)$ of Theorem 7.4. By Theorem 7.4, exp maps B onto M. Hence, M is compact. QED.

Remark. Theorems 7.3 and 7.5 imply that M is compact if and only if the cut locus $\tilde{C}(x_0)$ of x_0 in $T_{x_0}(M)$ is homeomorphic with a sphere of dimension $n - 1$, where $n = \dim M$.

Example 7.1. Let M be an n-dimensional unit sphere and x its north pole. The cut locus of x in $T_x(M)$ is the sphere of radius π with center at the origin of $T_x(M)$. The cut locus of x in M reduces to the north pole.

Example 7.2. Let S^n be the unit sphere in $\mathbf{R}^{n+1}$. Identifying each point of S^n with its antipodal point, we obtain the n-dimensional real projective space, which will be denoted by M. The Riemannian metric of S^n induces a Riemannian metric on M in a natural manner so that the projection of S^n onto M is a local isometry. The cut locus of a point $x \in M$ in $T_x(M)$ is the sphere of radius $\pi/2$ in $T_x(M)$ with center at the origin. If x corresponds to the north and south poles of S^n, then the cut locus of x in M is the image of the equator of S^n under the projection $S^n \to M$. The cut locus $C(x)$ is therefore a naturally imbedded $(n - 1)$-dimensional projective space.

Example 7.3. In the Euclidean plane $\mathbf{R}^2$ with coordinate system (x, y), consider the closed square given by $0 \leq x, y \leq 1$. By identifying $(x, 0)$ with $(x, 1)$ for all $0 \leq x \leq 1$ and $(0, y)$ with $(1, y)$ for all $0 \leq y \leq 1$, we obtain a torus, which we denote by M. The natural Riemannian metric on $\mathbf{R}^2$ induces a Riemannian metric on M. Let p be the point with coordinate $(\frac{1}{2}, \frac{1}{2})$. Then the cut locus of p in $T_p(M)$ can be identified with the boundary of the square in $\mathbf{R}^2$ under the natural identification of $T_p(M)$ with $\mathbf{R}^2$. The cut locus of p in M consists of two closed curves which form a basis for $H_1(M, \mathbf{Z}) \approx \mathbf{Z} + \mathbf{Z}$.

A list of papers on conjugate and cut loci can be found in an introductory article by Kobayashi [27]. For more recent results, see Warner [3], Weinstein [1], and Wong [4].

8. Spaces of non-positive curvature

We have already seen (cf. Corollary 2.4) that a Riemannian manifold with non-positive curvature has no conjugate point. In this section, we shall prove more.

THEOREM 8.1. *Let M be a complete Riemannian manifold with non-positive curvature. Let x be an arbitrary point of M. Then the exponential map $\exp_x: T_x(M) \to M$ at x is a covering map. In particular, if M is simply connected, then $\exp_x$ is a diffeomorphism of $T_x(M)$ onto M.*

Proof. In general, let f be a map from a Riemannian manifold N with metric g_N into a Riemannian manifold M with metric g_M. We say that f *increases* (resp. *decreases*) *the distance* if

$$f^* g_M \geq g_N \quad (\text{resp. } f^* g_M \leq g_N),$$

that is, if

$$\|X\| \leq \|f_* X\| \quad (\text{resp. } \|X\| \geq \|f_* X\|)$$

for every tangent vector X of N. We first prove the following general lemma.

LEMMA 1. *Let f be a map from a connected complete Riemannian manifold N onto another connected Riemannian manifold M of the same dimension. If f increases the distance, then f is a covering map and M is also complete.*

Proof. Let g_M and g_N be the metrics for M and N respectively. Since f increases the distance, $f^* g_M$ is also a Riemannian metric on N. For every curve τ in N, its arc length with respect to g_N is less than or equal to the one with respect to $f^* g_M$. It follows that if d and d' denote the distance functions for g_N and $f^* g_M$ respectively, then

$$d'(x, y) \geq d(x, y) \quad \text{for } x, y \in N.$$

It is then clear that every Cauchy sequence for d' is a Cauchy

VIII. VARIATIONS OF THE LENGTH INTEGRAL   103

sequence for d and hence is convergent as N is complete with respect to g_N. Since the topology of N is compatible with any Riemannian metric on N, N is complete with respect to f^*g_M.

Replacing g_N by f^*g_M we may assume that $f: N \to M$ is an isometric immersion, i.e., that $g_N = f^*g_M$. Lemma 1 now follows from Theorem 4.6 of Chapter IV.

LEMMA 2. *Let M be an arbitrary Riemannian manifold and let $x \in M$. Let ρ be a ray in $T_x(M)$ emanating from the origin and $\tau = \exp_x(\rho)$ the corresponding geodesic in M starting from x. Then every vector tangent to $T_x(M)$ and perpendicular to ρ is mapped by $\exp_x$ into a vector tangent to M and perpendicular to τ.*

Proof. Let $\rho = tX$, $0 \leq t \leq 1$, where $X \in T_x(M)$. Let W^* be a vector tangent to $T_x(M)$ at X and perpendicular to ρ. Let $X^s \in T_x(M)$, $-\varepsilon \leq s \leq \varepsilon$, be a 1-parameter family of vectors such that

(a) $X^0 = X$;
(b) each X^s has the same length as X;
(c) W^* is the tangent vector to the curve X^s, $-\varepsilon \leq s \leq \varepsilon$, at X^0.

For each s, let ρ^s be the ray given by tX^s, $0 \leq t \leq 1$. The 1-parameter family of rays ρ^s, $-\varepsilon \leq s \leq \varepsilon$, induces in a natural way a vector field Y^* along $\rho = \rho^0$ which is perpendicular to ρ; for each fixed t, Y^* at tX is the tangent vector to the curve described by tX^s, $-\varepsilon \leq s \leq \varepsilon$, at tX^0. The vector field Y^* vanishes at the origin of $T_x(M)$ and coincides with the vector W^* at X. If we set

$$\tau^s = \exp_x(\rho^s) \quad \text{and} \quad Y = (\exp_x)_*(Y^*),$$

then Y is the Jacobi field along τ^0 induced by the variation τ^s of the geodesic τ^0. We apply Theorem 5.1 to the variation τ^s. Since the length $L(\tau^s)$ is independent of s, the left hand side of the formula in Theorem 5.1 vanishes. Since Y vanishes at x and τ^0 is a geodesic, it follows that Y is perpendicular to τ^0 at the point $\exp_x(X)$. This completes the proof of Lemma 2.

LEMMA 3. *Let M be a complete Riemannian manifold with nonpositive curvature and x a point of M. Then $\exp_x: T_x(M) \to M$ increases the distance.*

Proof. Let W^* be a vector tangent to $T_x(M)$ at a point X of $T_x(M)$. We set
$$W = (\exp_x)_*(W^*).$$

Then W is a tangent vector of M at $\exp_x X$. To show that the length $\|W\|$ of W is not less than the length $\|W^*\|$ of W^*, we decompose W^* into a sum of two mutually perpendicular vectors at X:
$$W^* = W_1^* + W_2^*,$$
where W_1^* is in the direction of the ray ρ joining the origin of $T_x(M)$ to X and W_2^* is perpendicular to the ray ρ. We set
$$W_1 = (\exp_x)_*(W_1^*) \quad \text{and} \quad W_2 = (\exp_x)_*(W_2^*).$$
Since W_1^* is in the direction of ρ, it follows from the definition of $\exp_x$ that
$$\|W_1^*\| = \|W_1\|.$$
As W_1 is perpendicular to W_2 by Lemma 2, we have
$$\|W\|^2 = \|W_1 + W_2\|^2 = \|W_1\|^2 + \|W_2\|^2.$$
Hence,
$$\|W\|^2 - \|W^*\|^2 = \|W_1\|^2 + \|W_2\|^2 - \|W_1^*\|^2 - \|W_2^*\|^2$$
$$= \|W_2\|^2 - \|W_2^*\|^2.$$

The problem is thus reduced to proving the inequality $\|W_2\| \geq \|W_2^*\|$. In other words, it suffices to prove the inequality $\|W\| \geq \|W^*\|$ under the assumption that W^* is perpendicular to ρ.

We define vector fields Y^* and Y along ρ and τ respectively as in the proof of Lemma 2. Since Y^* and Y are induced by 1-parameter families of geodesics ρ^s and $\tau^s = \exp_x(\rho^s)$ respectively, they are Jacobi fields along ρ and τ respectively vanishing at $t = 0$. We shall now apply the comparison theorem of Rauch (Theorem 4.1) to Y^* and Y. Assumptions (1) and (4) in Theorem 4.1 are obviously satisfied. By Corollary 2.4, assumption (3) is also satisfied. To complete the proof of Lemma 3, it is therefore sufficient to verify assumption (2) for the vector fields Y^* and Y. Take a Euclidean coordinate system in $T_x(M)$ with origin at the origin of $T_x(M)$ and the corresponding normal coordinate system in a neighborhood of x. From the construction of Y, it follows that Y^* and Y have the same components with respect to the coordinate systems chosen above. On the other hand, the Christoffel symbols vanish at x (cf. Proposition 8.4 of Chapter III) and hence the covariant differentiation and the ordinary differentiation coincide at x (cf. Corollary 8.5 of Chapter III). These two facts imply assumption (2) of Theorem 4.1.

Combining Lemma 1 and Lemma 3, we obtain Theorem 8.1.

QED.

Remark 1. If M has vanishing curvature in Lemma 3, then we may apply the comparison theorem of Rauch in both directions to obtain $\|W^*\| \leq \|W\|$ and $\|W^*\| \geq \|W\|$. Hence we have:

If M has vanishing curvature, then $\exp_x\colon T_x(M) \to M$ is an isometric immersion.

Combining this with Theorem 8.1 we obtain

COROLLARY 8.2. *If M is a complete, simply connected Riemannian manifold with vanishing curvature and $x \in M$, then $\exp_x\colon T_x(M) \to M$ is an isometry.*

Remark 2. Kobayashi [16] strengthened Theorem 8.1 as follows: *If M is a connected complete Riemannian manifold and a point x of M has no conjugate point, then $\exp_x\colon T_x(M) \to M$ is a covering map.*

The proof may be achieved by replacing Lemma 3 by the following:

If a point $x \in M$ has no conjugate point, there is a complete Riemannian metric on $T_x(M)$ which makes $\exp_x$ distance-increasing.

For the proof we refer the reader to Kobayashi [16]; it is more elementary than that of Lemma 3.

Theorem 8.1 is due to Hadamard [1] and E. Cartan [8]. The proof presented here is based on Ambrose's lectures at MIT in 1957–58. Theorem 8.1 and its generalization as stated in Remark 2 may be proved by means of the theory of Morse (cf. Klingenberg [8] and Milnor [3]).

THEOREM 8.3. *Let M be a connected homogeneous Riemannian manifold with non-positive sectional curvature and negative-definite Ricci tensor. Then M is simply connected and, for every $x \in M$, $\exp_x\colon T_x(M) \to M$ is a diffeomorphism.*

Proof. An isometry of a Riemannian manifold is called a *Clifford translation* if the distance between a point and its image under the isometry is the same for every point. The following lemma is due to Wolf [1].

LEMMA 1. *Let M be a homogeneous Riemannian manifold and $\tilde{M}$ a covering manifold with the induced metric so that the covering projection $p \colon \tilde{M} \to M$ is an isometric immersion. Then a diffeomorphism f of $\tilde{M}$ onto itself satisfying $p \circ f = p$ is a Clifford translation of $\tilde{M}$.*

Proof of Lemma 1. Let G be a connected Lie group of isometries acting transitively on M, and $\mathfrak{g}$ the Lie algebra of G. Considering every element $X \in \mathfrak{g}$ as an infinitesimal isometry of M, let X^* be the lift of X to $\tilde{M}$. Then the set of these vector fields X^* generates a transitive Lie group G^* of isometries of $\tilde{M}$ whose Lie algebra is isomorphic to $\mathfrak{g}$. Since f induces the identity transformation of M, it leaves every X^* invariant. Hence f commutes with every element of G^*. For any two points $y, y' \in \tilde{M}$, let ψ be an element of G^* such that $y' = \psi(y)$. Then

$$d(y', f(y')) = d(\psi(y), f \circ \psi(y)) = d(\psi(y), \psi \circ f(y)) = d(y, f(y)),$$

where d denotes the distance between two points. This completes the proof of Lemma 1.

LEMMA 2. *Let M, $\tilde{M}$, and f be as in Lemma 1. Let $y_0 \in \tilde{M}$ and let $\tau^* = y_t$, $0 \leq t \leq a$, be a minimizing geodesic from y_0 to $f(y_0)$ so that $y_a = f(y_0)$. Set $x_t = p(y_t)$ for $0 \leq t \leq a$. Then $\tau = x_t$, $0 \leq t \leq a$, is a smoothly closed geodesic, that is, the outgoing direction of τ at x_0 coincides with the incoming direction of τ at x_a.*

Proof of Lemma 2. Let r be a small positive number such that the r-neighborhoods $V(y_0; r)$ and $V(y_a; r)$ of y_0 and y_a are homeomorphic with the r-neighborhood $U(x_0; r)$ of $x_0 = x_a$ under the projection p. Assume that τ is not smoothly closed at $x_0 = x_a$. Then there is a small positive number δ such that the points $x_{a-\delta}$ and x_δ can be joined by a curve σ in $U(x_0; r)$ whose length is less than 2δ (where 2δ is equal to the length of τ from $x_{a-\delta}$ through $x_a = x_0$ to x_δ). Let σ^* be the curve in $V(y_a; r)$ such that $p(\sigma^*) = \sigma$. Let y^* be the end point of σ^*. Then $y^* = f(y_\delta)$. Then

$$\begin{aligned} d(y_\delta, f(y_\delta)) = d(y_\delta, y^*) &\leq d(y_\delta, y_{a-\delta}) + d(y_{a-\delta}, y^*) \\ &\leq (a - 2\delta) + (\text{length of } \sigma^*) \\ &= (a - 2\delta) + (\text{length of } \sigma) \\ &< (a - 2\delta) + 2\delta = a = d(y_0, f(y_0)). \end{aligned}$$

This contradicts Lemma 1, thus completing the proof of Lemma 2.

We shall now complete the proof of Theorem 8.3. Assuming that M is not simply connected, let $\tilde{M}$ be the universal covering manifold of M. Let f be a covering transformation of $\tilde{M}$ different from the identity transformation. Let $\tau = x_t$, $0 \leq t \leq a$, be a smoothly closed geodesic of M given in Lemma 2. Let X be any infinitesimal isometry of M. We define a non-negative function $h(t)$, $-\infty < t < +\infty$, as follows:

$$h(t) = g(X, X)_{x_t} \quad \text{for } 0 \leq t \leq a,$$

and then extend h to a periodic function of period a. By Lemma 2, $h(t)$ is differentiable at every point t, $-\infty < t < +\infty$. We have

$$\frac{d}{dt} h(t) = 2g(X', X)_{x_t} \quad \text{for } 0 \leq t \leq a,$$

$$\frac{d^2}{dt^2} h(t) = 2g(X', X')_{x_t} - 2g(R(X, \dot{\tau})\dot{\tau}, X)_{x_t} \quad \text{for } 0 \leq t \leq a.$$

Since the sectional curvature is non-positive, $h''(t) \geq 0$ for $0 \leq t \leq a$. Since $h(t)$ is periodic, $h(t)$ is a constant function. Hence $h''(t) = 0$. In particular, $g(X', X') = 0$ and $g(R(X, \dot{\tau})\dot{\tau}, X) = 0$.

On the other hand, since M is a homogeneous Riemannian manifold with negative-definite Ricci tensor, there exists an infinitesimal isometry X of M such that $g(R(X, \dot{\tau})\dot{\tau}, X)_{x_0} < 0$. This contradiction comes from the assumption that M is not simply connected. The fact that $\exp_x \colon T_x(M) \to M$ is a diffeomorphism follows from Theorem 8.1. QED.

Theorem 8.3 is due to Kobayashi [19].

THEOREM 8.4. *Let M be a connected homogeneous Riemannian manifold with non-positive sectional curvature and negative-definite Ricci tensor. If a Lie subgroup G of the group $I(M)$ of isometries of M is transitive on M, then G has trivial center.*

Proof. Let C be the center of G. Let $\bar{G}$ be the closure of G in $I(M)$. Then every element of C commutes with every element of $\bar{G}$ and hence lies in the center of $\bar{G}$. We may therefore assume that G is a closed subgroup of $I(M)$.

We first prove that C is discrete. Suppose that X is an infinitesimal isometry of M which generates a 1-parameter group belonging to the center C. Since X is invariant under the action of G,

108   FOUNDATIONS OF DIFFERENTIAL GEOMETRY

the length $g(X, X)^{1/2}$ is constant on M. Since the Ricci tensor of M is negative-definite, Theorem 5.3 of Chapter VI implies that X vanishes identically.

Let φ be any element of C. Then it commutes with each element of G and hence is a Clifford translation in the sense defined in the proof of Theorem 8.3. In fact, for $x, x' \in M$ we choose an element ψ of G such that $x' = \psi(x)$. Then we have

$$d(x', \varphi(x')) = d(\psi(x), \varphi \circ \psi(x)) = d(\psi(x), \psi \circ \varphi(x)) = d(x, \varphi(x)).$$

It follows that the action of C on M is free. We shall show that C is properly discontinuous on M. Let H be the isotropy subgroup of G at a point of M so that $M = G/H$. Then H is compact (cf. Corollary 4.8 of Chapter I). By Proposition 4.5 of Chapter I, C is discontinuous on M. By Proposition 4.4 of Chapter I, C is properly discontinuous on M. Then the quotient space M/C is a manifold (cf. Proposition 4.3 of Chapter I) and M is a covering manifold of M/C with the natural projection $p: M \to M/C$ as a covering projection (cf. pp. 61–62 of Chapter I). Since C is the center of G, the action of G on M induces an action of G on M/C. It follows that with respect to the induced Riemannian metric M/C is also a homogeneous Riemannian manifold with non-positive sectional curvature and negative-definite Ricci tensor. By Theorem 8.3, M/C is simply connected. Hence C reduces to the identity element.   QED.

9. Center of gravity and fixed points of isometries

Let A be a compact topological space and $C(A)$ the algebra of real-valued continuous functions f on A. With the norm $\|f\|$ defined by

$$\|f\| = \sup_{a \in A} |f(a)|,$$

$C(A)$ is a Banach algebra. A *Radon measure* (or simply, *measure*) on A is a continuous linear mapping $\mu: C(A) \to R$. For each $f \in C(A)$, $\mu(f)$ is called the *integral* of f with respect to the measure μ and will be denoted by $\int_A f(a) \, d\mu(a)$ or simply by $\int_A f \, d\mu$. A measure μ is *positive* if $\mu(f) \geq 0$ for all non-negative $f \in C(A)$ and if $\mu \neq 0$.

VIII. VARIATIONS OF THE LENGTH INTEGRAL

We need the following theorem to prove the existence of a fixed point of a compact group of isometries of a complete, simply connected Riemannian manifold with non-positive curvature.

THEOREM 9.1. *Let μ be a positive measure on a compact topological space A. Let f be a continuous mapping from A into a complete, simply connected Riemannian manifold M with non-positive curvature. We set*

$$J(x) = \int_A d(x, f(a))^2 \, d\mu(a) \quad \text{for } x \in M,$$

where $d(x, f(a))$ is the distance between x and $f(a)$. Then J attains its minimum at precisely one point.

The point where J takes its minimum will be called the *center of gravity* of $f(A)$ with respect to μ.

Proof. By normalizing the measure if necessary, we may assume that the total measure of A, i.e., $\mu(1)$, is 1.

(1) The case where $M = R^n$. Let $x^1, \ldots, x^n$ be a Euclidean coordinate system in $\mathbf{R}^n$ and let $f: A \to \mathbf{R}^n$ be given by

$$x^i = f^i(a), \quad a \in A, i = 1, \ldots, n.$$

Then

$$\begin{aligned}
J(x) &= \int_A \sum_i (x^i - f^i(a))^2 \, d\mu(a) \\
&= \sum_i (x^i)^2 \int_A d\mu(a) - 2 \sum_i x^i \int_A f^i(a) \, d\mu(a) \\
&\quad + \sum_i \int_A (f^i(a))^2 \, d\mu(a) \\
&= \sum_i ((x^i)^2 - 2 b^i x^i + c^i),
\end{aligned}$$

where

$$b^i = \int_A f^i(a) \, d\mu(a) \quad \text{and} \quad c^i = \int_A (f^i(a))^2 \, d\mu(a).$$

Hence, we have

$$J(x) = \sum_i ((x^i - b^i)^2 + c^i - (b^i)^2),$$

showing that J takes its minimum at $b = (b^1, \ldots, b^n)$ and only at b.

(2) Existence of the center of gravity in the general case. Since J is continuous, it suffices to show the existence of a compact subset K of M and a positive number r such that

$$J(y_0) \leq r^2 \quad \text{for some } y_0 \text{ in } K$$

and

$$J(x) > r^2 \quad \text{for all } x \text{ not in } K.$$

Choose an arbitrary point y_0 of M and take a positive number r such that

$$d(y_0, f(a)) \leq r \quad \text{for all } a \in A.$$

We set

$$K = \{x \in M; \, d(x, f(A)) \leq r\}.$$

Since $f(A)$ is compact, K is bounded and closed. Being a bounded, closed subset of a complete Riemannian manifold M, K is compact by Theorem 4.1 of Chapter IV. Evidently, $y_0 \in K$. We have

$$J(y_0) \leq \int_A r^2 \, d\mu(a) = r^2.$$

If x is not in K, then $d(x, f(a)) > r$ for all $a \in A$. Hence,

$$J(x) > \int_A r^2 \, d\mu(a) = r^2 \quad \text{for } x \in K.$$

(3) The uniqueness of the center of gravity in the general case. We shall reduce the problem to the Euclidean case. Let o be a point of M where J takes its minimum. Since $\exp_o \colon T_o(M) \to M$ is a diffeomorphism by Theorem 8.1, there is a unique mapping $F \colon A \to T_o(M)$ such that

$$f = \exp_o \circ F.$$

We set

$$J'(X) = \int_A d'(X, F(a))^2 \, d\mu(a), \quad X \in T_o(M),$$

where d' denotes the Euclidean distance in $T_o(M)$. Let 0 be the origin of $T_o(M)$. We shall prove the following relations which imply obviously that o is the only point where J takes its minimum:

$$J(o) = J'(0) < J'(X) \leq J(x) \quad \text{if } \exp_o X = x \neq o.$$

Since $\exp_o$ preserves the length of a ray emanating from the origin 0, we have
$$d'(0, X) = d(o, x) \quad \text{for } x = \exp_o X.$$
Hence,
$$d'(0, F(a)) = d(o, f(a)) \quad \text{for } a \in A.$$
This implies the equality $J(o) = J'(0)$. Since $T_o(M)$ is a Euclidean space, (1) implies that 0 is the only point where J' takes its minimum, i.e.,
$$J'(0) < J'(X) \quad \text{for } 0 \neq X \in T_o(M).$$
Since $\exp_o$ is a distance-increasing mapping (cf. Lemma 3 in the proof of Theorem 8.1), we have
$$d'(X, F(a)) \leq d(x, f(a)) \quad \text{for } x = \exp_o X \text{ and } a \in A.$$
This implies the inequality
$$J'(X) \leq J(x) \quad \text{for } x = \exp_o X.$$
QED.

As an important application of Theorem 9.1, we prove

THEOREM 9.2. *Every compact group G of isometries of a complete, simply connected Riemannian manifold M with non-positive curvature has a fixed point.*

Proof. We choose any point, say x_0, of M and define a mapping $f: G \to M$ as follows:
$$f(a) = a(x_o) \quad \text{for } a \in G.$$
Let μ be a left invariant measure on G. (It is known that every compact group admits a bi-invariant measure; see for example, Nachbin [1; p. 81].) By setting $A = G$, we apply Theorem 9.1 and claim that the center of gravity of $f(G)$ with respect μ is a fixed point of G. Evidently, it suffices to show that $J(b(x)) = J(x)$ for $x \in M$ and for $b \in G$. Since the distance function d is invariant by G and μ is left invariant, we have
$$J(b(x)) = \int_{a \in G} d(b(x), a(x_o))^2 \, d\mu(a) = \int_{a \in G} d(x, b^{-1}a(x_o))^2 \, d\mu(a)$$
$$= \int_{a \in G} d(x, b^{-1}a(x_o))^2 \, d\mu(b^{-1}a) = J(x).$$
QED.

We owe the formulation of Theorem 9.1 to Iwahori [1]. Theorem 9.2 was originally proved by E. Cartan [16]. (See also Borel [4].)

As an immediate consequence of Theorem 9.2 we have

COROLLARY 9.3. *Let M be a connected, simply connected homogeneous Riemannian manifold of non-positive sectional curvature and let G be a closed subgroup of the group $I(M)$ of isometries of M. Assume that G is transitive so that $M = G/H$, where H is the isotropy subgroup of G at a point of M. Then*

(1) *H is a maximal compact subgroup of G and every maximal compact subgroup of G is conjugate to H;*

(2) *If G is connected, so is H.*

Proof. (1) Let K be any maximal compact subgroup of G. By Theorem 9.2, K is contained in the isotropy subgroup of G at a point say x, of M. Since the isotropy subgroup at x is compact (cf. Corollary 4.8 of Chapter I), it coincides with K. Since G is transitive, the isotropy subgroups are all conjugate to each other. Hence H is conjugate to K and is a maximal compact subgroup of G.

(2) Since M is simply connected, a simple homotopy argument shows that if G is connected, so is H. QED.

The following theorem will not be used except in §11 of Chapter XI.

THEOREM 9.4. *Let M be a connected, simply connected homogeneous Riemannian manifold with non-positive sectional curvature. Let G be a closed subgroup of the group of isometries of M which is transitive on M so that $M = G/H$, where H is the isotropy subgroup of G at a point, say o, of M. Assume that the linear isotropy representation of H leaves no non-zero vector of $T_o(M)$ invariant. Then we have*

(1) *M is in one-to-one correspondence with the set of all maximal compact subgroups of G under the correspondence which assigns to each point $x \in M$ the isotropy subgroup G_x of G at x;*

(2) *If α is an automorphism of G of prime period, then there is a maximal compact subgroup of G which is invariant by α.*

Proof. (1) By Corollary 9.3, G_x is a maximal compact subgroup of G and, conversely, every maximal compact subgroup of G coincides with G_x for some $x \in M$. It remains to prove that, for every $x \in M$, x is the only fixed point of G_x. Because of homogeneity, it suffices to show that o is the only fixed point of $H = G_o$. Assume that there is another fixed point x of H. By Theorem 8.1 there is a unique geodesic τ from o to x. By uniqueness of τ, H leaves τ pointwise invariant. Hence the tangent vector to τ at o is invariant by H, in contradiction to our assumption.

(2) Let γ be any automorphism of G. Then γ permutes the maximal compact subgroups of G. Let $\bar{\gamma}$ denote the corresponding transformation of M; if $\gamma(G_x) = G_y$, then $\bar{\gamma}(x) = y$ by definition. Let Γ be the cyclic group generated by α and assume that the order of Γ is prime. Then every element $\gamma \neq 1$ of Γ is a generator of Γ, and a fixed point of $\bar{\gamma}$ for any $\gamma \neq 1$ is a fixed point of $\bar{\alpha}$. If $\bar{\alpha}$ has no fixed point on M, then Γ acts freely on M. But this is impossible since M is homeomorphic to a Euclidean space by Theorem 8.1 of Chapter VIII. If Γ acts freely on M, then

$$H^k(\Gamma; \mathbf{Z}) = H^k(M/\Gamma; \mathbf{Z})$$

by a theorem of Eckmann, Eilenberg-MacLane and Hopf (see Cartan-Eilenberg [1; p. 356]). On the other hand, $H^k(M/\Gamma, \mathbf{Z}) = 0$ for $k > \dim M$. On the other hand, $H^k(\Gamma; \mathbf{Z}) = Z_p$ for all even k (cf. Cartan-Eilenberg [1; p. 251]). This shows that Γ cannot act freely on M. In other words, $\bar{\alpha}$ has a fixed point, say $x \in M$. Then α leaves G_x invariant. QED.

CHAPTER IX

Complex Manifolds

1. *Algebraic preliminaries*

The linear-algebraic results on real and complex vector spaces obtained in this section will be applied to tangent spaces of manifolds in subsequent sections.

A *complex structure* on a real vector space V is a linear endomorphism J of V such that $J^2 = -1$, where 1 stands for the identity transformation of V. A real vector space V with a complex structure J can be turned into a complex vector space by defining scalar multiplication by complex numbers as follows:

$$(a + ib)X = aX + bJX \quad \text{for } X \in V \text{ and } a, b \in \mathbf{R}.$$

Clearly, the real dimension m of V must be even and $\tfrac{1}{2}m$ is the complex dimension of V.

Conversely, given a complex vector space V of complex dimension n, let J be the linear endomorphism of V defined by

$$JX = iX \quad \text{for } X \in V.$$

If we consider V as a real vector space of real dimension $2n$, then J is a complex structure of V.

PROPOSITION 1.1. *Let J be a complex structure on a $2n$-dimensional real vector space V. Then there exist elements $X_1, \ldots, X_n$ of V such that $\{X_1, \ldots, X_n, JX_1, \ldots, JX_n\}$ is a basis for V.*

Proof. We turn V into an n-dimensional complex vector space as above. Let $X_1, \ldots, X_n$ be a basis for V as a complex vector space. It is easy to see that $\{X_1, \ldots, X_n, JX_1, \ldots, JX_n\}$ is a basis for V as a real vector space. QED.

Let $\mathbf{C}^n$ be the complex vector space of n-tuples of complex numbers $Z = (z^1, \ldots, z^n)$. If we set

$$z^k = x^k + iy^k, \qquad x^k, y^k \in \mathbf{R}, \qquad k = 1, \ldots, n,$$

then $\mathbf{C}^n$ can be identified with the real vector space $\mathbf{R}^{2n}$ of $2n$-tuples of real numbers $(x^1, \ldots, x^n, y^1, \ldots, y^n)$. In the following, unless otherwise stated, the identification of $\mathbf{C}^n$ with $\mathbf{R}^{2n}$ will always be done by means of the correspondence $(z^1, \ldots, z^n) \to (x^1, \ldots, x^n, y^1, \ldots, y^n)$. The complex structure of $\mathbf{R}^{2n}$ induced from that of $\mathbf{C}^n$ maps $(x^1, \ldots, x^n, y^1, \ldots, y^n)$ into $(y^1, \ldots, y^n, -x^1, \ldots, -x^n)$ and is called the *canonical complex structure* of $\mathbf{R}^{2n}$. In terms of the natural basis for $\mathbf{R}^{2n}$, it is given by the matrix

$$J_0 = \begin{pmatrix} 0 & I_n \\ -I_n & 0 \end{pmatrix},$$

where I_n denotes the identity matrix of degree n.

PROPOSITION 1.2. *Let J and J' be complex structures on real vector spaces V and V', respectively. If we consider V and V' as complex vector spaces in a natural manner, then a real linear mapping f of V into V' is complex linear when and only when $J' \circ f = f \circ J$.*

Proof. This follows from the fact that J or J' is the multiplication by i when V or V' is considered as a complex vector space. QED.

In particular, the complex general linear group $GL(n; \mathbf{C})$ of degree n can be identified with the subgroup of $GL(2n; \mathbf{R})$ consisting of matrices which commute with the matrix

$$J_0 = \begin{pmatrix} 0 & I_n \\ -I_n & 0 \end{pmatrix}.$$

A simple calculation shows that this representation of $GL(n; \mathbf{C})$ into $GL(2n; \mathbf{R})$, called the *real representation* of $GL(n; \mathbf{C})$, is given by

$$A + iB \to \begin{pmatrix} A & B \\ -B & A \end{pmatrix} \qquad \text{for} \qquad A + iB \in GL(n; \mathbf{C}),$$

where both A and B are real $n \times n$ matrices.

PROPOSITION 1.3. *There is a natural one-to-one correspondence between the set of complex structures on $\mathbf{R}^{2n}$ and the homogeneous space $GL(2n; \mathbf{R})/GL(n; \mathbf{C})$; the coset represented by an element $S \in GL(2n; \mathbf{R})$ corresponds to the complex structure SJ_0S^{-1}, where J_0 is (the matrix representing) the canonical complex structure of $\mathbf{R}^{2n}$.*

Proof. Every element S of $GL(2n; \mathbf{R})$ sends every complex structure J of $\mathbf{R}^{2n}$ into a complex structure SJS^{-1} of $\mathbf{R}^{2n}$; we consider $GL(2n; \mathbf{R})$ as a transformation group acting on the set of complex structures of $\mathbf{R}^{2n}$. It suffices to prove that this action is transitive and that the isotropy subgroup of $GL(2n; \mathbf{R})$ at J_0 is $GL(n; \mathbf{C})$. To prove the transitivity of the action, let J and J' be two complex structures of $\mathbf{R}^{2n}$. By Proposition 1.1, there are bases $\{e_1, \ldots, e_n, Je_1, \ldots, Je_n\}$ and $\{e'_1, \ldots, e'_n, J'e'_1, \ldots, J'e'_n\}$ for $\mathbf{R}^{2n}$. If we define an element S of $GL(2n; \mathbf{R})$ by

$$Se_k = e'_k, \qquad SJe_k = J'e'_k \qquad \text{for } k = 1, \ldots, n,$$

then $J' = SJS^{-1}$, thus proving the transitivity. On the other hand, an element S of $GL(2n; \mathbf{R})$ is in $GL(n; \mathbf{C})$ if and only if it commutes with J_0, that is, $J_0 = SJ_0S^{-1}$ (cf. the argument following Proposition 1.2). QED.

PROPOSITION 1.4. *Let J be a complex structure on a real vector space V. Then a real vector subspace V' of V is invariant by J if and only if V' is a complex subspace of V when V is considered as a complex vector space.*

Proof. As in the case of Proposition 1.2, this also follows from the fact that J is the multiplication by i when V is considered as a complex vector space. QED.

Let V be a real vector space and V^* its dual space. A complex structure J on V induces a complex structure on V^*, denoted also by J, as follows:

$$\langle JX, X^* \rangle = \langle X, JX^* \rangle \qquad \text{for } X \in V \text{ and } X^* \in V^*.$$

Let V be a real m-dimensional vector space and V^c the complexification of V, i.e., $V^c = V \otimes_{\mathbf{R}} \mathbf{C}$. Then V is a real subspace of V^c in a natural manner. More generally, the tensor space $\mathbf{T}^r_s(V)$ of type (r, s) over V can be considered as a real subspace of the tensor space $\mathbf{T}^r_s(V^c)$ in a natural manner. The *complex conjugation* in V^c is the real linear endomorphism defined by

$$Z = X + iY \to \bar{Z} = X - iY \qquad \text{for } X, Y \in V.$$

The complex conjugation of V^c extends in a natural manner to that of $\mathbf{T}_s^r(V^c)$.

Assume now that V is a real $2n$-dimensional vector space with a complex structure J. Then J can be uniquely extended to a complex linear endomorphism of V^c, and the extended endomorphism, denoted also by J, satisfies the equation $J^2 = -1$. The eigenvalues of J are therefore i and $-i$. We set

$$V^{1,0} = \{Z \in V^c; JZ = iZ\}, \qquad V^{0,1} = \{Z \in V^c; JZ = -iZ\}.$$

The following proposition is evident.

PROPOSITION 1.5. (1) $V^{1,0} = \{X - iJX; X \in V\}$ and $V^{0,1} = \{X + iJX; X \in V\}$;
(2) $V^c = V^{1,0} + V^{0,1}$ (complex vector space direct sum);
(3) The complex conjugation in V^c defines a real linear isomorphism between $V^{1,0}$ and $V^{0,1}$.

Let V^* be the dual space of V. Its complexification V^{*c} is the dual space of V^c. With respect to the eigenvalues $\pm i$ of the complex structure J on V^*, we have a direct sum decomposition as above:

$$V^{*c} = V_{1,0} + V_{0,1}.$$

The proof of the following proposition is also trivial.

PROPOSITION 1.6.
$$V_{1,0} = \{X^* \in V^{*c}; \langle X, X^* \rangle = 0 \quad \text{for all } X \in V^{0,1}\},$$
$$V_{0,1} = \{X^* \in V^{*c}; \langle X, X^* \rangle = 0 \quad \text{for all } X \in V^{1,0}\}.$$

The tensor space $\mathbf{T}_s^r(V^c)$ may be decomposed into a direct sum of tensor products of vector spaces each of which is identical with one of the spaces $V^{1,0}$, $V^{0,1}$, $V_{1,0}$, and $V_{0,1}$. We shall study the decomposition of the exterior algebra $\bigwedge V^{*c}$ more closely. The exterior algebras $\bigwedge V_{1,0}$ and $\bigwedge V_{0,1}$ can be considered as subalgebras of $\bigwedge V^{*c}$ in a natural manner. We denote by $\bigwedge^{p,q} V^{*c}$ the subspace of $\bigwedge V^{*c}$ spanned by $\alpha \wedge \beta$, where $\alpha \in \bigwedge^p V_{1,0}$ and $\beta \in \bigwedge^q V_{0,1}$. The following proposition is evident.

PROPOSITION 1.7. *The exterior algebra* $\bigwedge V^{*c}$ *may be decomposed as follows:*

$$\bigwedge V^{*c} = \sum_{r=0}^{n} \bigwedge^r V^{*c} \quad \text{with} \quad \bigwedge^r V^{*c} = \sum_{p+q=r} \bigwedge^{p,q} V^{*c},$$

and the complex conjugation in V^{*c}, extended to $\bigwedge V^{*c}$ in a natural manner, gives a real linear isomorphism between $\bigwedge^{p,q} V^{*c}$ and $\bigwedge^{q,p} V^{*c}$.

If $\{e^1, \ldots, e^n\}$ is a basis for the complex vector space $V_{1,0}$, then $\{\bar{e}^1, \ldots, \bar{e}^n\}$, where $\bar{e}^k = \overline{e^k}$, is a basis for $V_{0,1}$ (cf. Proposition 1.5) and the set of elements $e^{j_1} \wedge \cdots \wedge e^{j_p} \wedge \bar{e}^{k_1} \wedge \cdots \wedge \bar{e}^{k_q}$, $1 \leq j_1 < \cdots < j_p \leq n$ and $1 \leq k_1 < \cdots < k_q \leq n$, forms a basis for $\bigwedge^{p,q} V^{*c}$ over the field of complex numbers.

A *Hermitian inner product* on a real vector space V with a complex structure J is an inner product h such that

$$h(JX, JY) = h(X, Y) \qquad \text{for } X, Y \in V.$$

It follows that $h(JX, X) = 0$ for every $X \in V$.

PROPOSITION 1.8. *Let h be a Hermitian inner product in a $2n$-dimensional real vector space V with a complex structure J. Then there exist elements $X_1, \ldots, X_n$ of V such that $\{X_1, \ldots, X_n, JX_1, \ldots, JX_n\}$ is an orthonormal basis for V with respect to the inner product h.*

Proof. We use induction on $\dim V$. If X_1 is a unit vector, then $\{X_1, JX_1\}$ is orthonormal. Let W be the subspace spanned by X_1 and JX_1 and let $W^\perp$ be the orthogonal complement so that $V = W + W^\perp$. Then $W^\perp$ is invariant by J. By the inductive assumption, $W^\perp$ has an orthonormal basis of the form $\{X_2, \ldots, X_n, JX_2, \ldots, JX_n\}$. QED.

If h_0 is the canonical inner product in $\mathbf{R}^{2n}$, i.e., the inner product with respect to which the natural basis of $\mathbf{R}^{2n}$ is orthonormal, then h_0 is a Hermitian inner product with respect to the canonical complex structure J_0 of $\mathbf{R}^{2n}$.

PROPOSITION 1.9. *There is a natural one-to-one correspondence between the set of Hermitian inner products in $\mathbf{R}^{2n}$ with respect to the canonical complex structure J_0 and the homogeneous space $GL(n; \mathbf{C})/U(n)$; the coset represented by an element $S \in GL(n; \mathbf{C})$ corresponds to the Hermitian inner product h defined by*

$$h(X, Y) = h_0(SX, SY) \qquad \text{for } X, Y \in \mathbf{R}^{2n},$$

where h_0 is the canonical Hermitian inner product in $\mathbf{R}^{2n}$.

Proof. The proof is similar to that of Proposition 1.3. An element S of $GL(n; \mathbf{C})$ sends a Hermitian inner product h of $\mathbf{R}^{2n}$ (with respect to J_0) into a Hermitian inner product h' as follows:

$$h'(X, Y) = h(SX, SY) \qquad \text{for } X, Y \in \mathbf{R}^{2n}.$$

(It should be remarked that, considered as a subgroup of $GL(2n;\mathbf{R})$, the group $GL(n;\mathbf{C})$ acts on $\mathbf{R}^{2n}$.) We consider $GL(n;\mathbf{C})$ as a transformation group acting on the set of Hermitian inner products in $\mathbf{R}^{2n}$ (with respect to J_0) in the manner just described. It suffices to prove that this action is transitive and that the isotropy subgroup of $GL(n;\mathbf{C})$ at h_0 is $U(n)$. Given two Hermitian inner products h and h' of $\mathbf{R}^{2n}$, by Proposition 1.7 there are orthonormal bases $\{e_1, \ldots, e_n, J_0 e_1, \ldots, J_0 e_n\}$ with respect to h and $\{e'_1, \ldots, e'_n, J_0 e'_1, \ldots, J_0 e'_n\}$ with respect to h' for $\mathbf{R}^{2n}$. The element S of $GL(2n;\mathbf{R})$ defined by

$$Se'_k = e_k, \quad SJ_0 e'_k = J_0 e_k \quad \text{for } k = 1, \ldots, n,$$

is an element of $GL(n;\mathbf{C})$ and sends h into h', thus proving the transitivity of the action. On the other hand, the isotropy subgroup of $GL(n;\mathbf{C})$ at J_0 is evidently the intersection $GL(n;\mathbf{C}) \cap O(2n)$, where both $GL(n;\mathbf{C})$ and $O(2n)$ are considered as subgroups of $GL(2n;\mathbf{R})$. It is easy to see that $U(n)$ consists of elements of $GL(n;\mathbf{C})$ whose real representations are in $O(2n)$. QED.

The proof of the following proposition is straightforward.

PROPOSITION 1.10. *Let h be a Hermitian inner product in a real vector space V with a complex structure J. Then h can be extended uniquely to a complex symmetric bilinear form, denoted also by h, of V^c, and it satisfies the following conditions:*

(1) $h(\bar{Z}, \bar{W}) = \overline{h(Z, W)}$ *for $Z, W \in V^c$;*
(2) $h(Z, \bar{Z}) > 0$ *for all non-zero $Z \in V^c$;*
(3) $h(Z, \bar{W}) = 0$ *for $Z \in V^{1,0}$ and $W \in V^{0,1}$.*

Conversely, every complex symmetric bilinear form h on V^c satisfying (1), (2), (3) is the natural extension of a Hermitian inner product of V.

To each Hermitian inner product h on V with respect to a complex structure J, we associate an element φ of $\wedge^2 V^*$ as follows:
$$\varphi(X, Y) = h(X, JY) \quad \text{for } X, Y \in V.$$

We have to verify that φ is skew-symmetric:

$$\varphi(Y, X) = h(Y, JX) = h(JX, Y) = h(JX, -J^2 Y)$$
$$= h(X, -JY) = -\varphi(X, Y).$$

It can be easily seen that φ is also invariant by J. Since $\wedge^2 V^*$ can be considered as a subspace of $\wedge^2 V^{*c}$, φ may be considered as an element of $\wedge^2 V^{*c}$. In other words, φ may be uniquely extended to a skew-symmetric bilinear form on V^c, denoted also by φ. By Propositions 1.5, 1.6, and 1.10, we have

PROPOSITION 1.11. *Let φ be the skew-symmetric bilinear form on V^c associated to a Hermitian inner product h of V. Then $\varphi \in \wedge^{1,1} V^{*c}$.*

We prove

PROPOSITION 1.12. *Let h and φ be as in Proposition* 1.11, $\{Z_1, \ldots, Z_n\}$ *a basis for $V^{1,0}$ over $\mathbf{C}$ and $\{\xi^1, \ldots, \xi^n\}$ the dual basis for $V_{1,0}$. We set*

$$h_{j\bar{k}} = h(Z_j, \bar{Z}_k) \quad \text{for } j, k = 1, \ldots, n.$$

Then

(1) $h_{j\bar{k}} = \overline{h_{k\bar{j}}}$ *for $j, k = 1, \ldots, n$;*

(2) $\varphi = -2i \sum_{j,k=1}^n h_{j\bar{k}} \xi^j \wedge \bar{\xi}^k$.

Proof. (1) follows from (1) of Proposition 1.10. As for (2), given $Z, W \in V^c$, we may write

$$Z = \sum_{j=1}^n (\xi^j(Z) Z_j + \bar{\xi}^j(Z) \bar{Z}_j), \quad W = \sum_{j=1}^n (\xi^j(W) Z_j + \bar{\xi}^j(W) \bar{Z}_j).$$

A simple calculation then shows

$$\varphi(Z, W) = -i \sum_{j,k=1}^n h_{j\bar{k}} (\xi^j(Z) \bar{\xi}^k(W) - \xi^j(W) \bar{\xi}^k(Z)).$$

QED.

Example 1.1. Let $\mathfrak{g}$ be a Lie algebra over $\mathbf{C}$. Considering $\mathfrak{g}$ as a real vector space we have a complex structure J defined by $JX = iX$. The complex structure J satisfies $[JX, Y] = J[X, Y] = [X, JY]$ for all $X, Y \in \mathfrak{g}$, that is, $J \circ \text{ad}(X) = \text{ad}(X) \circ J$ for every $X \in \mathfrak{g}$. Conversely, suppose that $\mathfrak{g}$ is a Lie algebra over $\mathbf{R}$ with a complex structure J satisfying $J \circ \text{ad}(X) = \text{ad}(X) \circ J$ for every $X \in \mathfrak{g}$. Then, defining $(a + ib)X = aX + bJX$, where $a, b \in \mathbf{R}$, we get a complex Lie algebra; we may verify complex bilinearity of the bracket operation as follows:

$$[(a + ib)X, Y] = [aX, Y] + [bJX, Y] = a[X, Y] + b[JX, Y]$$
$$= a[X, Y] + bJ[X, Y] = (a + ib)[X, Y].$$

2. Almost complex manifolds and complex manifolds

A definition of complex manifold was given in Chapter I. For the better understanding of complex manifolds, we shall define the notion of almost complex manifolds and apply the results of §1 to tangent spaces of almost complex manifolds.

An *almost complex structure* on a real differentiable manifold M is a tensor field J which is, at every point x of M, an endomorphism of the tangent space $T_x(M)$ such that $J^2 = -1$, where 1 denotes the identity transformation of $T_x(M)$. A manifold with a fixed almost complex structure is called an *almost complex manifold*.

PROPOSITION 2.1. *Every almost complex manifold is of even dimensions and is orientable.*

Proof. An almost complex structure J on M defines a complex structure in each tangent space $T_x(M)$. As we have shown at the beginning of §1, dim $T_x(M)$ is even. Let $2n = $ dim M. In each tangent space $T_x(M)$ we fix a basis $X_1, \ldots, X_n, JX_1, \ldots, JX_n$. The existence of such a basis was proved in Proposition 1.1 and it is easy to see that any two such bases differ from each other by a linear transformation with positive determinant. To give an orientation to M, we consider the family of all local coordinate systems $x^1, \ldots, x^{2n}$ of M such that, at each point x where the coordinate system $x^1, \ldots, x^{2n}$ is valid, the basis $(\partial/\partial x^1)_x, \ldots, (\partial/\partial x^{2n})_x$ of $T_x(M)$ differs from the above chosen basis $X_1, \ldots, X_n, JX_1, \ldots, JX_n$ by a linear transformation with positive determinant. It is a simple matter to verify that the family of local coordinate systems thus obtained gives a complete atlas compatible with the pseudogroup of orientation-preserving transformations of $\mathbf{R}^{2n}$. QED.

The orientation of an almost complex manifold M given in the proof above is called the *natural orientation*.

To show that every complex manifold carries a natural almost complex structure, we consider the space $\mathbf{C}^n$ of n-tuples of complex numbers $(z^1, \ldots, z^n)$ with $z^j = x^j + iy^j$, $j = 1, \ldots, n$. With respect to the coordinate system $(x^1, \ldots, x^n, y^1, \ldots, y^n)$ we define an almost complex structure J on $\mathbf{C}^n$ by

$$J(\partial/\partial x^j) = \partial/\partial y^j, \qquad J(\partial/\partial y^j) = -(\partial/\partial x^j), \qquad j = 1, \ldots, n.$$

PROPOSITION 2.2. *A mapping f of an open subset of $\mathbf{C}^n$ into $\mathbf{C}^m$ preserves the almost complex structures of $\mathbf{C}^n$ and $\mathbf{C}^m$, i.e., $f_* \circ J = J \circ f_*$, if and only if f is holomorphic.*

Proof. Let $(w^1, \ldots, w^m)$ with $w^k = u^k + iv^k$, $k = 1, \ldots, m$, be the natural coordinate system in $\mathbf{C}^m$. If we express f in terms of these coordinate systems in $\mathbf{C}^n$ and $\mathbf{C}^m$:

$$u^k = u^k(x^1, \ldots, x^n, y^1, \ldots, y^n),$$
$$v^k = v^k(x^1, \ldots, x^n, y^1, \ldots, y^n), \quad k = 1, \ldots, m,$$

then f is holomorphic when and only when the following Cauchy-Riemann equations hold:

$$\partial u^k/\partial x^j - \partial v^k/\partial y^j = 0,$$
$$\partial u^k/\partial y^j + \partial v^k/\partial x^j = 0, \quad j = 1, \ldots, n; \quad k = 1, \ldots, m.$$

On the other hand, we have always (whether f is holomorphic or not)

$$f_*(\partial/\partial x^j) = \sum_{k=1}^{m} (\partial u^k/\partial x^j)(\partial/\partial u^k) + \sum_{k=1}^{m} (\partial v^k/\partial x^j)(\partial/\partial v^k),$$

$$f_*(\partial/\partial y^j) = \sum_{k=1}^{m} (\partial u^k/\partial y^j)(\partial/\partial u^k) + \sum_{k=1}^{m} (\partial v^k/\partial y^j)(\partial/\partial v^k),$$

for $j = 1, \ldots, n$.

From these formulas and the definition of J in $\mathbf{C}^n$ and $\mathbf{C}^m$ given above, we see that $f_* \circ J = J \circ f_*$ if and only if f satisfies the Cauchy-Riemann equations. QED.

To define an almost complex structure on a complex manifold M, we transfer the almost complex structure of $\mathbf{C}^n$ to M by means of charts. Proposition 2.2 implies that an almost complex structure can be thus defined on M independently of the choice of charts. An almost complex structure J on a manifold M is called a *complex structure* if M is an underlying differentiable manifold of a complex manifold which induces J in the way just described.

Let M and M' be almost complex manifolds with almost complex structures J and J', respectively. A mapping $f: M \to M'$ is said to be *almost complex* if $J' \circ f_* = f_* \circ J$. From Proposition 2.2 we obtain

PROPOSITION 2.3. *Let M and M' be complex manifolds. A mapping $f\colon M \to M'$ is holomorphic if and only if f is almost complex with respect to the complex structures of M and M'.*

In particular, two complex manifolds with the same underlying differentiable manifold are identical if the corresponding almost complex structures coincide.

Given an almost complex structure J on a manifold M, the tensor field $-J$ is also an almost complex structure which is said to be *conjugate* to J. If M is a complex manifold with atlas $\{(U_j, \varphi_j)\}$ then the family of charts $(U_j, \bar{\varphi}_j)$, where $\bar{\varphi}_j$ is the complex conjugate of φ_j, is an atlas of the topological space underlying M which is compatible with the pseudogroup of holomorphic transformations of $\mathbf{C}^n$. The atlas $\{(U_j, \bar{\varphi}_j)\}$ defines a complex manifold whose underlying differentiable manifold is the same as that of M; this new complex manifold is said to be *conjugate* to M and will be denoted by $\bar{M}$. It is easy to verify that if J is the complex structure of a complex manifold M, then $-J$ is the complex structure of $\bar{M}$.

PROPOSITION 2.4. *Let M be a $2n$-dimensional orientable manifold and $L(M)$ the bundle of linear frames over M. Then the set of almost complex structures on M are in one-to-one correspondence with the set of cross-sections of the associated bundle $B = L(M)/GL(n; \mathbf{C})$ with fibre $GL(2n; \mathbf{R})/GL(n; \mathbf{C})$, where $GL(n; \mathbf{C})$ is considered as a subgroup of $GL(2n; \mathbf{R})$ by its real representation.*

Proof. This follows from Proposition 5.6 of Chapter I (see also the remark following it) and Proposition 1.3. QED.

In general, given two tensor fields A and B of type $(1, 1)$ on a manifold M, we can construct the torsion of A and B, which is a tensor field of type $(1, 2)$ (cf. Proposition 3.12 of Chapter I). Specializing to the case where both A and B are an almost complex structure J, we define the *torsion* of J to be the tensor field N of type $(1, 2)$ given by

$$N(X, Y) = 2\{[JX, JY] - [X, Y] - J[X, JY] - J[JX, Y]\}$$
for $X, Y \in \mathfrak{X}(M)$.

Let $x^1, \ldots, x^{2n}$ be a local coordinate system in M. By setting $X = \partial/\partial x^j$ and $Y = \partial/\partial x^k$ in the equation defining N, we see that the components N^i_{jk} of N with respect to $x^1, \ldots, x^{2n}$ may be expressed in terms of the components J^i_j of J and its partial derivatives as follows:

$$N^i_{jk} = 2 \sum_{h=1}^{2n} (J^h_j \partial_h J^i_k - J^h_k \partial_h J^i_j - J^i_h \partial_j J^h_k + J^i_h \partial_k J^h_j),$$

where ∂_h denotes the partial differentiation $\partial/\partial x^h$. An almost complex structure is said to be *integrable* if it has no torsion.

THEOREM 2.5. *An almost complex structure is a complex structure if and only if it has no torsion.*

Proof. We shall only prove here that a complex structure has no torsion. The converse will be proved in Appendix 8 only in the case where the manifold and its almost complex structure are real analytic. Let $z^1, \ldots, z^n$, $z^j = x^j + iy^j$, be a complex local coordinate system in a complex manifold M. From the construction of the complex structure J given before Proposition 2.2, it is clear that the components of J with respect to the local coordinate system $x^1, \ldots, x^n, y^1, \ldots, y^n$ are constant functions in the coordinate neighborhood and hence have vanishing partial derivatives. By the expression above for N^i_{jk} it is clear that the torsion N is zero. QED.

The *complex tangent space* $T^c_x(M)$ of a manifold M at x is the complexification of the tangent space $T_x(M)$ and its elements are called *complex tangent vectors* at x. If we denote by $\mathfrak{D}^r(M)$ the space of r-forms on M, then an element of the complexification $\mathfrak{C}^r(M)$ of $\mathfrak{D}^r(M)$ is called a *complex r-form* on M. Every complex r-form ω may be written uniquely as $\omega' + i\omega''$, where ω' and ω'' are (real) r-forms. If we denote by T^{*c}_x the complexification of the dual space of $T_x(M)$, then a complex r-form ω on M gives an element of $\bigwedge^r T^{*c}_x$ at each point x of M; in other words, a skew-symmetric r-linear mappings $T^c_x(M) \times \cdots \times T^c_x(M) \to \mathbf{C}$ at each point x of M. More generally, we can define the space of complex tensor fields on M as the complexification of the space of (real) tensor fields. Such operations as contractions, brackets, exterior differentiation, Lie differentiations, interior products, etc. (cf. §3 of Chapter I) can be extended by linearity to complex tensor fields or complex differential forms.

If M is an almost complex manifold with almost complex structure J, then by Proposition 1.5

$$T^c_x(M) = T^{1,0}_x + T^{0,1}_x,$$

IX. COMPLEX MANIFOLDS 125

where $T_x^{1,0}$ and $T_x^{0,1}$ are the eigenspaces of J corresponding to the eigenvalues i and $-i$ respectively. A complex tangent vector (field) is of type $(1, 0)$ (resp. $(0, 1)$) if it belongs to $T_x^{1,0}$ (resp. $T_x^{0,1}$). By Proposition 1.5 we have

PROPOSITION 2.6. *A complex tangent vector Z of an almost complex manifold M is of type $(1, 0)$ (resp. $(0, 1)$) if and only if $Z = X - iJX$ (resp. $Z = X + iJX$) for some real tangent vector X.*

By Proposition 1.7, the space $\mathfrak{C} = \mathfrak{C}(M)$ of complex differential forms on an almost complex manifold M of dimension $2n$ may be bigraded as follows:

$$\mathfrak{C} = \sum_{p,q=0}^{n} \mathfrak{C}^{p,q}.$$

An element of $\mathfrak{C}^{p,q}$ is called a (complex) *form of degree* (p, q). By Proposition 1.6, a complex 1-form ω is of degree $(1, 0)$ (resp. $(0, 1)$) if and only if $\omega(Z) = 0$ for all complex vector fields Z of type $(0, 1)$ (resp. $(1, 0)$). If $\omega^1, \ldots, \omega^n$ is a local basis for $\mathfrak{C}^{1,0}$, then its complex conjugate $\bar{\omega}^1, \ldots, \bar{\omega}^n$ is a local basis for $\mathfrak{C}^{0,1}$ (cf. Proposition 1.5). It follows that the set of forms $\omega^{j_1} \wedge \cdots \wedge \omega^{j_p} \wedge \bar{\omega}^{k_1} \wedge \cdots \wedge \bar{\omega}^{k_q}$, $1 \leq j_1 < \cdots < j_p \leq n$ and $1 \leq k_1 < \cdots < k_q \leq n$, is a local basis for $\mathfrak{C}^{p,q}$. Consequently, if $\omega \in \mathfrak{C}^{p,q}$, then $\omega(Z_1, \ldots, Z_{p+q}) = 0$ for complex vector fields $Z_1, \ldots, Z_{p+q}$ of which more than p are of type $(1, 0)$ or more than q are of type $(0, 1)$. This property of $\mathfrak{C}^{p,q}$ may be used to define the bigrading of $\mathfrak{C}$.

PROPOSITION 2.7. *If $\mathfrak{C}^{p,q}$ is the space of forms of degree (p, q) on an almost complex manifold, then*

$$d\mathfrak{C}^{p,q} \subset \mathfrak{C}^{p+2,q-1} + \mathfrak{C}^{p+1,q} + \mathfrak{C}^{p,q+1} + \mathfrak{C}^{p-1,q+2}.$$

Proof. This follows from the fact that $\mathfrak{C}$ is locally generated by $\mathfrak{C}^{0,0}$, $\mathfrak{C}^{1,0}$ and $\mathfrak{C}^{0,1}$ and from the following obvious inclusions:

$$d\mathfrak{C}^{0,0} \subset \mathfrak{C}^{1,0} + \mathfrak{C}^{0,1}, \quad d\mathfrak{C}^{1,0} \subset \mathfrak{C}^{2,0} + \mathfrak{C}^{1,1} + \mathfrak{C}^{0,2},$$
$$d\mathfrak{C}^{0,1} \subset \mathfrak{C}^{2,0} + \mathfrak{C}^{1,1} + \mathfrak{C}^{0,2}.$$

QED.

THEOREM 2.8. *For an almost complex manifold M, the following conditions are equivalent:*

(a) *If Z and W are complex vector fields of type $(1, 0)$, so is $[Z, W]$;*
(b) *If Z and W are complex vector fields of type $(0, 1)$, so is $[Z, W]$;*
(c) $d\mathfrak{C}^{1,0} \subset \mathfrak{C}^{2,0} + \mathfrak{C}^{1,1}, \quad d\mathfrak{C}^{0,1} \subset \mathfrak{C}^{1,1} + \mathfrak{C}^{0,2}$;
(d) $d\mathfrak{C}^{p,q} \subset \mathfrak{C}^{p+1,q} + \mathfrak{C}^{p,q+1} \quad \text{for } p, q = 0, 1, \ldots, n$;
(e) *The almost complex structure has no torsion.*

Proof. The equivalence of (a) and (b) follows from the fact that, for any complex vector fields Z and W, the complex conjugate of $[Z, W]$ is $[\bar{Z}, \bar{W}]$ and the fact that $\bar{Z}$ is of type $(0, 1)$ if and only if Z is of type $(1, 0)$ (cf. Proposition 1.5). To prove that (a) and (b) imply (c), let $\omega \in \mathfrak{C}^{1,0}$. If Z and W are complex vector fields of type $(0, 1)$, then (b) implies (cf. Proposition 3.11 of Chapter I)

$$2\,d\omega(Z, W) = Z(\omega(W)) - W(\omega(Z)) - \omega([Z, W]) = 0,$$

showing that $d\omega$ does not have the component of degree $(0, 2)$. Using vector fields of type $(1, 0)$ and (a), we see that if ω is a form of degree $(0, 1)$, then $d\omega$ does not have the component of degree $(2, 0)$. To prove the converse (c) → (a), let Z and W be vector fields of type $(1, 0)$. The calculation above shows that $\omega([Z, W]) = 0$ for all forms of ω of degree $(0, 1)$. Hence, $[Z, W]$ is of type $(1, 0)$. The proof of (c) → (b) is similar. Since $\mathfrak{C}$ is locally generated by $\mathfrak{C}^{0,0}$, $\mathfrak{C}^{1,0}$, and $\mathfrak{C}^{0,1}$, (c) implies (d). The converse (d) → (c) is trivial. To prove the equivalence of (a) and (e), let X and Y be (real) vector fields and set $Z = [X - iJX, Y - iJY]$. Proposition 2.6 implies that (a) holds if and only if Z is of type $(1, 0)$ for all X and Y. On the other hand, a simple calculation shows

$$2(Z + iJZ) = -N(X, Y) - iJ(N(X, Y)).$$

Since $Z + iJZ = 0$ if and only if Z is of type $(1, 0)$, (a) holds if and only if $N(X, Y) = 0$ for all X and Y. QED.

Given an almost complex structure without torsion, we can define $\partial: \mathfrak{C}^{p,q} \to \mathfrak{C}^{p+1,q}$ and $\bar{\partial}: \mathfrak{C}^{p,q} \to \mathfrak{C}^{p,q+1}$ by (cf. the equivalence of (d) and (e) in Theorem 2.8)

$$d\omega = \partial\omega + \bar{\partial}\omega \quad \text{for } \omega \in \mathfrak{C}^{p,q}.$$

Since $d^2 = 0$, we obtain

$$\partial^2 = 0, \quad \bar{\partial}^2 = 0, \quad \partial \circ \bar{\partial} + \bar{\partial} \circ \partial = 0.$$

Given two manifolds M and M' and a mapping $f \colon M \to M'$, the differential f_* extends to a complex linear mapping of $T_x^c(M)$ into $T_y^c(M')$, $y = f(x)$, which will be denoted by the same symbol f_*. Similarly f^* maps each complex differential form of M' into a complex differential form of M.

PROPOSITION 2.9. *For a mapping f of an almost complex manifold M into another almost complex manifold M', the following conditions are equivalent:*

(a) *If Z is a complex vector of type $(1, 0)$ of M, so is the vector $f_*(Z)$ of M';*

(b) *If Z is a complex vector of type $(0, 1)$ of M, so is the vector $f_*(Z)$ of M';*

(c) *If ω is a form of degree (p, q) on M', then $f^*(\omega)$ is a form of degree (p, q) on M for all p and q;*

(d) *f is almost complex.*

Proof. The equivalence of (a) and (b) follows from the fact that $\bar{Z}$ is of type $(0, 1)$ if and only if Z is of type $(1, 0)$ (cf. Proposition 1.5) and the fact that f_* commutes with the complex conjugation. Since (a) (resp. (b)) is equivalent to the condition that f^* maps every form of degree $(0, 1)$ (resp. $(1, 0)$) into a form of the same degree (cf. Proposition 1.6) and since the algebra of complex forms is locally generated by functions, forms of degree $(1, 0)$ and forms of degree $(0, 1)$, we may conclude that (c) is equivalent to (a) and (b). To prove the equivalence of (a) and (d), let J and J' denote the almost complex structures of M and M', respectively. Assume (a). If X is a real tangent vector of M, then $X - iJX$ is of type $(1, 0)$ by Proposition 2.6. Since $f_*(X - iJX) = f_*(X) - i(f_*(JX))$, the vector $f_*(X) - i(f_*(JX))$ is of type $(1, 0)$ by (a). Hence, $f_*(JX) = J'(f_*(X))$, showing that f is almost complex. Assume (d). If Z is a complex tangent vector of type $(1, 0)$ of M, then $Z = X - iJX$ for some real tangent vector X by Proposition 2.6. From $f_*(JX) = J'(f_*(X))$, we obtain

$$f_*(X - iJX) = f_*(X) - i(f_*(JX)) = f_*(X) - iJ'(f_*(X)),$$

showing that Z is mapped by f into a vector of type $(1, 0)$.
QED.

An *infinitesimal automorphism* of an almost complex structure J

on M is a vector field X such that $L_X J = 0$, where L_X denotes the Lie differentiation with respect to X. By Corollary 3.7 of Chapter I, a vector field X is an infinitesimal automorphism of J if and only if it generates a local 1-parameter group of local almost complex transformations.

PROPOSITION 2.10. *A vector field X is an infinitesimal automorphism of an almost complex structure J on a manifold M if and only if*

$$[X, JY] = J([X, Y]) \quad \text{for all vector fields } Y \text{ on } M.$$

Proof. Let X and Y be any vector fields on M. By Proposition 3.2 of Chapter I, we have

$$[X, JY] = L_X(JY) = (L_X J)Y + J(L_X Y)$$
$$= (L_X J)Y + J([X, Y]).$$

Hence, $L_X J = 0$ if and only if $[X, JY] = J([X, Y])$ for all Y.
QED.

If X is an infinitesimal automorphism of J, JX need not be. In fact, if X is an infinitesimal automorphism of J and Y is arbitrary, then by Proposition 2.10 we have

$$N(X, Y) = \tfrac{1}{2}([JX, JY] - J[JX, Y]),$$

showing that JX is also an infinitesimal automorphism of J if and only if $N(X, Y) = 0$ for all Y. If $N = 0$, then the Lie algebra $\mathfrak{a}$ of infinitesimal automorphisms of J is stable under J and, by Proposition 2.10, $[X, JY] = J([X, Y])$ for $X, Y \in \mathfrak{a}$. Consequently, $\mathfrak{a}$ is a complex Lie algebra (possibly of infinite dimensions), the complex structure being defined by J (cf. Example 1.1).

Let M be a complex manifold and $z^1, \ldots, z^n$, $z^j = x^j + iy^j$, a complex local coordinate system of M. Then $dz^j = dx^j + i\, dy^j$ and $d\bar{z}^j = dx^j - i\, dy^j$. We set

$$\partial/\partial z^j = \tfrac{1}{2}(\partial/\partial x^j - i(\partial/\partial y^j)), \quad \partial/\partial \bar{z}^j = \tfrac{1}{2}(\partial/\partial x^j + i(\partial/\partial y^j)).$$

From the construction of the complex structure J given before Proposition 2.2, it follows that $\partial/\partial z^1, \ldots, \partial/\partial z^n$ (resp. $\partial/\partial \bar{z}^1, \ldots, \partial/\partial \bar{z}^n$) form a basis for $T_x^{1,0}$ (resp. $T_x^{0,1}$) at each point x of the coordinate neighborhood and $dz^1, \ldots, dz^n$ (resp. $d\bar{z}^1, \ldots, d\bar{z}^n$) form a local basis for $\mathfrak{C}^{1,0}$ (resp. $\mathfrak{C}^{0,1}$). Consequently, the set of forms $dz^{j_1} \wedge \cdots \wedge dz^{j_p} \wedge d\bar{z}^{k_1} \wedge \cdots \wedge d\bar{z}^{k_q}$, $1 \leq j_1 < \cdots < j_p \leq n$ and $1 \leq k_1 < \cdots < k_q \leq n$, forms a local basis for $\mathbb{C}^{p,q}$. A form ω of degree $(p, 0)$ is said to be *holomorphic* if $\bar{\partial}\omega = 0$. If we express ω in terms of $z^1, \ldots, z^n$:

$$\omega = \sum_{1 \leq j_1 < \cdots < j_p \leq n} f_{j_1 \cdots j_p} \, dz^{j_1} \wedge \cdots \wedge dz^{j_p},$$

then $\bar{\partial}\omega = 0$ if and only if $\bar{\partial} f_{j_1 \cdots j_p} = 0$. In general, if f is a complex-valued function, then $\bar{\partial} f = 0$ if and only if $\partial f/\partial \bar{z}^j = 0$ for $j = 1, \ldots, n$. On the other hand, the definition of $\partial/\partial \bar{z}^j$ given above shows that the equations $\partial f/\partial \bar{z}^j = 0$ are nothing but the Cauchy-Riemann equations. It follows that ω is holomorphic if and only if the coefficient functions $f_{j_1 \cdots j_p}$ are all holomorphic.

A *holomorphic vector field* on a complex manifold M is a complex vector field Z of type $(1, 0)$ such that Zf is holomorphic for every locally defined holomorphic function f. If we write

$$Z = \sum_{j=1}^{n} f^j (\partial/\partial z^j)$$

in terms of $z^1, \ldots, z^n$, then Z is holomorphic if and only if the components f^j are all holomorphic functions.

PROPOSITION 2.11. *On a complex manifold M, the Lie algebra of infinitesimal automorphisms of the complex structure J is isomorphic with the Lie algebra of holomorphic vector fields, the isomorphism being given by $X \to Z = \frac{1}{2}(X - iJX)$.*

Proof. Let X and Y be vector fields on M and, in terms of a local coordinate system $z^1, \ldots, z^n$, we write (cf. Proposition 2.6)

$$\tfrac{1}{2}(X - iJX) = \sum_{j=1}^{n} f^j (\partial/\partial z^j), \qquad \tfrac{1}{2}(Y - iJY) = \sum_{j=1}^{n} g^j (\partial/\partial z^j).$$

A simple calculation shows that the equation $[X, JY] = J([X, Y])$ is equivalent to the system of equations $\sum_{k=1}^{n} \bar{g}^k (\partial f^j/\partial \bar{z}^k) = 0$ for $j = 1, \ldots, n$. From Proposition 2.10 we may conclude that X is an infinitesimal automorphism of J if and only if $\partial f^j/\partial \bar{z}^k = 0$ for $j, k = 1, \ldots, n$, that is, if and only if $\frac{1}{2}(X - iJX)$ is holomorphic.

Let X and Y be infinitesimal automorphisms of J. Since J has no torsion, both JX and JY are also infinitesimal automorphisms of J as we saw just after Proposition 2.10. By Proposition 2.10, we have $J[X, Y] = [JX, Y] = [X, JY]$ and $[X, Y] = -[JX, JY]$.

Using these identities we may easily verify that the mapping $\theta\colon X \to \frac{1}{2}(X - iJX)$ satisfies $\theta([X, Y]) = [\theta(X), \theta(Y)]$. Clearly, θ is one-to-one. Finally, given a holomorphic vector field Z we show that there is an infinitesimal automorphism X of J such that $\theta(X) = Z$. In any coordinate neighborhood U, we write $Z = \sum_{j=1}^n f^j(\partial/\partial z^j)$ and set $f^j = \xi^j + i\eta^j$. Then we have

$$Z = (\tfrac{1}{2}) \sum_j \{\xi^j(\partial/\partial x^j) + \eta^j(\partial/\partial y^j)\}$$
$$- (\tfrac{1}{2})i \sum_j \{-\eta^j(\partial/\partial x^j) + \xi^j(\partial/\partial y^j)\}.$$

If we take $X_U = \sum_{j=1}^n \{\xi^j(\partial/\partial x^j) + \eta^j(\partial/\partial y^j)\}$, then $Z = (\tfrac{1}{2}) \times (X_U - iJX_U)$. If X_V is a vector field on another coordinate neighborhood V such that $Z = (\tfrac{1}{2})(X_V - iJX_V)$, then it is clear that $X_U = X_V$ on $U \cap V$. Thus there is a vector field X on M such that $Z = \theta(X)$. We have shown that θ is an isomorphism of the Lie algebra of infinitesimal automorphisms of J onto the Lie algebra of holomorphic vector fields on M.   QED.

We shall give a few examples of complex and almost complex manifolds.

Example 2.1. A *complex Lie group* G is a group which is at the same time a complex manifold such that the group operations $(a, b) \in G \times G \to ab^{-1} \in G$ is a holomorphic mapping of $G \times G$ into G. Here $G \times G$ is provided with the product complex structure. (More generally, if M and M' have almost complex structures J and J', respectively, then $M \times M'$ has a product almost complex structure $J \times J'$ whose definition is rather obvious. If J and J' are integrable (or complex structures), so is $J \times J'$.) The group $GL(n; \mathbf{C})$ is a complex Lie group; its complex structure is obtained by considering it as an open subset of $\mathbf{C}^{n^2}$. Its Lie algebra $\mathfrak{gl}(n; \mathbf{C})$ consists of all $n \times n$ complex matrices.

If G is a complex Lie group, its complex structure J is invariant by the mappings $L_a\colon x \to ax$, $R_a\colon x \to xa$, ad $(a)\colon x \to axa^{-1}$, and $\psi\colon x \to x^{-1}$. If X is a left invariant vector field, so is JX. Thus J induces a complex structure in the Lie algebra $\mathfrak{g}$ of G. Since J is invariant by ad (a), $a \in G$, it follows that $J \circ \text{ad}\,(X) = \text{ad}\,(X) \circ J$ for every $X \in \mathfrak{g}$. Thus $\mathfrak{g}$ is a complex Lie algebra (cf. Example 1.1). Conversely, let G be a Lie group whose Lie algebra $\mathfrak{g}$ is a complex

Lie algebra (with complex structure J). Assuming first that G is connected, we get from $J \circ \text{ad}(X) = \text{ad}(X) \circ J$, $X \in \mathfrak{g}$, that $J \circ \text{ad}(a) = \text{ad}(a) \circ J$ for every $a \in G$. If we transfer J to each tangent space $T_a(G)$ by L_a, we get a left invariant almost complex structure J on G which is also right invariant. Since $[JX, Y] = J[X, Y]$ for all left invariant vector fields X and Y, we see that the torsion $N(X, Y)$ is 0 for such X and Y. Thus $N = 0$. Since J is real analytic (as a left invariant tensor field on a Lie group), we conclude that G is a complex manifold. Since L_a and R_a, $a \in G$, preserve J, the Leibniz formula (Proposition 1.4 of Chapter I) shows that the mapping $(x, y) \in G \times G \to xy \in G$ is holomorphic. The mapping $\psi \colon x \to x^{-1}$ is holomorphic at e (since its differential is -1, which commutes with J) and hence everywhere, because $\psi \circ L_a = R_{a^{-1}} \circ \psi$ implies that $(\psi_*)_a = (R_{a^{-1}})_{*e} \circ (\psi_*)_e \circ (L_a)_{*e}$ preserves J. Thus G is a complex Lie group. The case where G is not connected is left to the reader.

Example 2.2. Consider $\mathbf{C}^n$ as a vector space and let $\tau_1, \ldots, \tau_{2n}$ be any basis of $\mathbf{C}^n$ over the field of real numbers. Let D be the subgroup of $\mathbf{C}^n$ generated by $\tau_1, \ldots, \tau_{2n}$: $D = \{\Sigma_{j=1}^{2n} m_j \tau_j ; m_j$ integers$\}$. The action of D on $\mathbf{C}^n$ is properly discontinuous and the quotient manifold $\mathbf{C}^n/D$ is called an n-dimensional *complex torus* (cf. Proposition 4.3 of Chapter I). In contrast to the case of real tori, two complex tori of the same dimension need not be holomorphically isomorphic. A complex torus $\mathbf{C}^n/D$ is called an *abelian variety* if there is a *real* skew-symmetric bilinear form E on $\mathbf{C}^n$ such that

(1) $E(iX, Y) = E(iY, X)$ for $X, Y \in \mathbf{C}^n$;
(2) $E(iX, X) > 0$ for every nonzero $X \in \mathbf{C}^n$;
(3) $E(X, Y)$ is an integer if $X, Y \in D$.

For $n = 1$, every complex torus is an abelian variety.

As a quotient group of $\mathbf{C}^n$, $\mathbf{C}^n/D$ is a connected compact complex Lie group. Conversely, *every connected compact complex Lie group G is a complex torus*. In fact, the adjoint representation of G is a holomorphic mapping of G into $GL(n; \mathbf{C}) \subset \mathbf{C}^{n^2}$, where $n = \dim G$. Since a holomorphic function on a compact complex manifold must be constant, the adjoint representation of G is trivial, that is, G is abelian. Thus G is an even-dimensional torus (Pontrjagin [1]) and hence a complex torus.

If $z^1, \ldots, z^n$ is the natural coordinate system in $\mathbf{C}^n$, then the holomorphic 1-forms $dz^1, \ldots, dz^n$ can be considered as forms on a complex torus $\mathbf{C}^n/D$. *Every holomorphic 1-form on $\mathbf{C}^n/D$ is a linear combination of $dz^1, \ldots, dz^n$ with constant coefficients.* In fact, every holomorphic 1-form on $\mathbf{C}^n/D$ is a linear combination of $dz^1, \ldots, dz^n$ with holomorphic functions as coefficients and since $\mathbf{C}^n/D$ is a compact complex manifold, these coefficient functions are constant functions.

Let $\mathbf{C}^m/D'$ be another complex torus and $w^1, \ldots, w^m$ the natural coordinate system in $\mathbf{C}^m$. A homomorphism $\mathbf{C}^n/D \to \mathbf{C}^m/D'$ is induced by a complex linear transformation of $\mathbf{C}^n$ into $\mathbf{C}^m$ which sends D into D'. If $f\colon \mathbf{C}^n/D \to \mathbf{C}^m/D'$ is a holomorphic mapping, then

$$f^*(dw^k) = \sum_{j=1}^n a_j^k\, dz^j, \quad k = 1, \ldots, m, \quad a_j^k \in \mathbf{C},$$

showing that f is induced by a mapping $\mathbf{C}^n \to \mathbf{C}^m$ of the form $w^k = \sum_{j=1}^n a_j^k z^j + b^k$, $k = 1, \ldots, m$, with $b^k \in \mathbf{C}$. Thus, *every holomorphic mapping of $\mathbf{C}^n/D$ into $\mathbf{C}^m/D'$ is a homomorphism modulo a translation in $\mathbf{C}^m/D'$.*

Example 2.3. An n-dimensional complex manifold M is said to be *complex parallelizable* if there exist n holomorphic vector fields $Z_1, \ldots, Z_n$ which are linearly independent at every point of M. Every complex torus is complex parallelizable. More generally, let G be a complex Lie group of complex dimension n. Taking n linearly independent complex vectors of type $(1, 0)$ at the identity element of G and extending them by left translations, we obtain n left invariant holomorphic vector fields $Z_1, \ldots, Z_n$ on G which are linearly independent at every point of G. If D is a discrete subgroup of G, then $Z_1, \ldots, Z_n$ induce n holomorphic vector fields on the quotient complex manifold G/D which are linearly independent at every point of G/D, showing that G/D is complex parallelizable. H. C. Wang [5] proved that, conversely, *every compact complex parallelizable manifold may be written as a quotient space G/D of a complex Lie group G by a discrete subgroup D.* In fact, if $Z_1, \ldots, Z_n$ are everywhere linearly independent holomorphic vector fields on a compact complex manifold M, then $[Z_j, Z_k] = \sum_{h=1}^n c_{jk}^h Z_h$, where, being holomorphic functions on a compact complex manifold M, c_{jk}^h are all constant functions on M. Let

IX. COMPLEX MANIFOLDS 133

$X_1, \ldots, X_n$ be the corresponding infinitesimal automorphisms of the complex structure of M, i.e., $X_j = Z_j + \bar{Z}_j$ for $j = 1, \ldots, n$ (cf. Proposition 2.11). Let G be the universal covering manifold of M and X_j^* the natural lift of X_j to G. Then, (i) $X_1^*, \ldots, X_n^*$ are infinitesimal automorphisms of the complex structure of G; (ii) $[X_j^*, X_k^*] = \Sigma_{h=1}^n c_{jk}^h X_h^*$ for $j, k = 1, \ldots, n$; (iii) $X_1^*, \ldots, X_n^*$ are complete vector fields on G. We point out that (iii) follows from the fact that all vector fields, in particular $X_1, \ldots, X_n$, on a compact manifold M are complete (cf. Proposition 1.6 of Chapter I). From (i), (ii), (iii), and the simple connectedness of G, it follows that G can be given a complex Lie group structure such that $X_1^*, \ldots, X_n^*$ are left invariant infinitesimal automorphisms of the complex structure of G. Let D be the group of covering transformations of G which gives M. Each element of D leaves $X_1^*, \ldots, X_n^*$ invariant and hence is a left translation of G. Thus D can be identified with a discrete subgroup of G acting on the left.

Example 2.4. The complex Grassmann manifold $G_{p,q}(\mathbf{C})$ of p-planes in $\mathbf{C}^{p+q}$ is the set of p-dimensional complex subspaces in $\mathbf{C}^{p+q}$ with the structure of a complex manifold defined as follows. Let $z^1, \ldots, z^{p+q}$ be the natural coordinate system in $\mathbf{C}^{p+q}$, each z^j being considered as a complex linear mapping $\mathbf{C}^{p+q} \to \mathbf{C}$. For each set $\alpha = \{\alpha_1, \ldots, \alpha_p\}$ of integers such that $1 \leq \alpha_1 < \cdots < \alpha_p \leq p + q$, let U_α be the subset of $G_{p,q}(\mathbf{C})$ consisting of p-dimensional subspaces S such that $z^{\alpha_1} |S, \ldots, z^{\alpha_p}| S$ are linearly independent. We shall define a mapping φ_α of U_α into the space $M(q, p; \mathbf{C})$ of $p \times q$ complex matrices, which may be identified with $\mathbf{C}^{pq}$. Let $\{\alpha_{p+1}, \ldots, \alpha_{p+q}\}$ be the complement of $\{\alpha_1, \ldots, \alpha_p\}$ in $\{1, \ldots, p + q\}$ in the increasing order. Since, for each $S \in U_\alpha$, $z^{\alpha_1} |S, \ldots, z^{\alpha_p}| S$ form a basis of the dual space of S, we may write

$$z^{\alpha_{p+k}} | S = \sum_{j=1}^p s_j^k (z^{\alpha_j} | S), \quad k = 1, \ldots, q.$$

We set
$$\varphi_\alpha(S) = (s_j^k) \in M(q, p; \mathbf{C}).$$

It is easy to see that φ_α maps U_α one-to-one onto $M(q, p; \mathbf{C})$ and that the family of $\binom{p+q}{p}$ charts $(U_\alpha, \varphi_\alpha)$ turns $G_{p,q}(\mathbf{C})$ into a complex manifold of complex dimension pq.

The group $GL(p+q; \mathbf{C})$ acting in $\mathbf{C}^{p+q}$ sends every p-dimensional subspace into a p-dimensional subspace and hence can be considered as a transformation group acting on $G_{p,q}(\mathbf{C})$. The action is holomorphic and transitive. If S_0 denote the p-dimensional subspace spanned by the first p elements of the natural basis of $\mathbf{C}^{p+q}$, then the isotropy subgroup at S_0 is given by

$$H = \left\{ \begin{pmatrix} * & * \\ 0 & * \end{pmatrix} \in GL(p+q; \mathbf{C}) \right\},$$

where 0 denotes the zero matrix with p columns and q rows. Thus $G_{p,q}(\mathbf{C})$ is a quotient space $GL(p+q; \mathbf{C})/H$ of a complex Lie group $GL(p+q; \mathbf{C})$ by a closed complex Lie subgroup H and the natural projection $GL(p+q; \mathbf{C}) \to G_{p,q}(\mathbf{C})$ is holomorphic. A similar argument applied to the unitary group $U(p+q)$ acting in $\mathbf{C}^{p+q}$ shows that $G_{p,q}(\mathbf{C})$ may be written also as a quotient space $U(p+q)/U(p) \times U(q)$, where

$$U(p) \times U(q) = \left\{ \begin{pmatrix} A & 0 \\ 0 & B \end{pmatrix}; A \in U(p), B \in U(q) \right\}.$$

Since $U(p+q)$ is compact, $G_{p,q}(\mathbf{C})$ is also compact.

The Grassmann manifold $G_{n,1}(\mathbf{C})$ is called an n-dimensional *complex projective space*, denoted by $P_n(\mathbf{C})$. The multiplicative group $\mathbf{C}^*$ of non-zero complex numbers acts freely on $\mathbf{C}^{n+1} - \{0\}$ by $(c, z) \in \mathbf{C}^* \times (\mathbf{C}^{n+1} - \{0\}) \to cz \in \mathbf{C}^{n+1} - \{0\}$. Let $z^0, z^1, \ldots, z^n$ be the natural coordinate system in $\mathbf{C}^{n+1}$. For each $j, j = 0, 1, \ldots, n$, let U_j^* be the set of points of $\mathbf{C}^{n+1} - \{0\}$ where $z^j \neq 0$ and let U_j be the image of U_j^* under the natural projection $\mathbf{C}^{n+1} - \{0\} \to (\mathbf{C}^{n+1} - \{0\})/\mathbf{C}^*$. It is easy to see that considering $z^0/z^j, \ldots, z^{j-1}/z^j, z^{j+1}/z^j, \ldots, z^n/z^j$ as functions defined on U_j we may identify $P_n(\mathbf{C})$ with $(\mathbf{C}^{n+1} - \{0\})/\mathbf{C}^*$ whose complex manifold structure is defined by the family of coordinate neighborhoods U_j with local coordinate system $z^0/z^j, \ldots, z^{j-1}/z^j, z^{j+1}/z^j, \ldots, z^n/z^j$, called the *inhomogeneous coordinate system* of $P_n(\mathbf{C})$ in U_j. The coordinate system $z^0, z^1, \ldots, z^n$ of $\mathbf{C}^{n+1}$ is called a *homogeneous coordinate system* of $P_n(\mathbf{C})$; homogeneous coordinates of a point of $P_n(\mathbf{C}) \approx (\mathbf{C}^{n+1} - \{0\})/\mathbf{C}^*$ is by definition the coordinates of a point of $\mathbf{C}^{n+1} - \{0\}$ representing it. Thus, homogeneous coordinates are defined up to a non-zero constant factor.

IX. COMPLEX MANIFOLDS 135

What we have just said may be rephrased more geometrically as follows. $\mathbf{C}^{n+1} - \{0\}$ *is a principal fibre bundle over* $P_n(\mathbf{C}) = (\mathbf{C}^{n+1} - \{0\})/\mathbf{C}^*$ *with group* $\mathbf{C}^*$. If we denote by π the projection $\mathbf{C}^{n+1} - \{0\} \to P_n(\mathbf{C})$, then local triviality $\psi_j \colon \pi^{-1}(U_j) \approx U_j \times \mathbf{C}^*$ is given by

$$\psi_j(z) = (\pi(z), z^j) \in U_j \times \mathbf{C}^*$$

for $z = (z^0, \ldots, z^n) \in \mathbf{C}^{n+1} - \{0\}$.

The transition functions $\psi_{kj} \colon U_j \cap U_k \to \mathbf{C}^*$ are given by

$$\psi_{kj} = z^k/z^j,$$

where $z^0, \ldots, z^n$ is considered as a homogeneous coordinate system in $P_n(\mathbf{C})$.

Let S^{2n+1} be the unit sphere in $\mathbf{C}^{n+1}$ defined by $|z^0|^2 + \cdots + |z^n|^2 = 1$ and S^1 the multiplicative group of complex numbers of absolute value 1. Then S^{2n+1} *is a principal fibre bundle over* $P_n(\mathbf{C})$ *with group* S^1; indeed, it is a subbundle of $\mathbf{C}^{n+1} - \{0\}$ in a natural manner. If we denote by π the projection $S^{2n+1} \to P_n(\mathbf{C})$, then local triviality $\psi_j \colon \pi^{-1}(U_j) \approx U_j \times S^1$ is given by

$$\psi_j(z) = (\pi(z), z^j/|z^j|) \in U_j \times S^1$$

for $z = (z^0, \ldots, z^n) \in S^{2n+1}$.

The transition functions $\psi_{kj} \colon U^j \cap U^k \to S^1$ are given by

$$\psi_{kj} = z^k |z^j|/z^j |z^k|.$$

It is sometimes necessary to identify the group S^1 with the additive group $\mathbf{R}/\mathbf{Z}$ (the real numbers modulo 1); the isomorphism is given by $\lambda = e^{2\pi i \theta} \in S^1 \to (1/2\pi i) \log \lambda = \theta \in \mathbf{R}/\mathbf{Z}$. If we consider S^{2n+1} as a principal fibre bundle over $P_n(\mathbf{C})$ with group $\mathbf{R}/\mathbf{Z}$, then the transition functions are given by

$$\frac{1}{2\pi i} \log \psi_{kj} = \frac{1}{2\pi i} (\log z^k/z^j - \log |z^k|/|z^j|).$$

Example 2.5. Let M be a complex manifold with an open covering $\{U_j\}$ and G a complex Lie group. Given a family of holomorphic mappings $\psi_{jk} \colon U_j \cap U_k \to G$ such that

$$\psi_{lj}(x) = \psi_{lk}(x)\psi_{kj}(x), \quad x \in U_j \cap U_k \cap U_l,$$

we can construct a principal fibre bundle P over M with group

136   FOUNDATIONS OF DIFFERENTIAL GEOMETRY

G and transition functions ψ_{kj} (cf. Proposition 5.2 of Chapter I). From the proof of Proposition 5.2 of Chapter I we see that P has a natural complex manifold structure such that the projection π is holomorphic and $\pi^{-1}(U_j) \approx U_j \times G$ holomorphically. We shall now apply this construction to the case where $M = P_p(\mathbf{C}) \times P_q(\mathbf{C})$ and G is a 1-dimensional complex torus $\mathbf{C}/D$. Let $z^0, \ldots, z^p$ and $w^0, \ldots, w^q$ be the homogeneous coordinate systems for $P_p(\mathbf{C})$ and $P_q(\mathbf{C})$, respectively. For each α, $\alpha = 0, \ldots, p$ (resp. λ, $\lambda = 0, \ldots, q$), let U_α (resp. V_λ) be the open subset of $P_p(\mathbf{C})$ (resp. $P_q(\mathbf{C})$) defined by $z^\alpha \neq 0$ (resp. $w^\lambda \neq 0$). We take $\{U_\alpha \times V_\lambda\}$ as an open covering of $P_p(\mathbf{C}) \times P_q(\mathbf{C})$. Let τ_1 and τ_2 generate D. We then define holomorphic mappings $\psi_{\beta\mu,\alpha\lambda}$: $(U_\alpha \times V_\lambda) \cap (U_\beta \times V_\mu) \to \mathbf{C}/D$ by

$$\psi_{\beta\mu,\alpha\lambda} = \frac{1}{2\pi i} \left(\tau_1 \log z^\beta/z^\alpha + \tau_2 \log w^\mu/w^\lambda \right) \mod D.$$

Then

$$\psi_{\gamma\nu,\alpha\lambda} = \psi_{\gamma\nu,\beta\mu} + \psi_{\beta\mu,\alpha\lambda}$$

on $(U_\alpha \times V_\lambda) \cap (U_\beta \times V_\mu) \cap (U_\gamma \times V_\nu)$.

We denote by $M^{p,q}_{\tau_1\tau_2}$ the bundle over $P_p(\mathbf{C}) \times P_q(\mathbf{C})$ with group $\mathbf{C}/D$ constructed by the transition functions $\{\psi_{\beta\mu,\alpha\lambda}\}$. We shall show that *the complex manifold $M^{p,q}_{\tau_1\tau_2}$ is diffeomorphic with the product $S^{2p+1} \times S^{2q+1}$ of a $(2p+1)$-sphere S^{2p+1} and a $(2q+1)$-sphere S^{2q+1}.* Define a family of mappings $\varphi_{\alpha\lambda}$: $U_\alpha \times V_\lambda \to \mathbf{C}/D$ by

$$\varphi_{\alpha\lambda} = \frac{1}{2\pi i} \left(\tau_1 \log |z^\alpha| + \tau_2 \log |w^\lambda| \right) \mod D$$

and set

$$\psi'_{\beta\mu,\alpha\lambda} = \varphi_{\beta\mu} + \psi_{\beta\mu,\alpha\lambda} - \varphi_{\alpha\lambda}.$$

The transition functions $\{\psi'_{\beta\mu,\alpha\lambda}\}$ define a bundle equivalent to $M^{p,q}_{\tau_1\tau_2}$. Define a family of mappings $f_{\beta\alpha}$: $U_\alpha \cap U_\beta \to \mathbf{R}/\mathbf{Z}$ by

$$f_{\beta\alpha} = \frac{1}{2\pi i} (\log z^\beta/z^\alpha - \log |z^\beta|/|z^\alpha|)$$

and a family of mappings $g_{\mu\lambda}$: $V_\lambda \cap V_\mu \to \mathbf{R}/\mathbf{Z}$ by

$$g_{\mu\lambda} = \frac{1}{2\pi i} (\log w^\mu/w^\lambda - \log |w^\mu|/|w^\lambda|).$$

IX. COMPLEX MANIFOLDS 137

The principal bundle over $P_p(\mathbf{C})$ with group $\mathbf{R}/\mathbf{Z}$ defined by the transition functions $\{f_{\beta\alpha}\}$ is a $(2p+1)$-sphere S^{2p+1} (cf. Example 2.4). Similarly, the principal bundle over $P_q(\mathbf{C})$ with group $\mathbf{R}/\mathbf{Z}$ defined by $\{g_{\mu\lambda}\}$ is a $(2q+1)$-sphere S^{2q+1}. The mapping $(a, b) \in \mathbf{R} \times \mathbf{R} \to a\tau_1 + b\tau_2 \in \mathbf{C}/D$ induces a group isomorphism of $(\mathbf{R}/\mathbf{Z}) \times (\mathbf{R}/\mathbf{Z})$ onto $\mathbf{C}/D$. Since

$$\psi'_{\beta\mu,\alpha\lambda} = \frac{1}{2\pi i}\{\tau_1(\log z^\beta/z^\alpha - \log |z^\beta|/|z^\alpha|) + \tau_2(\log w^\mu/w^\lambda - \log |w^\mu|/|w^\lambda|)\},$$

the principal bundle over $P_p(\mathbf{C}) \times P_q(\mathbf{C})$ with group $\mathbf{C}/D \in (\mathbf{R}/\mathbf{Z}) \times (\mathbf{R}/\mathbf{Z})$ defined by $\{\psi'_{\beta\mu,\alpha\lambda}\}$ is isomorphic with the product $S^{2p+1} \times S^{2q+1}$ of the two bundles

$$S^{2p+1}(P_p(\mathbf{C}), \mathbf{R}/\mathbf{Z}) \quad \text{and} \quad S^{2q+1}(P_q(\mathbf{C}), \mathbf{R}/\mathbf{Z}).$$

The fact that $S^{2p+1} \times S^1$ admits a complex structure was discovered by Hopf [5], and $M^{p,0}_{\tau_1\tau_2}$ is called a *Hopf manifold*. Calabi and Eckmann [1] later discovered a complex structure on $S^{2p+1} \times S^{2q+1}$ for all $p, q \geq 0$. ($M^{0,0}_{\tau_1\tau_2}$ is nothing but $\mathbf{C}/D$.) In Hopf's paper, $M^{p,0}_{\tau_1\tau_2}$ was described as follows. Let λ be a non-zero complex number with $|\lambda| \neq 1$ and Δ_λ the cyclic group of linear transformations of $\mathbf{C}^{p+1}$ generated by the transformation

$$z \in \mathbf{C}^{p+1} \to \lambda z \in \mathbf{C}^{q+1}.$$

Since Δ_λ is a properly discontinuous group acting freely on $\mathbf{C}^{p+1} - \{0\}$, $(\mathbf{C}^{p+1} - \{0\})/\Delta_\lambda$ is a complex manifold in a natural way. We shall show that $M^{p,0}_{\tau_1\tau_2} = (\mathbf{C}^{p+1} - 0)/\Delta_\lambda$ with $\lambda = e^{2\pi i \tau_2/\tau_1}$. For each α, $\alpha = 0, \ldots, p$, let h_α be the mapping from the set $U^*_\alpha = \{z = (z^0, \ldots, z^p) \in \mathbf{C}^{p+1} - \{0\}; z^\alpha \neq 0\}$ into $U_\alpha \times \mathbf{C}/D$ defined by

$$h_\alpha(z) = \left(\pi(z), \frac{1}{2\pi i} \tau_1 \log z^\alpha\right)$$

where $\pi \colon (\mathbf{C}^{p+1} - \{0\}) \to P_p(\mathbf{C})$ is the projection described in Example 2.4 and $(1/2\pi i)\tau_1 \log z^\alpha$ defines an element of $\mathbf{C}/D$. From the way $M^{p,0}_{\tau_1\tau_2}$ was constructed from $U_\alpha(\{\alpha\} \times U_\alpha \times \mathbf{C}/D)$ (cf. Proposition 5.2 of Chapter I), we see that the family of mappings $\{h_\alpha\}$ defines a holomorphic mapping h of $\mathbf{C}^{p+1} - \{0\}$ into $M^{p,0}_{\tau_1\tau_2}$. It is easy to verify that h is a covering projection and induces a holomorphic diffeomorphism of $(\mathbf{C}^{p+1} - 0)/\Delta_\lambda$ onto $M^{p,0}_{\tau_1\tau_2}$, where $\lambda = e^{2\pi i \tau_2/\tau_1}$.

We now define an action of $GL(p + 1; \mathbf{C}) \times GL(q + 1; \mathbf{C})$ on $M^{p,q}_{\tau_1\tau_2}$. Let $A \in GL(p + 1; \mathbf{C})$ and $B \in GL(q + 1; \mathbf{C})$. The group $GL(p + 1; \mathbf{C})$ (resp. $GL(q + 1; \mathbf{C})$) acts on $P_p(\mathbf{C})$ (resp. $P_q(\mathbf{C})$) in a natural way with respect to the homogeneous coordinate system). Let $(z, w, t) \in U_\alpha \times V_\lambda \times \mathbf{C}/D$, where z and w are represented by homogeneous coordinates $(z^0, \ldots, z^p)$ and $(w^0, \ldots, w^q)$ respectively so that $z^\alpha \neq 0$ and $w^\lambda \neq 0$. Let

$z' = Az$ with homogeneous coordinates $(z'^0, \ldots, z'^p)$,

$w' = Bw$ with homogeneous coordinates $(w'^0, \ldots, w'^q)$.

Assume that $(z', w') \in U_\beta \times V_\mu$, i.e., $z'^\beta \neq 0$ and $w'^\mu \neq 0$. Then we set

$$t' = t + \frac{1}{2\pi i}(\tau_1 \log z'^\beta/z^\alpha + \tau_2 \log w'^\mu/w^\lambda) \mod D.$$

It is straightforward to verify that the point x' of $M^{p,q}_{\tau_1\tau_2}$ represented by $(z', w', t') \in U_\alpha \times V_\lambda \times \mathbf{C}/D$ depends only on the point x of $M^{p,q}_{\tau_1\tau_2}$ represented by $(z, w, t) \in U_\alpha \times V_\lambda \times \mathbf{C}/D$ and on the element (A, B) of the group $GL(p + 1; \mathbf{C}) \times GL(q + 1; \mathbf{C})$. It is easy to see that *the action of $GL(p + 1; \mathbf{C}) \times GL(q + 1; \mathbf{C})$ on $M^{p,q}_{\tau_1\tau_2}$ defined by $((A, B), x) \to x'$ is holomorphic and transitive*. It should be pointed out that the action is not effective but is fibre-preserving with respect to the fibring $M^{p,q}_{\tau_1\tau_2} \to P_p(\mathbf{C}) \times P_q(\mathbf{C})$. The class of complex manifolds $M^{p,q}_{\tau_1\tau_2}$ ($p, q > 0$) constructed above is contained in the class of C-spaces of H. C. Wang [4] consisting of simply connected, compact homogeneous complex manifolds (see Note 24). (See also Ise [2] for Hopf manifolds.)

Example 2.6. Let S^n denote the unit sphere in $\mathbf{R}^{n+1}$. Kirchhoff [1, 2] has shown that *if S^n admits an almost complex structure, then S^{n+1} admits an absolute parallelism.* On the other hand, Borel and Serre [1] have proved that S^n, *for $n \neq 2, 6$, does not admit almost complex structures.* Later, Adams [1] proved that S^{n+1} admits an absolute parallelism only for $n + 1 = 1, 3,$ *and* 7. The result of Adams combined with that of Kirchhoff implies that of Borel and Serre. The proofs of the theorems of Adams and Borel-Serre are beyond the scope of this book. We shall prove only the elementary result of Kirchhoff. Let J be an almost complex structure on S^n.

IX. COMPLEX MANIFOLDS 139

We fix $\mathbf{R}^{n+1}$ as a subspace of $\mathbf{R}^{n+2}$ and a unit vector e of $\mathbf{R}^{n+2}$ perpendicular to $\mathbf{R}^{n+1}$. We shall construct a field σ of linear frames on S^{n+1}. Let $x \in S^{n+1}$. If $x \neq e$, then we may write uniquely as follows:

$$x = ae + by, \quad a, b \in \mathbf{R}, \quad b > 0 \quad \text{and} \quad y \in S^n.$$

Let V_y be the n-dimensional vector subspace of $\mathbf{R}^{n+2}$ parallel to the tangent space $T_y(S^n)$ in $\mathbf{R}^{n+2}$ and J_y the linear endomorphism of V_y corresponding to the linear endomorphism of $T_y(S^n)$ given by J. We define a linear frame $\sigma_x: \mathbf{R}^{n+1} \to T_x(S^{n+1})$ as follows:

$$\sigma_x(y) = ay - be,$$
$$\sigma_x(z) = az + bJ_y(z) \quad \text{for } z \in V_y.$$

(Note that $\mathbf{R}^{n+1}$ is spanned by y and V_y and that both $\sigma_x(y)$ and $\sigma_x(z)$ are perpendicular to x and hence can be considered as elements of $T_x(S^{n+1})$.) We define σ_e to be the identity transformation: $\mathbf{R}^{n+1} \to \mathbf{R}^{n+1} = T_e(S^{n+1})$. It is easy to see that σ is a continuous field.

Since the underlying differentiable manifold of $P_1(\mathbf{C})$ is S^2, S^2 admits a complex structure. We shall construct an almost complex structure on S^6 using Cayley numbers. A *Cayley number* $x = (q_1, q_2)$ is an ordered pair of quaternions. The set of Cayley numbers forms an 8-dimensional non-associative algebra over $\mathbf{R}$ with the addition and the multiplication defined as follows:

$$(q_1, q_2) \pm (q_1', q_2') = (q_1 \pm q_1', q_2 \pm q_2')$$
$$(q_1, q_2)(q_1', q_2') = (q_1 q_1' - \bar{q}_2' q_2, q_2' q_1 + q_2 \bar{q}_1'),$$

where $\bar{q}$ denotes the quaternion conjugate of q. Define the conjugate of a Cayley number $x = (q_1, q_2)$ to be $\bar{x} = (\bar{q}_1, -q_2)$. Then $x\bar{x} = (q_1\bar{q}_1 + \bar{q}_2 q_2, 0)$ and we set $|x|^2 = q_1\bar{q}_1 + \bar{q}_2 q_2$. Evidently $|x| > 0$ unless $x = 0$. It can be verified by direct calculation that $|xx'| = |x|\,|x'|$. Therefore $xx' = 0$ implies $x = 0$ or $x' = 0$. Although the associative law does not hold, the so-called alternative law is valid:

$$x(xx') = (xx)x', \quad (x'x)x = x'(xx).$$

A Cayley number $x = (q_1, q_2)$ is *real* if q_1 is real and $q_2 = 0$. It is called *purely imaginary* if q_1 is a purely imaginary quaternion. Let

140   FOUNDATIONS OF DIFFERENTIAL GEOMETRY

U_7 be the 7-dimensional real vector space formed by the purely imaginary Cayley numbers. We define an *inner product* (,) and a *vector product* $\times$ in U_7 as follows:

$$-(x, x') = \text{the real part of } xx', \quad x, x' \in U_7,$$

$$x \times x' = \text{the purely imaginary part of } xx', \quad x, x' \in U_7.$$

(This generalizes the inner product and the vector product in $\mathbf{R}^3$ which are defined by considering $\mathbf{R}^3$ as the space of purely imaginary quaternions.) It can be verified that if $x, x', x'' \in U_7$, then (i) $xx = -(x, x) = -|x|^2$, (ii) $x \times x' = -x' \times x$, (iii) $(x \times x', x'') = (x, x' \times x'')$. (For more details and references on Cayley numbers, see Jacobson [2].) Let S^6 be the unit sphere in U_7 defined by $S^6 = \{x \in U_7; |x| = 1\}$. We identify the tangent space $T_x(S^6)$ with the subspace $V_x = \{y \in U_7; (x, y) = 0\}$ of U_7 parallel to it. Define a linear endomorphism J_x of V_x such that $J_x \circ J_x = -1$ by

$$J_x(y) = x \times y, \quad y \in V_x.$$

By (ii) and (iii), $y \in V_x$ implies $x \times y \in V_x$ and hence J_x is an endomorphism of V_x. It follows also that $J_x(J_x(y)) = x \times (x \times y) = x(x \times y) - (x, x \times y) = x(x \times y) = x(xy) - x(x,y) = x(xy) = (xx)y = -|x|^2 y = -y$, showing that $J_x \circ J_x = -1$. The family of endomorphisms J_x, $x \in S^6$, defines an almost complex structure on S^6. It may be verified by direct calculation that this almost complex structure has non-vanishing torsion (cf. Frölicher [1]). The group of automorphisms of the algebra of Cayley numbers is an exceptional simple Lie group G_2 (see Jacobson [1, 2]) and the group G_2 acts transitively on the unit sphere S^6 in U_7, leaving the almost complex structure defined above invariant. It is not known whether S^6 admits any complex structure. Invariant (almost) complex structures on homogeneous spaces will be discussed later in §6 of Chapter X.

Example 2.7.   Let M be a 6-dimensional orientable manifold immersed in $\mathbf{R}^7$. From each almost complex structure J on S^6 we shall induce an almost complex structure on M. (When J is the almost complex structure on S^6 defined from Cayley numbers as in Example 2.6, the induced almost complex structure on M coincides with the one constructed by Calabi [5].) Let $g: M \to S^6$ be the spherical map of Gauss (cf. §2 of Chapter VII). The tangent spaces $T_x(M)$ and $T_{g(x)}(S^6)$ are parallel in $\mathbf{R}^7$ and can be identified in a natural manner. Hence every almost complex structure on S^6 induces an almost complex structure on M^6 in a natural manner.

Example 2.8. We show that *every almost complex structure J on a 2-dimensional orientable manifold M has vanishing torsion*. For any vector field X on M, we have

$$N(X, JX) = 2(-[JX, X] - [X, JX] + J[X, X] - J[JX, JX])$$
$$= 0.$$

On the other hand, in a neighborhood of a point where $X \neq 0$, every vector field Y is a linear combination of X and JX. Hence $N = 0$, proving our assertion. We show also that *every Riemannian metric on a 2-dimensional oriented manifold M defines an almost complex structure J in a natural manner*. In fact, every Riemannian metric on M defines a reduction to $SO(2)$ of the structure group $GL(2; \mathbf{R})$ of the bundle $L(M)$ of linear frames. Since $SO(2)$ is contained in the real representation of $GL(1; \mathbf{C})$, i.e., $SO(2) \subset GL(1; \mathbf{C}) \subset GL(2; \mathbf{R})$, our assertion follows from Proposition 2.4.

3. Connections in almost complex manifolds

On each almost complex manifold M, we shall construct the bundle $C(M)$ of complex linear frames and study connections in $C(M)$ and their torsions.

Let M be an almost complex manifold of dimension $2n$ with almost complex structure J and let J_0 be the canonical complex structure of the vector space $\mathbf{R}^{2n}$ defined in §1. Then a *complex linear frame* at a point x of M is a non-singular linear mapping $u: \mathbf{R}^{2n} \to T_x(M)$ such that $u \circ J_0 = J \circ u$. In §1 we showed that J defines the structure of a complex vector space in $T_x(M)$ and, by Proposition 1.2, $u: \mathbf{R}^{2n} \to T_x(M)$ is a complex linear frame at x if and only if it is a non-singular complex linear mapping of $\mathbf{C}^n = \mathbf{R}^{2n}$ onto $T_x(M)$. The set of complex linear frames forms a principal fibre bundle over M with group $GL(n; \mathbf{C})$; it is called the *bundle of complex linear frames* and is denoted by $C(M)$. The proof of this fact is almost identical with the case of bundles of linear frames (see Example 5.2 of Chapter I), except perhaps local triviality of $C(M)$. To prove local triviality of $C(M)$, let $x^1, \ldots, x^{2n}$

be a local coordinate system valid in a neighborhood of a point o of M. By a change of ordering if necessary, we may assume that $\partial/\partial x^1, \ldots, \partial/\partial x^n$ form a basis for $T_o(M)$ as a complex vector space and hence for $T_x(M)$ for all x in a neighborhood U of o. Let $e_1, \ldots, e_n$ be the natural basis for $\mathbf{R}^{2n} = \mathbf{C}^n$ as a complex vector space. To each complex linear frame u at x, we assign an element $(x, a) \in U \times GL(n; \mathbf{C})$, where $a = (a_j^i)$ is defined by

$$u(e_k) = \sum_{j=1}^{n} a_k^j (\partial/\partial x^j), \quad k = 1, \ldots, n.$$

It is an easy matter to verify that $u \to (x, a)$ gives the local triviality of $C(M)$. We remark here that, although we may obtain an open covering $\{U_\alpha\}$ of M and transition functions $\psi_{\alpha\beta}\colon U_\alpha \cap U_\beta \to GL(n; \mathbf{C})$ as explained in §5 of Chapter I, these transition functions are in general not almost complex.

Since the bundle $C(M)$ is a subbundle of the bundle $L(M)$ of linear frames, each almost complex structure gives rise to a reduction of the structure group $GL(2n; \mathbf{R})$ of $L(M)$ to $GL(n; \mathbf{C})$.

PROPOSITION 3.1. *Given a 2n-dimensional manifold M, there is a natural one-to-one correspondence between the almost complex structures and the reductions of the structure group of $L(M)$ to $GL(n; \mathbf{C})$.*

Proof. Since we have already defined a mapping from the set of almost complex structures of M into the set of reductions of the structure group of $L(M)$ to $GL(n; \mathbf{C})$, we shall just construct its inverse mapping. Let P be a subbundle of $L(M)$ with structure group $GL(n; \mathbf{C})$. For each point x of M, choose a linear frame $u \in P$ at x and then transfer the canonical complex structure of $\mathbf{R}^{2n}$ onto $T_x(M)$ by the linear transformation $u\colon \mathbf{R}^{2n} \to T_x(M)$ to obtain a complex structure on the vector space $T_x(M)$. Since any other frame $u' \in P$ at x differs from u by right multiplication by an element of $GL(n; \mathbf{C})$, the complex structure defined on $T_x(M)$ is independent of choice of u. QED.

From Proposition 3.1 above and Proposition 5.6 of Chapter I, we obtain

PROPOSITION 3.2. *Given a 2n-dimensional manifold M, there is a natural one-to-one correspondence between the almost complex structures of M and the cross-sections of the associated bundle $L(M)/GL(n; \mathbf{C})$ over M.*

IX. COMPLEX MANIFOLDS   143

We know that, given a Riemannian manifold M with metric tensor g, a linear connection Γ of M is a metric connection, i.e., Γ comes from a connection in the bundle $O(M)$ of orthonormal frames if and only if g is parallel with respect to Γ (cf. Proposition 2.1 of Chapter IV as well as Proposition 1.5 of Chapter III). The proof of the following proposition is analogous to that of Proposition 1.5 of Chapter III and is left to the reader.

PROPOSITION 3.3. *For a linear connection Γ on an almost complex manifold M, the following conditions are equivalent:*
(a) *Γ is a connection in the bundle $C(M)$ of complex linear frames;*
(b) *The almost complex structure J is parallel with respect to Γ.*

A linear (or affine) connection on M is said to be *almost complex* if it satisfies any one (and hence both) of the conditions above. From the general theory of connections (cf. Theorem 2.1 of Chapter II) we know that every almost complex manifold admits an almost complex affine connection (provided it is paracompact). We now prove the existence of a connection of more special type.

THEOREM 3.4. *Every almost complex manifold M admits an almost complex affine connection such that its torsion T is given by*

$$N = 8T,$$

where N is the torsion of the almost complex structure J of M.

Proof. Consider an arbitrary torsion-free affine connection on M with covariant differentiation ∇, and let Q be the tensor field of type $(1, 2)$ defined by

$$4Q(X, Y) = (\nabla_{JY} J)X + J((\nabla_Y J)X) + 2J((\nabla_X J)Y),$$

where X and Y are vector fields. Consider an affine connection whose covariant differentiation $\tilde{\nabla}$ is defined by

$$\tilde{\nabla}_X Y = \nabla_X Y - Q(X, Y).$$

By Proposition 7.5 of Chapter III, $\tilde{\nabla}$ is really covariant differentiation of an affine connection. We shall show that this is a desired connection.

To prove that the connection given by $\tilde{\nabla}$ is almost complex, we compare $\tilde{\nabla}_X(JY)$ with $J(\tilde{\nabla}_X Y)$. Then

$$\tilde{\nabla}_X(JY) = \nabla_X(JY) - Q(X, JY)$$
$$= (\nabla_X J)Y + J(\nabla_X Y) - Q(X, JY),$$
$$J(\tilde{\nabla}_X Y) = J(\nabla_X Y) - J(Q(X, Y)).$$

To prove that $\tilde{\nabla}_X(JY) = J(\tilde{\nabla}_X Y)$, we shall establish the equality

$$Q(X, JY) - J(Q(X, Y)) = (\nabla_X J)Y.$$

We have

$$4Q(X, JY) = -(\nabla_Y J)X + J((\nabla_{JY} J)X) + 2J((\nabla_X J) \circ JY)$$
$$4J(Q(X, Y)) = J((\nabla_{JY} J)X) - (\nabla_Y J)X - 2(\nabla_X J)Y.$$

On the other hand, from $0 = \nabla_X(J^2) = (\nabla_X J)J + J(\nabla_X J)$, we obtain

$$2J((\nabla_X J) \circ JY) = -2J(J \circ (\nabla_X J)Y) = 2(\nabla_X J)Y.$$

This establishes the desired equality, thus showing that $\tilde{\nabla}$ commutes with J, i.e., J is parallel with respect to the connection given by $\tilde{\nabla}$.

The torsion T of $\tilde{\nabla}$ is given by (cf. Theorem 5.1 of Chapter III)

$$T(X, Y) = \tilde{\nabla}_X Y - \tilde{\nabla}_Y X - [X, Y]$$
$$= \nabla_X Y - \nabla_Y X - [X, Y] - Q(X, Y) + Q(Y, X).$$

Since ∇ has no torsion, we obtain

$$T(X, Y) = -Q(X, Y) + Q(Y, X).$$

From the defining equation of Q, we obtain

$$4(Q(Y, X) - Q(X, Y)) = (\nabla_{JX} J)Y + J((\nabla_Y J)X)$$
$$- (\nabla_{JY} J)X - J((\nabla_X J)Y).$$

The four terms on the right may be written as follows:

$$(\nabla_{JX} J)Y = \nabla_{JX}(JY) - J(\nabla_{JX} Y),$$
$$J((\nabla_Y J)X) = J(\nabla_Y(JX)) + \nabla_Y X,$$
$$(\nabla_{JY} J)X = \nabla_{JY}(JX) - J(\nabla_{JY} X),$$
$$J((\nabla_X J)Y) = J(\nabla_X(JY)) + \nabla_X Y$$

IX. COMPLEX MANIFOLDS 145

Hence,

$$4(Q(Y, X) - Q(X, Y))$$
$$= (\nabla_{JX}(JY) - \nabla_{JY}(JX)) - (\nabla_X Y - \nabla_Y X)$$
$$- J(\nabla_{JX} Y - \nabla_Y(JX)) - J(\nabla_X(JY) - \nabla_{JY} X).$$

Since ∇ has no torsion, the four terms on the right may be replaced by the ordinary brackets and we have

$$4(Q(Y, X) - Q(X, Y))$$
$$= [JX, JY] - [X, Y] - J[JX, Y] - J[X, JY]$$
$$= \tfrac{1}{2} N(X, Y).$$

QED.

COROLLARY 3.5. *An almost complex manifold M admits a torsion-free almost complex affine connection if and only if the almost complex structure has no torsion.*

Proof. Assume that M admits a torsion-free almost complex affine connection and denote its covariant differentiation by ∇. If we use this ∇ in the proof of Theorem 3.4, then from $\nabla J = 0$ we obtain $Q = 0$ and hence $N = 0$. The converse is a special case of Theorem 3.4. QED.

PROPOSITION 3.6. *Let M be an almost complex manifold with almost complex structure J. Then the torsion T and the curvature R of an almost complex affine connection satisfy the following identities:*

(1) $T(JX, JY) - J(T(JX, Y)) - J(T(X, JY)) - T(X, Y)$
$$= -\tfrac{1}{2} N(X, Y)$$

for any vector fields X and Y, where N is the torsion of J;

(2) $R(X, Y) \circ J = J \circ R(X, Y)$

for any vector fields X and Y.

Proof. This is an immediate consequence of the two formulas

$$T(X, Y) = \nabla_X Y - \nabla_Y X - [X, Y]$$

and

$$R(X, Y) = [\nabla_X, \nabla_Y] - \nabla_{[X,Y]}$$

in Theorem 5.1 of Chapter III and $\nabla J = 0$. QED.

146 FOUNDATIONS OF DIFFERENTIAL GEOMETRY

We shall conclude this section by a simple remark on the structure equations of an almost complex affine connection. Let $C(M)$ be the bundle of complex linear frames on an almost complex manifold M, θ the canonical form on $C(M)$, i.e., the restriction of the canonical form of $L(M)$ to $C(M)$, and ω the connection form of an almost complex affine connection with torsion form Θ and curvature form Ω. Taking the natural bases in $\mathbf{R}^{2n}$ and $\mathfrak{gl}(2n;\mathbf{R})$, we write the structure equations as follows (cf. §2 of Chapter III):

$$d\theta^j = -\sum_{k=1}^{2n} \omega^j_k \wedge \theta^k + \Theta^j, \qquad j = 1, \ldots, 2n,$$

$$d\omega^j_k = -\sum_{h=1}^{2n} \omega^j_h \wedge \omega^h_k + \Omega^j_k, \qquad j, k = 1, \ldots, 2n.$$

Since the connection is almost complex, ω and Ω are $\mathfrak{gl}(n;\mathbf{C})$-valued on $C(M)$, where $\mathfrak{gl}(n;\mathbf{C})$ is considered as a subalgebra of $\mathfrak{gl}(2n;\mathbf{R})$ as explained in §1. If we therefore set

$$\varphi^\alpha = \theta^\alpha + i\theta^{n+\alpha}, \qquad \Phi^\alpha = \Theta^\alpha + i\Theta^{n+\alpha}, \qquad \alpha = 1, \ldots, n,$$
$$\psi^\alpha_\beta = \omega^\alpha_\beta + i\omega^\alpha_{n+\beta}, \qquad \Psi^\alpha_\beta = \Omega^\alpha_\beta + i\Omega^\alpha_{n+\beta}, \qquad \alpha, \beta = 1, \ldots, n,$$

then $\varphi = (\varphi^\alpha)$ and $\Phi = (\Phi^\alpha)$ are $\mathbf{C}^n$-valued and $\psi = (\psi^\alpha_\beta)$ and $\Psi = (\Psi^\alpha_\beta)$ are $\mathfrak{gl}(n;\mathbf{C})$-valued, where $\mathfrak{gl}(n\colon\mathbf{C})$ is now considered as the Lie algebra of $n \times n$ complex matrices. The structure equations on $C(M)$ may be now written as follows:

$$d\varphi^\alpha = -\sum_{\beta=1}^{n} \psi^\alpha_\beta \wedge \varphi^\beta + \Phi^\alpha, \qquad \alpha = 1, \ldots, n,$$

$$d\psi^\alpha_\beta = -\sum_{\gamma=1}^{n} \psi^\alpha_\gamma \wedge \psi^\gamma_\beta + \Psi^\alpha_\beta, \qquad \alpha, \beta = 1, \ldots, n.$$

4. Hermitian metrics and Kaehler metrics

A *Hermitian metric* on an almost complex manifold M is a Riemannian metric g invariant by the almost complex structure J, i.e.,

$$g(JX, JY) = g(X, Y) \qquad \text{for any vector fields } X \text{ and } Y.$$

A Hermitian metric thus defines a Hermitian inner product (cf. §1) on each tangent space $T_x(M)$ with respect to the complex structure defined by J. An almost complex manifold (resp. a complex manifold) with a Hermitian metric is called an *almost Hermitian manifold* (resp. a *Hermitian manifold*).

PROPOSITION 4.1. *Every almost complex manifold admits a Hermitian metric provided it is paracompact.*

Proof. Given an almost complex manifold M, take any Riemannian metric g (which exists provided M is paracompact; see Example 5.7 of Chapter I). We obtain a Hermitian metric h by setting

$$h(X, Y) = g(X, Y) + g(JX, JY) \quad \text{for any vector fields } X \text{ and } Y.$$
QED.

Remark. By Proposition 1.10, every Hermitian metric g on an almost complex manifold M can be extended uniquely to a complex symmetric tensor field of covariant degree 2, also denoted by g, such that
(1) $g(\bar{Z}, \bar{W}) = \overline{g(Z, W)}$ for any complex vector fields Z and W;
(2) $g(Z, \bar{Z}) > 0$ for any non-zero complex vector Z;
(3) $g(Z, \bar{W}) = 0$ for any vector field Z of type $(1, 0)$ and any vector field W of type $(0, 1)$.
Conversely, every complex symmetric tensor field g with the properties (1), (2), and (3) is the natural extension of a Hermitian metric on M.

In §1 we associated to each Hermitian inner product on a vector space V a skew-symmetric bilinear form on V. Applying the same construction to a Hermitian metric of an almost complex manifold M we obtain a 2-form on M. More explicitly, the *fundamental 2-form* Φ of an almost Hermitian manifold M with almost complex structure J and metric g is defined by

$$\Phi(X, Y) = g(X, JY) \quad \text{for all vector fields } X \text{ and } Y.$$

Since g is invariant by J, so is Φ, i.e.,

$$\Phi(JX, JY) = \Phi(X, Y) \quad \text{for all vector fields } X \text{ and } Y.$$

The almost complex structure J is not, in general, parallel with respect to the Riemannian connection defined by the Hermitian metric g. Indeed we have the following formula.

PROPOSITION 4.2. *Let M be an almost Hermitian manifold with almost complex structure J and metric g. Let Φ be the fundamental 2-form, N the torsion of J and ∇ the covariant differentiation of the Riemannian connection defined by g. Then for any vector fields X, Y, and Z on M we have*

$$4g((\nabla_X J)Y, Z) = 6\, d\Phi(X, JY, JZ) - 6\, d\Phi(X, Y, Z)$$
$$+ g(N(Y, Z), JX).$$

Proof. First we have

$$g((\nabla_X J)Y, Z) = g(\nabla_X(JY), Z) - g(J(\nabla_X Y), Z)$$
$$= g(\nabla_X(JY), Z) + g(\nabla_X Y, JZ).$$

To the two terms on the right we apply the following formula (cf. Proposition 2.3 of Chapter IV):

$$2g(\nabla_X Y, Z) = X(g(Y, Z)) + Y(g(X, Z)) - Z(g(X, Y))$$
$$+ g([X, Y], Z) + g([Z, X], Y) + g(X, [Z, Y]).$$

Our proposition now follows by direct calculation from the definitions of Φ and N and from the following formula (cf. Proposition 3.11 of Chapter I):

$$3\, d\Phi(X, Y, Z) = X(\Phi(Y, Z)) + Y(\Phi(Z, X)) + Z(\Phi(X, Y))$$
$$- \Phi([X, Y], Z) - \Phi([Z, X], Y)$$
$$- \Phi([Y, Z], X).$$

QED.

As an application, we have

THEOREM 4.3. *For an almost Hermitian manifold M with almost complex structure J and metric g, the following conditions are equivalent:*
(1) *The Riemannian connection defined by g is almost complex;*
(2) *The almost complex structure has no torsion and the fundamental 2-form Φ is closed.*

Proof. Assume (1). Since the Riemannian connection is almost complex and has no torsion, J has no torsion by Corollary 3.5 (or by Proposition 3.6). Since both g and J are parallel with respect to the Riemannian connection, so is Φ, as easily seen from the definition of Φ. In particular, Φ is closed (by Corollary 8.6 of Chapter III).

IX. COMPLEX MANIFOLDS 149

Assume (2). By Proposition 4.2, we have $\nabla_X J = 0$ for all vector fields X. QED.

COROLLARY 4.4. *For a Hermitian manifold M, the following conditions are equivalent:*
(1) *The Riemannian connection defined by the Hermitian metric is almost complex;*
(2) *The fundamental 2-form Φ is closed.*

A Hermitian metric on an almost complex manifold is called a *Kaehler metric* if the fundamental 2-form is closed. An almost complex manifold (resp. a complex manifold) with a Kaehler metric is called an *almost Kaehler manifold* (resp. a *Kaehler manifold*). An almost Hermitian manifold with $d\Phi = 0$ and $N = 0$ used to be called a *pseudo-Kaehler manifold*. In view of the theorem of Newlander-Nirenberg to the effect that an almost complex manifold with $N = 0$ is a complex manifold (cf. Theorem 2.5), a pseudo-Kaehler manifold is necessarily a Kaehler manifold.

Since a Hermitian metric g is non-degenerate (more strongly, positive-definite) and J is non-singular at each point, it follows that the fundamental 2-form Φ is non-degenerate at each point, that is, $\Phi^n = \Phi \wedge \cdots \wedge \Phi$ (n times, where $2n = \dim M$) is non-zero at each point. In general, a $2n$-dimensional manifold with a 2-form (resp. a closed 2-form) Φ which is non-degenerate at each point of M is called an *almost symplectic* or *almost Hamiltonian manifold* (resp. a *symplectic* or *Hamiltonian manifold*).

PROPOSITION 4.5. *The curvature R and the Ricci tensor S of a Kaehler manifold possess the following properties:*
(1) $R(X, Y) \circ J = J \circ R(X, Y)$ *and* $R(JX, JY) = R(X, Y)$ *for all vector fields X and Y;*
(2) $S(JX, JY) = S(X, Y)$ *and*
$$S(X, Y) = \tfrac{1}{2}\{\text{trace of } J \circ R(X, JY)\}$$
for all vector fields X and Y.

Proof. (1) For a Kaehler manifold, the Riemannian connection is almost complex by Corollary 4.4. The first formula follows therefore from Proposition 3.6. To prove the second formula, we recall the following formula (cf. Proposition 2.1 of Chapter V):
$$g(R(X, Y)V, U) = g(R(U, V)Y, X)$$
for all vector fields U, V, X, Y.

150 FOUNDATIONS OF DIFFERENTIAL GEOMETRY

Hence we have
$$g(R(JX, JY)V, U) = g(R(U, V)JY, JX) = g(J(R(U, V)Y), JX)$$
$$= g(R(U, V)Y, X) = g(R(X, Y)V, U),$$
where the second equality is a consequence of the first formula and the third equality follows from the fact that g is Hermitian. We may now conclude $R(JX, JY) = R(X, Y)$.

(2) We recall (cf. §5 of Chapter VI) that the Ricci tensor S of a Riemannian manifold was defined by
$$S(X, Y) = \text{trace of the map } V \to R(V, X)Y.$$
The first formula may be proved as follows:
$$\begin{aligned} S(JX, JY) &= \text{trace of } V \to R(V, JX)JY \\ &= \text{trace of } JV \to R(JV, JX)JY \\ &= \text{trace of } JV \to R(V, X)JY \\ &= \text{trace of } JV \to J(R(V, X)Y) \\ &= \text{trace of } V \to R(V, X)Y \\ &= S(X, Y), \end{aligned}$$
where the third and the fourth equalities follow from the second and the first formulas of (1), respectively. We shall now prove the second formula. Using the first formula in (1) and Bianchi's first identity (cf. Theorem 5.3 of Chapter III), we obtain
$$\begin{aligned} S(X, Y) &= \text{trace of } V \to R(V, X)Y \\ &= \text{trace of } V \to -J(R(V, X)JY) \\ &= \text{trace of } V \to (J(R(X, JY)V) + J(R(JY, V)X)). \end{aligned}$$
On the other hand, using the second formula in (1), we have
$$\begin{aligned} \text{trace of } V \to J(R(JY, V)X) &= \text{trace of } JV \to J(R(JY, JV)X) \\ &= \text{trace of } JV \to J(R(Y, V)X) \\ &= \text{trace of } V \to R(Y, V)X \\ &= \text{trace of } V \to -R(V, Y)X \\ &= -S(Y, X) = -S(X, Y). \end{aligned}$$
Hence,
$$S(X, Y) = (\text{trace of } V \to J(R(X, JY)V)) - S(X, Y),$$
which establishes the second formula. QED.

IX. COMPLEX MANIFOLDS 151

The following theorem is originally due to Iwamoto [3].

THEOREM 4.6. *For a Kaehler manifold M of complex dimension n, the restricted linear holonomy group is contained in $SU(n)$ if and only if the Ricci tensor vanishes identically.*

Proof. The following lemma is immediate from Theorem 8.1 of Chapter II.

LEMMA 1. *Consider a connection in a principal fibre bundle P over a manifold M with group G. Let $\mathfrak{g}$ be the Lie algebra of G and $\mathfrak{h}$ an ideal of $\mathfrak{g}$. Then the Lie algebra of the holonomy group is contained in $\mathfrak{h}$ if and only if the curvature form takes its values in $\mathfrak{h}$ on P.*

Since the holonomy groups with respect to different reference points are all conjugate to each other in G, the condition that the Lie algebra of the holonomy group is contained in an *ideal* $\mathfrak{h}$ is independent of reference points.

We apply Lemma 1 to the bundle $C(M)$ of complex linear frames of an almost complex manifold M of real dimension $2n$, where G is the real representation of $GL(n; \mathbf{C})$ and $\mathfrak{h}$ is the Lie algebra of $SL(n; \mathbf{C})$. Since the Lie algebra of $SL(n; \mathbf{C})$ consists of complex $n \times n$ matrices with trace 0, under the real representation it consists of real $2n \times 2n$ matrices A satisfying the following two conditions (cf. §1):

$$\text{trace } A = 0 \quad \text{and} \quad \text{trace } J_0 A = 0,$$

where J_0 is the matrix representing the canonical complex structure of $\mathbf{R}^{2n}$. From Lemma 1 we may now conclude

LEMMA 2. *For an almost complex linear connection with curvature tensor R on a $2n$-dimensional almost complex manifold M, the restricted linear holonomy group is contained in (the real representation of) $SL(n; \mathbf{C})$ if and only if*

$$\text{trace } R(X, Y) = 0 \quad \text{and} \quad \text{trace } J \circ R(X, Y) = 0$$

for all vector fields X and Y, where J denotes the almost complex structure.

We apply Lemma 2 to a Kaehler manifold M of complex dimension n. Its holonomy group, which is a subgroup of $O(2n) \cap GL(n; \mathbf{C}) = U(n)$, is contained in $O(2n) \cap SL(n; \mathbf{C}) = SU(n)$ if and only if the curvature tensor R satisfies the two conditions in Lemma 2. However, the first condition is automatically satisfied by every Riemannian manifold, for $R(X, Y)$ gives a skew-symmetric linear transformation at each point with respect to the metric tensor (cf. Proposition 2.1 of Chapter V). On the other hand, by (2) of Proposition 4.5 the second condition is satisfied if and only if the Ricci tensor vanishes identically. QED.

For an almost Hermitian manifold and particularly for a Kaehler manifold M, the natural bundle to consider is the intersection of the two bundles $C(M)$ and $O(M)$. A *unitary frame* of an almost Hermitian manifold M is a complex linear frame which is at the same time orthonormal with respect to the Hermitian metric. The set of unitary frames of M forms a principal fibre bundle over M with group $U(n)$ (more precisely, the real representation of $U(n)$), where $2n$ is the real dimension of M. This bundle, denoted by $U(M)$, is called the *bundle of unitary frames* over M. The proof of the following proposition is similar to those of Propositions 3.1 and 3.2 and hence is left to the reader.

PROPOSITION 4.7. *Given an almost complex manifold M of dimension $2n$, there is a natural one-to-one correspondence between any two of the following three sets: the Hermitian metrics on M, the reductions of the structure group of $C(M)$ to $U(n)$, and the cross sections of the associated bundle $C(M)/U(n)$ over M.*

Since $U(M) = O(M) \cap C(M)$, every connection in $U(M)$ is metric (cf. §2 of Chapter IV) and almost complex (cf. §3), and conversely. In other words, an affine connection on an almost Hermitian manifold M is a connection in the bundle $U(M)$ if and only if both the almost complex structure and the metric tensor are parallel with respect to it. Now Theorem 4.3 may be reformulated as follows.

THEOREM 4.8. *An almost Hermitian manifold M is a Kaehler manifold if and only if the bundle $U(M)$ of unitary frames admits a torsion-free connection (which is necessarily unique).*

Proof. If M is a Kaehler manifold, then the Riemannian connection is a connection in $U(M)$ by Theorem 4.3. Since a torsion-free connection in $U(M)$ is also a torsion-free connection in $O(M)$, it is the Riemannian connection determined by the Hermitian metric. Since it is also a connection in $C(M)$, it is almost complex by Proposition 3.3. Now by Theorem 4.3, the existence of a torsion-free connection in $U(M)$ implies that the almost complex structure has no torsion (and hence is a complex structure by Theorem 2.5) and that the fundamental 2-form is closed. QED.

Using the bundle $U(M)$, we shall study the structure equations and the curvature form of a Kaehler manifold M. Let $\theta = (\theta^1, \ldots, \theta^{2n})$ be the canonical form on $U(M)$ and let $\omega = (\omega_j^i)_{i,j=1,\ldots,2n}$ be the connection form on $U(M)$ which defines the Riemannian connection of M. We denote by $\Omega = (\Omega_j^i)$ the curvature form and write

$$\Omega_j^i = \tfrac{1}{2} \sum_{k,l=1}^{2n} R_{jkl}^i \theta^k \wedge \theta^l, \qquad i,j = 1, \ldots, 2n.$$

As in §3 we set

$$\varphi^\alpha = \theta^\alpha + i\theta^{n+\alpha}, \qquad \alpha = 1, \ldots, n,$$
$$\psi_\beta^\alpha = \omega_\beta^\alpha + i\omega_{n+\beta}^\alpha, \qquad \Psi_\beta^\alpha = \Omega_\beta^\alpha + i\Omega_{n+\beta}^\alpha, \qquad \alpha, \beta = 1, \ldots, n.$$

Since $\omega = (\omega_j^i)$ has values in the real representation of $\mathfrak{u}(n)$, we have

$$\omega_\beta^\alpha = \omega_{n+\beta}^{n+\alpha}, \qquad \omega_{n+\beta}^\alpha = -\omega_\beta^{n+\alpha}, \qquad \omega_\beta^\alpha = -\omega_\alpha^\beta, \qquad \omega_{n+\beta}^\alpha = \omega_{n+\alpha}^\beta.$$

The analogous relations are satisfied by Ω_j^i. We have therefore

$$\psi_\beta^\alpha = -\bar{\psi}_\alpha^\beta \quad \text{and} \quad \Psi_\beta^\alpha = -\bar{\Psi}_\alpha^\beta, \qquad \alpha, \beta = 1, \ldots, n.$$

We shall relate the curvature form (Ψ_β^α) with the Ricci tensor. In the same way as we associate the fundamental 2-form to each Kaehler metric, we associate a 2-form ρ, called the *Ricci form*, to the Ricci tensor S of a Kaehler manifold M in the following manner:

$$\rho(X, Y) = S(X, JY) \qquad \text{for all vector fields } X \text{ and } Y.$$

The fact that ρ is skew-symmetric and hence is a 2-form follows from the first formula in (2) of Proposition 4.5.

PROPOSITION 4.9. *For a Kaehler manifold M, the Ricci form ρ is closed and is related with the curvature form (Ψ_β^α) as follows:*

$$\pi^*(\rho) = -2i \sum_{\alpha=1}^n \Psi_\alpha^\alpha,$$

where π is the projection of $U(M)$ onto M.

154   FOUNDATIONS OF DIFFERENTIAL GEOMETRY

Proof. By the second formula in (2) of Proposition 4.5, $\rho(X, Y)$ is equal to $-\frac{1}{2}(\text{trace of } J \circ R(X, Y))$. We recall that R and Ω are related as follows (cf. §5 of Chapter III):

$$R(X, Y) = u \circ (2\Omega(X^*, Y^*)) \circ u^{-1} \quad \text{for } X, Y \in T_x(M),$$

where $u \in U(M)$ is a unitary frame at $x \in M$ and X^* and Y^* are the lifts of X and Y to u respectively. Denote by J_0 the canonical complex structure of $\mathbf{R}^{2n}$ (cf. §1). Since the frame $u \colon \mathbf{R}^{2n} \to T_x(M)$ preserves the complex structure, we obtain

$$J \circ R(X, Y) = u \circ J_0 \circ (2\Omega(X^*, Y^*)) \circ u^{-1}.$$

Since the trace of the left hand side is equal to that of

$$J_0 \circ (2\Omega(X^*, Y^*)),$$

we obtain

$$\rho(X, Y) = -\tfrac{1}{2}(\text{trace of } J \circ R(X, Y)) = 2 \sum_{\alpha=1}^{n} \Omega^{\alpha}_{n+\alpha}(X^*, Y^*).$$

On the other hand, since (Ω^j_k) is skew-symmetric and, in particular, $\Omega^j_j = 0$ for $j = 1, \ldots, 2n$, we have

$$\sum_{\alpha=1}^{n} \Psi'^{\alpha}_{\alpha} = \sum_{\alpha=1}^{n} (\Omega^{\alpha}_{\alpha} + i\Omega^{\alpha}_{n+\alpha}) = i \sum_{\alpha=1}^{n} \Omega^{\alpha}_{n+\alpha}.$$

Hence,

$$\rho(X, Y) = -2i \sum_{\alpha=1}^{n} \Psi'^{\alpha}_{\alpha}(X^*, Y^*),$$

thus proving the second assertion. To prove that ρ is closed, we consider the structure equation:

$$d\psi^{\alpha}_{\beta} = -\sum_{\gamma=1}^{n} \psi^{\alpha}_{\gamma} \wedge \psi^{\gamma}_{\beta} + \Psi^{\alpha}_{\beta}, \quad \alpha, \beta = 1, \ldots, n.$$

We set $\alpha = \beta$ and sum over α. Then

$$d\left(\sum_{\alpha=1}^{n} \psi^{\alpha}_{\alpha} \right) = \sum_{\alpha=1}^{n} \Psi'^{\alpha}_{\alpha},$$

showing that the right hand side is also closed. From the second assertion of our proposition just established, it follows that ρ is also closed.   QED.

Remark. Since $SU(n)$ is a closed normal subgroup of $U(n)$, we can consider the principal fibre bundle $U(M)/SU(n)$ over M with group $U(n)/SU(n)$. The connection in the bundle $U(M)/SU(n)$ induced by the homomorphism $U(M) \to U(M)/SU(n)$ from the connection in $U(M)$ is given by a 1-form $\Sigma_{\alpha=1}^{n} \psi_{\alpha}^{\alpha}$ (cf. Proposition 6.1 of Chapter II). Then the equation $d(\Sigma_{\alpha=1}^{n} \psi_{\alpha}^{\alpha}) = \Sigma_{\alpha=1}^{n} \Psi_{\alpha}^{\alpha}$ is its structure equation and $\Sigma_{\alpha=1}^{n} \Psi_{\alpha}^{\alpha}$ is its curvature form. This curvature form vanishes if and only if the Ricci tensor of the Kaehler manifold M vanishes. In other words, the restricted holonomy group of the induced connection in $U(M)/SU(n)$ is trivial if and only if the Ricci tensor of M vanishes identically. It is not hard to see that this fact implies Theorem 4.6.

5. *Kaehler metrics in local coordinate systems*

We shall express here various tensor fields introduced in §4 in terms of complex local coordinate systems. Throughout this section, let M be an n-dimensional complex manifold and $z^1, \ldots, z^n$ a complex local coordinate system in M. Unless otherwise stated, Greek indices $\alpha, \beta, \gamma, \ldots$ run from 1 to n, while Latin capitals $A, B, C, \ldots$ run through $1, \ldots, n, \bar{1}, \ldots, \bar{n}$. We set

(1) $\quad Z_\alpha = \partial/\partial z^\alpha, \quad Z_{\bar\alpha} = \bar{Z}_\alpha = \partial/\partial \bar{z}^\alpha.$

Given a Hermitian metric g on M, we extend the Hermitian inner product in each tangent space $T_x(M)$ defined by g to a unique complex symmetric bilinear form in the complex tangent space $T_x^c(M)$ (cf. Proposition 1.10) and set

(2) $\quad g_{AB} = g(Z_A, Z_B).$

Then, by Proposition 1.10,

(3) $\quad g_{\alpha\beta} = g_{\bar\alpha\bar\beta} = 0$

and $(g_{\alpha\bar\beta})$ is an $n \times n$ Hermitian matrix. It is then customary to write

(4) $\quad ds^2 = 2 \sum_{\alpha,\beta} g_{\alpha\bar\beta} \, dz^\alpha \, d\bar{z}^\beta$

for the metric g. By Proposition 1.12, the fundamental 2-form is given by

(5) $\quad \Phi = -2i \sum_{\alpha,\beta} g_{\alpha\bar\beta} \, dz^\alpha \wedge d\bar{z}^\beta.$

A necessary and sufficient condition for g to be a Kaehler metric is given by

(6) $\partial g_{\alpha\bar{\beta}}/\partial z^{\gamma} = \partial g_{\gamma\bar{\beta}}/\partial z^{\alpha}$ or $\partial g_{\alpha\bar{\beta}}/\partial z^{\bar{\gamma}} = \partial g_{\alpha\bar{\gamma}}/\partial z^{\bar{\beta}}$.

Given any affine connection with covariant differentiation ∇ on M, we set

(7) $\nabla_{Z_B} Z_C = \sum_A \Gamma^A_{BC} Z_A$.

Notice that covariant differentiation, which is originally defined for real vector fields, is extended by complex linearity to act on complex vector fields.

Then

(8) $\overline{\Gamma^A_{BC}} = \Gamma^{\bar{A}}_{\bar{B}\bar{C}}$

with the convention that $\bar{\bar{\alpha}} = \alpha$. From the fact that $JZ_\alpha = iZ_\alpha$ and $JZ_{\bar{\alpha}} = -iZ_{\bar{\alpha}}$, it follows that the connection is almost complex if and only if

(9) $\Gamma^\alpha_{B\bar{\gamma}} = \Gamma^{\bar{\alpha}}_{B\gamma} = 0$.

By direct calculation we see that an almost complex connection has no torsion if and only if

(10) $\Gamma^\alpha_{\beta\gamma} = \Gamma^\alpha_{\gamma\beta}$, $\Gamma^{\bar{\alpha}}_{\bar{\beta}\bar{\gamma}} = \Gamma^{\bar{\alpha}}_{\bar{\gamma}\bar{\beta}}$

and

(11) other $\Gamma^A_{BC} = 0$.

In particular, (8), (9), (10), and (11) hold for every Kaehler manifold. For a Kaehler manifold, the Γ^A_{BC}'s are determined by the metric as follows:

(12) $\sum_\alpha g_{\alpha\bar{\varepsilon}} \Gamma^\alpha_{\beta\gamma} = \partial g_{\bar{\varepsilon}\beta}/\partial z^\gamma$, $\sum_\alpha g_{\bar{\alpha}\varepsilon} \Gamma^{\bar{\alpha}}_{\bar{\beta}\bar{\gamma}} = \partial g_{\varepsilon\bar{\beta}}/\partial z^{\bar{\gamma}}$.

The proof of (12) is similar to that of Corollary 2.4 of Chapter IV or may be derived from the corollary.

As in §7 of Chapter III, we set

(13) $R(Z_C, Z_D)Z_B = \sum_A K^A_{BCD} Z_A$.

IX. COMPLEX MANIFOLDS

We also define

(14) $K_{ABCD} = g(R(Z_C, Z_D)Z_B, Z_A)$

so that

(15) $K_{ABCD} = \sum_E g_{AE} K^E_{BCD}$

as in §2 of Chapter V. Just from the fact that the connection is almost complex so that $R(Z_C, Z_D)$ commutes with J, it follows that

(16) $K^\alpha_{\bar\beta CD} = K^{\bar\alpha}_{\beta CD} = 0.$

From (3), (15), and (16) we obtain

(17) $K_{\alpha\beta CD} = K_{\bar\alpha\bar\beta CD} = 0.$

From Proposition 2.1 of Chapter V, we obtain $K_{ABCD} = -K_{ABDC} = -K_{BACD} = K_{CDAB}$. This together with (16) and (17) yields

(18) $K^A_{B\gamma\delta} = K^A_{B\bar\gamma\bar\delta} = 0$

and

(19) $K_{AB\gamma\delta} = K_{AB\bar\gamma\bar\delta} = 0.$

Consequently, only the components of the following types can be different from zero:

$$K^\alpha_{\beta\gamma\bar\delta},\ K^\alpha_{\beta\bar\gamma\delta},\ K^{\bar\alpha}_{\beta\gamma\bar\delta},\ K^{\bar\alpha}_{\beta\bar\gamma\delta},$$
$$K_{\alpha\bar\beta\gamma\bar\delta},\ K_{\alpha\bar\beta\bar\gamma\delta},\ K_{\bar\alpha\beta\gamma\bar\delta},\ K_{\bar\alpha\beta\bar\gamma\delta}.$$

In the formula $R(X, Y)Z = [\nabla_X, \nabla_Y]Z - \nabla_{[X,Y]}Z$ we set $X = Z_\gamma$, $Y = Z_{\bar\delta}$ and $Z = Z_\beta$. Using (7), (9), (10), and (11), we obtain

(20) $K^\alpha_{\beta\gamma\bar\delta} = -\partial \Gamma^\alpha_{\beta\gamma}/\partial \bar z^\delta.$

From (3), (12), (15), (20) we obtain

(21) $K_{\alpha\bar\beta\gamma\bar\delta} = \dfrac{\partial^2 g_{\alpha\bar\beta}}{\partial z^\gamma\, \partial \bar z^\delta} - \sum_{\tau,\varepsilon} g^{\bar\varepsilon\tau} \dfrac{\partial g_{\alpha\bar\varepsilon}}{\partial z^\gamma} \dfrac{\partial g_{\beta\bar\tau}}{\partial \bar z^\delta},$

where $(g^{\alpha\bar\beta})$ is the inverse matrix to $(g_{\alpha\bar\beta})$ so that $\Sigma_\beta\, g^{\alpha\bar\beta} g_{\gamma\bar\beta} = \delta^\alpha_\gamma$.

Since the components K_{AB} of the Ricci tensor are given by $K_{AB} = \Sigma_C\, K^C_{ACB}$, we obtain from (16), (18), and (20)

(22) $K_{\alpha\bar\beta} = -\sum_\gamma \Gamma^\gamma_{\alpha\gamma}/\partial \bar z^\beta, \qquad K_{\bar\alpha\beta} = \bar K_{\alpha\bar\beta}, \qquad K_{\alpha\beta} = K_{\bar\alpha\bar\beta} = 0.$

If we denote by G the determinant of the matrix $(g_{\alpha\bar\beta})$ and apply the rule for the differentiation of a determinant and the definition of the inverse matrix $(g^{\alpha\bar\beta})$ by means of minors of $(g^{\alpha\bar\beta})$ and G, then we have

$$\frac{\partial G}{\partial z^\alpha} = G \sum_{\beta,\gamma} g^{\beta\bar\gamma} \frac{\partial g_{\beta\bar\gamma}}{\partial z^\alpha}.$$

Applying (12) we obtain

(23) $\quad \sum_\gamma \Gamma^\gamma_{\alpha\gamma} = \dfrac{\partial \log G}{\partial z^\alpha}.$

This together with (22) gives us

(24) $\quad K_{\alpha\bar\beta} = -\dfrac{\partial^2 \log G}{\partial z^\alpha \, \partial \bar z^\beta}.$

Remark. For a Riemannian manifold with metric

$$ds^2 = \Sigma \, g_{ij} \, dx^i \, dx^j,$$

(23) still holds if we set $G = (\det (g_{ij}))^{1/2}$. But the expression for the Ricci tensor is far more complicated than (24); see, for instance, Eisenhart [1; p. 21].

In the same way as we obtained the local coordinate expression (5) for Φ, we obtain the following expression for the Ricci form:

(25) $\quad \rho = -2i \sum_{\alpha,\beta} K_{\alpha\bar\beta} \, dz^\alpha \wedge d\bar z^\beta.$

By (24), ρ may be rewritten as follows:

(26) $\quad \rho = 2i \, \partial \bar\partial \log G.$

Let F be a real-valued function in a coordinate neighborhood of a complex manifold M and set

(27) $\quad g_{\alpha\bar\beta} = \dfrac{\partial^2 F}{\partial z^\alpha \, \partial \bar z^\beta}.$

Since F is real, $(g_{\alpha\bar\beta})$ is always Hermitian. If it is positive definite, then $2\Sigma_{\alpha,\beta} \, g_{\alpha\bar\beta} \, dz^\alpha \, d\bar z^\beta$ is a Kaehler metric by (6). It is known that every Kaehler metric can be locally written $ds^2 = 2\Sigma_{\alpha,\beta} \, g_{\alpha\bar\beta} \, dz^\alpha \, d\bar z^\beta$ with $(g_{\alpha\bar\beta})$ given by (27). Although this fact is rarely used, Kaehler metrics are usually constructed by (27) from certain functions F (cf. examples given in the next section).

When $(g_{\alpha\bar\beta})$ is given by (27), substituting (27) into (21) we obtain

$$(28) \quad K_{\alpha\bar\beta\gamma\bar\delta} = \frac{\partial^4 F}{\partial z^\alpha\, \partial \bar z^\beta\, \partial z^\gamma\, \partial \bar z^\delta} - \sum_{\varepsilon,\tau} g^{\bar\varepsilon\tau} \frac{\partial^3 F}{\partial \bar z^\varepsilon\, \partial z^\alpha\, \partial z^\gamma} \frac{\partial^3 F}{\partial z^\tau\, \partial \bar z^\beta\, \partial \bar z^\delta}.$$

In this section we followed fairly closely Chapter VIII of Yano-Bochner [1]. It should be noted, however, that their curvature tensor differs from ours by sign.

6. Examples of Kaehler manifolds

Example 6.1. The complex n-space $\mathbf{C}^n$ with the metric

$$ds^2 = \sum_{\alpha=1}^n dz^\alpha\, d\bar z^\alpha$$

(where $z^1, \ldots, z^n$ is the natural coordinate system) is a complete, flat Kaehler manifold. According to (5) in §5, the fundamental 2-form Φ is given by

$$\Phi = -i \sum_{\alpha=1}^n dz^\alpha \wedge d\bar z^\alpha.$$

Example 6.2. Let $\mathbf{C}^n/D$ be a complex torus as defined in Example 2.2 in §2. Since D is a discrete group of translations, the Kaehler metric in $\mathbf{C}^n$ defined in Example 6.1 induces a Kaehler metric on $\mathbf{C}^n/D$. The complex tori are the only compact complex parallelizable manifolds which admit Kaehler metrics, Wang [5].

Example 6.3. Let $P_n(\mathbf{C})$ be the n-dimensional complex projective space with homogeneous coordinate system $z^0, z^1, \ldots, z^n$ (cf. Example 2.4). For each index j, let U_j be the open subset of $P_n(\mathbf{C})$ defined by $z^j \neq 0$. We set

$$t_j^k = z^k/z^j, \qquad j, k = 0, \ldots, n.$$

On each U_j, we take $t_j^0, \ldots, \hat t_j^j, \ldots, t_j^n$ (where $\hat t_j^j$ indicates that t_j^j is deleted) as a local coordinate system and consider the function $f_j = \sum_{k=0}^n t_j^k\, \bar t_j^k$. Then

$$f_j = f_k\, t_j^k\, \bar t_j^k \qquad \text{on } U_j \cap U_k.$$

Since t_j^k is a holomorphic function in U_j and in particular in $U_j \cap U_k$, it follows that

$$\partial\bar\partial \log f_j = \partial\bar\partial \log f_k \qquad \text{on } U_j \cap U_k.$$

By setting
$$\Phi = -4i\, \partial\bar{\partial} \log f_j \quad \text{on } U_j,$$
we obtain a globally defined closed $(1, 1)$-form Φ on $P_n(\mathbf{C})$. We set
$$g(X, Y) = \Phi(JX, Y) \quad \text{for all vector fields } X \text{ and } Y.$$
To see that the symmetric form g is positive definite (so that it is a Kaehler metric with fundamental form Φ), we verify this fact on each U_j by a direct calculation. For instance, on U_0 we have

$$ds^2 = 4\, \frac{\left(1 + \sum_\alpha t^\alpha \bar{t}^\alpha\right)\left(\sum_\alpha dt^\alpha\, d\bar{t}^\alpha\right) - \left(\sum_\alpha \bar{t}^\alpha\, dt^\alpha\right)\left(\sum_\alpha t^\alpha\, d\bar{t}^\alpha\right)}{\left(1 + \sum_\alpha t^\alpha \bar{t}^\alpha\right)^2}$$

where we set $t_0^\alpha = t^\alpha$, $\alpha = 1, \ldots, n$, for the sake of simplicity. The Kaehler metric just constructed on $P_n(\mathbf{C})$ is sometimes called the *Fubini-Study metric* (cf. Fubini [1] and Study [1]). We can look at the preceding construction from a slightly different angle. Let $\pi\colon \mathbf{C}^{n+1} - \{0\} \to P_n(\mathbf{C})$ be the projection defined in Example 2.4. Using the natural complex coordinate system $z^0, z^1, \ldots, z^n$ in $\mathbf{C}^{n+1}$, we set
$$\tilde{\Phi} = -4i\, \partial\bar{\partial} \log \left(z^0 \bar{z}^0 + z^1 \bar{z}^1 + \cdots + z^n \bar{z}^n\right).$$
The form $\tilde{\Phi}$ projects onto Φ, i.e., $\pi^*(\Phi) = \tilde{\Phi}$. Since the action of $U(n + 1)$ on $\mathbf{C}^{n+1}$ preserves the function $z^0 \bar{z}^0 + z^1 \bar{z}^1 + \cdots + z^n \bar{z}^n$ as well as the complex structure, it preserves $\tilde{\Phi}$. Consequently, the induced action of $U(n + 1)$ on $P_n(\mathbf{C})$ (cf. Example 2.4) preserves Φ as well as the complex structure and hence the metric.

Example 6.4. Let $G_{p,q}(\mathbf{C})$ be the complex Grassmann manifold of p-planes in $\mathbf{C}^{p+q}$ (cf. Example 2.4). Let $M^*(p + q, p;\, \mathbf{C})$ be the space of $p \times (p + q)$ complex matrices (i.e., p columns and $(p + q)$ rows) of rank p; it may be considered as the set of p linearly independent vectors in $\mathbf{C}^{p+q}$. Each set of p linearly independent vectors in $\mathbf{C}^{p+q}$ determines a p-plane in $\mathbf{C}^{p+q}$. Thus we get the natural projection: $M^*(p + q, p;\, \mathbf{C}) \to G_{p,q}(\mathbf{C})$. Then $M^*(p + q, p;\, \mathbf{C})$ is a principal fibre bundle over $G_{p,q}(\mathbf{C})$ with group $GL(p;\, \mathbf{C})$; the group $GL(p;\, \mathbf{C})$ acts on $M^*(p + q, p;\, \mathbf{C})$ on the right by matrix multiplication. Representing the coordinate system in $M^*(p+q, p; \mathbf{C})$ by a $p \times (p+q)$ complex matrix Z. We define a closed (1, 1)-form $\tilde{\Phi}$ on $M^*(p+q, p; \mathbf{C})$ by

$$\tilde{\Phi} = -4i\, \partial\bar{\partial} \log \det({}^t\bar{Z}Z).$$

It is easy to verify that $\tilde{\Phi}$ projects onto a closed (1, 1)-form, say Φ, of the base space $G_{p,q}(\mathbf{C})$, i.e., $\pi^*(\Phi) = \tilde{\Phi}$. We set

$$g(X, Y) = \Phi(JX, Y) \qquad \text{for all vector fields } X \text{ and } Y.$$

Since the action of $U(p+q)$ on $M^*(p+q, p; \mathbf{C})$ on the left preserves the function ${}^t\bar{Z}Z$ as well as the complex structure, it preserves $\tilde{\Phi}$. Consequently, the induced action of $U(p+q)$ on $G_{p,q}(\mathbf{C})$ preserves Φ as well as the complex structure and hence g. Thus, g is invariant by the group $U(p+q)$, which acts transitively on $G_{p,q}(\mathbf{C})$. To show that g is a Kaehler metric with fundamental form Φ, we shall express g in terms of local coordinate systems of $G_{p,q}(\mathbf{C})$. Writing

$$Z = \begin{pmatrix} Z_0 \\ Z_1 \end{pmatrix},$$

where Z_0 is a $p \times p$ matrix and Z_1 is a $p \times q$ matrix, we consider for instance the open subset of $G_{p,q}(\mathbf{C})$ defined by $\det Z_0 \neq 0$. Then $Z \to Z_1 Z_0^{-1}$ gives a one-to-one mapping of U onto the space $M(q, p; \mathbf{C})$ of $p \times q$ matrices. Setting $T = Z_1 Z_0^{-1}$, we may use T as a local coordinate system in U. Since ${}^t\bar{Z}Z = {}^t\bar{Z}_0(I + {}^t\bar{T}T)Z_0$ where I denotes the $p \times p$ identity matrix, we obtain

$$\partial\bar{\partial} \log \det({}^t\bar{Z}Z) = \partial\bar{\partial} \log \det(I + {}^t\bar{T}T)$$

and hence

$$\Phi = -4i\, \partial\bar{\partial} \log \det(I + {}^t\bar{T}T).$$

A direct calculation shows that the form g coincides at $T = 0$ with

$$4 \text{ trace } (d^t\bar{T}\, dT),$$

which implies that g is positive definite at $T = 0$. Since g is invariant by $U(p+q)$ acting transitively on $G_{p,q}(\mathbf{C})$, g is positive definite everywhere. The forms $\tilde{\Phi}$ and Φ and the metric g generalize those given in Example 6.3.

Example 6.5. Let $D_{p,q}$ be the space of $p \times q$ complex matrices Z satisfying $I_p - {}^t\bar{Z}Z > 0$, where I_p is the $p \times p$ identity matrix and the symbol ">0" means "positive definite". Then $D_{p,q}$ is a bounded domain in $\mathbf{C}^{pq}$. Let $U(q, p)$ be the group of $(q + p) \times (q + p)$ matrices

$$\begin{pmatrix} A & B \\ C & D \end{pmatrix} \quad (A: q \times q \text{ matrix}; D: p \times p \text{ matrix})$$

such that

$$\begin{pmatrix} {}^t\bar{A} & {}^t\bar{C} \\ {}^t\bar{B} & {}^t\bar{D} \end{pmatrix} \begin{pmatrix} I_q & 0 \\ 0 & -I_p \end{pmatrix} \begin{pmatrix} A & B \\ C & D \end{pmatrix} = \begin{pmatrix} I_q & 0 \\ 0 & -I_p \end{pmatrix}.$$

If we set
$$W = (AZ + B)(CZ + D)^{-1},$$
then
$$I_p - {}^t\overline{W}W = {}^t(\bar{C}Z + \bar{D})^{-1}(I_p - {}^t\bar{Z}Z)(CZ + D)^{-1},$$

which shows that W is in $D_{p,q}$ if Z is in $D_{p,q}$. It is a simple exercise to show that the group $U(q, p)$ thus acts transitively on $D_{p,q}$. We set

$$\Phi = 4i \, \partial\bar{\partial} \log \det(I_p - {}^t\bar{Z}Z).$$

From the preceding equation it follows that Φ is invariant by the action of $U(q, p)$. In the same way as in Example 6.4 we see that Φ gives rise to a Kaehler metric g which coincides at $Z = 0$ with 4 trace $(d^t\bar{Z} \, dZ)$. In particular, $D_{1,n}$ is the interior of the unit ball in $\mathbf{C}^n$ and the metric g may be written in terms of the coordinate system $z^1, \ldots, z^n$ of $\mathbf{C}^n$ as follows:

$$ds^2 = 4 \frac{\left(1 - \sum_\alpha z^\alpha \bar{z}^\alpha\right)\left(\sum_\alpha dz^\alpha \, d\bar{z}^\alpha\right) - \left(\sum_\alpha \bar{z}^\alpha \, dz^\alpha\right)\left(\sum_\alpha z^\alpha \, d\bar{z}^\alpha\right)}{\left(1 - \sum_\alpha z^\alpha \bar{z}^\alpha\right)^2}.$$

Example 6.6. We shall sketch briefly how the construction of the metric in Example 6.5 may be generalized to the case of an arbitrary bounded domain in $\mathbf{C}^n$. Let M be an n-dimensional complex manifold and H the space of holomorphic n-forms φ (i.e., forms of degree $(n, 0)$ such that $\bar{\partial}\varphi = 0$) such that

$$\int_m i^{n^2} \varphi \wedge \bar{\varphi} < \infty.$$

IX. COMPLEX MANIFOLDS

If we define an inner product in H by

$$(\varphi, \psi) = \int_M i^{n^2} \varphi \wedge \bar{\psi}, \qquad \varphi, \psi \in H,$$

then H is a separable Hilbert space. Taking any orthonormal basis $\varphi_0, \varphi_1, \ldots$ of H, we set

$$K = i^{n^2} \sum_{h=0}^{\infty} \varphi_h \wedge \bar{\varphi}_h.$$

Then K is a form of degree (n, n) and is independent of the choice of orthonormal basis. Taking a complex local coordinate system $z^1, \ldots, z^n$ in M, we write

$$K = i^{n^2} k \, dz^1 \wedge \cdots \wedge dz^n \wedge d\bar{z}^1 \wedge \cdots \wedge d\bar{z}^n,$$

where k is a non-negative function defined in the coordinate neighborhood. Assuming that $K \neq 0$ so that $k > 0$ everywhere, we set

$$ds^2 = 2 \sum g_{\alpha\bar{\beta}} \, dz^\alpha \, d\bar{z}^\beta \quad \text{with } g_{\alpha\bar{\beta}} = \partial^2 \log k / \partial z^\alpha \, \partial \bar{z}^\beta.$$

Then ds^2 is in general positive semi-definite. For any bounded domain in $\mathbf{C}^n$, $K \neq 0$ everywhere and ds^2 is positive definite. Since every holomorphic transformation of M induces a linear transformation of H preserving the inner product (i.e., a unitary transformation of H), it preserves K and ds^2. The Kaehler metric of a bounded domain in $\mathbf{C}^n$ thus constructed is called the *Bergman metric* of M. Let $i^{n^2} G dz^1 \wedge \cdots \wedge dz^n \wedge d\bar{z}^1 \wedge \cdots \wedge d\bar{z}^n$ be the volume element of M with respect to the Kaehler metric constructed above (where G is given by $G = \det(g_{\alpha\bar{\beta}})$; cf. Appendix 6). Since both K and the volume element are forms of degree (n, n) which are invariant by the group of holomorphic (and hence isometric) transformations, they coincide up to a constant multiple if M is homogeneous, i.e., if the group of holomorphic transformations is transitive on M. It follows from (24) of §5 that $K_{\alpha\bar{\beta}} = -g_{\alpha\bar{\beta}}$, i.e., $S = -g$ (where S denotes the Ricci tensor and g denotes the metric tensor) if M is homogeneous. For the domain $D_{p,q}$ in Example 6.5, the function k (with respect to the natural coordinate system) is given by $c\{\det(I_p - {}^t\bar{Z}Z)\}^{-(p+q)}$, where

$$c = \frac{1! \, 2! \cdots (p+q-1)!}{1! \, 2! \cdots (p-1)! \, 1! \, 2! \cdots (q-1)! \, \pi^{pq}}.$$

For the detail of the construction of Bergman metrics, we refer the reader to Bergman [1], Helgason [2; pp. 293–300], Kobayashi [14] and Weil [2; pp. 55–65]. For the domain $D_{p,q}$, see for instance Hua [1; p. 83].

Example 6.7. Let M be a complex analytic submanifold of a Kaehler manifold M', i.e., the immersion $M \to M'$ is holomorphic. Then the Riemannian metric induced on M is Hermitian. It is easy to see that the fundamental 2-form associated with this Hermitian metric is the restriction of the fundamental 2-form of M' and hence is closed. This shows that every complex analytic submanifold M of a Kaehler manifold is also a Kaehler manifold with the induced metric. In particular, every closed complex analytic submanifold of the complex projective space $P_N(\mathbf{C})$ is a Kaehler manifold. It is a celebrated theorem of Chow that such a manifold is always algebraic, i.e., is given as the zeros of homogeneous polynomials in the homogeneous coordinate system z^0, z^1, ..., z^N of $P_N(\mathbf{C})$. (See Chow [1] and also the anonymous letter in Amer. J. Math. 78 (1956), p. 898 for a simpler proof in the non-singular case.)

Example 6.8. Let M be a 1-dimensional complex manifold. Then every Hermitian metric is a Kaehler metric since every 2-form and, in particular, the fundamental 2-form on a manifold of real dimension 2 is closed. From Theorem 2.5 and Example 2.8 we see that every Riemannian metric on an oriented 2-dimensional manifold is a Kaehler metric with respect to the naturally induced complex structure.

Example 6.9. We shall now give examples of complex manifolds which do not admit any Kaehler metric. For this purpose we shall first show that the even-dimensional Betti numbers of a compact Kaehler manifold M are all positive. If Φ is the fundamental 2-form, then $\Phi^k = \Phi \wedge \cdots \wedge \Phi$ (k times) is a closed $2k$-form and defines an element of the $2k$-th cohomology group $H^{2k}(M; \mathbf{R})$ in the de Rham theory. Since Φ^n, $n = \dim M$, coincides with the volume element up to a non-zero constant multiple, its integral over M is different from zero and hence it gives a non-zero element of $H^{2n}(M; \mathbf{R})$. It follows that each Φ^k, $1 \leq k \leq n$, defines a non-zero element of $H^{2k}(M; \mathbf{R})$, thus proving our assertion. The complex manifold $M^{p,q}_{r_1 r_2}$ defined in Example 2.5 is diffeomorphic with $S^{2p+1} \times S^{2q+1}$ but cannot carry any Kaehler metric except for $p = q = 0$.

Example 6.10. Let M be an arbitrary manifold of dimension n and $T^*(M)$ the space of covectors of M. We define a 1-form ω on $T^*(M)$ as follows. If X is a tangent vector of $T^*(M)$ at $\xi \in T^*(M)$, then we set $\omega(X) = \langle \xi, \pi(X) \rangle$, where $\pi: T^*(M) \to M$ is the projection. We show that the closed 2-form $d\omega$ defines a symplectic structure on $T^*(M)$ (cf. §4). If $x^1, \ldots, x^n$ is a local coordinate system in M, then we introduce a local coordinate system $x^1, \ldots, x^n, p^1, \ldots, p^n$ in $T^*(M)$ by setting

$$\xi = p^1(\xi)\, dx^1 + \cdots + p^n(\xi)\, dx^n \quad \text{for } \xi \in T^*(M).$$

It follows immediately that

$$\omega = p^1\, dx^1 + \cdots + p^n\, dx^n$$

in terms of this local coordinate system. Hence, $d\omega = dp^1 \wedge dx^1 + \cdots + dp^n \wedge dx^n$ and $(d\omega)^n \neq 0$, thus proving our assertion.

7. *Holomorphic sectional curvature*

We begin with the study of algebraic properties of the Riemannian curvature tensor of a Kaehler manifold.

Let V be a $2n$-dimensional real vector space with a complex structure J as in §1. Consider a quadrilinear mapping

$$R: V \times V \times V \times V \to \mathbf{R}$$

satisfying the following four conditions:

(a) $R(X, Y, Z, W) = -R(Y, X, Z, W) = -R(X, Y, W, Z)$
(b) $R(X, Y, Z, W) = R(Z, W, X, Y)$
(c) $R(X, Y, Z, W) + R(X, Z, W, Y) + R(X, W, Y, Z) = 0$
(d) $R(JX, JY, Z, W) = R(X, Y, JZ, JW) = R(X, Y, Z, W)$.

We know (cf. Proposition 1.1 of Chapter V) that (b) is a consequence of (a) and (c) and (cf. Proposition 2.1 of Chapter V) that the Riemannian curvature tensor satisfies (a), (b), and (c) at each point of the manifold. In addition to (a), (b), and (c), the Riemannian curvature of a Kaehler manifold satisfies (d). In fact, the second equality in (d) is nothing but the second equality of (1) in Proposition 4.5. The equality $R(JX, JY, Z, W) = R(X, Y, Z, W)$ follows either from (b) and the second equality of (d) or from the first equality of (1) in Proposition 4.5.

Sharpening Proposition 1.2 of Chapter V in the present situation, we have

PROPOSITION 7.1. *Let R and T be two quadrilinear mappings satisfying the above conditions* (a), (b), (c), *and* (d). *If*
$$R(X, JX, X, JX) = T(X, JX, X, JX) \quad \text{for all } X \in V,$$
then $R = T$.

Proof. We may assume that $T = 0$; consider $R - T$ and 0 instead of R and T. Consider the quadrilinear mapping which sends $(X, Y, Z, W) \in V \times V \times V \times V$ into
$$R(X, JY, Z, JW) + R(X, JZ, Y, JW) + R(X, JW, Y, JZ).$$
This quadrilinear mapping is symmetric in X, Y, Z and W by (a), (b), and (d). Since it vanishes for $X = Y = Z = W$ by the assumption that $T = 0$, it must vanish identically. Setting $X = Z$ and $Y = W$, we obtain

(1) $2R(X, JY, X, JY) + R(X, JX, Y, JY) = 0$.

On the other hand, by (c) we have
$$R(X, JX, Y, JY) + R(X, Y, JY, JX) + R(X, JY, JX, Y) = 0,$$
which is, by (a) and (d), equivalent to
$$R(X, JX, Y, JY) - R(X, Y, X, Y) - R(X, JY, X, JY) = 0.$$
This, combined with (1), yields

(2) $3R(X, JY, X, JY) + R(X, Y, X, Y) = 0$.

Replacing Y by JY in (2), we obtain

(3) $3R(X, Y, X, Y) + R(X, JY, X, JY) = 0$.

From (2) and (3) we obtain
$$R(X, Y, X, Y) = 0.$$
By Proposition 1.2 of Chapter V, we have $R = 0$. QED.

Besides a quadrilinear mapping R, we consider a Hermitian inner product g in V as in §1. We set
$$\begin{aligned}R_0(X, Y, Z, W) = \tfrac{1}{4}\{ & g(X, Z)g(Y, W) - g(X, W)g(Y, Z) \\ & + g(X, JZ)g(Y, JW) - g(X, JW)g(Y, JZ) \\ & + 2g(X, JY)g(Z, JW)\}.\end{aligned}$$

The verification of the following proposition is trivial.

PROPOSITION 7.2. *The quadrilinear mapping R_0 satisfies* (a), (b), (c), (d), *and the following relations:*

$$R_0(X, Y, X, Y) = \tfrac{1}{4}\{g(X, X)g(Y, Y) - g(X, Y)^2 + 3g(X, JY)^2\},$$
$$R_0(X, JX, X, JX) = g(X, X)^2.$$

Let p be a plane, that is, a 2-dimensional subspace, in V and let X, Y be an orthonormal basis for p. We know (cf. §1 of Chapter V) that $K(p)$ which is defined by

$$K(p) = R(X, Y, X, Y)$$

depends only on p and is independent of the choice of an orthonormal basis for p.

PROPOSITION 7.3. *Let R be a quadrilinear mapping satisfying* (a), (b), (c), *and* (d). *If $K(p) = c$ for all planes p which are invariant by J, then $R = cR_0$.*

Proof. A plane p is invariant by J if and only if X, JX is an orthonormal basis for p for any unit vector X in p. Our assumption is therefore equivalent to

$$R(X, JX, X, JX) = c \quad \text{for all unit vector } X \text{ in } V.$$

It follows that $R(X, JX, X, JX) = cR_0(X, JX, X, JX)$ for all vectors $X \in V$ by Proposition 7.2. Applying Proposition 7.1, we may conclude that $R = cR_0$. QED.

Given two planes p and p' in V, consider the angle $\widehat{XX'}$ between a vector X in p and a vector X' in p'. The infimum of $\widehat{XX'}$ over X and X' is called the angle between p and p'. If we denote by $\alpha(p)$ the angle between p and $J(p)$, then by analytic geometry we see that

$$\cos \alpha(p) = |g(X, JY)|,$$

where X, Y is any orthonormal basis for p. From Proposition 7.2 we obtain

PROPOSITION 7.4. *If X, Y is an orthonormal basis for a plane p and if we set $K_0(p) = R_0(X, Y, X, Y)$, then*

$$K_0(p) = \tfrac{1}{4}(1 + 3 \cos^2 \alpha(p)).$$

We shall now apply these algebraic results to the Riemannian curvature tensor R of a Kaehler manifold M with metric g. For each plane p in the tangent space $T_x(M)$, the sectional curvature $K(p)$ was defined in §2 of Chapter V by $K(p) = R(X, Y, X, Y)$, where X, Y is an orthonormal basis for p. If p is invariant by the complex structure J, then $K(p)$ is called the *holomorphic sectional curvature* by p. If p is invariant by J and X is a unit vector in p, then X, JX is an orthonormal basis for p and hence $K(p) = R(X, JX, X, JX)$. Proposition 7.1 implies that the holomorphic sectional curvatures $K(p)$ for all p in $T_x(M)$ invariant by J determine the Riemannian curvature tensor R at x.

If $K(p)$ is a constant for all planes p in $T_x(M)$ invariant by J and for all points $x \in M$, then M is called a *space of constant holomorphic sectional curvature*. The following is a Kaehlerian analogue of Schur's Theorem (Theorem 2.2 of Chapter V).

THEOREM 7.5. *Let M be a connected Kaehler manifold of complex dimension $n \geq 2$. If the holomorphic sectional curvature $K(p)$, where p is a plane in $T_x(M)$ invariant by J, depends only on x, then M is a space of constant holomorphic sectional curvature.*

Proof. Although it is possible to give a proof similar to that of Theorem 2.2 of Chapter V, we shall appeal here to Theorem 1 of Note 3, which states that if M is a Riemannian manifold of dimension ≥ 3 with metric g and the Ricci tensor S such that $S = \lambda g$ where λ is a function on M, then λ is a constant. We define a tensor field R_0 on M using g and J as above (before Proposition 7.2). By Proposition 7.3, we have $R = \lambda R_0$, where λ is a function on M. By a direct calculation using the definition of R_0, we see that the Ricci tensor S is given by

$$S = \tfrac{1}{2}(n+1)\lambda g.$$

Applying Theorem 1 of Note 3 mentioned above, we may conclude that λ is a constant. QED.

Remark. In the course of the proof we established that if M is a Kaehler manifold of constant holomorphic sectional curvature c, then its Ricci tensor S is given by $S = \tfrac{1}{2}(n+1)cg$, where n is the complex dimension of M.

We have seen also that $R = cR_0$ holds for a space of constant holomorphic sectional curvature c. We shall now express this equality in terms of a complex local coordinate system $z^1, \ldots, z^n$ of M. Among the components K_{ABCD} of the Riemannian curvature defined by (14) of §5, the only components which can be different from zero are $K_{\alpha\bar{\beta}\gamma\bar{\delta}}$ and those which differ from $K_{\alpha\bar{\beta}\gamma\bar{\delta}}$ by complex conjugation and sign. With our notations in §5, we have

PROPOSITION 7.6. *A Kaehler manifold M is a space of constant holomorphic sectional curvature c if and only if*

$$K_{\alpha\bar{\beta}\gamma\bar{\delta}} = -\tfrac{1}{2}c(g_{\alpha\bar{\beta}}\, g_{\gamma\bar{\delta}} + g_{\alpha\bar{\delta}}\, g_{\beta\bar{\gamma}}).$$

Proof. We calculate $K_{\alpha\bar{\beta}\gamma\bar{\delta}} = cR_0(Z_\alpha, Z_{\bar{\beta}}, Z_\gamma, Z_{\bar{\delta}})$ using the definition of R_0, (3) of §5 and the fact that $JZ_\alpha = iZ_\alpha$ and $JZ_{\bar{\alpha}} = -iZ_{\bar{\alpha}}$ (cf. §2). QED.

With our notations in §4, we have

PROPOSITION 7.7. *If M is a Kaehler manifold of constant holomorphic sectional curvature c, then the curvature form (Ψ^α_β) is given by*

$$\Psi^\alpha_\beta = \tfrac{1}{2}c(\varphi^\alpha \wedge \bar{\varphi}^\beta + \delta_{\alpha\beta}\Sigma_\gamma\, \varphi^\gamma \wedge \bar{\varphi}^\gamma) \quad \text{on } U(M).$$

Since the proof is similar to that of Proposition 2.4 of Chapter V, it is left to the reader.

We shall now establish the existence of a space of constant holomorphic sectional curvature.

THEOREM 7.8. (1) *For any positive number c, the complex projective space $P_n(\mathbf{C})$ carries a Kaehler metric of constant holomorphic sectional curvature c. With respect to an inhomogeneous coordinate system $z^1, \ldots, z^n$ it is given by*

$$ds^2 = \frac{4}{c} \cdot \frac{(1 + \sum z^\alpha \bar{z}^\alpha)(\sum dz^\alpha\, d\bar{z}^\alpha) - (\sum \bar{z}^\alpha\, dz^\alpha)(\sum z^\alpha\, d\bar{z}^\alpha)}{(1 + \sum z^\alpha \bar{z}^\alpha)^2}.$$

(2) *For any negative number c, the open unit ball $D_n = \{(z^1, \ldots, z^n);\ \Sigma z^\alpha \bar{z}^\alpha < 1\}$ in $\mathbf{C}^n$ carries a complete Kaehler metric of constant holomorphic sectional curvature c. With respect to the coordinate system $z^1, \ldots, z^n$ of $\mathbf{C}^n$ it is given by*

$$ds^2 = -\frac{4}{c} \cdot \frac{(1 - \sum z^\alpha \bar{z}^\alpha)(\sum dz^\alpha\, d\bar{z}^\alpha) - (\sum \bar{z}^\alpha\, dz^\alpha)(\sum z^\alpha\, d\bar{z}^\alpha)}{(1 - \sum z^\alpha \bar{z}^\alpha)^2}.$$

Proof. By 4 of §5, $(g_{\alpha\bar{\beta}})$ is given by

$$\tfrac{1}{2}c(1 \pm \sum z^\gamma \bar{z}^\gamma)^2 g_{\alpha\bar{\beta}} = (1 \pm \sum z^\gamma \bar{z}^\gamma)\, \delta_{\alpha\beta} - \bar{z}^\alpha z^\beta,$$

where the plus sign is taken in the case (1) and the minus sign in the case (2). We differentiate this identity with respect to $\partial/\partial z^\gamma$ and $\partial^2/\partial z^\gamma \, \partial \bar{z}^\delta$ and set $z^1 = \cdots = z^n = 0$. Then we obtain

$$g_{\alpha\beta} = \frac{2}{c} \delta_{\alpha\beta}, \qquad \partial g_{\alpha\beta}/\partial z^\gamma = 0,$$

$$\partial^2 g_{\alpha\beta}/\partial z^\gamma \, \partial \bar{z}^\delta = -\frac{2}{c} (\delta_{\alpha\beta} \, \delta_{\gamma\delta} + \delta_{\alpha\delta} \, \delta_{\beta\gamma})$$

at the origin. Applying (21) of §5 and Proposition 7.6, we see that the metric ds^2 is of constant holomorphic sectional curvature c at the origin $z^1 = \cdots = z^n = 0$. Since we know (cf. Example 6.3 and Example 6.4) that both $P_n(\mathbf{C})$ and D_n admit a transitive group of holomorphic isometric transformations, we may conclude that ds^2 is of constant holomorphic sectional curvature c everywhere and is complete (by Theorem 4.5 of Chapter IV).
QED.

Since both g and J are parallel tensor fields (cf. Corollary 4.4) so is the tensor field R_0 constructed before Proposition 7.2. It follows that the curvature tensor of a Kaehler manifold of constant holomorphic sectional curvature is also parallel. We are now in a position to prove a Kaehlerian analogue of Theorem 7.10 of Chapter VI.

THEOREM 7.9. *Any two simply connected complete Kaehler manifolds of constant holomorphic sectional curvature c are holomorphically isometric to each other.*

Proof. Let M and M' be two such manifolds and choose a point o in M and a point o' in M'. Then any linear isomorphism $F: T_o(M) \to T_{o'}(M')$ preserving both the metric and the almost complex structure necessarily maps the curvature tensor at o into the curvature tensor at o' (cf. Proposition 7.3). Then by Theorem 7.8 of Chapter VI, there exists a unique affine isomorphism f of M such that $f(o) = o'$ and that the differential of f at o is F. Let x be any point of M and τ a curve from o to x. We set $x' = f(x)$ and $\tau' = f(\tau)$. Since the parallel displacement along τ' corresponds to that along τ under f and since the metric tensors and the almost complex structures of M and M' are all parallel, the affine isomorphism f maps the metric tensor and the almost complex structure of M into those of M'. By Proposition 2.3, f is holomorphic.
QED.

Theorem 7.9 together with Theorem 7.8 implies that a simply connected complete Kaehler manifold of constant holomorphic sectional curvature c can be identified with the complex projective space $P_n(\mathbf{C})$, the open unit ball D_n in $\mathbf{C}^n$ or $\mathbf{C}^n$ according as $c > 0$, $c < 0$ or $c = 0$. This has been proved independently by Hawley [1] and Igusa [1].

We remark here that a complete Kaehler manifold of positive constant holomorphic sectional curvature is necessarily simply connected (cf. Synge [2], Kobayashi [17], Note 23).

8. De Rham decomposition of Kaehler manifolds

According to the decomposition theorem of de Rham (Theorem 6.2 of Chapter IV), a simply connected, complete Kaehler manifold M is isometric to the direct product $M_0 \times M_1 \times \cdots \times M_k$, where M_0 is a Euclidean space and $M_1, \ldots, M_k$ are all simply connected, complete, irreducible Riemannian manifolds and, moreover, such a decomposition is unique. We shall show in this section that $M_0, M_1, \ldots, M_k$ are all Kaehler manifolds and M is holomorphically isometric to the product $M_0 \times M_1 \times \cdots \times M_k$.

We begin with the following lemma.

LEMMA. *Let M be a complex manifold and M' an almost complex manifold. If there exists an almost complex immersion $f\colon M' \to M$, then M' is also a complex manifold and f is holomorphic.*

Proof. Let $2n$ and $2n'$ be the real dimensions of M and M' respectively. Let (U, φ) be a chart of M, where U is an open subset of M and φ is a homeomorphism of U onto an open subset of $\mathbf{C}^n$. Since f is an immersion, there exist a linear mapping p of $\mathbf{C}^n$ onto $\mathbf{C}^{n'}$ and an open subset U' of M' such that $f(U') \subset U$ and that the mapping $\varphi'\colon U' \to \mathbf{C}^{n'}$ defined by $\varphi' = p \circ \varphi \circ f$, is a diffeomorphism of U' onto an open subset $\varphi'(U')$ of $\mathbf{C}^{n'}$. We shall show that the atlas consisting of these (U', φ') defines the structure of a complex manifold on M'. Since p, φ, and f are almost complex mappings, φ' is also almost complex. Let (V', ψ') be constructed from a chart (V, ψ) of M as above. Then the mapping $\psi' \circ \varphi'^{-1}\colon \varphi'(U' \cap V') \to \psi'(U' \cap V')$ is almost complex. By Proposition 2.3, $\psi' \circ \varphi'^{-1}$ is holomorphic, thus showing that

172   FOUNDATIONS OF DIFFERENTIAL GEOMETRY

the family of (U', φ') is an atlas defining a complex structure on M'. Since φ' is an almost complex mapping of U' into $\mathbf{C}^{n'}$, the complex structure thus constructed is compatible with the almost complex structure originally given on M'. The assertion that f is holomorphic follows also from Proposition 2.3.   QED.

THEOREM 8.1. *Let $M_0 \times M_1 \times \cdots \times M_k$ be the de Rham decomposition of a simply connected, complete Kaehler manifold M. Then $M_0, M_1, \ldots, M_k$ are all Kaehler manifolds in a natural manner and the isometry between M and $M_0 \times M_1 \times \cdots \times M_k$ is holomorphic.*

Proof. We fix a point o and let $T_o(M) = \Sigma_{i=0}^k T_o^{(i)}$ be the canonical decomposition of $T_o(M)$; it is unique up to an order (cf. Theorem 5.4 of Chapter IV). Since the almost complex structure J of M is parallel, it commutes with the linear holonomy group $\Psi(o)$ of M at o. In particular, $\Psi(o)$ acts trivially on $J(T_o^{(0)})$ and hence $J(T_o^{(0)}) = T_o^{(0)}$. Let T_o' be the orthogonal complement of $T_o^{(0)}$ in $T_o(M)$. It is invariant by $\Psi(o)$ and $T_o' = \Sigma_{i=1}^k T_o^{(i)}$ is a direct sum decomposition into mutually orthogonal, invariant, and irreducible subspaces. Since J leaves the metric invariant and is parallel, it follows that $J(T_o') = T_o'$ and $T_o' = \Sigma_{i=1}^k J(T_o^{(i)})$ is also a direct sum decomposition into mutually orthogonal, invariant, and irreducible subspaces. Hence, $J(T_o^{(1)}), \ldots, J(T_o^{(k)})$ coincide with $T_o^{(1)}, \ldots, T_o^{(k)}$ up to an order. To show that they actually coincide including the order, we recall (cf. Theorem 5.4 of Chapter IV) that $\Psi(o)$ is the direct product $\Psi_0(o) \times \Psi_1(o) \times \cdots \times \Psi_k(o)$ of normal subgroups, where $\Psi_i(o)$ is trivial on $T_o^{(j)}$ if $i \neq j$ and is irreducible on $T_o^{(i)}$ for each $i = 1, \ldots, k$, and $\Psi^0(o)$ consists of the identity only. Assume now $J(T_o^{(i)}) = T_o^{(j)}$ with $i \neq j$. Choose a non-zero element $X \in T_o^{(i)}$ and let a be an element of $\Psi_i(o)$ such that $a(X) \neq X$. Since $J(X) \in T_o^{(j)}$ and a acts trivially on $T_o^{(j)}$, we have

$$J(X) = a(J(X)) = J(a(X)) \neq J(X),$$

which is a contradiction. Hence, $J(T_o^{(i)}) = T_o^{(i)}$ for all i. Since J is parallel, the involutive distribution $T^{(i)}$ obtained by parallel displacement of $T_o^{(i)}$ is invariant by J for each $i = 0, 1, \ldots, k$. Since M_i is the maximal integral submanifold of the distribution $T^{(i)}$ through o, the restriction of J to $T^{(i)}$ defines an almost complex structure J_i on M_i. By Lemma, J_i comes from a complex

IX. COMPLEX MANIFOLDS 173

structure and the imbedding $M_i \to M$ is holomorphic. It is straightforward to verify that the induced Riemannian metric on M_i is Hermitian with respect to J_i, and that J_i is parallel with respect to the Riemannian connection. By Corollary 4.4, M_i is a Kaehler manifold. QED.

Remark. In Theorem 8.1, M_0 is therefore isometric with a complex Euclidean space $\mathbf{C}^{n_0}$. The original proof of Theorem 8.1 by Hano and Matsushima [1] relied on Theorem 2.5.

We shall say that a Kaehler manifold M is *non-degenerate* if the restricted linear holonomy group $\Psi^0(x)$ contains the endomorphism J_x of $T_x(M)$ given by the almost complex structure J for an arbitrarily chosen reference point x. Note that the condition $J_x \in \Psi^0(x)$ is independent of the choice of a reference point x. For any point y, let τ denote the parallel displacement: $T_x(M) \to T_y(M)$ along an arbitrary curve from x to y. Since J is parallel, we have $J_y = \tau \circ J_x \circ \tau^{-1}$. On the other hand (cf. §4 of Chapter II), we have $\Psi^0(y) = \tau \circ \Psi^0(x) \circ \tau^{-1}$.

COROLLARY 8.2. *In Theorem 8.1, if M is non-degenerate, then M_0 does not appear (i.e., reduces to a point) and $M_1, \ldots, M_k$ are all non-degenerate.*

Proof. This is immediate from (3) of Theorem 5.4 of Chapter IV. QED.

The following theorem shows the relationship of the notion of non-degeneracy to the Ricci curvature.

THEOREM 8.3. *Let M be a Kaehler manifold of complex dimension n.*

(1) *If the restricted linear holonomy group of M is irreducible and the Ricci tensor is not zero, then M is non-degenerate.*

(2) *If M is non-degenerate and n is not divisible by 4, then the Ricci tensor is not zero;*

(3) *If the Ricci tensor is non-degenerate at some point of M, then M is non-degenerate.*

Proof. (1) Since the Ricci tensor is not zero, the restricted linear holonomy group $\Psi^0(x)$ is a subgroup of $U(n)$ not contained in $SU(n)$ by Theorem 4.6. Since $U(n)/SU(n)$ is of dimension 1, the natural homomorphism $U(n) \to U(n)/SU(n)$ maps $\Psi^0(x)$ onto $U(n)/SU(n)$. The kernel $SU(n) \cap \Psi^0(x)$ of the homomorphism

$\Psi^0(x) \to U(n)/SU(n)$ is therefore of codimension 1 in $\Psi^0(x)$. Let $\mathfrak{g}$ and $\mathfrak{s}$ be the Lie algebras of $\Psi^0(x)$ and $SU(n) \cap \Psi^0(x)$ respectively. Since $\Psi^0(x)$ is closed in $U(n)$ (cf. Theorem 5.5 of Chapter IV) and hence is compact, its Lie algebra $\mathfrak{g}$ is reductive (cf. Appendix 9). It follows that there exists an ideal $\mathfrak{c}$ complementary to $\mathfrak{s}$ in $\mathfrak{g}$. Hence,

$$\mathfrak{g} = \mathfrak{s} + \mathfrak{c}, \quad [\mathfrak{s}, \mathfrak{c}] = 0.$$

Since $\dim \mathfrak{c} = \operatorname{codim} \mathfrak{s} = 1$, we have

$$[\mathfrak{c}, \mathfrak{c}] = 0,$$

showing that $\mathfrak{c}$ is in the center of $\mathfrak{g}$. We have therefore shown that $\Psi^0(x)$ has a non-discrete center. Since $\Psi^0(x)$ is irreducible on $T_x(M) = \mathbf{R}^{2n}$ and hence on $T_x^{(1,0)} = \mathbf{C}^n$, we can apply Schur's lemma and see that its center is of dimension 1 and consists of elements $\lambda I_n \in U(n)$, where λ are complex numbers of absolute value 1. In particular, the center contains the transformation $\sqrt{-1}\, I_n$. In the real representation, this means that the center of $\Psi^0(x)$ contains J_x, that is, M is non-degenerate.

(2) If we consider $\Psi^0(x)$ as a subgroup of $U(n)$, then non-degeneracy of M means that $\Psi^0(x)$ contains the transformation $\sqrt{-1}\, I_n$ whose determinant, the n-th power of $\sqrt{-1}$, is not equal to 1 since n is not divisible by 4. Hence $\Psi^0(x)$ is not contained in $SU(n)$ and, by Theorem 4.6, the Ricci tensor is not zero.

(3) Let x be a point of M where the Ricci tensor is non-degenerate (as a symmetric bilinear form on $T_x(M)$). Let U be a properly chosen neighborhood of x which is holomorphically isometric with the direct product $U_0 \times U_1 \times \cdots \times U_k$, where U_0 is locally a complex Euclidean space and $U_1, \ldots, U_k$ are irreducible Kaehler manifolds. To see the existence of such a neighborhood U, we use Theorem 5.4 of Chapter IV and choose a neighborhood V which is isometric to the direct product $V_0 \times V_1 \times \cdots \times V_h$, where $V_0, V_1, \ldots, V_h$ are constructed from the canonical decomposition of $T_x(M)$ and V_0 is locally Euclidean. We apply Theorem 5.4 of Chapter IV in the same way to those which are reducible among $V_1, \ldots, V_h$. Repeating this process we obtain eventually a neighborhood U of x which is isometric with $U_0 \times U_1 \times \cdots \times U_k$ where U_0 is locally Euclidean and $U_1, \ldots, U_k$ are irreducible Riemannian manifolds. By the same

argument as in the proof of Theorem 8.1, we see that U_0 is locally a complex Euclidean space and $U_1, \ldots, U_k$ are irreducible Kaehler manifolds and U is holomorphically isometric with $U_0 \times U_1 \times \cdots \times U_k$. We may consider $U_0, U_1, \ldots, U_k$ as submanifolds of U passing through x. Then the Ricci tensor of U at x is the direct sum of the Ricci tensors of $U_0, U_1, \ldots, U_k$. Therefore the Ricci tensor of each U_i must be non-degenerate at x. It follows that U_0 does not appear in the decomposition and that $U_1, \ldots, U_k$ are non-degenerate by (1). Since the holonomy group $\Psi^0(x)$ of M contains the restricted linear holonomy group of U at x which is the direct product of those of $U_1, \ldots, U_k$, we see that $\Psi^0(x)$ contains J_x. QED.

Theorem 8.3 is found in Kobayashi-Nomizu [1].

Remark. Under the assumptions of (1) or (3) in Theorem 8.3 the Lie algebra of $\Psi'^0(x)$ also contains J. It is not known if irreducibility of $\Psi'_0(x)$ implies non-degeneracy of M. In fact, we do not know of any irreducible Kaehler manifold with vanishing Ricci tensor.

9. *Curvature of Kaehler submanifolds*

Let M' be a Kaehler manifold of complex dimension $n + p$ with complex structure J and Kaehler metric g and M a complex submanifold of M' of complex dimension n. Then M is a Kaehler manifold with the induced complex structure J and the induced metric g (cf. Example 6.7). This fact can be proved anew together with

PROPOSITION 9.1. *Let α be the second fundamental form of a complex submanifold M of a Kaehler manifold M'. Then*

$$\alpha(JX, Y) = \alpha(X, JY) = J(\alpha(X, Y))$$

for all vector fields X and Y on M.

Proof. For the Kaehlerian connection ∇' of M' and for vector fields X and Y on M, we write $\nabla'_X Y = \nabla_X Y + \alpha(X, Y)$ as in §3 of Chapter VII, where $\nabla_X Y$ is the tangential component and $\alpha(X, Y)$ is the normal component. We know (Proposition 3.1 of Chapter VII) that $\nabla_X Y$ is the Riemannian connection of M. We have

$$\nabla'_X(JY) = \nabla_X(JY) + \alpha(X, JY),$$

176   FOUNDATIONS OF DIFFERENTIAL GEOMETRY

where JY is also a vector field on M. Since ∇' is Kaehlerian, we have

$$\nabla'_X(JY) = J(\nabla'_X Y) = J(\nabla_X Y) + J(\alpha(X, Y)).$$

Both the tangent space $T_x(M)$ and the normal space $T_x(M)$ being invariant by J, we obtain

$$\nabla_X(JY) = J(\nabla_X Y) \quad \text{and} \quad \alpha(X, JY) = J(\alpha(X, Y)).$$

The first identity shows once more that ∇ is Kaehlerian. From the second identity and from symmetry of $\alpha(X, Y)$ in X and Y, we have

$$\alpha(JX, Y) = \alpha(Y, JX) = J(\alpha(Y, X)) = J(\alpha(X, Y)).$$

QED.

The following proposition is immediate from Proposition 4.1 (equation of Gauss) of Chapter VII and Proposition 9.1.

PROPOSITION 9.2. *Let M be a complex submanifold of a Kaehler manifold M' with second fundamental form α. Let R and R' be the Riemannian curvature tensor fields of M and M' respectively. Then*

$$R(X, JX, X, JX) = R'(X, JX, X, JX) - 2g(\alpha(X, X), \alpha(X, X))$$

for all vector fields X on M.

We see from Proposition 9.2 that the holomorphic sectional curvature of M does not exceed that of the ambient space M'. In particular, we have

PROPOSITION 9.3. *Let M' be a Kaehler manifold with non-positive holomorphic sectional curvature. Then every complex submanifold M of M' has non-positive holomorphic sectional curvature.*

Making use of Propositions 9.1 and 9.2 we can relate the Ricci tensor S of M with Ricci tensor S' of M'. Let $X \in T_x(M)$ and choose $V_1, \ldots, V_{n+p} \in T_x(M')$ in such a way that $V_1, \ldots, V_n$ are tangent to M and that $V_1, \ldots, V_{n+p}, JV_1, \ldots, JV_{n+p}$ form an orthonormal basis for $T_x(M')$. Then (cf. p. 249 of Volume I)

$$S(X, X) = \sum_{i=1}^{n} (R(V_i, X, V_i, X) + R(JV_i, X, JV_i, X)),$$

$$S'(X, X) = \sum_{i=1}^{n+p} (R'(V_i, X, V_i, X) + R'(JV_i, X, JV_i, X)).$$

From Propositions 9.1 and 9.2 we obtain easily the following formula:

$$S(X, X) = \sum_{i=1}^{n} (R'(V_i, X, V_i, X) + R'(JV_i, X, JV_i, X))$$
$$- 2 \sum_{i=1}^{n} g(\alpha(V_i, X), \alpha(V_i, X)).$$

Hence,

$$S(X, X) = S'(X, X) - \sum_{i=n+1}^{n+p} \{R'(V_i, X, V_i, X)$$
$$+ R'(JV_i, X, JV_i, X)\} - 2 \sum_{i=1}^{n} g(\alpha(V_i, X), \alpha(V_i, X)).$$

The following two propositions are now immediate. (For the proof of Proposition 9.5, consult Proposition 7.2.)

PROPOSITION 9.4. *Let M' be a Kaehler manifold of zero curvature Then every complex submanifold M of M' has negative semi-definite Ricci tensor S, i.e.,*

$$S(X, X) \leq 0 \quad \text{for all vector fields } X \text{ on } M.$$

PROPOSITION 9.5. *Let M' be a Kaehler manifold of constant holomorphic sectional curvature c. Then the Ricci tensor S of an n-dimensional complex submanifold M of M' is given by*

$$S(X, X) = \tfrac{1}{2} c(n + 1) g(X, X) - 2 \sum_{i=1}^{n} g(\alpha(V_i, X), \alpha(V_i, X))$$

for $X \in T_x(M)$, where $V_1, \ldots, V_n \in T_x(M)$ are chosen in such a way that $V_1, \ldots, V_n, JV_1, \ldots, JV_n$ form an orthonormal basis for $T_x(M)$.

Proposition 9.2 is due to Bochner [2]. See also O'Neill [2].

Remark. According to Remmert, every Stein manifold M of complex dimension n can be imbedded as a closed complex submanifold of $\mathbf{C}^{2n+1}$ (for the proof as well as the concept of Stein manifold, see Gunning and Rossi [1; pp. 219–226], E. Bishop [1], and Narasimhan [1]). Using Proposition 4.9 of Chapter IV and Propositions 9.3 and 9.5 above, we see that every Stein manifold admits a complete Kaehler metric with non-positive holomorphic sectional curvature and negative semi-definite Ricci tensor.

10. Hermitian connections in Hermitian vector bundles

A holomorphic vector bundle over a complex manifold with Hermitian fibre metric is called a *Hermitian vector bundle*. We shall generalize some of the basic formulas in Kaehlerian manifolds to Hermitian vector bundles.

Let M be an n-dimensional complex manifold and E a Hermitian vector bundle over M with fibre $\mathbf{C}^r$ and fibre metric h. Then there are two principal fibre bundles naturally associated to E. Let P be the principal fibre bundle with group $GL(r; \mathbf{C})$ associated to E. It is a holomorphic principal bundle. Let Q be the subbundle of P with group $U(r)$ defined by the Hermitian fibre metric h. If we take as an example of E the tangent bundle of a Hermitian manifold M, then P is the bundle $C(M)$ of complex linear frames and Q is the bundle $U(M)$ of unitary frames.

The unique connection in Q (or in P) defined in the following theorem is called the *Hermitian connection* of the Hermitian vector bundle E. In the special case where E is the tangent bundle of a Hermitian manifold M, it is called the Hermitian connection of M.

THEOREM 10.1. *There exists a unique connection in the bundle Q such that, when extended to a connection in P, its horizontal subspaces are complex subspaces of the tangent spaces of P.*

Proof. At each point u of Q, we consider the tangent space $T_u(Q)$ as a real subspace of $T_u(P)$ and set $H_u = T_u(Q) \cap J(T_u(Q))$. Then H_u is invariant by the complex structure J and hence is a complex subspace of $T_u(P)$. We shall show that H_u is the horizontal subspace at u defined by the desired connection.

Since the action of $GL(r; \mathbf{C})$ on P is holomorphic, every element $a \in U(r)$ maps H_u onto H_{ua}.

We shall show that $T_u(Q) = G_u + H_u$, where G_u is the vertical subspace of $T_u(Q)$. The following lemma can be easily verified.

LEMMA. *The Lie algebra $\mathfrak{u}(r)$ is a real form of $\mathfrak{gl}(r; \mathbf{C})$, i.e.,*

$$\mathfrak{gl}(r; \mathbf{C}) = \mathfrak{u}(r) + i\mathfrak{u}(r).$$

IX. COMPLEX MANIFOLDS   179

Since $G_u \cap H_u$ is isomorphic to $\mathfrak{u}(r) \cap i\mathfrak{u}(r) = 0$, it follows that $G_u \cap H_u = 0$. Also from the lemma it follows that $T_u(P) = T_u(Q) + J(T_u(Q))$ (not a direct sum) for every $u \in Q$. Since

$$\dim (T_u(Q) + J(T_u(Q))) + \dim (T_u(Q) \cap J(Q))$$
$$= \dim T_u(Q) + \dim J(T_u(Q))$$

and $H_u = T_u(Q) \cap J(T_u(Q))$, it follows that $\dim H_u = \dim M$. Hence, $T_u(Q) = G_u + H_u$. This proves that the distribution $\{H_u; u \in Q\}$ defines a connection in Q.

We extend this connection in P by setting $H_{ua} = R_a(H_u)$ for $u \in Q$ and $a \in GL(r; \mathbf{C})$, where R_a denotes right translation by a. For $u \in Q$, H_u is invariant by J by definition of H_u. Since R_a is a holomorphic transformation, H_{ua} is also invariant by J.

In order to prove the uniqueness of the connection, let $\{H'_u;$ $u \in Q\}$ be the horizontal distribution defined by another connection satisfying the conditions stated in the theorem. Since $J(H'_u) = H'_u$ and $H'_u \subset T_u(Q)$, we have $H'_u \subset T_u(Q) \cap J(T_u(Q)) = H_u$. Since $\dim H'_u = \dim M = \dim H_u$, we conclude $H'_u = H_u$.

QED.

Theorem 10.1 may be reformulated as follows.

THEOREM 10.1'. *There exists a unique connection form ψ in the bundle Q such that, when extended to a connection form ψ in P, it is of degree $(1, 0)$.*

Proof. We first prove

LEMMA. *If a complex Lie group G acts holomorphically on a complex manifold P to the right, then the homomorphism $A \in \mathfrak{g} \to A^* \in \mathfrak{X}(P)$, defined in Proposition 4.1 of Chapter I, satisfies $(iA)^* = J(A^*)$.*

Proof of Lemma. By assumption, the mapping $(u, a) \in P \times G \to ua \in P$ is holomorphic. For each $u \in P$, the mapping $\sigma_u: a \in G \to ua \in P$ is hence holomorphic so that $(\sigma_u)^*(iA_e) = J((\sigma_u)^*A_e)$ for each $A \in \mathfrak{g}$. Hence $(iA)^*_u = J(A^*_u)$, thus completing the proof of the lemma.

From this lemma applied to the structure group $GL(r; \mathbf{C})$ acting on P, we see that $(iA)^* = J(A^*)$ for each $A \in \mathfrak{gl}(r; \mathbf{C})$, where A^* is the fundamental vector field corresponding to A. Thus the connection form ψ on P satisfies $\psi(JA^*) = iA$. Since $J(H_u) = H_u$, we obtain $\psi(JX) = i\psi(X)$ for every vector field X on P. Thus

180   FOUNDATIONS OF DIFFERENTIAL GEOMETRY

$\psi(X + iJX) = \psi(X) + i\psi(JX) = \psi(X) - \psi(X) = 0$, that is, ψ is of degree $(1, 0)$. QED.

In the special case where M is a Hermitian manifold and Q is the bundle $U(M)$ of unitary frames, it is possible to characterize the Hermitian connection in Q without extending it to a connection in P.

PROPOSITION 10.2. *The Hermitian connection of a Hermitian manifold M is a unique affine connection such that both the metric tensor and the complex structure J are parallel and that the torsion T is pure in the following sense:*

$$T(JX, Y) = T(X, JY) \quad \text{for all vector fields } X, Y \text{ on } M.$$

Proof. An affine connection of M is a connection defined in the bundle $Q = U(M)$ of unitary frames if and only if both the metric tensor and the complex structure J are parallel. Extend such a connection to a connection in the bundle $P = C(M)$ of complex linear frames and write its first structure equation (in its complex form, cf. §5):

$$d\varphi^i = -\sum_{j=1}^{n} \psi^i_j \wedge \varphi^j + \Phi^i, \quad i = 1, \ldots, n.$$

Since the φ^i's are holomorphic, the $d\varphi^i$'s are of degree $(2, 0)$. Hence the ψ^i_j's are of degree $(1, 0)$ if and only if the Φ^i's are of degree $(2, 0)$. This is equivalent to the statement that the connection form (ψ^i_j) is of degree $(1, 0)$ if and only if the torsion tensor T is pure in the sense defined above. QED.

From Proposition 10.2 we see that a Hermitian manifold is a Kaehler manifold if and only if its Hermitian connection is torsion-free.

Returning to the general case, we shall show that the curvature form Ψ on P of the Hermitian connection is of degree $(1, 1)$, that is,

$$\Psi(X, Y) = \Psi(JX, JY) \quad \text{for all vector fields } X, Y \text{ on } P.$$

Since Ψ is a tensorial form of type ad $GL(r; \mathbf{C})$ and the connection is reducible to Q, it suffices to verify the equality $\Psi(X, Y) = \Psi(JX, JY)$ for horizontal vectors X, Y at each point u of Q. Given horizontal vectors X, Y at $u \in Q$, we extend them to horizontal

IX. COMPLEX MANIFOLDS

vector fields defined in a neighborhood of u in P. Then JX and JY are also horizontal. We then have

$$\Psi(X, Y) = d\psi(X, Y) = -\tfrac{1}{2}\psi([X, Y]),$$
$$\Psi(JX, JY) = d\psi(JX, JY) = -\tfrac{1}{2}\psi([JX, JY]).$$

Since the almost complex structure J on P has no torsion, we have (cf. §2)

$$\Psi(X, Y) - \Psi(JX, JY) = \tfrac{1}{2}\psi(J([X, JY] + [JX, Y]))$$
$$= \frac{i}{2}\,\psi([X, JY] + [JX, Y]).$$

Since X, Y, JX, and JY are all horizontal and hence tangent to Q at each point of Q, both $[X, JY]$ and $[JX, Y]$ are also tangent to Q at $u \in Q$. Hence $i\psi([X, JY] + [JX, Y])_u$ belongs to $i\mathfrak{u}(r)$. On the other hand, $\Psi(X, Y)_u - \Psi(JX, JY)_u$ belongs to $\mathfrak{u}(r)$. Since $\mathfrak{u}(r) \cap i\mathfrak{u}(r) = 0$, it follows that $\Psi(X, Y)_u - \Psi(JX, JY)_u = 0$. This proves our assertion.

In the same way as the curvature form of an affine connection defines the curvature tensor (cf. §5 of Chapter III), the curvature form Ψ of the Hermitian connection of a Hermitian vector bundle E gives rise to the curvature "tensor" R in the following sense. For any vector fields X, Y on M, $R(X, Y)$ is a field of linear endomorphisms of the bundle E, that is, at each point $x \in M$, $R(X, Y)_x$ is a complex linear endomorphism of the fibre E_x of E at x. The fact that Ψ is of degree $(1, 1)$ implies the following proposition.

PROPOSITION 10.3. *Let R be the curvature tensor of the Hermitian connection of a Hermitian vector bundle E. Then*

$$R(X, Y) = R(JX, JY) \quad \textit{for all vector fields } X, Y \textit{ on } M.$$

Given a holomorphic vector bundle E over M with fibre $\mathbf{C}^r$ and an integer k with $0 \leq k \leq r$, we define a holomorphic vector bundle $\bigwedge^k E$ over M with fibre $\bigwedge^k \mathbf{C}^r$ as follows. If E_x is the fibre of E at x, then $\bigwedge^k E_x$ is the fibre of $\bigwedge^k E$ at x. The group $GL(r; \mathbf{C})$ acting on $\mathbf{C}^r$ acts also on $\bigwedge^k \mathbf{C}^r$ in a natural manner. But the action of $GL(r; \mathbf{C})$ on $\bigwedge^k \mathbf{C}^r$ is in general not effective. Let N be the normal subgroup of $GL(r; \mathbf{C})$ consisting of elements which act trivially on $\bigwedge^k \mathbf{C}^r$. If $k = r$, then $N = SL(r; \mathbf{C})$. Let $\tilde{P}$ be the principal fibre bundle with group $GL(r; \mathbf{C})/N$ associated to $\bigwedge^k E$. A Hermitian fibre metric in E induces a Hermitian fibre metric in $\bigwedge^k E$; if $s_1, \ldots, s_r$ is an orthonormal basis for E_x, then $\{s_{i_1} \wedge \cdots \wedge s_{i_k}; i_1 < \cdots < i_k\}$ is an orthonormal basis for $\bigwedge^k E_x$. If $k = r$, then $s_1 \wedge \cdots \wedge s_r$ is a basis of $\bigwedge^r E_x$ with length 1.

From now on we shall consider the special case $k = r$. Then $\bigwedge^r E$ is a Hermitian vector bundle with fibre $\bigwedge^r \mathbf{C}^r = \mathbf{C}$ and the associated principal fibre bundle $\tilde{P} = P/SL(r; \mathbf{C})$ has structure group $GL(r; \mathbf{C})/SL(r; \mathbf{C}) = \mathbf{C}^*$. Let ψ be the Hermitian connection form on P. Since ψ is $\mathfrak{gl}(r; \mathbf{C})$-valued, trace ψ is a 1-form on P with values in the Lie algebra $\mathfrak{gl}(r; \mathbf{C})/\mathfrak{sl}(r; \mathbf{C}) = \mathbf{C}$. Let $f\colon P \to \tilde{P} = P/SL(r; \mathbf{C})$ be the natural homomorphism. Then the form $\tilde{\psi}$ on $\tilde{P}$ determined by

$$f^*\tilde{\psi} = \text{trace } \psi$$

defines a connection in $\tilde{P}$ (cf. Proposition 6.1 of Chapter II). Using Theorem 10.1' we shall show that $\tilde{\psi}$ defines the Hermitian connection of the Hermitian vector bundle $\bigwedge^r E$. Since ψ is of degree $(1, 0)$, so is $\tilde{\psi}$. If Q denotes the subbundle with group $U(r)$ of P as before, then $\tilde{Q} = Q/SU(r)$ is the subbundle of $\tilde{P}$ corresponding to the Hermitian fibre metric of $\bigwedge^r E$. Since the restriction of ψ to Q takes its values in $\mathfrak{u}(r)$, the restriction of $\tilde{\psi}$ to $\tilde{Q}$ takes its values in $\mathfrak{u}(r)/\mathfrak{su}(r)$ and defines a connection in $\tilde{Q}$. Hence $\tilde{\psi}$ is the Hermitian connection form for $\bigwedge^r E$. Since the structure group $GL(r; \mathbf{C})/SL(r; \mathbf{C})$ of $\tilde{P}$ is abelian, the structure equation for the Hermitian connection of $\bigwedge^r E$ is given by

$$d\tilde{\psi} = \tilde{\Psi},$$

where $\tilde{\Psi}$ is the curvature form. By Proposition 6.1 of Chapter II, we have

$$f^*\tilde{\Psi} = \text{trace } \Psi.$$

If R is the curvature tensor of the Hermitian connection of E, then the curvature tensor $\tilde{R}$ of the Hermitian connection of $\bigwedge^r E$ is given by

$$\tilde{R}(X, Y) = \text{trace } R(X, Y) \quad \text{for all vector fields } X, Y \text{ on } M,$$

where the right hand side means the trace of the linear endomorphism $R(X, Y)_x: E_x \to E_x$ for each $x \in M$. Since, restricted to Q, Ψ is $\mathfrak{u}(r)$-valued, it follows that trace Ψ, restricted to Q, is purely imaginary. Hence $\tilde{\Psi}$, restricted to $\tilde{Q}$, is also purely imaginary. We can conclude that the curvature tensor $\tilde{R}$ is a 2-form on M whose values are purely imaginary. We set $\rho = -2iR$ and call it the *Ricci form* of the Hermitian vector bundle E. This is a generalization of the Ricci form of a Kaehler manifold (cf. Proposition 4.11).

We shall now express various quantities described above in terms of local coordinates. Let $z^1, \ldots, z^n$ be a local coordinate system for M and we set

$$Z_i = \partial/\partial z^i, \quad Z_{\bar{i}} = \bar{Z}_i = \partial/\partial \bar{z}^i \quad \text{for } i = 1, \ldots, n.$$

Let $s_1, \ldots, s_r$ be holomorphic local cross sections of E which are everywhere linearly independent. Denote by σ the corresponding holomorphic local cross section of the principal bundle P. We set

(1) $\quad \sigma^*(\psi^\alpha_\beta) = \sum_{i=1}^n \Gamma^\alpha_{i\beta} dz^i \quad$ for $\alpha, \beta = 1, \ldots, r$.

(It should be observed that the $\sigma^*(\psi^\alpha_\beta)$'s do not involve $d\bar{z}^i$ since they are of degree $(1, 0)$.) As in Proposition 7.4 of Chapter III we have

(2) $\quad \nabla_{Z_i} s_\beta = \sum_\alpha \Gamma^\alpha_{i\beta} s_\alpha$.

We set

(3) $\quad h_{\alpha\bar{\beta}} = h(s_\alpha, \bar{s}_\beta) \quad$ for $\alpha, \beta = 1, \ldots, r$,

and denote by $(h^{\alpha\bar{\beta}})$ the inverse matrix of $(h_{\alpha\bar{\beta}})$. From (2) and (3) we obtain (cf. the proof of Corollary 2.4 of Chapter IV)

(4) $\quad \Gamma^\alpha_{i\beta} = \sum_\gamma h^{\alpha\bar{\gamma}} \dfrac{\partial h_{\beta\bar{\gamma}}}{\partial z^i} \quad$ for $\alpha, \beta = 1, \ldots, r; i = 1, \ldots, n.$

If we set

(5) $\quad R(Z_i, Z_{\bar{j}}) s_\beta = \sum_\alpha K^\alpha_{\beta i\bar{j}} s_\alpha$,

then

(6) $\quad K^\alpha_{\beta i\bar{j}} = -\dfrac{\partial \Gamma^\alpha_{i\beta}}{\partial \bar{z}^j} = -\sum_\gamma h^{\alpha\bar{\gamma}} \dfrac{\partial^2 h_{\beta\bar{\gamma}}}{\partial z^i \partial \bar{z}^j} + \sum_{\gamma,\delta,\epsilon} h^{\alpha\bar{\delta}} h^{\epsilon\bar{\gamma}} \dfrac{\partial h_{\beta\bar{\gamma}}}{\partial z^i} \dfrac{\partial h_{\epsilon\bar{\delta}}}{\partial \bar{z}^j}.$

The proof of (6) is straightforward from (2) and from
$$R(Z_i, Z_{\bar{j}})s_\beta = ([\nabla_{Z_i}, \nabla_{Z_{\bar{j}}}] - \nabla_{[Z_i, Z_{\bar{j}}]})s_\beta.$$
If we set

(7) $K_{\alpha\bar\beta i\bar j} = h(R(Z_i, Z_{\bar j})\bar s_\beta, s_\alpha),$

then we obtain

(8) $K_{\alpha\bar\beta i\bar j} = \dfrac{\partial^2 h_{\alpha\bar\beta}}{\partial z^i\, \partial \bar z^j} - \sum\limits_{\gamma,\epsilon} h^{\epsilon\bar\gamma} \dfrac{\partial h_{\alpha\bar\gamma}}{\partial z^i} \dfrac{\partial h_{\epsilon\bar\beta}}{\partial \bar z^j}.$

If we set

(9) $K_{i\bar j} = \tilde R(Z_i, Z_{\bar j})$ for $i, j = 1, \ldots, n,$

then

(10) $K_{i\bar j} = \sum\limits_\alpha K^\alpha_{\alpha i \bar j}.$

The Ricci form ρ is given by

(11) $\rho = -2\sqrt{-1} \sum\limits_{i,j=1}^{n} K_{i\bar j}\, dz^i \wedge d\bar z^j.$

If we denote the induced Hermitian fibre metric in $\bigwedge^r E$ by the same letter h and set

(12) $H = h(s_1 \wedge \cdots \wedge s_r, \bar s_1 \wedge \cdots \wedge \bar s_r),$

then

(13) $H = \det(h_{\alpha\bar\beta}).$

To calculate the components of the curvature tensor R of the Hermitian vector bundle $\bigwedge^r E$, we specialize (6) to the case where the fibre dimension is equal to 1. Then we obtain
$$K_{i\bar j} = -\frac{1}{H} \frac{\partial^2 H}{\partial z^i\, \partial \bar z^j} + \frac{1}{H^2} \frac{\partial H}{\partial z^i} \frac{\partial H}{\partial \bar z^j}.$$
Hence,

(14) $K_{i\bar j} = -\dfrac{\partial^2 \log H}{\partial z^i\, \partial \bar z^j}.$

From (14) and (11) we obtain

(15) $\rho = -2\sqrt{-1}\, \bar\partial \partial \log H.$

Remark. For a Hermitian manifold, the Hermitian connection was first defined by Chern [10]. For an interpretation of the Hermitian connection from the viewpoint of the theory of G-structures, see Klingenberg [1]. For a Hermitian vector bundle, the Hermitian connection was considered by Nakano [1]. The proof of Theorem 10.1 is due to Singer [1]. As we see from the proof, he showed that if P is a holomorphic principal bundle with a complex Lie group G and if Q is a real subbundle of P with structure group U such that $\mathfrak{u}$ is a real form of $\mathfrak{g}$, then there is a unique connection in Q such that, when extended to a connection in P, its horizontal subspaces are complex subspaces of the tangent spaces of P.

CHAPTER X

Homogeneous Spaces

1. Invariant affine connections

Let M be a manifold of dimension n and G a Lie subgroup of $GL(n; \mathbf{R})$. We recall (cf. p. 288 of Volume I) that a G-structure on M is a principal subbundle P of the bundle $L(M)$ with structure group $G \subset GL(n; \mathbf{R})$. Examples of G we have in mind for later applications are $GL(n; \mathbf{R})$, $O(n)$, $GL(m; \mathbf{C})$ and $U(m)$, where $n = 2m$ for the last two groups.

A transformation f of M induces in a natural manner an automorphism $\tilde{f}$ of the bundle $L(M)$; $\tilde{f}$ maps a linear frame $u = (X_1, \ldots, X_n)$ at $x \in M$ into the linear frame $\tilde{f}(u) = (fX_1, \ldots, fX_n)$ at $f(x) \in M$. If $\tilde{f}$ maps a G-structure P into itself, then f is called an *automorphism of the G-structure* P. An infinitesimal transformation X of M is called an *infinitesimal automorphism of* P if the local 1-parameter group of local transformations f_t generated by X consists of local automorphisms of P, i.e., if each $\tilde{f}_t$ maps P into itself. With the notion of natural lift defined on p. 230 of Volume I, we can characterize an infinitesimal automorphism of P as follows.

PROPOSITION 1.1. *Let $\tilde{X}$ be the natural lift to $L(M)$ of an infinitesimal transformation X of M. Then X is an infinitesimal automorphism of a G-structure P if and only if $\tilde{X}$ is tangent to P at each point of P.*

Proof. We recall how $\tilde{X}$ was constructed in the proof of Proposition 2.1 of Chapter VI. Let f_t be the local 1-parameter group of local transformations of M generated by X. Then $\tilde{X}$ is the vector field on $L(M)$ induced by the local 1-parameter group of local automorphisms $\tilde{f}_t$ of $L(M)$. It is clear that every $\tilde{f}_t$ maps P into itself if and only if $\tilde{X}$ is tangent to P at each point of P.
QED.

X. HOMOGENEOUS SPACES 187

Let $M = K/H$ be a homogeneous space, where K is a Lie group and H is a closed subgroup of K (cf. Proposition 4.2 of Chapter I). The coset H is called the origin of M and will be denoted by o. The group K acts transitively on M in a natural manner; an element $f \in K$ maps a coset $f'H$ into the coset $ff'H$. The set N of elements of K which act as the identity transformation on M forms a closed normal subgroup of K contained in H; in fact, N is the largest such subgroup of K. The action of K on M is said to be *effective* if N consists of the identity element only (cf. p. 42 of Volume I) and it is said to be *almost effective* if N is a discrete subgroup of K. The *linear isotropy representation* is by definition the homomorphism of H into the group of linear transformations of $T_o(M)$ which assigns to each $h \in H$ the differential of h at o. The image of H by the linear isotropy representation is called the *linear isotropy group* at o and is denoted by $\tilde{H}$. It is clear that if the linear isotropy representation is faithful, then K is effective on M. The action of K on M induces an action of K on $L(M)$ and, as it can be seen easily, the linear isotropy representation of H is faithful if and only if K acts freely on $L(M)$, i.e., if and only if no element of K except the identity element leaves a linear frame of M fixed.

Unless otherwise stated we shall assume throughout this section that *M is a homogeneous space K/H with connected Lie group K and faithful linear isotropy representation of H and that P is a G-structure on M invariant by K*, i.e., K acts on P as an automorphism group. Under these assumptions we shall study K-invariant connections in P. (The necessity of assuming that the linear isotropy representation is faithful will be explained at the end of this section.) We shall also fix a linear frame $u_o \in P$ at o throughout. If we identify $T_o(M)$ with $\mathbf{R}^n$ by the linear isomorphism $u_o \colon \mathbf{R}^n \to T_o(M)$, then the linear isotropy representation of H may be identified with the homomorphism $\lambda \colon H \to G$ defined by

$$\lambda(h) = u_o^{-1} \circ h_* \circ u_o \quad \text{for } h \in H,$$

where $h_* \colon T_o(M) \to T_o(M)$ denotes the differential of h at o.

For the sake of convenience we shall restate the theorem of Wang on invariant connections specialized to the present situation (cf. Theorem 11.5 of Chapter II).

THEOREM 1.2. *Let P be a K-invariant G-structure on $M = K/H$. Then there is a one-to-one correspondence between the set of K-invariant connections in P and the set of linear mappings $\Lambda \colon \mathfrak{k} \to \mathfrak{g}$ such that*

(1) $\Lambda(X) = \lambda(X) \quad \text{for } X \in \mathfrak{h}$,
(2) $\Lambda(\mathrm{ad}\,(h))(X) = \mathrm{ad}\,(\lambda(h))(\Lambda(X)) \quad \text{for } h \in H \text{ and } X \in \mathfrak{k}$,

where λ denotes not only the linear isotropy representation $H \to G$ but also the induced Lie algebra homomorphism $\mathfrak{h} \to \mathfrak{g}$, $\mathrm{ad}\,h$ denotes the adjoint representation of H in $\mathfrak{k}$ and $\mathrm{ad}\,(\lambda(h))$ denotes the adjoint representation of G in $\mathfrak{g}$.

To a K-invariant connection in P with connection form ω there corresponds the linear mapping defined by

$$\Lambda(X) = \omega_{u_o}(\tilde{X}) \quad \text{for } X \in \mathfrak{k},$$

where $\tilde{X}$ denotes the natural lift to P of a vector field $X \in \mathfrak{k}$ of M.

We shall now express the one-to-one correspondence in Theorem 1.2 in terms of covariant differentiation. If ∇ is the covariant differentiation with respect to an affine connection on M and if X is a vector field on M, then the tensor field A_X of type $(1, 1)$ on M defined by (cf. p. 235 of Volume I)

$$A_X = L_X - \nabla_X$$

is characterized by (cf. Proposition 2.5 of Chapter VI)

$$A_X Y = -\nabla_Y X - T(X, Y) \quad \text{for all vector fields } Y \text{ on } M,$$

where T denotes the torsion tensor field.

COROLLARY 1.3. *In Theorem 1.2, the one-to-one correspondence between the set of linear mappings $\tilde{X} \colon \mathfrak{k} \to \mathfrak{g}$ satisfying (1) and (2) and the set of K-invariant connections in P can be given by*

$$u_o \circ (\Lambda(X)) \circ u_o^{-1} = -(A_X)_o \quad \text{for } X \in \mathfrak{k}.$$

Proof. We fix a connection in P. For each vector field Y on M, let Y^* be the horizontal lift of Y to P (cf. p. 64 of Volume I). For each $X \in \mathfrak{k}$, the 1-parameter group $\exp tX$ of K acts on M as well as on P. We denote by X and $\tilde{X}$ the vector fields induced by $\exp tX$ on M and P, respectively. By Lemma in the proof of Proposition 1.1 of Chapter III, we have

(1) $u_o^{-1}((\nabla_Y X)_o) = (Y^*(\theta(\tilde{X})))_{u_o}.$

X. HOMOGENEOUS SPACES

On the other hand, by Theorem 1.2 we have

(2) $\Lambda(X)(\theta(Y^*)_{u_o}) = (\omega(\tilde{X})_{u_o})(\theta(Y^*)_{u_o})$.

By Proposition 2.1 of Chapter VI, Proposition 3.10 of Chapter I and the first structure equation (Theorem 2.4 of Chapter III), we have

(3) $0 = \iota_{Y*} \circ L_{\tilde{X}}\theta = \iota_{Y*}(d \circ \iota_{\tilde{X}}\theta + \iota_{\tilde{X}} \circ d\theta)$
$= \iota_{Y*} \circ d(\theta(\tilde{X})) + 2 \, d\theta(\tilde{X}, Y^*)$
$= Y^*(\theta(\tilde{X})) - \omega(\tilde{X})\theta(Y^*) + 2\Theta(\tilde{X}, Y^*)$.

By (1), (2), (3), and the definition of T on p. 132 of Volume I, we have

(4) $0 = (\nabla_Y X)_o - u_o \circ (\Lambda(X)) \circ u_o^{-1}(Y_o) + T(X, Y)_o$.

Hence,

(5) $u_o \circ (\Lambda(X)) \circ u_o^{-1}(Y_o) = (\nabla_Y X)_o + T(X, Y)_o = -(A_X Y)_o$.
QED.

Remark. Condition (2) in Theorem 1.2 implies

(2′) $\Lambda([Y, X]) = [\lambda(Y), \Lambda(X)]$ for $Y \in \mathfrak{h}$ and $X \in \mathfrak{k}$,

and, conversely, if H is connected, then (2′) implies (2).

We can express the torsion and the curvature of an invariant connection in terms of Λ as follows.

PROPOSITION 1.4. *In Theorem 1.2, the torsion tensor T and the curvature tensor R of the invariant connection corresponding to Λ can be expressed at the origin as follows:*

(1) $T(X, Y)_o = u_o \circ (\Lambda(X)) \circ u_o^{-1}(Y_o) - u_o \circ (\Lambda(Y)) \circ u_o^{-1}(X_o)$
$- [X, Y]_o$ *for $X, Y \in \mathfrak{k}$*;

(2) $R(X, Y)_o = u_o \circ \{[\Lambda(X), \Lambda(Y)] - \Lambda([X, Y])\} \circ u_o^{-1}$
for $X, Y \in \mathfrak{k}$.

Proof. Let $X, Y \in \mathfrak{k}$. Let $\tilde{X}$ and $\tilde{Y}$ be the natural lifts to P of the vector fields on M induced by X and Y. Then $[\tilde{X}, \tilde{Y}]$ is the natural lift of $-[X, Y]$.† Since $X_o = u_o(\theta(\tilde{X})u_o)$ by definition of θ and

† See correction for p. 106 of Volume I.

$\Lambda(X) = \omega(X)_{u_o}$ by Theorem 1.2, the proof of our proposition reduces to showing the following two formulas:

$$2\Theta(\tilde{X}, \tilde{Y}) = \theta([\tilde{X}, \tilde{Y}]) + \omega(\tilde{X})\theta(\tilde{Y}) - \omega(\tilde{X})\theta(\tilde{X}),$$
$$2\Omega(\tilde{X}, \tilde{Y}) = \omega([\tilde{X}, \tilde{Y}]) + \omega(\tilde{X})\omega(\tilde{Y}) - \omega(\tilde{Y})\omega(\tilde{X}).$$

The second formula has been established in the course of the proof of Proposition 11.4 of Chapter II and the first formula can be proved in the same way. QED.

Theorem 1.2 does not completely answer the question: when does a homogeneous space K/H admit a K-invariant affine connection? A simple sufficient condition will be given in the next section. A simple necessary condition is that the linear isotropy representation of H must be faithful, provided that K is effective on K/H. (More generally, given a manifold M with an affine connection, an affine transformation f of M is the identity transformation if it leaves one linear frame fixed. This is a special case of Lemma 4, p. 254 of Volume I.)

2. Invariant connections on reductive homogeneous spaces

As in §1, let $M = K/H$ be a homogeneous space on which a connected Lie group K acts transitively and effectively and let P be a K-invariant G-structure on M. We fix a frame $u_o \in P$ at the origin $o \in M$.

We say that a homogeneous space K/H is *reductive* if the Lie algebra $\mathfrak{k}$ of K may be decomposed into a vector space direct sum of the Lie algebra $\mathfrak{h}$ of H and an ad (H)-invariant subspace $\mathfrak{m}$, that is, if

(1) $\mathfrak{k} = \mathfrak{h} + \mathfrak{m}, \quad \mathfrak{h} \cap \mathfrak{m} = 0;$

(2) ad $(H)\mathfrak{m} \subset \mathfrak{m}.$

Condition (2) implies

(2′) $[\mathfrak{h}, \mathfrak{m}] \subset \mathfrak{m},$

and, conversely, if H is connected, then (2′) implies (2). We shall see shortly that if K/H is reductive, then P admits a K-invariant connection and the linear isotropy representation of H is necessarily faithful.

Specializing Theorem 11.7 of Chapter II to the present situation, we may state

THEOREM 2.1. *Let P be a K-invariant G-structure on a reductive homogeneous space $M = K/H$ with decomposition $\mathfrak{k} = \mathfrak{h} + \mathfrak{m}$. Then there is a one-to-one correspondence between the set of K-invariant connections in P and the set of linear mappings $\Lambda_\mathfrak{m} : \mathfrak{m} \to \mathfrak{g}$ such that*

$$\Lambda_\mathfrak{m}(\operatorname{ad} h(X)) = \operatorname{ad}(\lambda(h))(\Lambda_\mathfrak{m}(X)) \quad \text{for } X \in \mathfrak{m} \text{ and } h \in H,$$

where λ denotes the linear isotropy representation $H \to G$.

The correspondence is given in the manner of Theorem 1.2 by

$$\Lambda(X) = \begin{cases} \lambda(X) & \text{if } X \in \mathfrak{h}, \\ \Lambda_\mathfrak{m}(X) & \text{if } X \in \mathfrak{m}. \end{cases}$$

From Theorem 2.1 and Corollary 1.3, we obtain immediately

COROLLARY 2.2. *The one-to-one correspondence in Theorem 2.1 may be also given by*

$$u_o \circ (\Lambda_m(X)) \circ u_o^{-1} = -(A_X)_o \quad \text{for } X \in \mathfrak{m}.$$

From Proposition 1.4 and Theorem 2.1 we obtain the torsion T and the curvature R expressed in terms of $\Lambda_\mathfrak{m}$. To give simpler expressions to the formulas in Proposition 1.4, we make the following identification:

$$\mathfrak{m} = T_o(M) = \mathbf{R}^n.$$

The identification $\mathfrak{m} = T_o(M)$ is given by $X \in \mathfrak{m} \to X_o \in T_o(M)$, i.e., by the evaluation of $X \in \mathfrak{m}$ at o as a vector field on M. The identification $T_o(M) = \mathbf{R}^n$ is given by the linear frame $u_o \in P$ at o. Consequently, we consider $\Lambda_\mathfrak{m}(X)$ as a linear transformation of $\mathfrak{m}$. We define the notations $[\ ,\ ]_\mathfrak{h}$ and $[\ ,\ ]_\mathfrak{m}$ by

$$[X, Y] = [X, Y]_\mathfrak{h} + [X, Y]_\mathfrak{m}, \quad [X, Y]_\mathfrak{h} \in \mathfrak{h}, \quad [X, Y]_\mathfrak{m} \in \mathfrak{m}.$$

Then Proposition 1.4 may be restated in the following form.

PROPOSITION 2.3. *In Theorem 2.1, the torsion tensor T and the curvature tensor R of the invariant connection corresponding to $\Lambda_\mathfrak{m}$ can be* *expressed at the origin o as follows:*

(1) $T(X, Y)_o = \Lambda_m(X)Y - \Lambda_m(Y)X - [X, Y]_m$
$\qquad\qquad\qquad\qquad\qquad\qquad\qquad$ *for* $X, Y \in \mathfrak{m}$;

(2) $R(X, Y)_o = [\Lambda_m(X), \Lambda_m(Y)] - \Lambda_m([X, Y]_m) - \lambda([X, Y]_\mathfrak{h})$
$\qquad\qquad\qquad\qquad\qquad\qquad\qquad$ *for* $X, Y \in \mathfrak{m}$.

It is clear that if we set $\Lambda_m(X) = 0$ for all $X \in \mathfrak{m}$, then Λ_m satisfies the condition of Theorem 2.1. This shows that if K/H is reductive, then P admits a K-invariant connection. The K-invariant connection in P defined by $\Lambda_m = 0$ is called the *canonical connection* (with respect to the decomposition $\mathfrak{k} = \mathfrak{h} + \mathfrak{m}$) as in p. 110 of Volume I. The canonical connection can be given the following geometric interpretation.

PROPOSITION 2.4. *Let P be a K-invariant G-structure on a reductive homogeneous space $M = K/H$ with decomposition $\mathfrak{k} = \mathfrak{h} + \mathfrak{m}$. Then the canonical connection in P is the unique K-invariant connection in P possessing the following property:*

Let $f_t = \exp tX$ be the 1-parameter subgroup of K generated by an arbitrary element $X \in \mathfrak{m}$ and $\tilde{f}_t$ the 1-parameter group of transformations of P induced by f_t. Then the orbit $\tilde{f}_t(u_o)$ is horizontal.

Proof. For $X \in \mathfrak{m}$, let $\tilde{X}$ denote the vector field induced by $\tilde{f}_t$ on P. We fix a K-invariant connection in P. Since $\Lambda_m(X) = \omega(\tilde{X})_{u_o}$ by Theorems 1.2 and 2.1, it follows that $\Lambda_m(X) = 0$ if and only if $\tilde{X}$ is horizontal at u_o. Since $\tilde{f}_t$ leaves $\tilde{X}$ invariant because of $\tilde{f}_t = \exp t\tilde{X}$, $\tilde{X}$ is horizontal at each $\tilde{f}_t(u_o)$ if it is horizontal at u_o. Hence $\Lambda_m(X) = 0$ if and only if $\tilde{X}$ is horizontal at each $\tilde{f}_t(u_o)$. In other words, $\Lambda_m(X) = 0$ if and only if the integral curve $\tilde{f}_t(u_o)$ of $\tilde{X}$ through u_o is horizontal. $\qquad$ QED.

COROLLARY 2.5. *Let P and K/H be as in Proposition 2.4 and consider the canonical connection in P. Then we have*

(1) *For each $X \in \mathfrak{m}$, set $f_t = \exp tX$ in K and $x_t = f_t(o)$ in K/H. Then the parallel displacement of tangent vectors at o along the curve x_t, $0 \leq t \leq s$, coincides with the differential of f_s acting on K/H.*

(2) *For each $X \in \mathfrak{m}$, the curve x_t defined in (1) is a geodesic. Conversely, every geodesic starting from o is of the form $f_t(o)$ for some $X \in \mathfrak{m}$.*

(3) *The canonical connection is complete.*

X. HOMOGENEOUS SPACES   193

Proof. (1) follows almost immediately from Proposition 2.4. Indeed, since $f_t(u_o)$, $0 \leq t \leq s$, is a horizontal curve which projects on the curve x_t, $0 \leq t \leq s$, we see that, for any $Y \in T_o(K/H)$,

$$(f_s)_*(Y) = f_s(u_o) \cdot u_o^{-1} Y$$

is parallel to Y along the curve x_t, $0 \leq t \leq s$.

(2) follows from (1), since the tangent vector $\dot{x}_t$ is equal to $(f_t)_* X_o$. (3) is immediate from (2), since $\exp tX$ is defined for all t.
                                                                QED.

Remark. Let P be a K-invariant G-structure on a reductive homogeneous space $M = K/H$ as in Proposition 2.4. Let P' be any K-invariant subbundle of P with structure group $G' \subset G$. Assume for the sake of simplicity that u_o is also contained in P'. Then the canonical connection in P' defined by $\Lambda_\mathfrak{m} = 0$ is evidently (the restriction of) the canonical connection in P (cf. Theorem 2.1). In particular, if we set $P' = \{f(u_o); f \in K\}$, then P' is a subbundle of P with group H (or more precisely, $\lambda(H) \subset G$) which is isomorphic with the bundle $K(K/H, H)$. The canonical connection in P' corresponds to the invariant connection in $K(K/H, H)$ defined in Theorem 11.1 of Chapter II.

THEOREM 2.6. *As in Theorem 2.1, let P be a K-invariant G-structure on a reductive homogeneous space $M = K/H$ with decomposition $\mathfrak{k} = \mathfrak{h} + \mathfrak{m}$. For the torsion tensor T and the curvature tensor R of the canonical connection in P, we have*

(1) $T(X, Y)_o = -[X, Y]_\mathfrak{m}$ *for $X, Y \in \mathfrak{m}$;*
(2) $(R(X, Y)Z)_o = -[[X, Y]_\mathfrak{h}, Z]$ *for $X, Y, Z \in \mathfrak{m}$;*
(3) $\nabla T = 0$;
(4) $\nabla R = 0$.

Proof. (1) Set $\Lambda_\mathfrak{m} = 0$ in (1) of Proposition 2.3.

(2) Set $\Lambda_\mathfrak{m} = 0$ in (2) of Proposition 2.3 and use the fact that the linear isotropy representation λ of H corresponds to the restriction to $\mathfrak{m}$ of the adjoint representation of H in $\mathfrak{k}$.

(3) and (4) follow from the invariance of T and R by K and from the following proposition.

PROPOSITION 2.7. *Let P be a K-invariant G-structure on a reductive homogeneous space $M = K/H$ as in Theorem 2.1. If a tensor field on M*

is invariant by K, then it is parallel with respect to the canonical connection in P.

Proof. Let L be a K-invariant tensor field on K/H. From (1) of Corollary 2.5 and from the definition of covariant derivative in terms of parallel displacement it follows that $\nabla L = 0$ at o. Since the connection is invariant, it follows that $\nabla L = 0$ on K/H.
QED.

We recall (cf. p. 262 of Volume I) that, in general, an affine connection on a manifold M is said to be *invariant by parallelism* if, for arbitrary points x and y of M and for an arbitrary curve τ from x to y, there exists a (unique) local affine transformation f such that $f(x) = y$ and that the differential of f at x coincides with the parallel displacement $\tau\colon T_x(M) \to T_y(M)$ along τ. We know (cf. Corollary 7.6 of Chapter VI) that a connection is invariant by parallelism if and only if $\nabla T = 0$ and $\nabla R = 0$. Theorem 2.6 implies therefore that the canonical connection in P is invariant by parallelism. But we can see from the remark made after Corollary 2.5 that for the canonical connection in P over a reductive homogeneous space $M = K/H$, every parallel displacement $\tau\colon T_x(M) \to T_y(M)$ is actually given by the differential at x of a *global* affine transformation $f \in K$. In fact, we have considered in the remark the bundle $K(K/H, H)$ as a subbundle of P by sending $f \in K$ into $f(u_o) \in P$ and have shown that the canonical connection is already defined in this subbundle $K(K/H, H)$, i.e., $K(K/H, H)$ contains the holonomy bundle through u_o of the canonical connection.

Conversely, we have

THEOREM 2.8. *Let M be a manifold with an affine connection. Choose a point $o \in M$ and a linear frame u_o at o and let $P(u_o)$ be the holonomy bundle through u_o.*

(1) *If K is a connected group of affine transformations of M such that every $u \in P(u_o)$ is the image of u_o by some element $f \in K$, i.e., $u = \tilde{f}(u_o)$ and if H is the isotropy subgroup of K at o, i.e., $H = \{f \in K; f(o) = o\}$, then M is a reductive homogeneous space and the connection is the canonical connection in $L(M)$;*

(2) *If the connection is complete and satisfies $\nabla T = 0$ and $\nabla R = 0$ and if M is simply connected, then M admits an affine transformation*

X. HOMOGENEOUS SPACES 195

group K satisfying the condition in (1) *and, moreover, the smallest such group is simply transitive on the holonomy bundle $P(u_o)$.*

Proof. (1) Let K be as described in (1) and consider a bundle isomorphism of $K(K/H, H)$ into $L(M)$ which maps $f \in K$ into $\tilde{f}(u_o)$. Let $K(u_o)$ be the image of K under this isomorphism; it is a subbundle of $L(M)$. By our assumption on K, we have

$$P(u_o) \subset K(u_o) \subset L(M).$$

Let Q_{u_o} be the horizontal subspace of $T_{u_o}(P(u_o))$. Then

$$Q_{u_o} \subset T_{u_o}(P(u_o)) \subset T_{u_o}(K(u_o)).$$

If we consider the Lie algebra $\mathfrak{k}$ as the tangent space of K at the identity, then the isomorphism $K \to K(u_o)$ induces a linear isomorphism $\mathfrak{k} \to T_{u_o}(K(u_o))$. Let $\mathfrak{m}$ be the inverse image of Q_{u_o} under this linear isomorphism. It is clear that the Lie algebra $\mathfrak{h}$ of H is the inverse image of the vertical subspace under the same linear isomorphism. It follows that

$$\mathfrak{k} = \mathfrak{h} + \mathfrak{m} \quad \text{(vector space direct sum)}.$$

To prove $\mathrm{ad}(H)\mathfrak{m} \subset \mathfrak{m}$, it suffices to verify that, for each $h \in H$, the mapping $f \in K \to hfh^{-1} \in K$ sends the subspace $\mathfrak{m}$ of $\mathfrak{k}$ into itself. But this amounts to showing that for each $h \in H$ the mapping $u = \tilde{f}(u_o) \in K(u_o) \to \tilde{h}\tilde{f}\tilde{h}^{-1}(u_o) \in K(u_o)$ sends Q_{u_o} into itself. If we denote by λ the linear isotropy representation $H \to GL(n; \mathbf{R})$ defined by

$$\tilde{h}(u_o) = u_o \cdot \lambda(h),$$

then the mapping $u \to \tilde{h}\tilde{f}\tilde{h}^{-1}(u_o)$ may be written as $u \to \tilde{h}(u) \cdot \lambda(h)^{-1}$. This mapping sends u_o into itself. Since h is an affine transformation, $\tilde{h}$ sends every horizontal subspace into a horizontal subspace. The right translation by $\lambda(h)^{-1}$ maps also every horizontal subspace into a horizontal subspace. It follows that the mapping $u \to \tilde{h}(u) \cdot \lambda(h)^{-1}$ sends Q_{u_o} into itself. From our definition of $\mathfrak{m}$ it is clear that $\tilde{X}_{u_o}$ is horizontal for every $X \in \mathfrak{m}$. By Theorems 1.2 and 2.1, we have $\Lambda_\mathfrak{m}(X) = \omega(X)_{u_o} = 0$ for all $X \in \mathfrak{m}$, showing that the given connection is the canonical connection in $L(M)$.

(2) Let $A(M)$ be the group of affine transformations of M. It acts also on the bundle $L(M)$ in a natural way. We first show that

the orbit of $A(M)$ through u_o contains $P(u_o)$ under the assumptions stated in (2). Since $\nabla T = 0$ and $\nabla R = 0$, every $u \in P(u_o)$ is of the form $\tilde{f}(u_o)$ for some local affine transformation f by Corollary 7.6 of Chapter VI. By Corollary 6.2 of Chapter VI and Theorem 7.7 of Chapter VI, such an f can be extended to a (unique) global affine transformation $f \in A(M)$. Let $A^0(M)$ be the identity component of the group $A(M)$ (cf. Theorem 1.5 of Chapter VI). Since $P(u_o)$ is connected, the orbit of $A^0(M)$ through u_o already contains $P(u_o)$. We may therefore take $A^0(M)$ as K. Let K^* be the subgroup of K consisting of elements f such that $\tilde{f}$ maps $P(u_o)$ into itself. We claim that an element $f \in K$ is in K^* if (and only if) $\tilde{f}(u_o) \in P(u_o)$. In fact, if $u \in P(u_o)$ is obtained from u_o by the parallel displacement along a curve τ in M, then $\tilde{f}(u)$ is obtained from $\tilde{f}(u_o)$ by the parallel displacement along $f(\tau)$. This shows that if $\tilde{f}(u_o)$ is in $P(u_o)$, then $\tilde{f}(u)$ is also in $P(u_o)$. We have thus shown that K^* is transitive on $P(u_o)$. As we have shown after Proposition 1.4, if f is an affine transformation of M other than the identity transformation, then $\tilde{f}$ acts on $L(M)$ without fixed point. Hence K^* is simply transitive on $P(u_o)$. From the construction of K^* it is clear that K^* is the smallest subgroup of $A(M)$ satisfying the condition in (1). QED.

We have shown (cf. Corollary 2.5) that for the canonical connection every geodesic starting from o is of the form $f_t(o)$, where $f_t = \exp tX$ for some $X \in \mathfrak{m}$. We shall now determine all K-invariant connections with the same family of geodesics as the canonical connection.

PROPOSITION 2.9. *Let P be a K-invariant G-structure on a reductive homogeneous space $M = K/H$ with decomposition $\mathfrak{k} = \mathfrak{h} + \mathfrak{m}$. Then the invariant connection defined by a linear mapping $\Lambda_\mathfrak{m} : \mathfrak{m} \to \mathfrak{g}$ of Theorem 2.1 has the same geodesics as the canonical connection if and only if*

$$\Lambda_\mathfrak{m}(X)X = 0 \quad \text{for all } X \in \mathfrak{m}.$$

Proof. Since we are considering only the invariant connections we have only to compare the geodesics starting from o. Let ∇ be the covariant differentiation defined by a K-invariant connection in P. Let $X \in \mathfrak{m}$ and set $f_t = \exp tX$. Then $f_t(o)$ is a geodesic with

respect to ∇ if and only if
$$(\nabla_X X)_{f_t(o)} = 0 \quad \text{for all } t.$$
Since f_t is an affine transformation for each t, it follows (cf. Proposition 1.4 of Chapter VI) that
$$f_t((\nabla_X X)_o) = (\nabla_{f_t X}(f_t X))_{f_t(o)}.$$
Since f_t is the 1-parameter group generated by X, we have $f_t X = X$. Hence
$$f_t((\nabla_X X)_o) = (\nabla_X X)_{f_t(o)}.$$
It follows that $f_t(o)$ is a geodesic with respect to ∇ if and only if
$$(\nabla_X X)_o = 0.$$
On the other hand, Corollary 1.3 and Theorem 2.1 imply (after the identification of $T_o(M)$, $\mathbf{R}^n$ and $\mathfrak{m}$)
$$\Lambda_\mathfrak{m}(X)X = -(A_X)_o X_o = -(L_X X)_o + (\nabla_X X)_o = (\nabla_X X)_o.$$
Hence, $f_t(o)$ is a geodesic with respect to ∇ if and only if
$$\Lambda_\mathfrak{m}(X)X = 0. \qquad \text{QED.}$$

Now we wish to consider a K-invariant torsion-free connection having the same geodesics as the canonical connection. As we shall see from the proof of the next theorem, we may not be able to find such a connection in a given K-invariant G-structure if G is too small. In the next theorem we take therefore $L(M)$ as P and we shall speak of an affine connection on M instead of a connection in $L(M)$.

THEOREM 2.10. *Every reductive homogeneous space $M = K/H$ admits a unique torsion-free K-invariant affine connection having the same geodesics as the canonical connection. It is defined by*
$$\Lambda_\mathfrak{m}(X)Y = \tfrac{1}{2}[X, Y]_\mathfrak{m} \quad \text{for } X, Y \in \mathfrak{m}.$$
Proof. By polarization we see that $\Lambda_\mathfrak{m}(X)Y + \Lambda_\mathfrak{m}(Y)X = 0$ for $X, Y \in \mathfrak{m}$ if and only if $\Lambda_\mathfrak{m}(X)X = 0$ for $X \in \mathfrak{m}$. Theorem 2.10 now follows from Propositions 2.3 and 2.9. QED.

We shall call the invariant connection defined in Theorem 2.10 the *natural torsion-free connection* on K/H (with respect to the decomposition $\mathfrak{k} = \mathfrak{h} + \mathfrak{m}$).

From Corollary 2.5 and Theorem 2.10 we obtain

COROLLARY 2.11. *The natural torsion-free connection on a reductive homogeneous space K/H is complete.*

Remark. In the foregoing discussion, we assumed for the sake of simplicity that K acts effectively on K/H, i.e., H contains no non-trivial normal subgroup of K. This made it possible for instance to imbed $K(K/H, H)$ as a subbundle of $L(M)$. But, otherwise, the results obtained in this section are all valid without the assumption that K is effective on K/H. In fact, if K is not effective on K/H, let N be the largest normal subgroup of K contained in H and set $K^* = K/N$ and $H^* = H/N$. Then the natural homomorphism $K \to K^*$ induces a diffeomorphism $K/H \to K^*/H^*$ and K^* is effective on K^*/H^*. Moreover, if $\mathfrak{k} = \mathfrak{h} + \mathfrak{m}$ is an ad (H)-invariant decomposition of $\mathfrak{k}$, then the homomorphism $\mathfrak{k} \to \mathfrak{k}^*$ induces an ad (H^*)-invariant decomposition $\mathfrak{k}^* = \mathfrak{h}^* + \mathfrak{m}^*$ of $\mathfrak{k}^*$. The natural isomorphism $\mathfrak{m} \to \mathfrak{m}^*$ makes it possible to transfer the results proven for K^*/H^* to K/H. The remark we have just made is useful in applying the results of this section to concrete examples. Many important homogeneous spaces are often written in the form K/H where K and H are (products of) classical groups but K may not be effective on K/H.

As an example of homogeneous space, we shall consider a group manifold. Let M be a connected Lie group and let the group $K = M \times M$ act on M by

$$((x,y), z) \in (M \times M) \times M \to xyz^{-1} \in M.$$

Then the isotropy subgroup H of K at the identity element $o \in M$ coincides with the diagonal $\Delta(M \times M) = \{(x, x); x \in M\}$. Let $\mathfrak{m}$ be the Lie algebra of M so that $\mathfrak{k} = \mathfrak{m} + \mathfrak{m}$ and $\mathfrak{h} = \Delta(\mathfrak{m} + \mathfrak{m}) = \{(X, X); X \in \mathfrak{m}\}$. The homogeneous space $M = K/H$ is reductive and there are three natural decompositions of $\mathfrak{k}$:

$\mathfrak{k} = \mathfrak{h} + \mathfrak{m}_+,$ where $\mathfrak{m}_+ = \{(0, X); X \in \mathfrak{m}\};$

$\mathfrak{k} = \mathfrak{h} + \mathfrak{m}_-,$ where $\mathfrak{m}_- = \{(X, 0); X \in \mathfrak{m}\};$

$\mathfrak{k} = \mathfrak{h} + \mathfrak{m}_0,$ where $\mathfrak{m}_0 = \{(X, -X); X \in \mathfrak{m}\}.$

The canonical connections defined by these three decompositions are called, respectively, the $(+)$-*connection*, the $(-)$-*connection*, and the (0)-*connection*; these connections were first introduced by Cartan and Schouten [1].

PROPOSITION 2.12. *Let M be a connected Lie group and let X, Y, and Z be left invariant vector fields. Then*

For the $(+)$-connection: $T(X, Y) = [X, Y]$ and $R = 0$;
For the $(-)$-connection: $T(X, Y) = -[X, Y]$ and $R = 0$;
For the (0)-connection: $T = 0$ and $R(X, Y)Z = -\frac{1}{4}[[X, Y], Z]$.

This proposition will not be used later and the proof is left to the reader.

The following example of a non-reductive homogeneous space with an invariant affine connection is due to Wang (cf. Nomizu [6]).

Example 2.1 Let M be the affine space $\mathbf{R}^n$, $n \geq 2$, deprived of the origin. The largest group of affine transformations of M is easily seen to be $GL(n; \mathbf{R})$. Let K be the identity component of $GL(n; \mathbf{R})$ and H the isotropy subgroup of K at a point, say $(1, 0, \ldots, 0) \in M$. A simple calculation using matrices shows that K/H is not reductive.

But many a homogeneous space is reductive. In any of the following cases, a homogeneous space K/H is reductive:

(a) H is compact;

(b) H is connected and $\mathfrak{h}$ is reductive in $\mathfrak{k}$ in the sense that ad $(\mathfrak{h})$ in $\mathfrak{k}$ is completely reducible. This is the case if H is connected and semi-simple;

(c) H is a discrete subgroup of K.

To prove that K/H is reductive in case (a), let $(\ ,\ )'$ be an arbitrary inner product on $\mathfrak{k}$. Define a new inner product $(\ ,\ )$ on $\mathfrak{k}$ by

$$(X, Y) = \int_H (\operatorname{ad} h(X), \operatorname{ad} h(Y))' \, dh,$$

where dh denotes the Haar measure on H. The new inner product is invariant by ad (H). If we denote by $\mathfrak{m}$ the orthogonal complement of $\mathfrak{h}$ with respect to the inner product $(\ ,\ )$, then we have an ad (H)-invariant decomposition $\mathfrak{k} = \mathfrak{h} + \mathfrak{m}$.

Most of the results in this section are due to Nomizu [2]. (We note that the canonical connection in Proposition 2.4 was called the *canonical connection of the second kind*, and the natural torsion-free connection in Theorem 2.10 was called the *canonical connection of the first kind.*) Theorem 2.8 is a technical improvement of Kobayashi [3]. Invariant connections on reductive homogeneous spaces have been studied independently by Raševskiĭ [1], [2], Kurita [1], and Vinberg [1], [2].

3. Invariant indefinite Riemannian metrics

Let $M = K/H$ be a homogeneous space, where K is a connected Lie group and, unless otherwise stated, acts effectively on M. We shall often identify the tangent space $T_o(M)$ at the origin o with the quotient space $\mathfrak{k}/\mathfrak{h}$ in a natural manner. If K/H is reductive with an ad (H)-invariant decomposition $\mathfrak{k} = \mathfrak{h} + \mathfrak{m}$, then both $T_o(M)$ and $\mathfrak{k}/\mathfrak{h}$ will be identified with $\mathfrak{m}$ also in a natural manner. Since, for each $h \in H$, ad $h \colon \mathfrak{k} \to \mathfrak{k}$ maps the subalgebra $\mathfrak{h}$ into itself, it induces a linear transformation of $\mathfrak{k}/\mathfrak{h}$ which will be also denoted by ad h. As in the preceding sections, we shall identify each element $X \in \mathfrak{k}$ with the vector field on M induced by X.

To express some of the basic properties of an invariant metric on K/H in the Lie algebraic language, we first prove

PROPOSITION 3.1. *There is a natural one-to-one correspondence between the K-invariant indefinite Riemannian metrics g on $M = K/H$ and the ad (H)-invariant non-degenerate symmetric bilinear forms B on $\mathfrak{k}/\mathfrak{h}$. The correspondence is given by*

$$B(\bar{X}, \bar{Y}) = g(X, Y)_o \quad \text{for } X, Y \in \mathfrak{k},$$

where $\bar{X}$ and $\bar{Y}$ are the elements of $\mathfrak{k}/\mathfrak{h}$ represented by X and Y respectively. A form B is positive definite if and only if the corresponding metric g is positive definite.

Proof. We shall only show how to construct g from B and shall leave the rest of the proof to the reader. Any point of M is of the form $f(o)$ for some $f \in K$ and any vector at $f(o) \in M$ is of the form $f(X_o)$ for some $X \in \mathfrak{k}$. It is straightforward to verify that the following equality defines a K-invariant metric g on M:

$$g(f(X_o), f(Y_o)) = B(\bar{X}, \bar{Y}) \quad \text{for } X, Y \in \mathfrak{k}. \qquad \text{QED.}$$

COROLLARY 3.2. *If $M = K/H$ is reductive with an ad (H)-invariant decomposition $\mathfrak{k} = \mathfrak{h} + \mathfrak{m}$, then there is a natural one-to-one* correspondence between the K-invariant indefinite Riemannian metrics g on $M = K/H$ and the ad (H)-invariant non-degenerate symmetric bilinear forms B on $\mathfrak{m}$. The correspondence is given by

$$B(X, Y) = g(X, Y)_o \quad \text{for } X, Y \in \mathfrak{m}.$$

The invariance of B by ad (H) implies

$$B([Z, X], Y) + B(X, [Z, Y]) = 0$$

for $X, Y \in \mathfrak{m}$ and $Z \in \mathfrak{h}$,

and the converse holds if H is connected.

Remark. Proposition 3.1 is a special case of a natural one-to-one correspondence between the K-invariant tensor fields L of type (r, s) on K/H and the ad (H)-invariant tensors of type (r, s) on $\mathfrak{k}/\mathfrak{h}$. The same remark applies to Corollary 3.2.

THEOREM 3.3. *Let $M = K/H$ be a reductive homogeneous space with an ad (H)-invariant decomposition $\mathfrak{k} = \mathfrak{h} + \mathfrak{m}$ and an ad (H)-invariant non-degenerate symmetric bilinear form B on $\mathfrak{m}$. Let g be the K-invariant metric corresponding to B. Then*

(1) *The Riemannian connection for g is given by*

$$\Lambda_\mathfrak{m}(X)Y = (1/2)[X, Y]_\mathfrak{m} + U(X, Y),$$

where $U(X, Y)$ is the symmetric bilinear mapping of $\mathfrak{m} \times \mathfrak{m}$ into $\mathfrak{m}$ defined by

$$2B(U(X, Y), Z) = B(X, [Z, Y]_\mathfrak{m}) + B([Z, X]_\mathfrak{m}, Y)$$

for all $X, Y, Z \in \mathfrak{m}$.

(2) *The Riemannian connection for g coincides with the natural torsion-free connection if and only if B satisfies*

$$B(X, [Z, Y]_\mathfrak{m}) + B([Z, X]_\mathfrak{m}, Y) = 0 \quad \text{for } X, Y, Z \in \mathfrak{m}.$$

Proof. Identifying $\mathfrak{m}$ and $T_o(M)$ we have $\Lambda_\mathfrak{m}(X) = -(A_X)_o$ by Corollary 2.2. Since A_X is skew-symmetric with respect to g (cf. Proposition 3.2 of Chapter VI), $\Lambda_\mathfrak{m}(X)$ is skew-symmetric with respect to B, that is, $B(\Lambda_\mathfrak{m}(X)Y, Z) + B(Y, \Lambda_\mathfrak{m}(X)Z) = 0$ for $Y, Z \in \mathfrak{m}$. We have also $\Lambda_\mathfrak{m}(X)Y - \Lambda_\mathfrak{m}(Y)X = [X, Y]_\mathfrak{m}$ by Proposition 2.3. If we set

$$U(X, Y) = \Lambda_\mathfrak{m}(X)Y - (1/2)[X, Y]_\mathfrak{m},$$

then $U(X, Y)$ is symmetric in X and Y and satisfies

$$B(U(X, Y), Z) + B(Y, U(X, Z))$$
$$= (1/2)\{B([Y, X]_\mathfrak{m}, Z) + B(Y, [Z, X]_\mathfrak{m})\}.$$

From this and from the two identities resulting by cyclic permutations of X, Y, Z, we obtain by using symmetry of U

$$2B(U(X, Y), Z) = B(X, [Z, Y]_\mathfrak{m}) + B([Z, X]_\mathfrak{m}, Y)$$

for $X, Y, Z \in \mathfrak{m}$, proving the first part. The second part follows immediately. QED.

Theorem 3.3 was proved by Nomizu [2], where the formula (13.1), p. 51, has an error of the sign (as was pointed out by H. Wu).

A homogeneous space $M = K/H$ with a K-invariant indefinite Riemannian metric g is said to be *naturally reductive* if it admits an ad (H)-invariant decomposition $\mathfrak{k} = \mathfrak{h} + \mathfrak{m}$ satisfying the condition

$$B(X, [Z, Y]_\mathfrak{m}) + B([Z, X]_\mathfrak{m}, Y) = 0 \quad \text{for } X, Y, Z \in \mathfrak{m}$$

in Theorem 3.3.

PROPOSITION 3.4. *Let $M = K/H$ be a naturally reductive homogeneous space with an ad (H)-invariant decomposition $\mathfrak{k} = \mathfrak{h} + \mathfrak{m}$ and a K-invariant indefinite Riemannian metric g. Let B be the bilinear form on $\mathfrak{m}$ which corresponds to g. Then the curvature tensor R of the Riemannian connection satisfies*

$$g(R(X, Y)Y, X)_o = \tfrac{1}{4}B([X, Y]_\mathfrak{m}, [X, Y]_\mathfrak{m}) - B([[X, Y]_\mathfrak{h}, Y], X)$$
for $X, Y \in \mathfrak{m}$.

Proof. From Proposition 2.3 and Theorem 2.10 we obtain

$$(R(X, Y)Z)_o = \tfrac{1}{4}[X, [Y, Z]_\mathfrak{m}]_\mathfrak{m} - \tfrac{1}{4}[Y, [X, Z]_\mathfrak{m}]_\mathfrak{m}$$
$$- \tfrac{1}{2}[[X, Y]_\mathfrak{m}, Z]_\mathfrak{m} - [[X, Y]_\mathfrak{h}, Z]$$
for $X, Y, Z \in \mathfrak{m}$.

From this formula and Theorem 3.3 we obtain Proposition 3.4 easily. QED.

The following theorem furnishes a very simple case where Theorem 3.3 and Proposition 3.4 may be applied.

THEOREM 3.5. *Let K/H be a homogeneous space. Assume that the Lie algebra $\mathfrak{k}$ of K admits an* ad (K)-*invariant non-degenerate symmetric bilinear form B such that its restriction $B_\mathfrak{h}$ to $\mathfrak{h}$ is non-degenerate. Then*

(1) *The decomposition $\mathfrak{k} = \mathfrak{h} + \mathfrak{m}$ defined by*

$$\mathfrak{m} = \{X \in \mathfrak{k}; B(X, Y) = 0 \quad \text{for all } Y \in \mathfrak{h}\}$$

is ad (H)-*invariant and the restriction $B_\mathfrak{m}$ of B to $\mathfrak{m}$ is also non-degenerate and* ad (H)-*invariant;*

(2) *The homogeneous space K/H is naturally reductive with respect to the decomposition $\mathfrak{k} = \mathfrak{h} + \mathfrak{m}$ defined above and the K-invariant metric g defined by $B_\mathfrak{m}$;*

(3) *The curvature tensor R defined by the metric g satisfies*

$$g(R(X, Y)Y, X)_\mathfrak{h} = \tfrac{1}{4} B_\mathfrak{m}([X, Y]_\mathfrak{m}, [X, Y]_\mathfrak{m}) + B_\mathfrak{h}([X, Y]_\mathfrak{h}, [X, Y]_\mathfrak{h})$$

for $X, Y \in \mathfrak{m}$.

Proof. The proof of (1) is trivial. To prove (2), let $X, Y, Z \in \mathfrak{m}$. Since B is ad (K)-invariant, we have

$$B([Z, X], Y) + B(X, [Z, Y]) = 0.$$

Since $\mathfrak{h}$ and $\mathfrak{m}$ are perpendicular with respect to B, we obtain

$$B([Z, X]_\mathfrak{m}, Y) + B(X, [Z, Y]_\mathfrak{m}) = 0,$$

thus proving (2). To prove (3), let $X, Y \in \mathfrak{m}$. Then

$$\begin{aligned}B_\mathfrak{m}([[X, Y]_\mathfrak{h}, Y], X) &= B([[X, Y]_\mathfrak{h}, Y], X) \\ &= -B([X, Y]_\mathfrak{h}, [X, Y]) \\ &= -B([X, Y]_\mathfrak{h}, [X, Y]_\mathfrak{h}) \\ &= -B_\mathfrak{h}([X, Y]_\mathfrak{h}, [X, Y]_\mathfrak{h}).\end{aligned}$$

Now (3) follows from Proposition 3.4. QED.

COROLLARY 3.6. *Let K/H be a homogeneous space such that the Lie algebra $\mathfrak{k}$ of K admits an* ad (K)-*invariant positive definite symmetric bilinear form B. Then K/H is naturally reductive with respect to the decomposition $\mathfrak{k} = \mathfrak{h} + \mathfrak{m}$ and the K-invariant Riemannian metric g defined in Theorem 3.5. The sectional curvature of g is non-negative.*

Remark. Samelson [2] gave a more direct proof of the last assertion of Corollary 3.6.

Example 3.1. As a special case of Corollary 3.6 we take a compact Lie group K, considered as a homogeneous space $K/\{e\}$. For an ad (K)-invariant positive definite symmetric bilinear form B on the Lie algebra $\mathfrak{k}$, we get a left invariant Riemannian metric g on K, which is also right invariant. The curvature tensor R is given by

$$R(X, Y) = -\tfrac{1}{4}\,\mathrm{ad}\,([X, Y]).$$

In fact, this Riemannian connection coincides with the (0)-connection of K in Proposition 2.12.

Example 3.2. The Killing-Cartan form φ on the Lie algebra $\mathfrak{k}$ of a Lie group K is negative semi-definite if K is compact (cf. Appendix 9). We shall prove that, conversely, *if φ is negative definite, then K is compact* (a theorem of Weyl). Starting from $B = -\varphi$, which is an ad (K)-invariant positive definite symmetric bilinear form on $\mathfrak{k}$, we obtain a bi-invariant Riemannian metric g on K in the manner of Example 3.1. Since $R(X, Y) = (-\tfrac{1}{4})\,\mathrm{ad}\,([X, Y])$, the Ricci tensor S is given by

$$\begin{aligned}S(X, Y) &= \text{trace of } \{Z \to R(Z, X)Y\} \\ &= \text{trace of } \{Z \to -\tfrac{1}{4}[[Z, X], X]\} \\ &= -\tfrac{1}{4}\text{ trace ad }(X)\,\mathrm{ad}\,(Y) = -\tfrac{1}{4}\varphi(X, Y) \\ &= \tfrac{1}{4}\,g(X, Y).\end{aligned}$$

Thus the metric g is Einstein, and Theorem 3.4 of Chapter VIII implies that K is compact. The theorem of Weyl implies the following. *If K is a compact semi-simple Lie group* (so that φ is negative definite; cf. Appendix 9), *then the universal covering group $\tilde{K}$ is also compact and the fundamental group $\pi_1(K)$ is finite.*

4. Holonomy groups of invariant connections

For the sake of convenience we shall restate the theorem of Wang on the holonomy algebra of an invariant connection specialized to the case of an invariant affine connection (cf. Theorem 11.8 of Chapter II).

THEOREM 4.1. *Let P be a K-invariant G-structure on $M = K/H$ and $\Lambda\colon \mathfrak{k} \to \mathfrak{g}$ a linear mapping defining a K-invariant connection in P as in Theorem 1.2. Then the Lie algebra of the holonomy group $\Psi(u_o)$ of the invariant connection defined by Λ is given by*

$$\mathfrak{m}_0 + [\Lambda(\mathfrak{k}), \mathfrak{m}_0] + [\Lambda(\mathfrak{k}), [\Lambda(\mathfrak{k}), \mathfrak{m}_0]] + \cdots,$$

where $\mathfrak{m}_0$ is the subspace of $\mathfrak{g}$ spanned by
$$\{[\Lambda(X), \Lambda(Y)] - \Lambda([X, Y]) ; X, Y \in \mathfrak{k}\}.$$
In the reductive case, we obtain

COROLLARY 4.2. *In Theorem 4.1, assume further that K/H is reductive with an $\mathrm{ad}\,(H)$-invariant decomposition $\mathfrak{k} = \mathfrak{h} + \mathfrak{m}$ and let $\Lambda_\mathfrak{m}: \mathfrak{m} \to \mathfrak{g}$ be the restriction of Λ to $\mathfrak{m}$. Then the Lie algebra of the holonomy group $\Psi(u_0)$ is given by*
$$\mathfrak{m}_0 + [\Lambda_\mathfrak{m}(\mathfrak{m}), \mathfrak{m}_0] + [\Lambda_\mathfrak{m}(\mathfrak{m}), [\Lambda_\mathfrak{m}(\mathfrak{m}), \mathfrak{m}_0]] + \cdots,$$
where $\mathfrak{m}_0$ is the subspace of $\mathfrak{g}$ spanned by
$$\{[\Lambda_\mathfrak{m}(X), \Lambda_\mathfrak{m}(Y)] - \Lambda_\mathfrak{m}([X, Y]_\mathfrak{m}) - \lambda([X, Y]_\mathfrak{h}) ; X, Y \in \mathfrak{m}\}.$$

Proof. We shall first show that $\mathfrak{m}_0$ in Corollary 4.2 coincides with $\mathfrak{m}_0$ in Theorem 4.1. Let $X, Y \in \mathfrak{k}$ and decompose them according to the decomposition $\mathfrak{k} = \mathfrak{h} + \mathfrak{m}$:
$$X = X_\mathfrak{h} + X_\mathfrak{m}, \qquad Y = Y_\mathfrak{h} + Y_\mathfrak{m}.$$
Using the property of $\Lambda_\mathfrak{m}$ stated in Theorem 2.1 and the fact that λ is a homomorphism of $\mathfrak{h}$ into $\mathfrak{g}$, we obtain
$$[\Lambda(X), \Lambda(Y)] - \Lambda([X, Y])$$
$$= [\lambda(X_\mathfrak{h}) + \Lambda_\mathfrak{m}(X_\mathfrak{m}), \lambda(Y_\mathfrak{h}) + \Lambda_\mathfrak{m}(Y_\mathfrak{m})]$$
$$\quad - \lambda([X_\mathfrak{h}, Y_\mathfrak{h}] + [X_\mathfrak{m}, Y_\mathfrak{m}]_\mathfrak{h})$$
$$\quad - \Lambda_\mathfrak{m}([X_\mathfrak{h}, Y_\mathfrak{m}] + [X_\mathfrak{m}, Y_\mathfrak{h}] + [X_\mathfrak{m}, Y_\mathfrak{m}]_\mathfrak{m})$$
$$= [\Lambda_\mathfrak{m}(X_\mathfrak{m}), \Lambda_\mathfrak{m}(Y_\mathfrak{m})] - \Lambda_\mathfrak{m}([X_\mathfrak{m}, Y_\mathfrak{m}]_\mathfrak{m}) - \lambda([X_\mathfrak{m}, Y_\mathfrak{m}]_\mathfrak{h}),$$
thus proving our assertion.

Let $X, Y \in \mathfrak{m}$ and set
$$A(X, Y) = [\Lambda_\mathfrak{m}(X), \Lambda_\mathfrak{m}(Y)] - \Lambda_\mathfrak{m}([X, Y]_\mathfrak{m}) - \lambda([X, Y]_\mathfrak{h}).$$
Let $Z = Z_\mathfrak{h} + Z_\mathfrak{m} \in \mathfrak{k}$. By a simple calculation, we obtain
$$[\Lambda(Z), A(X, Y)] = A([Z_\mathfrak{h}, X], Y) + A(X, [Z_\mathfrak{h}, Y])$$
$$\quad + [\Lambda_\mathfrak{m}(Z_\mathfrak{m}), A(X, Y)],$$
thus proving
$$[\Lambda(\mathfrak{k}), \mathfrak{m}_0] \subset \mathfrak{m}_0 + [\Lambda_\mathfrak{m}(\mathfrak{m}), \mathfrak{m}_0].$$

Hence we obtain also

$$[\Lambda(Z), [\Lambda_{\mathfrak{m}}(\mathfrak{m}), \mathfrak{m}_0]]$$
$$\subset [\lambda(Z_{\mathfrak{h}}), [\Lambda_{\mathfrak{m}}(\mathfrak{m}), \mathfrak{m}_0]] + [\Lambda_{\mathfrak{m}}(Z_{\mathfrak{m}}), [\Lambda_{\mathfrak{m}}(\mathfrak{m}), \mathfrak{m}_0]]$$
$$\subset [[\lambda(Z_{\mathfrak{h}}), \Lambda_{\mathfrak{m}}(\mathfrak{m})], \mathfrak{m}_0] + [\Lambda_{\mathfrak{m}}(\mathfrak{m}), [\lambda(Z_{\mathfrak{h}}), \mathfrak{m}_0]]$$
$$\quad + [\Lambda_{\mathfrak{m}}(Z_{\mathfrak{m}}), [\Lambda_{\mathfrak{m}}(\mathfrak{m}), \mathfrak{m}_0]]$$
$$\subset [\Lambda_{\mathfrak{m}}([Z_{\mathfrak{h}}, \mathfrak{m}]), \mathfrak{m}_0] + [\Lambda_{\mathfrak{m}}(\mathfrak{m}), \mathfrak{m}_0 + [\Lambda_{\mathfrak{m}}(\mathfrak{m}), \mathfrak{m}_0]]$$
$$\quad + [\Lambda_{\mathfrak{m}}(Z_{\mathfrak{m}}), [\Lambda_{\mathfrak{m}}(\mathfrak{m}), \mathfrak{m}_0]]$$
$$\subset [\Lambda_{\mathfrak{m}}(\mathfrak{m}), \mathfrak{m}_0] + [\Lambda_{\mathfrak{m}}(\mathfrak{m}), [\Lambda_{\mathfrak{m}}(\mathfrak{m}), \mathfrak{m}_0]],$$

thus proving

$$[\Lambda(\mathfrak{k}), [\Lambda_{\mathfrak{m}}(\mathfrak{m}), \mathfrak{m}_0]] \subset [\Lambda_{\mathfrak{m}}(\mathfrak{m}), \mathfrak{m}_0] + [\Lambda_{\mathfrak{m}}(\mathfrak{m}), [\Lambda_{\mathfrak{m}}(\mathfrak{m}), \mathfrak{m}_0]].$$

Continuing this way, we obtain finally

$$\mathfrak{m}_0 + [\Lambda(\mathfrak{k}), \mathfrak{m}_0] + [\Lambda(\mathfrak{k}), [\Lambda(\mathfrak{k}), \mathfrak{m}_0]] + \cdots$$
$$\subset \mathfrak{m}_0 + [\Lambda_{\mathfrak{m}}(\mathfrak{m}), \mathfrak{m}_0] + [\Lambda_{\mathfrak{m}}(\mathfrak{m}), [\Lambda_{\mathfrak{m}}(\mathfrak{m}), \mathfrak{m}_0]] + \cdots.$$

Since the reversed inclusion holds trivially, Corollary 4.2 follows now from Theorem 4.1. QED.

Setting $\Lambda_{\mathfrak{m}} = 0$ in Corollary 4.2, we obtain (Nomizu [2])

COROLLARY 4.3. *In Theorem 4.1, assume that K/H is reductive with an ad (H)-invariant decomposition $\mathfrak{k} = \mathfrak{h} + \mathfrak{m}$. Then the Lie algebra of the holonomy group $\Psi(u_o)$ of the canonical connection is spanned by*

$$\{\lambda([X, Y]_{\mathfrak{h}}) ; X, Y \in \mathfrak{m}\}.$$

We remark here that $\{[X, Y]_{\mathfrak{h}} ; X, Y \in \mathfrak{m}\}$ spans an ideal of $\mathfrak{h}$ and, consequently, the restricted linear holonomy group of the canonical connection is a normal subgroup of the linear isotropy group $\tilde{H} = \lambda(H)$.

As we have already remarked (cf. Remark after Corollary 2.5), the canonical connection is related to the invariant connection in $K(K/H, H)$ defined in Theorem 11.1 of Chapter II. Corollary 4.3 corresponds to the statement (4) in Theorem 11.1 there.

In spite of the fact that, for the natural torsion-free connection, $\Lambda_{\mathfrak{m}}$ is explicitly written by means of the bracket, Corollary 4.2 does not give a particularly simple expression for the holonomy algebra in this case.

The following reformulation of Theorem 4.1 and Corollary 4.2 is sometimes more useful for applications.

THEOREM 4.4 *In Theorem 4.1, the Lie algebra of the holonomy group* $\Psi'(u_o)$ *is equal to the smallest subalgebra* $\mathfrak{h}^*$ *of* $\mathfrak{g}$ *such that* (1) $R(X, Y)_o \in \mathfrak{h}^*$ *for all* $X, Y \in \mathfrak{k}$ *and* (2) $[\Lambda(X), \mathfrak{h}^*] \in h^*$ *for all* $X \in \mathfrak{k}$.

As usual $T_o(M)$ is identified with $\mathbf{R}^n$ and hence $R(X, Y)_o$ above really means $u_o^{-1} \circ (R(X, Y)_o) \circ u_o$.

Proof. By Theorem 4.1, the Lie algebra of $\Psi'(u_o)$ is equal to the smallest subalgebra $\mathfrak{h}^*$ of $\mathfrak{g}$ such that (1) $\mathfrak{m}_0 \subset \mathfrak{h}^*$, where $\mathfrak{m}_0$ is defined in Theorem 4.1 and (2) $[\Lambda(X), \mathfrak{h}^*] \subset \mathfrak{h}^*$ for all $X \in \mathfrak{k}$. On the other hand, $\mathfrak{m}_0$ is generated by $R(X, Y)_o$, $X, Y \in \mathfrak{k}$, by Proposition 1.4. QED.

The same reasoning using Corollary 4.2 and Proposition 2.3 yields

COROLLARY 4.5. *In Corollary 4.2, the Lie algebra of the holonomy group* $\Psi'(u_o)$ *is equal to the smallest subalgebra* $\mathfrak{h}^*$ *of* $\mathfrak{g}$ *such that* (1) $R(X, Y)_o \in \mathfrak{h}^*$ *for all* $X, Y \in \mathfrak{m}$ *and* (2) $[\Lambda_\mathfrak{m}(X), \mathfrak{h}^*] \subset \mathfrak{h}^*$ *for all* $X \in \mathfrak{m}$.

In Nomizu [3], Corollary 4.5 was derived more directly from a result of Nijenhuis (cf. Theorem 9.2 of Chapter III).

For an invariant connection on a homogeneous space, we shall sharpen results in §4 of Chapter VI. Let P be a K-invariant G-structure on a homogeneous space K/H and $\Lambda: \mathfrak{k} \to \mathfrak{g}$ a linear mapping defining a K-invariant connection in P (cf. Theorem 1.2). We set

$\mathfrak{a}_\mathfrak{k}$ = the subalgebra of $\mathfrak{g}$ generated by $\{\Lambda(X); X \in \mathfrak{k}\}$.

By Corollary 1.3, $\mathfrak{a}_\mathfrak{k}$ may be considered also as the Lie algebra of linear transformations of $T_o(M)$ generated by $\{(A_X)_o; X \in \mathfrak{k}\}$, where $A_X = L_X - \nabla_X$. Originally, $\mathfrak{a}_\mathfrak{k}$ was introduced as such in the Riemannian case by Kostant [1], [2], and has been used by Lichnerowicz [3] and Wang [1] under more general circumstances.

The basic properties of $\mathfrak{a}_\mathfrak{k}$ are given by

PROPOSITION 4.6. *Let P be a K-invariant G-structure on $M = K/H$ and $\Lambda: \mathfrak{k} \to \mathfrak{g}$ a linear mapping defining a K-invariant connection in P. Let $\mathfrak{h}^*$ be the Lie algebra of the holonomy group* $\Psi'(u_o)$. *Then*

(1) $\mathfrak{h}^* \subset \mathfrak{a}_\mathfrak{k} \subset \mathfrak{n}_\mathfrak{g}(\mathfrak{h}^*)$,

where $\mathfrak{n}_\mathfrak{g}(\mathfrak{h}^*)$ is the normalizer of $\mathfrak{h}^*$ in $\mathfrak{g}$, i.e.,

$$\mathfrak{n}_\mathfrak{g}(\mathfrak{h}^*) = \{A \in \mathfrak{g}; [A, \mathfrak{h}^*] \subset \mathfrak{h}^*\};$$

(2) $\lambda(\mathfrak{h}) \subset \mathfrak{a}_\mathfrak{f}$.

Proof. (1) By Theorem 4.1, $\mathfrak{a}_\mathfrak{f}$ clearly contains $\mathfrak{h}^*$. Also by Theorem 4.1, $[\Lambda(X), \mathfrak{h}^*] \subset \mathfrak{h}^*$ and hence $[\mathfrak{a}_\mathfrak{f}, \mathfrak{h}^*] \subset \mathfrak{h}^*$.

(2) Since $\Lambda(X) = \lambda(X)$ for $X \in \mathfrak{h}$ by definition, $\mathfrak{a}_\mathfrak{f}$ contains $\lambda(\mathfrak{h}) = \{\Lambda(X); X \in \mathfrak{h}\}$. QED.

We shall give several cases in which the holonomy algebra $\mathfrak{h}^*$ coincides with $\mathfrak{a}_\mathfrak{f}$. We shall say that a K-invariant connection in P is *normal* if $\mathfrak{h}^* = \mathfrak{a}_\mathfrak{f}$.

THEOREM 4.7. *Let* $M = K/H$ *be a homogeneous space with a K-invariant Riemannian metric. Then its Riemannian connection is normal in any of the following cases:*

(a) *M is compact;*

(b) *M does not admit a non-zero parallel 2-form.*

Proof. As we have shown in the proof of Theorem 4.5 of Chapter VI, every infinitesimal isometry X of M gives rise to a parallel tensor field B_X of type $(1, 1)$ which is skew-symmetric with respect to the Riemannian metric and the vanishing of B_X implies $(A_X)_o \in \mathfrak{h}^*$. If (a) is satisfied, then $\mathfrak{a}_\mathfrak{f} \subset \mathfrak{h}^*$ by Theorem 4.5 of Chapter VI. Let β_X be the 2-form corresponding to B_X by the duality defined by the metric tensor g, i.e.,

$$\beta_X(Y, Z) = g(B_X Y, Z) \quad \text{for all vector fields } Y, Z \text{ on } M.$$

If (b) is satisfied, then $\beta_X = 0$ for all $X \in \mathfrak{k}$ and hence $B_X = 0$ for all $X \in \mathfrak{k}$. It follows that $\mathfrak{a}_\mathfrak{f} \subset \mathfrak{h}^*$. QED.

Remark. The proof of Theorem 4.5 of Chapter VI actually yields the following slightly more general result:

(a') *If $M = K/H$ is a compact homogeneous space with a K-invariant indefinite Riemannian metric, then its Riemannian connection is normal;*

(b') *For a homogeneous space $M = K/H$ with a K-invariant indefinite Riemannian metric, every K-invariant metric connection (i.e., every K-invariant connection for which the metric tensor is parallel) which does not admit a non-zero parallel 2-form is normal.*

Theorem 4.7 is due to Kostant [1]. As we see from Proposition 4.6, if $\mathfrak{h}^* = \mathfrak{n}_\mathfrak{g}(\mathfrak{h}^*)$ then $\mathfrak{h}^* = \mathfrak{a}_\mathfrak{f}$.

X. HOMOGENEOUS SPACES   209

The following theorem gives a geometric interpretation to the notion of normal connection.

THEOREM 4.8. *Let P be a K-invariant G-structure on a homogeneous space $M = K/H$. Fixing a K-invariant connection in P, let $P(u_o)$ be the holonomy bundle through a frame $u_o \in P$. Then the connection is normal if and only if every element of K maps $P(u_o)$ into itself.*

Proof. Let ω be the connection form for the given K-invariant connection. For each $f \in K$, the induced transformation $\tilde{f}$ of P maps every horizontal curve into a horizontal curve. It follows that $\tilde{f}$ maps $P(u_o)$ into itself if and only if $\tilde{f}(u_o) \in P(u_o)$. Since K is connected by assumption, $\tilde{f}(u_o) \in P(u_o)$ for all $f \in K$ if and only if $\tilde{X}$ is tangent to $P(u_o)$ at u_o for all $X \in \mathfrak{k}$. (Here, $\tilde{X}$ denotes as before the natural lift of X to P.) Since the horizontal component of $\tilde{X}$ at u_o is always tangent to $P(u_o)$, $\tilde{X}$ is tangent to $P(u_o)$ at u_o if and only if its vertical component is. But the latter holds if and only if $\omega(\tilde{X})_{u_o}$ is in the Lie algebra $\mathfrak{h}^*$ of the holonomy group $\Psi(u_o)$. Hence, $\tilde{f}$ maps $P(u_o)$ into itself for all $f \in K$ if and only if $\omega(\tilde{X})_{u_o} \in \mathfrak{h}^*$ for all $X \in \mathfrak{k}$. On the other hand, Corollary 1.3 implies that $\omega(\tilde{X})_{u_o} \in \mathfrak{h}^*$ if and only if $X \in \mathfrak{a}_\mathfrak{f}$.   QED.

We have already seen in Chapter II that, by virtue of the reduction theorem (cf. Theorem 7.1 of Chapter II), for certain types of problems concerning a connection in a principal bundle P we can assume that P is the holonomy bundle. If an automorphism group K of P is involved, such a simplification is not in general available unless K maps the holonomy bundle into itself. Theorem 4.8 means that if an invariant connection on a homogeneous space is normal, then the reduction theorem can be still used advantageously. The proof of the following corollary will illustrate the point.

COROLLARY 4.9. *Let P be a K-invariant G-structure on a homogeneous space $M = K/H$. If a K-invariant connection in P is normal, then every parallel tensor field on M is invariant by K.*

Proof. By Theorem 4.8 we may assume that P itself is the holonomy bundle. Let S be a parallel tensor field of type (r, s) on M and $\tilde{S}$ the corresponding mapping of P into the tensor space $\mathbf{T}_s^r$, i.e.,

$$\tilde{S}(u) = u^{-1}(S_{\pi(u)}),$$

where $\pi\colon P \to M$ is the projection and u^{-1} denotes the isomorphism $\mathbf{T}^r_s(\pi(u)) \to \mathbf{T}^r_s$ induced by $u^{-1}\colon T_x(M) \to \mathbf{R}^n$. (More intuitively, $\tilde{S}(u)$ may be considered as the components of S with respect to the frame u.) From the definition of covariant differentiation given in §1 of Chapter III, the assumption that S is parallel means that $\tilde{S}$ is a constant map of P into $\mathbf{T}^r_s$. Hence, $\tilde{S}$ is invariant by $\tilde{f}$ for all $f \in K$. This means that S is invariant by K. QED.

Remark. Corollary 4.9 has been proved by Lichnerowicz [3] along the following line of argument. For each $x \in M = K/H$, let $\mathfrak{a}_\mathfrak{k}(x)$ be the Lie algebra of linear transformations of $T_x(M)$ generated by $\{(A_X)_x;\ X \in \mathfrak{k}\}$. Then a tensor field S on M satisfying any two of the following three conditions satisfies necessarily the third:

(a) S is invariant by K;
(b) S is parallel;
(c) S_x is invariant by the linear transformation group generated by $\mathfrak{a}_\mathfrak{k}(x)$ for all $x \in M$.

(This fact easily follows from $A_X = L_X - \nabla_X$ without the assumption that the given invariant connection is normal.) If the connection is normal, then (b) implies (c) and hence (a) also.

From Theorem 4.7 and Corollary 4.9 it follows that on a compact homogeneous Riemannian manifold M every parallel tensor field S on M is invariant by the largest connected group of isometries of M. This is a special case of the result of Wang proved in Volume I (cf. Theorem 4.6 of Chapter VI) and was originally obtained by Kostant [2].

5. The de Rham decomposition and irreducibility

Let M be a simply connected complete Riemannian manifold. Then (cf. Theorem 6.2 of Chapter IV), M is isometric to the direct product $M_0 \times M_1 \times \cdots \times M_r$, where M_0 is a Euclidean space (possibly of dimension 0) and $M_1, \ldots, M_r$ are all simply connected, complete, irreducible Riemannian manifolds, and such a decomposition is unique up to an order. By Theorem 3.5 of Chapter VI, the largest connected group $I^0(M)$ of isometries of M is naturally isomorphic to the direct product of the largest connected groups $I^0(M_i)$ of isometries of the factors M_i:

$$I^0(M) \approx I^0(M_0) \times I^0(M_1) \times \cdots \times I^0(M_r).$$

From this, it is evident that M is a homogeneous Riemannian manifold if and only if all factors $M_0, M_1, \ldots, M_r$ are. The following theorem (due to Nomizu [3]) gives a more precise result.

THEOREM 5.1. *Let $M = K/H$ be a simply connected homogeneous space with an invariant Riemannian metric. Then there exist connected closed subgroups $K_0, K_1, \ldots, K_r$ of K, all containing H, such that*

$$K/H = K_0/H \times K_1/H \times \cdots \times K_r/H$$

is the de Rham decomposition, where each factor K_i/H is provided with an invariant Riemannian metric. (K_i may not be effective on K_i/H.)

Proof. We first remark that, being homogeneous, M is complete (cf. Theorem 4.5 of Chapter IV) and hence the de Rham decomposition theorem stated above may be applied to M. Let $M = M_0 \times M_1 \times \cdots \times M_r$ be the de Rham decomposition of M. We may identify each M_i with a totally geodesic submanifold $\overline{M}_i$ of M through the origin o of $M = K/H$ (cf. §6 of Chapter IV). Since K is assumed to be connected, it is contained in $I^0(M)$. For each fixed i, set

$$K_i = \{f \in K; f(\overline{M}_i) \subset \overline{M}_i\},$$

that is,

$$K_i = K \cap I^0(M_i).$$

Since, for each $x \in \overline{M}_i$, there exists an element $f \in K$ such that $f(o) = x$ and since such an element f is necessarily in $I^0(M_i)$, K_i is transitive on M_i and the isotropy subgroup of K_i at o coincides with H. Hence, $M_i = K_i/H$. QED.

Although Theorem 5.1 holds for all simply connected, homogeneous Riemannian manifolds, it is more desirable to have, even under a stronger assumption, the decomposition of the following type:

$$K/H = K_0/H_0 \times K_1/H_1 \times \cdots \times K_r/H_r$$

with

$$K = K_0 \times K_1 \times \cdots \times K_r, \qquad H = H_0 \times H_1 \times \cdots \times H_r.$$

The following theorem of Kostant [2] gives results in this direction.

THEOREM 5.2. *Let $M = K/H$ be a simply connected naturally reductive homogeneous space with an $\mathrm{ad}\,(H)$-invariant decomposition $\mathfrak{k} = \mathfrak{h} + \mathfrak{m}$ and a K-invariant Riemannian metric g. Let*

$$T_o(M) = T_o^{(0)} + T_o^{(1)} + \cdots + T_o^{(r)}$$

be the de Rham decomposition of the tangent space $T_o(M)$ and

$$\mathfrak{m} = \mathfrak{m}_0 + \mathfrak{m}_1 + \cdots + \mathfrak{m}_r$$

the corresponding decomposition of $\mathfrak{m}$ under the natural identification $T_o(M) = \mathfrak{m}$. If we set

$$\mathfrak{k}(\mathfrak{m}) = \mathfrak{m} + [\mathfrak{m}, \mathfrak{m}],$$

$$\mathfrak{k}_i = \mathfrak{m}_i + [\mathfrak{m}_i, \mathfrak{m}_i], \qquad \mathfrak{h}_i = \mathfrak{k}_i \cap \mathfrak{h} \qquad \text{for } i = 0, 1, \ldots, r,$$

then $\mathfrak{k}(\mathfrak{m})$, $\mathfrak{k}_0, \mathfrak{k}_1, \ldots, \mathfrak{k}_r$ are ideals of $\mathfrak{k}$ and

$$\mathfrak{k}(\mathfrak{m}) = \mathfrak{k}_0 + \mathfrak{k}_1 + \cdots + \mathfrak{k}_r \qquad \textit{(direct sum of Lie algebras)}$$

and

$$\mathfrak{h} = \mathfrak{h}_0 + \mathfrak{h}_1 + \cdots + \mathfrak{h}_r \qquad \textit{(direct sum of Lie algebras)}.$$

Proof. We first prove the following general fact.

LEMMA 1. *Let $T_x(M) = \Sigma_{i=0}^r T_x^{(i)}$ be the de Rham decomposition of the tangent space $T_x(M)$ at x of a simply connected Riemannian manifold M. Let N be the normalizer of the (restricted) linear holonomy group $\Psi'(x)$ (it consists of linear transformations s of $T_x(M)$ such that $s^{-1}\Psi'(x)s \subset \Psi'(x)$). Then the subspaces $T_x^{(i)}$ are all invariant by the identity component N^0 of N.*

Proof of Lemma 1. Let $s \in N$. Since for each $a \in \Psi'(x)$ there exists an element $a' \in \Psi'(x)$ such that

$$as(T_x^{(i)}) = sa'(T_x^{(i)}) = s(T_x^{(i)}) \qquad \text{for } i = 0, 1, \ldots, r,$$

each $s(T_x^{(i)})$ is invariant by $\Psi'(x)$. On the other hand, $T_x(M) = \Sigma_{i=0}^r s(T_x^{(i)})$. From the uniqueness of the de Rham decomposition (cf. Theorem 5.4 of Chapter IV), it follows that each s permutes the subspaces $T_x^{(0)}, T_x^{(1)}, \ldots, T_x^{(r)}$. This gives rise to a homomorphism of N into the group of permutations of the set $\{0, 1, \ldots, r\}$. Evidently, the kernel of this homomorphism contains the identity component N^0. This completes the proof of Lemma 1.

X. HOMOGENEOUS SPACES   213

LEMMA 2. *The following relations hold:*

(1) $\mathfrak{a}_{\mathfrak{k}}(\mathfrak{m}_i) \subset \mathfrak{m}_i$, $[\mathfrak{h}, \mathfrak{m}_i] \subset \mathfrak{m}_i$ *for* $i = 0, 1, \ldots, r$;

(2) $[\mathfrak{m}_i, \mathfrak{m}_i] \subset \mathfrak{m}_i + \mathfrak{h}$ *for* $i = 0, 1, \ldots, r$;

(3) $[\mathfrak{m}_i, \mathfrak{m}_j] = 0$ *for* $i \neq j$, $i, j = 0, 1, \ldots, r$.

Proof of Lemma 2. Since $\mathfrak{a}_{\mathfrak{k}}$ is contained in the normalizer of the holonomy algebra $\mathfrak{h}^*$ by Proposition 4.6, Lemma 1 implies the first relation in (1). Since $[\mathfrak{h}, \mathfrak{m}_i] \subset \mathfrak{a}_{\mathfrak{k}}(\mathfrak{m}_i)$ also by Proposition 4.6, the second relation in (1) holds. Since K/H is naturally reductive, the Riemannian connection coincides with the natural torsion-free connection. Hence (cf. Theorem 2.10),

$$[X, Y]_{\mathfrak{m}} = -2(A_X Y)_o \quad \text{for } X, Y \in \mathfrak{m}.$$

It follows that

$$[X, Y]_{\mathfrak{m}} \subset \mathfrak{a}_{\mathfrak{k}}(\mathfrak{m}_i) \subset \mathfrak{m}_i \quad \text{for } X \in \mathfrak{m}, Y \in \mathfrak{m}_i,$$

that is,

(4) $[\mathfrak{m}, \mathfrak{m}_i]_{\mathfrak{m}} \subset \mathfrak{m}_i$,

which implies (2) and also

(5) $[\mathfrak{m}_i, \mathfrak{m}_j]_{\mathfrak{m}} \subset \mathfrak{m}_i \cap \mathfrak{m}_j = 0$ for $i \neq j$.

By Proposition 2.3, the curvature tensor R is given by

$$(R(X, Y)Z)_o = \tfrac{1}{4}[X, [Y, Z]_{\mathfrak{m}}]_{\mathfrak{m}} - [Y, [X, Z]_{\mathfrak{m}}]_{\mathfrak{m}}$$
$$- \tfrac{1}{2}[[X, Y]_{\mathfrak{m}}, Z]_{\mathfrak{m}} - [[X, Y]_{\mathfrak{h}}, Z]$$

for $X, Y, Z \in \mathfrak{m}$.
If $X \in \mathfrak{m}_i$, $Y \in \mathfrak{m}_j$ and $i \neq j$, then (4) and (5) imply

$$(R(X, Y)Z)_o = -[[X, Y]_{\mathfrak{h}}, Z] \quad \text{for } Z \in \mathfrak{m}.$$

But the left hand side vanishes since X and Y belong to different factors in the de Rham decomposition. Since the linear isotropy representation is faithful, from $[[X, Y]_{\mathfrak{h}}, \mathfrak{m}] = 0$ we obtain $[X, Y]_{\mathfrak{h}} = 0$. This together with (5) implies (3), thus completing the proof of Lemma 2.

We shall now complete the proof of Theorem 5.2 with the aid of Lemmas 1 and 2. Using the Jacobi identity we obtain from Lemma 2

(6) $[\mathfrak{k}, \mathfrak{m}_i] \subset [\mathfrak{h}, \mathfrak{m}_i] + [\mathfrak{m}_0, \mathfrak{m}_i] + \cdots + [\mathfrak{m}_r, \mathfrak{m}_i]$
$\subset \mathfrak{m}_i + [\mathfrak{m}_i, \mathfrak{m}_i]$.

By the Jacobi identity and (6), we obtain

(7) $[\mathfrak{k},[\mathfrak{m}_i, \mathfrak{m}_i]] \subset [\mathfrak{m}_i + [\mathfrak{m}_i, \mathfrak{m}_i], \mathfrak{m}_i]$
$\subset [\mathfrak{k}, \mathfrak{m}_i] \subset \mathfrak{m}_i + [\mathfrak{m}_i, \mathfrak{m}_i]$.

By (6) and (7), each $\mathfrak{k}_i$ is an ideal of $\mathfrak{k}$. By Lemma 2,

(8) $[\mathfrak{k}_i, \mathfrak{k}_j] = 0 \quad$ for $i \neq j$.

By Lemma 2,

$\mathfrak{k}(\mathfrak{m}) = \mathfrak{k}_0 + \mathfrak{k}_1 + \cdots + \mathfrak{k}_r \quad$ (not necessarily a direct sum).

To show that the right hand side is actually a direct sum, let $X \in \mathfrak{k}_i \cap \mathfrak{k}_j$ for some $i \neq j$. By (8),

(9) $[X, \mathfrak{m}] \subset [X, \mathfrak{k}_0] + [X, \mathfrak{k}_1] + \cdots + [X, \mathfrak{k}_r] = 0$.

On the other hand, $\mathfrak{k}_i \supset \mathfrak{m}_i$ and $\mathfrak{k} = \mathfrak{h} + \mathfrak{m}$ imply

$\mathfrak{k}_i = \mathfrak{h}_i + \mathfrak{m}_i \quad$ where $\mathfrak{h}_i = \mathfrak{k}_i \cap \mathfrak{h}$.

Since $\mathfrak{k}_i \cap \mathfrak{k}_j \subset \mathfrak{h}_i \cap \mathfrak{h}_j \subset \mathfrak{h}$ for $i \neq j$, X is in $\mathfrak{h}$. From (9) and the faithfulness of the linear isotropy representation, it follows that $X = 0$. The rest of the theorem is now evident. QED.

Remark. If K is connected and K/H is simply connected, a simple homotopy argument shows that H is connected. By taking the universal covering group of K, we may assume in Theorem 5.2 that K is simply connected; K remains to be almost effective on K/H although it may no longer be effective. Since K is simply connected, the normal subgroups $K(\mathfrak{m})$, K_0, K_1, ..., K_r of K generated by $\mathfrak{k}(\mathfrak{m})$, $\mathfrak{k}_0$, $\mathfrak{k}_1$, ..., $\mathfrak{k}_r$, respectively, are closed and simply connected, and moreover

$$K(\mathfrak{m}) = K_0 \times K_1 \times \cdots \times K_r.$$

If we set $H(\mathfrak{m}) = K(\mathfrak{m}) \cap H$, $H_i = K_i \cap H$ for $i = 0, 1, \ldots, r$, then $K(\mathfrak{m})/H(\mathfrak{m})$, K_i/H_i, $i = 0, 1, \ldots, r$, are naturally reductive, and

$$K/H = K(\mathfrak{m})/H(\mathfrak{m}) = K_0/H_0 \times K_1/H_1 \times \cdots \times K_r/H_r$$

coincides with the de Rham decomposition of $M = K/H$.

It is now quite natural to look for cases in which $\mathfrak{k} = \mathfrak{k}(\mathfrak{m})$.

X. HOMOGENEOUS SPACES 215

COROLLARY 5.3. *In Theorem 5.2, if the* ad (H)-*invariant inner product* $B_\mathfrak{m}$ *on* $\mathfrak{m}$ *corresponding to the metric* g *can be extended to an* ad (K)-*invariant non-degenerate symmetric bilinear form* B *on* $\mathfrak{k}$ *such that* $B(\mathfrak{m}, \mathfrak{h}) = 0$, *then*

$$\mathfrak{k} = \mathfrak{k}_0 + \mathfrak{k}_1 + \cdots + \mathfrak{k}_r.$$

Proof. It suffices to prove $\mathfrak{k} = \mathfrak{k}(\mathfrak{m})$. But this holds more generally as follows.

LEMMA. *Under the same assumption as in Theorem 3.5, we have*

$$\mathfrak{k} = \mathfrak{m} + [\mathfrak{m}, \mathfrak{m}].$$

Proof of Lemma. Since $\mathfrak{k}(\mathfrak{m}) = \mathfrak{m} + [\mathfrak{m}, \mathfrak{m}]$ is an ideal of $\mathfrak{k}$, its orthogonal complement $\mathfrak{n}$ (in $\mathfrak{k}$ with respect to B) is also an ideal of $\mathfrak{k}$. Since $\mathfrak{n}$ is perpendicular to $\mathfrak{m}$, it is contained in $\mathfrak{h}$. Since K is (almost) effective on K/H, $\mathfrak{n}$ reduces to 0.   QED.

COROLLARY 5.4. *Let* $M = K/H$ *be a simply connected, naturally reductive homogeneous space with a* K-*invariant Riemannian metric* g. *If* K *is simple, then* M *is irreducible (as a Riemannian manifold)*.
Proof. Since $\mathfrak{k}$ has no non-trivial ideal, $\mathfrak{k} = \mathfrak{k}(\mathfrak{m})$. By Theorem 5.2, M is either irreducible or a Euclidean space. The following lemma whose proof is taken from Lichnerowicz [1; p. 113] completes the proof.

LEMMA. *If a connected Lie group* K *acts transitively on a Euclidean space* $\mathbf{R}^n$ *as a group of Euclidean motions, then* K *is not semi-simple*.

Proof of Lemma. Assume that K is semi-simple. Every element of K can be uniquely written as rt, where r is a rotation at the origin and t is a translation. Let $\rho\colon K \to SO(n)$ be the homomorphism which sends rt into r. Its kernel N is an abelian normal subgroup of K and hence must be discrete. On the other hand, the image of ρ, being a connected semi-simple subgroup of $SO(n)$, must be closed (cf. p. 279 of Volume I) and hence compact. Hence, K/N (which is isomorphic to the image of ρ) is compact and semi-simple. Its covering group K is also compact by the theorem of Weyl (cf. Example 3.2). Being compact, K has a fixed point in $\mathbf{R}^n$ (cf. the last 6 lines on p. 193 of Volume I or Theorem 9.2 of Chapter VIII), which is a contradiction.   QED.

6. Invariant almost complex structures

Let $M = K/H$ be a homogeneous space of a Lie group K. In this section we shall first give an algebraic characterization of an invariant almost complex structure on M and its integrability.

We denote by o the origin of M. We choose once and for all a vector space direct sum decomposition of the Lie algebra $\mathfrak{k} = \mathfrak{h} + \mathfrak{m}$, where we are not assuming ad $(H)\mathfrak{m} \subset \mathfrak{m}$ as in the reductive case. Accordingly we write $X = X_\mathfrak{h} + X_\mathfrak{m}$ for $X \in \mathfrak{k}$. We identify $\mathfrak{m}$ with the tangent space $T_o(M)$ by the restriction of $\pi_*: T_e(K) = \mathfrak{k} \to T_o(M)$ to the subspace $\mathfrak{m}$. Each element X of $\mathfrak{k}$ generates a 1-parameter subgroup $\exp tX$ and this induces a vector field on M, which we denote by X again.

Let J be an almost complex structure on M invariant by K. Restricting J to $T_o(M) = \mathfrak{m}$ we obtain a linear endomorphism $I: \mathfrak{m} \to \mathfrak{m}$ such that $I^2 X = -X$ for $X \in \mathfrak{m}$. Since J is invariant by K, $J_o: T_o(M) \to T_o(M)$ commutes with the linear isotropy representation λ of H. But under the identification $T_0(M) = \mathfrak{m}$, λ is given by

$$\lambda(a)X = (\text{ad }(a)X)_\mathfrak{m} \quad \text{for } a \in H \text{ and } X \in \mathfrak{m},$$

where ad (a) denotes the adjoint action of a in $\mathfrak{k}$. Thus we obtain

$$I(\text{ad }(a)X)_\mathfrak{m} = (\text{ad }(a)IX)_\mathfrak{m} \quad \text{for } a \in H \text{ and } X \in \mathfrak{m}.$$

PROPOSITION 6.1. *The invariant almost complex structures on* $M = K/H$ *are in a natural one-to-one correspondence with the set of linear endomorphisms I of $\mathfrak{m}$ such that*

(1) $I^2 X = -X$ *for* $X \in \mathfrak{m}$;

(2) $I(\text{ad }(a)X)_\mathfrak{m} = (\text{ad }(a)IX)_\mathfrak{m}$ *for* $a \in H$ *and* $X \in \mathfrak{m}$.

The proof is straightforward and is left to the reader. The following proposition is equally easy; (2)' below is an infinitesimal version of (2) above.

PROPOSITION 6.2. *If H is connected, then the invariant almost complex structures on $M = K/H$ are in a natural one-to-one correspondence with*

X. HOMOGENEOUS SPACES

the set of linear endomorphisms I of $\mathfrak{m}$ such that
(1) $I^2 X = -X$ *for* $X \in \mathfrak{m}$;
(2)' $I[Y, X]_\mathfrak{m} = [Y, IX]_\mathfrak{m}$ *for* $X \in \mathfrak{m}$ *and* $Y \in \mathfrak{h}$.

We extend $I: \mathfrak{m} \to \mathfrak{m}$ to a linear endomorphism $\tilde{I}: \mathfrak{k} \to \mathfrak{k}$ by setting $\tilde{I}X = 0$ for $X \in \mathfrak{h}$. Then conditions (1), (2), and (2)' imply
(a) $\tilde{I}X = 0$ for $X \in \mathfrak{h}$;
(b) $\tilde{I}^2 X = -X$ mod $\mathfrak{h}$ for $X \in \mathfrak{k}$;
(c) $\tilde{I}(\mathrm{ad}\,(a)X) = \mathrm{ad}\,(a)\tilde{I}X$ mod $\mathfrak{h}$ for $a \in H$ and $X \in \mathfrak{k}$;
(c)' $\tilde{I}[Y, X] = [Y, \tilde{I}X]$ mod $\mathfrak{h}$ for $X \in \mathfrak{k}$ and $Y \in \mathfrak{h}$.

It is obvious that the almost complex structure J on K/H and $\tilde{I}$ are related by
(d) $\pi^*(\tilde{I}X) = J_o(\pi_* X)$ for $X \in \mathfrak{k}$.

Identifying $\tilde{I}$ and $\tilde{I}'$ such that $\tilde{I}(X) - \tilde{I}'(X) \in \mathfrak{h}$ for $X \in \mathfrak{k}$ we have

PROPOSITION 6.3. *The invariant almost complex structures J on $M = K/H$ are in a natural one-to-one correspondence with the linear endomorphisms $\tilde{I}$ of $\mathfrak{k}$ (mod $\mathfrak{h}$) satisfying* (a), (b), *and* (c). *When H is connected,* (c) *may be replaced by* (c)'. *The correspondence is given by* (d).

Proof. Given J on K/H, we have indicated how to obtain $\tilde{I}$ on $\mathfrak{k}$. Conversely, given $\tilde{I}$ on $\mathfrak{k}$ satisfying (a), (b), and (c) (or (c')), we may choose a subspace $\mathfrak{m}$ such that $\mathfrak{g} = \mathfrak{h} + \mathfrak{m}$ and define I on $\mathfrak{m}$ by $IX = (\tilde{I}X)_\mathfrak{m}$ for $X \in \mathfrak{m}$. We may easily verify that I satisfies (1) and (2) (or (2)') and thus determines an invariant almost complex structure J on K/H satisfying (d). It remains to show that the correspondence $J \to \tilde{I}$ is one-to-one. If J and J' correspond to the same $\tilde{I}$, then (d) implies $J_o(\pi_* X) = J'_o(\pi_* X)$ for every $X \in \mathfrak{k}$. Since $\pi_*(\mathfrak{k}) = T_o(K/H)$, this means that J_o and J'_o coincide. Since J and J' are invariant, they coincide everywhere on K/H.
QED.

As for the integrability we prove

THEOREM 6.4. *An invariant almost complex structure J on $M = K/H$ has no torsion if and only if the corresponding linear endomorphism $\tilde{I}$ of $\mathfrak{k}$ satisfies*
$$[\tilde{I}X, \tilde{I}Y] - [X, Y] - \tilde{I}[X, \tilde{I}Y] - \tilde{I}[\tilde{I}X, Y] \in \mathfrak{h}$$
for all $X, Y \in \mathfrak{k}$.

Proof. Since $\mathfrak{k}$ is the Lie algebra of left invariant vector fields on K, $\tilde{I}$ defines in a natural manner a left invariant tensor field of type $(1, 1)$ on K, which we shall denote by $\tilde{I}$. We define a tensor field S of type $(1, 2)$ on K by

$$S(U, V) = [\tilde{I}U, \tilde{I}V] + \tilde{I}^2[U, V] - \tilde{I}[U, \tilde{I}V] - \tilde{I}[\tilde{I}U, V]$$

for all vector fields U and V on K. (The tensor field $2S$ is what we called the torsion of two tensor fields A and B of type $(1, 1)$ when $A = B = \tilde{I}$ (pp. 37–38 of Volume I).) S is also a left invariant tensor field. Our problem is to show that J is integrable if and only if $S(X, Y)$ is tangent to H at e for all $X, Y \in \mathfrak{k}$.

We prepare the following terminology. We say that a vector field U on K is *projectable* if there is a vector field U' on K/H such that U is π-related to U' (cf. p. 10 of Volume I), namely, $(\pi_*)U_x = U'_{\pi(x)}$ for every $x \in K$. In this case we shall denote U' by $\pi_* U$. If U and V are projectable, $U + V$, cU (c constant), and $[U, V]$ are projectable and

$$\pi_*(U + V) = \pi_*(U) + \pi_*(V), \quad \pi_*(cU) = c\pi_*(U),$$

and

$$\pi_*([U, V]) = [\pi_*(U), \pi_*(V)].$$

Thus the set of all projectable vector fields on K is a subalgebra, say, $\mathfrak{P}(K)$, of the Lie algebra $\mathfrak{X}(M)$ of all vector fields on K and π_* gives a Lie algebra homomorphism of $\mathfrak{P}(K)$ onto the Lie algebra $\mathfrak{X}(K/H)$.

We claim that if U is projectable, so is $\tilde{I}U$ and $\pi_*(\tilde{I}U) = J(\pi_* U)$. It suffices to show that $\pi_*(\tilde{I}U) = J(\pi_* U)$ when U is a vector at a point a of K. This equality is obvious from the relationship (d) of $\tilde{I}$ at e and J at o. If we denote by L_a the left translation of K by a as well as the transformation of K/H by a, then U is of the form $U = L_a V$ for some vector V at e and

$$\pi_*(\tilde{I}U) = \pi_*(\tilde{I}L_a V) = \pi_*(L_a \tilde{I} V) = L_a \pi_*(\tilde{I} V)$$
$$= L_a J(\pi_* V) = JL_a \pi_*(V) = J(\pi_*(L_a V)) = J(\pi_* U),$$

proving our assertion. If U and V are projectable vector fields on

X. HOMOGENEOUS SPACES   219

K, then $S(U, V)$ is also projectable and

$$\pi_*(S(U, V)) = [J\pi_*U, J\pi_*V] + J^2[\pi_*U, \pi_*V]$$
$$- J[\pi_*U, J\pi_*V] - J[J\pi_*U, \pi_*V]$$
$$= [J\pi_*U, J\pi_*V] - [\pi_*U, \pi_*V]$$
$$- J[\pi_*U, J\pi_*V] - J[J\pi_*U, \pi_*V],$$

which is half of the value of the torsion tensor of J for the vector fields π_*U and π_*V on K/H. It follows that J is integrable if and only if $\pi_*(S(U, V)) = 0$ for all projectable vector fields, since $U \in \mathfrak{P}(K) \to \pi_*U \in \mathfrak{X}(K/H)$ is onto. Since S is a left invariant tensor field, $\pi_*(S(U, V))$ is equivalent to the condition that $S(U, V)$ is tangent to H at e for all vectors $X, Y \in T_e(K)$, that is, $S(U, V) \in \mathfrak{h}$ for all $X, Y \in \mathfrak{k}$.   QED.

Proposition 6.3 and Theorem 6.4 are due to Koszul [2]. (See also Frölicher [1].)

Specializing to reductive homogeneous spaces we obtain

PROPOSITION 6.5. *Let* $M = K/H$ *be a reductive homogeneous space with a decomposition* $\mathfrak{k} = \mathfrak{h} + \mathfrak{m}$, *where* ad (H) $\mathfrak{m} \subset \mathfrak{m}$. *Then*

(i) *There is a natural one-to-one correspondence between the set of all invariant almost complex structures J on K/H and the set of linear endomorphisms I of $\mathfrak{m}$ satisfying*

(1) $I^2 = -1$;
(2) $I \circ$ ad $(a) =$ ad $(a) \circ I$ *for every* $a \in H$.

When H is connected, (2) *can be replaced by a weaker condition*

(2)′ $I \circ$ ad $(Y) =$ ad $(Y) \circ I$ *for every* $Y \in \mathfrak{h}$.

(ii) *An invariant almost complex structure J is integrable if and only if the corresponding endomorphism I of $\mathfrak{m}$ satisfies*

$$[IX, IY]_\mathfrak{m} - [X, Y]_\mathfrak{m} - I[X, IY]_\mathfrak{m} - I[IX, Y]_\mathfrak{m} = 0$$

for all $X, Y \in \mathfrak{m}$.

Proof. The first part follows from Propositions 6.1 and 6.2 if we note that ad $(a)\mathfrak{m} \subset \mathfrak{m}$ and ad $(Y)\mathfrak{m} \subset \mathfrak{m}$ for $a \in H$ and $Y \in \mathfrak{h}$. To prove the second assertion we check the condition given in Theorem 6.4. The condition is trivially satisfied if $X, Y \in \mathfrak{h}$. It is satisfied for $X \in \mathfrak{m}$ and $Y \in \mathfrak{h}$, since it reduces to $[X, Y] + I[IX, Y] \in \mathfrak{h}$, which is satisfied by virtue of $I \circ$ ad $(Y) =$ ad $(Y) \circ I$.

Thus J is integrable if and only if

$$[IX, IY] - [X, Y] - I[X, IY]_\mathfrak{m} - I[X, IY]_\mathfrak{m} \in \mathfrak{h}$$

for all $X, Y \in \mathfrak{m}$, that is,

$$[IX, IY]_\mathfrak{m} - [X, Y]_\mathfrak{m} - I[X, IY]_\mathfrak{m} - I[X, IY]_\mathfrak{m} = 0$$

for all $X, Y \in \mathfrak{m}$. QED.

A homogeneous space $M = K/H$ with an invariant complex structure is called a *complex homogeneous space*. If an invariant almost complex structure J on K/H is integrable, then it is an invariant complex structure (by the real analytic version of Theorem 2.5 of Chapter IX; see Appendix 8).

Example 6.1. Let K be an even-dimensional Lie group. Considering K as a homogeneous space $K/\{e\}$ with the left action of K, we may obtain a left invariant almost complex structure on K by simply taking an arbitrary complex structure on $\mathfrak{k} = T_e(K)$, namely, a linear endomorphism I of $\mathfrak{k}$ such that $I^2 = -1$. This does not mean, however, that K is a complex Lie group (cf. Example 2.1 of Chapter IX).

Example 6.2. Let K be a complex Lie group and let H be a closed complex Lie subgroup. In terms of Lie algebras, this means that there is a complex structure J on $\mathfrak{k}$ such that $\mathrm{ad}\,(X) \circ J = J \circ \mathrm{ad}\,(X)$ for every $X \in \mathfrak{k}$ and that the subalgebra $\mathfrak{h}$ is stable by J. In the homogeneous space $M = K/H$, we have an invariant complex structure, as can be proved by following the analogy of the proof for the existence of a differentiable structure on an arbitrary homogeneous space of a Lie group (cf. Chevalley [1, p. 110]). Here we indicate how our results can be applied to this situation. In $\mathfrak{k}$ we may take a subspace $\mathfrak{m}$ such that $\mathfrak{k} = \mathfrak{h} + \mathfrak{m}$ and $J\mathfrak{m} \subset \mathfrak{m}$ (cf. the proof of Proposition 1.1 of Chapter IX). Define $\tilde{I}$ on $\mathfrak{k}$ by $\tilde{I}X = JX$ for $X \in \mathfrak{m}$ and $\tilde{I}X = 0$ for $X \in \mathfrak{h}$. Then $\tilde{I}$ satisfies conditions (a), (b), and (c) of Proposition 6.3. We may also verify the integrability condition in Theorem 6.4. For example, for $X, Y \in \mathfrak{m}$, we have

$$[\tilde{I}X, \tilde{I}Y] = [JX, JY] = J^2[X, Y] = -[X, Y].$$

Also, $[X, JY] = J[X, Y]$ implies $[X, JY]_\mathfrak{m} = J[X, Y]_\mathfrak{m}$ so that

X. HOMOGENEOUS SPACES 221

$\tilde{I}[X, \tilde{I}Y] = J[X, JY]_m = J^2[X, Y]_m = -[X, Y]_m$ and, similarly, $\tilde{I}[\tilde{I}X, Y] = -[X, Y]_m$. Thus the m-component of $[\tilde{I}X, \tilde{I}Y] - [X, Y] - \tilde{I}[X, \tilde{I}Y] - \tilde{I}[\tilde{I}X, Y]$ is 0.

Remark. In Chapter XI we shall reexamine many examples given in Chapter IX from the point of view of complex homogeneous spaces (in fact, Hermitian symmetric spaces).

CHAPTER XI

Symmetric Spaces

1. Affine locally symmetric spaces

Let M be an n-dimensional manifold with an affine connection. The *symmetry* s_x at a point $x \in M$ is a diffeomorphism of a neighborhood U onto itself which sends $\exp X$, $X \in T_x(M)$, into $\exp(-X)$. Since the symmetry at x defined in one neighborhood U of x and the symmetry at x defined in another neighborhood V of x coincide in $U \cap V$, we can legitimately speak of *the* symmetry at x. If $\{x^1, \ldots, x^n\}$ is a normal coordinate system with origin at x, then s_x sends $(x^1, \ldots, x^n)$ into $(-x^1, \ldots, -x^n)$. The differential of s_x at x is equal to $-I_x$, where I_x is the identity transformation of $T_x(M)$. The symmetry s_x is involutive in the sense that $s_x \circ s_x$ is the identity transformation of a neighborhood of x. If s_x is an affine transformation for every $x \in M$, then M is said to be *affine locally symmetric*.

THEOREM 1.1. *A manifold M with an affine connection is affine locally symmetric if and only if $T = 0$ and $\nabla R = 0$, where T and R are the torsion and the curvature tensors respectively.*

Proof. Assume that M is affine locally symmetric. Since s_x is an affine transformation, it preserves T and ∇R. On the other hand, T is a tensor field of degree 3 and ∇R is a tensor field of degree 5. (The degree of a tensor field of type (r, s) is, by definition, $r + s$.)

LEMMA. *On an affine locally symmetric space M, a tensor field of odd degree which is invariant by s_x vanishes at x.*

Proof of Lemma. Since the differential of s_x at x is $-I_x$, s_x sends a tensor K of degree p at x into $(-1)^p K$.

From the lemma it follows that $T = 0$ and $\nabla R = 0$.

XI. SYMMETRIC SPACES 223

Conversely, assume $T = 0$ and $\nabla R = 0$. Since R is a tensor field of degree 4, $-I_x$ preserves R_x. By Theorem 7.4 of Chapter VI, for each fixed x there exists a local affine transformation f such that $f(x) = x$ and the differential of f at x coincides with $-I_x$. Being an affine transformation, f commutes with exp (cf. Proposition 1.1 of Chapter VI), i.e.,

$$f(\exp X) = \exp(-I_x X) \quad \text{for } X \in T_x(M).$$

Hence, $f = s_x$.   QED.

Theorem 1.1 enables us to apply a number of results obtained in §7 of Chapter VI to affine locally symmetric spaces. In particular, every affine locally symmetric space is a real analytic manifold with a real analytic connection with respect to the atlas consisting of normal coordinate systems (cf. Theorem 7.7 of Chapter VI).

A manifold M with an affine connection is said to be *affine symmetric* if, for each $x \in M$, the symmetry s_x can be extended to a global affine transformation of M. From Corollary 7.9 of Chapter VI we obtain

THEOREM 1.2. *A complete, simply connected, affine locally symmetric space is affine symmetric.*

The following theorem shows that the assumption of completeness in Theorem 1.2 is necessary.

THEOREM 1.3. *Every affine symmetric space is complete.*
Proof. Let $\tau = x_t$, $0 \leq t \leq a$, be a geodesic from x to y. Using the symmetry s_y, we can extend τ beyond y as follows. Set

$$x_{a+t} = s_y(x_{a-t}) \quad \text{for } 0 \leq t \leq a.$$

QED.

THEOREM 1.4. *On every affine symmetric space, the group of affine transformations is transitive.*
Proof. Given any two points x and y, there exists a finite sequence of convex neighborhoods $U_1, U_2, \ldots, U_k$ such that $x \in U_1$, $y \in U_k$, and $U_i \cap U_{i+1} \neq \phi$ for $i = 1, \ldots, n-1$. Hence, x and y can be joined by a finite number of geodesic segments. (For the existence and the property of a convex neighborhood, see

Theorem 8.7 of Chapter III.) It suffices therefore to prove that, for a pair of points x and y which can be joined by a single geodesic segment, there exists an affine transformation sending x into y. Let z be the midpoint on the geodesic segment from x to y (with respect to the affine parameter). Then the symmetry s_z maps x into y. QED.

The group $\mathfrak{A}(M)$ of affine transformations of M is known to be a Lie group (cf. Theorem 1.5 of Chapter VI). Let G denote the identity component $\mathfrak{A}^0(M)$ of $\mathfrak{A}(M)$ for the sake of simplicity. Since, as is easily seen, the identity component of a Lie group acting transitively on a manifold M is itself transitive on M, an affine symmetric space M may be written as a homogeneous space G/H. We shall later see that $M = G/H$ is reductive and that the given connection on M coincides with the canonical connection as well as the natural torsion-free connection defined in §2 of Chapter X. As a preparation, we prove

THEOREM 1.5. *Let G be the largest connected group of affine transformations of an affine symmetric space M and H the isotropy subgroup of G at a fixed point o of M so that $M = G/H$. Let s_o be the symmetry of M at o and σ the automorphism of G defined by*

$$\sigma(g) = s_o \circ g \circ s_o^{-1} \quad \text{for } g \in G.$$

Let G_σ be the closed subgroup of G consisting of elements fixed by σ. Then H lies between G_σ and the identity component of G_σ.

Proof. Let $h \in H$ and consider $\sigma(h) = s_o \circ h \circ s_o^{-1}$. Since the differential of s_o at o coincides with $-I_o$, the differential of $\sigma(h)$ at o coincides with that of h at o. Since two affine transformations with the same differential at one point coincide with each other (see the end of §1 of Chapter X), $\sigma(h)$ is equal to h, thus proving $H \subset G_\sigma$. Let g_t be an arbitrary 1-parameter subgroup of G_σ. Then

$$s_o \circ g_t(o) = g_t \circ s_o(o) = g_t(o),$$

which shows that the orbit $g_t(o)$ is left fixed pointwise by s_o. Since o is an isolated fixed point of s_o, the orbit $g_t(o)$ must reduce to the point o, that is $g_t(o) = o$. Hence $g_t \in H$. Since a connected Lie group is generated by its 1-parameter subgroups, we may conclude that the identity component of G_σ is contained in H. QED.

2. Symmetric spaces

Theorem 1.5 suggests the following definition. A *symmetric space* is a triple (G, H, σ) consisting of a connected Lie group G, a closed subgroup H of G and an involutive automorphism σ of G such that H lies between G_σ and the identity component of G_σ, where G_σ denotes the closed subgroup of G consisting of all the elements left fixed by σ. A symmetric space (G, H, σ) is said to be *effective* (resp. *almost effective*) if the largest normal subgroup N of G contained in H reduces to the identity element only (resp. is discrete). If (G, H, σ) is a symmetric space, then $(G/N, H/N, \sigma^*)$ is an effective symmetric space, where σ^* is the involutive automorphism of G/N induced by σ. Here it is more convenient not to assume that (G, H, σ) is effective, i.e., G acts effectively on G/H. The results in Chapter X we need here are all valid without the assumption that G be effective on G/H (cf. the remark following Corollary 2.11 of Chapter X).

Given a symmetric space (G, H, σ), we shall construct for each point x of the quotient space $M = G/H$ an involutive diffeomorphism s_x, called the *symmetry at x*, which has x as an isolated fixed point. For the origin o of G/H, s_o is defined to be the involutive diffeomorphism of G/H onto itself induced by the automorphism σ of G. To show that o is an isolated fixed point of s_o, let $g(o)$ be a fixed point of s_o, where $g \in G$. This means $\sigma(g) \in gH$. Set $h = g^{-1}\sigma(g) \in H$. Since $\sigma(h) = h$, we have

$$h^2 = h\sigma(h) = g^{-1}\sigma(g)\sigma(g^{-1}\sigma(g)) = g^{-1}\sigma(g)\sigma(g^{-1})g = 1,$$

thus showing that h^2 is the identity element. If g is sufficiently close to the identity element so that h is also near the identity element, then h itself must be the identity element and hence $\sigma(g) = g$. Being invariant by σ and near the identity element, g lies in the identity component of G_σ and hence in H. This implies $g(o) = o$, thus proving our assertion that o is an isolated fixed point of s_o. For $x = g(o)$, we set $s_x = g \circ s_o \circ g^{-1}$. Then s_x is independent of the choice of g such that $x = g(o)$.

We define now an infinitesimal version of a symmetric space. A *symmetric Lie algebra* (sometimes called an *involutive Lie algebra*) is a triple $(\mathfrak{g}, \mathfrak{h}, \sigma)$ consisting of a Lie algebra $\mathfrak{g}$, a subalgebra $\mathfrak{h}$ of $\mathfrak{g}$, and an involutive automorphism σ of $\mathfrak{g}$ such that $\mathfrak{h}$ consists of all elements of $\mathfrak{g}$ which are left fixed by σ. A symmetric Lie algebra $(\mathfrak{g}, \mathfrak{h}, \sigma)$ is said to be *effective* if $\mathfrak{h}$ contains no non-zero ideal of $\mathfrak{g}$.

Every symmetric space (G, H, σ) gives rise to a symmetric Lie algebra $(\mathfrak{g}, \mathfrak{h}, \sigma)$ in a natural manner; $\mathfrak{g}$ and $\mathfrak{h}$ are the Lie algebras of G and H, respectively, and the automorphism σ of $\mathfrak{g}$ is the one induced by the automorphism σ of G. Conversely, if $(\mathfrak{g}, \mathfrak{h}, \sigma)$ is a symmetric Lie algebra and if G is a connected, simply connected Lie group with Lie algebra $\mathfrak{g}$, then the automorphism σ of $\mathfrak{g}$ induces an automorphism σ of G (cf. Chevalley [1; p. 113]) and, for any subgroup H lying between G_σ and the identity component of G_σ, the triple (G, H, σ) is a symmetric space. Note that since G_σ is closed, so is H. If G is not simply connected, the Lie algebra automorphism σ may not induce an automorphism of G. It is clear that (G, H, σ) is almost effective if and only if $(\mathfrak{g}, \mathfrak{h}, \sigma)$ is effective. It is also easy to see that there is a natural one-to-one correspondence between the (effective) symmetric Lie algebras $(\mathfrak{g}, \mathfrak{h}, \sigma)$ and the (almost effective) symmetric spaces (G, H, σ) with G simply connected and H connected.

Let $(\mathfrak{g}, \mathfrak{h}, \sigma)$ be a symmetric Lie algebra. Since σ is involutive, its eigenvalues as a linear transformation of $\mathfrak{g}$ are 1 and -1, and $\mathfrak{h}$ is the eigenspace for 1. Let $\mathfrak{m}$ be the eigenspace for -1. The decomposition

$$\mathfrak{g} = \mathfrak{h} + \mathfrak{m}$$

is called the *canonical decomposition* of $(\mathfrak{g}, \mathfrak{h}, \sigma)$.

PROPOSITION 2.1. *If* $\mathfrak{g} = \mathfrak{h} + \mathfrak{m}$ *is the canonical decomposition of a symmetric Lie algebra* $(\mathfrak{g}, \mathfrak{h}, \sigma)$, *then*

$$[\mathfrak{h}, \mathfrak{h}] \subset \mathfrak{h}, \qquad [\mathfrak{h}, \mathfrak{m}] \subset \mathfrak{m}, \qquad [\mathfrak{m}, \mathfrak{m}] \subset \mathfrak{h}.$$

Proof. The first relation just expresses the fact that $\mathfrak{h}$ is a subalgebra. If $X \in \mathfrak{h}$ and $Y \in \mathfrak{m}$, then

$$\sigma([X, Y]) = [\sigma(X), \sigma(Y)] = [X, -Y] = -[X, Y],$$

which proves the second relation. If $X, Y \in \mathfrak{m}$, then

$$\sigma([X, Y]) = [\sigma(X), \sigma(Y)] = [-X, -Y] = [X, Y],$$

which proves the third relation. QED.

Remark. The inclusion relations given in Proposition 2.1 characterize a symmetric Lie algebra. Given a Lie algebra $\mathfrak{g}$ and a decomposition $\mathfrak{g} = \mathfrak{h} + \mathfrak{m}$ (vector space direct sum) satisfying the relations in Proposition 2.1, let σ be the linear transformation of $\mathfrak{g}$ defined by

$$\sigma(X) = X \quad \text{for } X \in \mathfrak{h} \quad \text{and} \quad \sigma(Y) = -Y \quad \text{for } Y \in \mathfrak{m}.$$

It is easy to verify that σ is an involutive automorphism of $\mathfrak{g}$ and $(\mathfrak{g}, \mathfrak{h}, \sigma)$ is a symmetric Lie algebra.

PROPOSITION 2.2. *Let (G, H, σ) be a symmetric space and $(\mathfrak{g}, \mathfrak{h}, \sigma)$ its symmetric Lie algebra. If $\mathfrak{g} = \mathfrak{h} + \mathfrak{m}$ is the canonical decomposition of $(\mathfrak{g}, \mathfrak{h}, \sigma)$, then*

$$\operatorname{ad}(H)\mathfrak{m} \subset \mathfrak{m}.$$

Proof. Let $X \in \mathfrak{m}$ and $h \in H$. Then

$$\sigma(\operatorname{ad} h \cdot X) = \operatorname{ad} \sigma(h) \cdot \sigma(X) = \operatorname{ad} h \cdot (-X) = -\operatorname{ad} h \cdot X.$$

QED.

A *homomorphism* of a symmetric space (G', H', σ') into a symmetric space (G, H, σ) is a homomorphism α of G' into G such that $\alpha(H') \subset H$ and that $\sigma \circ \alpha = \alpha \circ \sigma'$; it will be denoted by $\alpha \colon (G', H', \sigma') \to (G, H, \sigma)$. A homomorphism $\alpha \colon (G', H', \sigma') \to (G, H, \sigma)$ is called a *mono-* or *epimorphism* according as $\alpha \colon G' \to G$ is a mono- or epimorphism. It is called an *isomorphism* if $\alpha \colon G' \to G$ is an isomorphism and if $\alpha(H') = H$.

A triple (G', H', σ') is called a *symmetric subspace* (resp. *symmetric closed subspace* or *symmetric normal subspace*) of a symmetric space (G, H, σ) if G' is a Lie subgroup (resp. closed subgroup or normal subgroup) of G invariant by σ, if $H' = G' \cap H$ and if σ' is the restriction of σ to G'. Unless otherwise stated we shall assume that G' is also connected, although the definition above makes sense for a non-connected G'. Given a symmetric closed normal subspace (G', H', σ') of a symmetric space (G, H, σ), we obtain the symmetric quotient space $(G/G', H/H', \sigma'')$, where σ'' is the involutive automorphism of G/G' induced by σ.

For symmetric Lie algebras, we define the notions of homomorphism, monomorphism, epimorphism, isomorphism, symmetric subalgebra, and symmetric ideal in the same way. It should be remarked that, in the definition of a homomorphism $\alpha\colon (\mathfrak{g}', \mathfrak{h}', \sigma') \to (\mathfrak{g}, \mathfrak{h}, \sigma)$, the condition $\alpha(\mathfrak{h}') \subset \mathfrak{h}$ follows from $\sigma \circ \alpha = \alpha \circ \sigma'$.

Each homo-, mono-, epi-, or isomorphism $\alpha\colon (G', H', \sigma') \to (G, H, \sigma)$ induces in a natural manner a homo-, mono-, epi-, or isomorphism $\alpha\colon (\mathfrak{g}', \mathfrak{h}', \sigma') \to (\mathfrak{g}, \mathfrak{h}, \sigma)$, respectively. The converse holds under suitable assumptions. For instance, each homomorphism $\alpha\colon (\mathfrak{g}', \mathfrak{h}', \sigma') \to (\mathfrak{g}, \mathfrak{h}, \sigma)$ generates a homomorphism $\alpha\colon (G', H', \sigma') \to (G, H, \sigma)$ if G' is simply connected and H' is connected.

Each homomorphism $\alpha\colon (G', H', \sigma') \to (G, H, \sigma)$ induces a mapping $\bar{\alpha}\colon G'/H' \to G/H$ which commutes with the symmetries, i.e.,

$$\bar{\alpha} \circ s_{x'} = s_{\bar{\alpha}(x')} \circ \bar{\alpha} \quad \text{for } x' \in G'/H',$$

where $s_{x'}$ denotes the symmetry of G'/H' at x' and $s_{\bar{\alpha}(x')}$ the symmetry of G/H at $\bar{\alpha}(x')$.

Each homomorphism $\alpha\colon (\mathfrak{g}', \mathfrak{h}', \sigma') \to (\mathfrak{g}, \mathfrak{h}, \sigma)$ is compatible with canonical decompositions: if $\mathfrak{g}' = \mathfrak{h}' + \mathfrak{m}'$ and $\mathfrak{g} = \mathfrak{h} + \mathfrak{m}$ are the canonical decompositions, then $\alpha\colon \mathfrak{g}' \to \mathfrak{g}$ sends $\mathfrak{m}'$ into $\mathfrak{m}$.

Given two symmetric spaces (G, H, σ) and (G', H', σ'), their *direct product* is a symmetric space given by $(G \times G', H \times H', \sigma \times \sigma')$. Similarly, the *direct sum* of two symmetric Lie algebras $(\mathfrak{g}, \mathfrak{h}, \sigma)$ and $(\mathfrak{g}', \mathfrak{h}', \sigma')$ is given by $(\mathfrak{g} + \mathfrak{g}', \mathfrak{h} + \mathfrak{h}', \sigma + \sigma')$.

The following example will show that the concept of symmetric space is a generalization of that of Lie group. We shall see more examples in due course.

Example 2.1. Let G be a connected Lie group and ΔG the diagonal of $G \times G$, i.e., $\Delta G = \{(g, g) \in G \times G; g \in G\}$. Define $\sigma\colon G \times G \to G \times G$ by $\sigma(g, g') = (g', g)$. Then $(G \times G, \Delta G, \sigma)$ is a symmetric space. The quotient space $(G \times G)/\Delta G$ is diffeomorphic with G, the diffeomorphism being induced from $g \in G \to (g, e) \in G \times G$, where e is the identity element of G. Given two connected Lie groups G and G' with a homomorphism $\alpha\colon G' \to G$, the homomorphism $\alpha \times \alpha\colon G' \times G' \to G \times G$ gives rise to a homomorphism $\alpha \times \alpha\colon (G' \times G', \Delta G', \sigma') \to (G \times G, \Delta G, \sigma)$. If G' is a (closed or normal) subgroup of G, then $(G' \times G', \Delta G', \sigma')$ is a symmetric (closed or normal) subspace of $(G \times G, \Delta G, \sigma)$. Similarly, every Lie algebra $\mathfrak{g}$ gives rise to a symmetric Lie algebra $(\mathfrak{g} + \mathfrak{g}, \Delta \mathfrak{g}, \sigma)$.

Example 2.2. Let M be an n-dimensional affine locally symmetric space. At a point x of M, consider the tangent space $\mathfrak{m} = T_x(M)$ and the curvature tensor R_x. We denote by $\mathfrak{h}$ the set of all linear endomorphisms U of $T_x(M)$ which, when extended to a derivation of the tensor algebra at x, map R_x into 0, that is, (writing R for R_x)

$$(U \cdot R)(X, Y) \equiv UR(X, Y) - R(UX, Y) - R(X, UY) - R(X, Y)U$$
$$= 0$$

for all $X, Y \in \mathfrak{m}$. Then $\mathfrak{h}$ is a Lie algebra under the usual bracket operation (cf. Proposition 2.13 of Chapter I). Note that $R(X, Y) \in \mathfrak{h}$ for any $X, Y \in \mathfrak{m}$, since if we extend X and Y to vector fields, then $R(X, Y) = [\nabla_X, \nabla_Y] - \nabla_{[X,Y]}$ maps R_x into 0 by virtue of $\nabla R = 0$. In the direct sum $\mathfrak{g} = \mathfrak{m} + \mathfrak{h}$, we define

$[X, Y] = -R(X, Y) \quad$ for $X, Y \in \mathfrak{m}$,

$[U, X] = UX \quad$ for $U \in \mathfrak{h}, X \in \mathfrak{m}$,

$[U, V] = [U, V] \quad$ for $U, V \in \mathfrak{h}$, as already defined in $\mathfrak{h}$.

We may verify that the Jacobi identity is satisfied in $\mathfrak{g}$ as follows: If $X, Y, Z \in \mathfrak{m}$, then $[[X, Y], Z] = -R(X, Y)Z$ so that

$$\mathfrak{S}[[X, Y], Z] = 0$$

by Bianchi's first identity (Theorem 5.3 of Chapter III). If $X, Y \in \mathfrak{m}$ and $U \in \mathfrak{h}$, then

$$[[X, Y], U] = -[R(X, Y), U],$$
$$[[Y, U], X] = [-UY, X] = R(UY, X)$$

and
$$[[U, X], Y] = [UX, Y] = -R(UX, Y)$$

so that $\mathfrak{S}[[X, Y], U] = 0$ by virtue of the fact that $U \cdot R = 0$. Finally, for $U, V \in \mathfrak{h}$ and $X \in \mathfrak{m}$, $\mathfrak{S}[[U, V], X] = 0$ follows immediately from $[U, V] = UV - VU$. Hence $\mathfrak{g} = \mathfrak{m} + \mathfrak{h}$ is a Lie algebra, and obviously $[\mathfrak{m}, \mathfrak{m}] \subset \mathfrak{h}, [\mathfrak{m}, \mathfrak{h}] \subset \mathfrak{m}$, and $[\mathfrak{h}, \mathfrak{h}] \subset \mathfrak{h}$. If we define $\sigma(X) = -X$ for $X \in \mathfrak{m}$ and $\sigma(U) = U$ for $U \in \mathfrak{h}$, then σ is an involutive automorphism of $\mathfrak{g}$. Thus $(\mathfrak{g}, \mathfrak{h}, \sigma)$ is a symmetric Lie algebra. It is effective, for if U is an element of an ideal $\mathfrak{n}$ of $\mathfrak{g}$ contained in $\mathfrak{h}$, then for $X \in \mathfrak{m}$ we have $[U, X] \in \mathfrak{n}$ and also $[U, X] = UX \in \mathfrak{m}$, showing $UX = 0$ for every $X \in \mathfrak{m}$, that is, $U = 0$.

3. The canonical connection on a symmetric space

We have shown in §1 that a simply connected manifold with a complete affine connection such that $T = 0$ and $\nabla R = 0$ gives rise to a symmetric space (G, H, σ) such that $M = G/H$. We shall show now that, conversely, if (G, H, σ) is a symmetric space, then the homogeneous space G/H admits an invariant affine connection with $T = 0$ and $\nabla R = 0$. Let

$$\mathfrak{g} = \mathfrak{h} + \mathfrak{m}$$

be the canonical decomposition of the symmetric Lie algebra $(\mathfrak{g}, \mathfrak{h}, \sigma)$. According to Proposition 2.2, G/H is reductive with respect to the canonical decomposition in the sense defined in §2 of Chapter X. Theorem 2.1 of Chapter X states that there is a one-to-one correspondence between the set of G-invariant affine connections on G/H and the set of linear mappings $\Lambda_\mathfrak{m}: \mathfrak{m} \to \mathfrak{gl}(n; \mathbf{R})$ such that

$$\Lambda_\mathfrak{m}(\mathrm{ad}\ h(X)) = \mathrm{ad}\ (\lambda(h))(\Lambda_\mathfrak{m}(X)) \qquad \text{for } X \in \mathfrak{m} \text{ and } h \in H,$$

where λ denotes the linear isotropy representation $H \to GL(n; \mathbf{R})$. (We recall that both $\mathfrak{m}$ and the tangent space $T_o(M)$ at the origin $o \in M = G/H$ are identified with $\mathbf{R}^n$.) The invariant connection corresponding to $\Lambda_\mathfrak{m} = 0$ is called the *canonical connection* of (G, H, σ) or G/H. Since $[\mathfrak{m}, \mathfrak{m}] \subset \mathfrak{h}$ for a symmetric space, the canonical connection coincides with the natural torsion-free connection defined in §2 of Chapter X. We have $T = 0$ and $\nabla R = 0$.

THEOREM 3.1. *Let (G, H, σ) be a symmetric space. The canonical connection is the only affine connection on $M = G/H$ which is invariant by the symmetries of M.*

Proof. We shall first prove that the canonical connection is invariant by the symmetries s_x. Since $s_x = g \circ s_o \circ g^{-1}$ if $x = g(o)$ as was shown in §2 and since the canonical connection is invariant by G, it suffices to show that the canonical connection is invariant by s_o. From the way s_o was constructed by σ in §2, it follows that

$$g \circ s_o = s_o \circ \sigma(g) \quad \text{for } g \in G,$$

considered as transformations on $M = G/H$. This implies that s_o maps every G-invariant connection into a G-invariant connection. Let s_o map a G-invariant connection with $\Lambda_\mathfrak{m}$ into a G-invariant connection with $\Lambda'_\mathfrak{m}$. From Theorems 1.2 and 2.1 of Chapter X it follows that $\Lambda'_\mathfrak{m} = -\Lambda_\mathfrak{m}$. In particular, s_o maps the canonical connection (defined by $\Lambda_\mathfrak{m} = 0$) into itself.

We claim that the symmetry s_x constructed from σ coincides with the symmetry of M at x with respect to the canonical connection in the sense of §1. In fact, since σ maps $X \in \mathfrak{m}$ into $-X \in \mathfrak{m}$, the differential of s_x at x coincides with $-I_x$. On the other hand, since s_x is an affine transformation (with respect to the canonical connection), it commutes with the exponential maps, i.e.,

$$s_x(\exp X) = \exp(-X) \quad \text{for } X \in T_x(M),$$

(cf. Proposition 1.1 of Chapter VI), which proves our claim. By the lemma in the proof of Theorem 1.1, every tensor field of odd degree invariant by all s_x must vanish identically. We use this fact to prove the uniqueness of a connection invariant by s_x.

We know (cf. Proposition 7.10 of Chapter III) that the difference of two connections on a manifold is a tensor field of type $(1, 2)$. Considering a connection on $M = G/H$ invariant by all s_x, let S be its difference with the canonical connection. Being the difference of two connections invariant by all s_x, S is invariant by all s_x and hence vanishes identically. QED.

Specializing results in §2 of Chapter X to the canonical connection of a symmetric space, we state

THEOREM 3.2. *With respect to the canonical connection of a symmetric space* (G, H, σ), *the homogeneous space* $M = G/H$ *is a (complete) affine symmetric space with symmetries* s_x *and possesses the following properties:*

(1) $T = 0$, $\nabla R = 0$, *and* $R(X, Y)Z = -[[X, Y], Z]$ *for* $X, Y, Z \in \mathfrak{m}$, *where* $\mathfrak{m}$ *is identified with* $T_o(M)$, o *being the origin of* M;

(2) *For each* $X \in \mathfrak{m}$, *the parallel displacement along* $\pi(\exp tX)$ *coincides with the differential of the transformation* $\exp tX$ *on* M;

(3) *For each* $X \in \mathfrak{m}$, $\pi(\exp tX) = (\exp tX) \cdot o$ *is a geodesic starting from o and, conversely, every geodesic from o is of this form;*
(4) *Every G-invariant tensor field on M is parallel;*
(5) *The Lie algebra of the linear holonomy group (with reference point o) is spanned by* $\{R(X, Y) = -\mathrm{ad}_{\mathfrak{m}}([X, Y]), X, Y \in \mathfrak{m}\}$.

We note that (5) follows from Corollary 4.3 of Chapter X or more directly from Theorem 9.2 of Chapter III.

Remark. If a tensor field F on $M = G/H$ is invariant by G, then each symmetry s_x maps F into F or $-F$ according as the degree of F is even or odd. Since $s_x = g \circ s_o \circ g^{-1}$ for some $g \in G$, it suffices to prove the assertion above for s_o. Since $g \circ s_o = s_o \circ \sigma(g)$ for every $g \in G$ as in the proof of Theorem 3.1, $s_o(F)$ is also invariant by G. But clearly $s_o(F) = \pm F$ at o depending on the parity of the degree of F. Since both F and $s_o(F)$ are invariant by G, we have $s_o(F) = \pm F$ according as the degree of F is even or odd.

THEOREM 3.3. *Let* (G, H, σ) *be a symmetric space. A G-invariant (indefinite) Riemannian metric on* $M = G/H$, *if there exists any, induces the canonical connection on* M.

Proof. Such a metric is parallel with respect to the canonical connection by (4) of Theorem 3.2. Since the canonical connection is also torsion-free by Theorem 3.2, it must be the Riemannian connection (cf. Theorem 2.2 of Chapter IV). This theorem may be also derived from Theorem 3.3 of Chapter X. QED.

We also remark here that if (G, H, σ) is a symmetric space, then G/H is naturally reductive (in the sense of §3 of Chapter X) with respect to an arbitrary G-invariant indefinite Riemannian metric (if there exists any).

There is a large class of symmetric spaces which admit invariant indefinite Riemannian metrics.

THEOREM 3.4. *Let* (G, H, σ) *be a symmetric space with G semi-simple and let* $\mathfrak{g} = \mathfrak{h} + \mathfrak{m}$ *be the canonical decomposition. Then the restriction of the Killing-Cartan form of* $\mathfrak{g}$ *to* $\mathfrak{m}$ *defines a G-invariant (indefinite) Riemannian metric on G/H in the manner described in Corollary 3.2 of Chapter* X.

Proof. Since the Killing-Cartan form of a semi-simple Lie algebra is non-degenerate and invariant by all automorphisms (cf. Appendix 9), it suffices to prove the following

XI. SYMMETRIC SPACES   233

LEMMA. *Let $(\mathfrak{g}, \mathfrak{h}, \sigma)$ be a symmetric Lie algebra and $\mathfrak{g} = \mathfrak{h} + \mathfrak{m}$ the canonical decomposition. If B is a symmetric bilinear form on $\mathfrak{g}$ invariant by σ, then $B(\mathfrak{h}, \mathfrak{m}) = 0$. If B is moreover non-degenerate, so are its restriction $B_\mathfrak{h}$ to $\mathfrak{h}$ and its restriction $B_\mathfrak{m}$ to $\mathfrak{m}$.*

Proof of Lemma. If $X \in \mathfrak{h}$ and $Y \in \mathfrak{m}$, then

$$B(X, Y) = B(\sigma(X), \sigma(Y)) = B(X, -Y) = -B(X, Y),$$

which proves our assertion. It is clear that B is non-degenerate if and only if both $B_\mathfrak{h}$ and $B_\mathfrak{m}$ are so.   QED.

Let M be an affine symmetric space and G the largest connected group of affine transformations of M. Choosing an origin $o \in M$, we obtain a symmetric space (G, H, σ) such that $M = G/H$ (cf. Theorem 1.5). In general, there might exist a symmetric subspace (G', H', σ') of (G, H, σ) such that $M = G/H = G'/H'$ in a natural manner. To see this, let $\mathfrak{g} = \mathfrak{h} + \mathfrak{m}$ be the canonical decomposition. It is a simple matter to verify that $\mathfrak{h}(\mathfrak{m}) = [\mathfrak{m}, \mathfrak{m}]$ is an ideal of $\mathfrak{h}$ and $\mathfrak{g}(\mathfrak{m}) = \mathfrak{h}(\mathfrak{m}) + \mathfrak{m}$ is an ideal of $\mathfrak{g}$. For each subalgebra $\mathfrak{g}'$ of $\mathfrak{g}$ containing $\mathfrak{g}(\mathfrak{m})$, we obtain a symmetric subspace (G', H', σ') such that $M = G/H = G'/H'$ by taking for G' the connected Lie subgroup of G generated by $\mathfrak{g}'$ and setting $H' = G' \cap H$ and $\sigma' = \sigma \mid G'$. Conversely, if (G', H', σ') is a symmetric space with $M = G'/H'$ (with the same origin as $M = G/H$) and if G' is effective on M, then (G', H', σ') is a symmetric subspace of (G, H, σ) and $\mathfrak{g}'$ contains $\mathfrak{g}(\mathfrak{m})$. If $M = \mathbf{R}^n$ with the natural flat affine connection, then $\mathfrak{h}(\mathfrak{m}) = 0$ and $\mathfrak{g}(\mathfrak{m}) = \mathfrak{m}$. Consequently, every connected Lie subgroup G' of G containing the group of translations of $\mathbf{R}^n$ yields a symmetric subspace (G', H', σ') of (G, H, σ) such that $\mathbf{R}^n = G'/H'$.

When M is an affine locally symmetric space, we may construct, for each point x of M, a symmetric Lie algebra $(\mathfrak{g}, \mathfrak{h}, \sigma)$ as in Example 2.2. Let (G, H, σ) be a symmetric space corresponding to $(\mathfrak{g}, \mathfrak{h}, \sigma)$, where G and G/H may be assumed simply connected. Let o be the origin of G/H. Then we have a linear isomorphism F of $T_x(M)$ onto $T_o(G/H)$ which maps the curvature tensor R_x of M upon the curvature tensor of the canonical connection of G/H. By Theorem 7.4 of Chapter VI we see that F is the differential of a certain affine isomorphism f of a neighborhood U of x in M onto a neighborhood V of o in G/H. We have thus shown that each point of an affine locally symmetric space M has a neighborhood on which the affine connection is isomorphic to the canonical connection of a certain symmetric space G/H restricted to a neighborhood of the origin. The proof above is a special case of a method which works for a manifold with an affine connection invariant by parallelism ($\nabla T = 0$ and $\nabla R = 0$) and which gives a local version of (2) of Theorem 2.8 of Chapter X.

4. Totally geodesic submanifolds

Let (G, H, σ) be a symmetric space. With respect to the canonical connection, $M = G/H$ is an affine symmetric space (cf. Theorem 3.2).

Let (G', H', σ') be a symmetric subspace of (G, H, σ), i.e., let G' be a connected Lie subgroup of G invariant by σ, $H' = G' \cap H$ and $\sigma' = \sigma \mid G'$. Then there is a natural imbedding of $M' = G'/H'$ into $M = G/H$. From the way the symmetries of M and M' are constructed from σ and σ' (see §2), we easily see that the symmetry s_x of M at a point x of M' restricts to the symmetry of M' at x. We may therefore denote the symmetry of M' at x by s_x.

THEOREM 4.1. *If (G', H', σ') is a symmetric subspace of a symmetric space (G, H, σ), then $M' = G'/H'$ is a totally geodesic submanifold of $M = G/H$ (with respect to the canonical connection of M). The canonical connection of M restricted to M' coincides with the canonical connection of M.*

Proof. Let $\mathfrak{g} = \mathfrak{h} + \mathfrak{m}$ and $\mathfrak{g}' = \mathfrak{h}' + \mathfrak{m}'$ be the canonical decompositions of $\mathfrak{g}$ and $\mathfrak{g}'$, respectively. For the moment, we consider only the canonical connection of M. A geodesic of M tangent to M' at the origin o is of the form $f_t(o)$, where $f_t = \exp tX$ with $X \in \mathfrak{m}' \subset \mathfrak{m}$. Obviously, the geodesic $f_t(o)$ lies in M'. Given any point $x = g'(o)$ of M' with $g' \in G'$, a geodesic of M tangent to M' at x is of the form $g'(f_t(o))$, which is obviously contained in M', thus proving that M' is a totally geodesic submanifold of M.

Since the canonical connection of M is invariant by the symmetries of M, its restriction to M' is also invariant by the symmetries of M'. By Theorem 3.1, the connection induced on M' by the canonical connection of M coincides with the canonical connection of M'. QED.

Conversely, we have

THEOREM 4.2. *Let (G, H, σ) be a symmetric space, $M = G/H$ the affine symmetric space with the canonical connection and M' a complete totally geodesic submanifold of M through the origin o. Let G' be the largest connected Lie subgroup of G leaving M' invariant, H' the intersection $G' \cap H$ and σ' the restriction of σ to G'. Then (G', H', σ') is a symmetric subspace of (G, H, σ) and $M' = G'/H'$.*

By a complete totally geodesic submanifold, we mean a totally geodesic submanifold of M which is complete with respect to the connection induced from the connection of M. The assumption that M' contains o is not restrictive because every totally geodesic submanifold can be translated by an element of G so as to contain o. Although not explicitly stated, the theorem asserts that G' is invariant by σ.

Proof. We need a few lemmas.

LEMMA 1. *The product of two symmetries s_x and s_y of M is a transformation belonging to G.*

Proof of Lemma 1. Write

$$s_x = g \circ s_o \circ g^{-1} \quad \text{where } x = g(o), \quad g \in G,$$
$$s_y = g' \circ s_o \circ g'^{-1} \quad \text{where } y = g'(o), \quad g' \in G.$$

Then
$$s_x \circ s_y = g \circ (s_o \circ g^{-1} \circ g' \circ s_o) \circ g'^{-1}.$$

On the other hand (cf. the proof of Theorem 3.1), we have

$$s_o \circ g^{-1} \circ g' \circ s_o = \sigma(g^{-1}g').$$

Hence,
$$s_x \circ s_y = g\sigma(g^{-1}g')g'^{-1}.$$

LEMMA 2. *Let M' and M'' be two complete totally geodesic submanifolds of M. If at one point of $M' \cap M''$ the tangent space of M' coincides with that of M'', then $M' = M''$.*

Proof of Lemma 2. Let $x \in M' \cap M''$ and $T_x(M') = T_x(M'')$. Any point y of M' can be joined to x by a broken geodesic (that is, a curve composed of a finite number of geodesic segments) of M'. This broken geodesic may be considered as a broken geodesic of M because M' is totally geodesic in M. Since M'' is totally geodesic and complete in M, M'' contains this broken geodesic. Hence, $y \in M''$, thus proving $M' \subset M''$. Similarly, $M'' \subset M'$.

LEMMA 3. *For every $x \in M'$, the symmetry s_x of M at x sends M' into itself*

Proof of LEMMA 3. Since s_x is an affine transformation of M, $s_x(M')$ is also totally geodesic in M. Applying Lemma 2 to M' and $s_x(M')$, we obtain Lemma 3.

We are now in a position to complete the proof of Theorem 4.2. We shall first prove that G' is transitive on M'. Given two points x and y of M', we have to show the existence of an element of G' which maps x into y. Since any two points of M' can be joined by a broken geodesic of M', we may assume that x and y can be joined by a geodesic segment τ of M'. Write

$$\tau = x_t, \quad 0 \leq t \leq 4a, \quad \text{with } x = x_0 \text{ and } y = x_{4a}$$

For each fixed t, consider the symmetries s_{x_t} and $s_{x_{3t}}$ and set

$$f_t = s_{x_{3t}} \circ s_{x_t}.$$

Then f_0 is the identity element and $f_a(x) = y$. Each f_t is a transformation belonging to G by Lemma 1 and sends M' into itself by Lemma 3. Hence, each f_t lies in G', proving that G' is transitive on M'.

We have shown earlier (cf. the proof of Theorem 3.1) that

$$g \circ s_o = s_o \circ \sigma(g) \quad \text{for } g \in G.$$

Since s_o leaves M' invariant by Lemma 3, $\sigma(g) = s_o^{-1} \circ g \circ s_o$ leaves M' invariant for $g \in G'$. Let g_t be a curve in G' starting from the identity element. Then $\sigma(g_t)$ leaves M' invariant. Since $\sigma(g_0)$ is the identity element of G', the curve $\sigma(g_t)$ lies in G'. This proves that σ sends G' into itself.

If we set $H' = G' \cap H$ and $\sigma' = \sigma \mid G'$, then (G', H', σ') is a symmetric subspace of (G, H, σ). The transitivity of G' on M' implies $M' = G'/H'$. QED.

Let τ be a geodesic from x to y in an affine symmetric space M. The product of two symmetries s_x and s_y is known as a *transvection* (along τ). Lemma 1 says that every transvection is contained in the connected group G although the symmetries may not be. The proof above shows that the set of transvections generates a transitive group.

XI. SYMMETRIC SPACES   237

Theorems 4.1 and 4.2 do not give a one-to-one correspondence between the complete totally geodesic submanifolds M' through o of M and the symmetric subspaces (G', H', σ') of (G, H, σ). Two different symmetric subspaces (G', H', σ') and (G'', H'', σ'') may give rise to the same totally geodesic submanifold (cf. Remark at the end of §3).

THEOREM 4.3. *Let (G, H, σ) be a symmetric space and $\mathfrak{g} = \mathfrak{h} + \mathfrak{m}$ the canonical decomposition. Then there is a natural one-to-one correspondence between the set of linear subspaces $\mathfrak{m}'$ of $\mathfrak{m}$ such that $[[\mathfrak{m}', \mathfrak{m}'], \mathfrak{m}'] \subset \mathfrak{m}'$ and the set of complete totally geodesic submanifolds M' through the origin o of the affine symmetric space $M = G/H$, the correspondence being given by $\mathfrak{m}' = T_o(M')$ (under the identification $\mathfrak{m} = T_o(M)$).*

Proof. Let M' be a complete totally geodesic submanifold through o of M and (G', H', σ') a symmetric subspace of (G, H, σ) such that $M' = G'/H'$. Let $\mathfrak{g}' = \mathfrak{h}' + \mathfrak{m}'$ be the canonical decomposition and $\mathfrak{m}' = T_o(M')$. Since $[\mathfrak{m}', \mathfrak{m}'] \subset \mathfrak{h}'$, we have $[[\mathfrak{m}', \mathfrak{m}'], \mathfrak{m}'] \subset \mathfrak{m}'$.

Conversely, let $\mathfrak{m}'$ be a subspace of $\mathfrak{m}$ such that $[[\mathfrak{m}', \mathfrak{m}'], \mathfrak{m}'] \subset \mathfrak{m}'$. Then set

$$\mathfrak{h}' = [\mathfrak{m}', \mathfrak{m}'], \qquad \mathfrak{g}' = \mathfrak{h}' + \mathfrak{m}', \qquad \sigma' = \sigma \mid \mathfrak{g}'.$$

Then $(\mathfrak{g}', \mathfrak{h}', \sigma')$ is a symmetric subalgebra of $(\mathfrak{g}, \mathfrak{h}, \sigma)$. Let G' be the connected Lie subgroup of G generated by $\mathfrak{g}'$ and set $H' = G' \cap H$. Since σ leaves $\mathfrak{g}'$ invariant, it leaves G' invariant. By setting $\sigma' = \sigma \mid G'$, we obtain a symmetric subspace (G', H', σ') of (G, H, σ). The totally geodesic submanifold $M' = G'/H'$ has the property $T_o(M') = \mathfrak{m}'$. Now, from Lemma 2 in the proof of Theorem 4.2, we may conclude that the correspondence $M' \to \mathfrak{m}'$ is one-to-one.   QED.

It would be of some interest to note that the symmetric subalgebra $(\mathfrak{g}', \mathfrak{h}', \sigma')$ constructed in the proof above by setting $\mathfrak{g}' = [\mathfrak{m}', \mathfrak{m}'] + \mathfrak{m}'$ is the smallest symmetric subalgebra of $(\mathfrak{g}, \mathfrak{h}, \sigma)$ with the desired property. In general, a subspace $\mathfrak{t}$ of a Lie algebra $\mathfrak{g}$ such that $[[\mathfrak{t}, \mathfrak{t}], \mathfrak{t}] \subset \mathfrak{t}$ is called a *Lie triple system*.

PROPOSITION 4.4. *Let (G, H, σ) be a symmetric space and M' a complete totally geodesic submanifold through the origin o of $M = G/H$.*

Then M' is flat (i.e., has vanishing curvature) if and only if the corresponding subspace $\mathfrak{m}'$ of $\mathfrak{m}$ in Theorem 4.3 satisfies

$$[[\mathfrak{m}', \mathfrak{m}'], \mathfrak{m}'] = 0.$$

Proof. Apply (1) of Theorem 3.2 to a symmetric subspace (G', H', σ') such that $M' = G'/H'$. QED.

5. Structure of symmetric Lie algebras

Let $\mathfrak{g}$ be a Lie algebra and $\mathfrak{r}$ its radical, i.e., the largest solvable ideal of $\mathfrak{g}$, so that $\mathfrak{g}/\mathfrak{r}$ is semi-simple. There exists a semi-simple subalgebra $\mathfrak{s}$ of $\mathfrak{r}$ which is complementary to $\mathfrak{r}$ (Levi's theorem; cf. for instance, Jacobson [1; p. 91]). In other words, the short exact sequence

$$0 \to \mathfrak{r} \to \mathfrak{g} \to \mathfrak{g}/\mathfrak{r} \to 0$$

splits. It is also a well known theorem (cf. Appendix 9 or Jacobson [1; p. 71]) that a semi-simple Lie algebra $\mathfrak{s}$ is the direct sum of its simple ideals. We shall extend these results to symmetric Lie algebras.

Since the radical $\mathfrak{r}$ is *the* largest solvable ideal of $\mathfrak{g}$, it is invariant by every automorphism of $\mathfrak{g}$. Hence (cf. §2), given a symmetric Lie algebra $(\mathfrak{g}, \mathfrak{h}, \sigma)$ we have a symmetric ideal $(\mathfrak{r}, \mathfrak{h} \cap \mathfrak{r}, \sigma')$, where $\sigma' = \sigma \mid \mathfrak{r}$. We obtain also a symmetric Lie algebra $(\mathfrak{g}/\mathfrak{r}, \mathfrak{h}/\mathfrak{h} \cap \mathfrak{r}, \sigma^*)$ together with a natural epimorphism $(\mathfrak{g}, \mathfrak{h}, \sigma) \to (\mathfrak{g}/\mathfrak{r}, \mathfrak{h}/\mathfrak{h} \cap \mathfrak{r}, \sigma^*)$.

THEOREM 5.1. *Let $(\mathfrak{g}, \mathfrak{h}, \sigma)$ be a symmetric Lie algebra and $\mathfrak{r}$ the radical of $\mathfrak{g}$. Then there exists a symmetric subalgebra $(\mathfrak{s}, \mathfrak{h} \cap \mathfrak{s}, \sigma'')$ of $(\mathfrak{g}, \mathfrak{h}, \sigma)$ such that $\mathfrak{s}$ is a semi-simple subalgebra of $\mathfrak{g}$ which is complementary to $\mathfrak{r}$. In other words, the short exact sequence*

$$0 \to (\mathfrak{r}, \mathfrak{h} \cap \mathfrak{r}, \sigma') \to (\mathfrak{g}, \mathfrak{h}, \sigma) \to (\mathfrak{g}/\mathfrak{r}, \mathfrak{h}/\mathfrak{h} \cap \mathfrak{r}, \sigma^*) \to 0$$

splits.

Proof. All we have to know is that in Levi's theorem there exists a semi-simple subalgebra $\mathfrak{s}$ which is invariant by σ. This fact is proved in Appendix 9. QED.

THEOREM 5.2. *Let* $(\mathfrak{g}, \mathfrak{h}, \sigma)$ *be a symmetric Lie algebra with* $\mathfrak{g}$ *semi-simple. Then* $(\mathfrak{g}, \mathfrak{h}, \sigma)$ *can be decomposed into the direct sum*

$$(\mathfrak{g}_1 + \mathfrak{g}_1', \mathfrak{h}_1, \sigma_1) + \cdots + (\mathfrak{g}_k + \mathfrak{g}_k', \mathfrak{h}_k, \sigma_k)$$
$$+ (\mathfrak{g}_{k+1}, \mathfrak{h}_{k+1}, \sigma_{k+1}) + \cdots + (\mathfrak{g}_r, \mathfrak{h}_r, \sigma_r),$$

where

(1) $\mathfrak{g}_1, \mathfrak{g}_1', \ldots, \mathfrak{g}_k, \mathfrak{g}_k', \mathfrak{g}_{k+1}, \ldots, \mathfrak{g}_r$ *are the simple ideals of* $\mathfrak{g}$;

(2) $\mathfrak{g}_1 + \mathfrak{g}_1', \ldots, \mathfrak{g}_k + \mathfrak{g}_k', \mathfrak{g}_{k+1}, \ldots, \mathfrak{g}_r$ *are invariant by* σ *and* $\sigma_i = \sigma \,|\, \mathfrak{g}_i + \mathfrak{g}_i'$ *for* $i = 1, \ldots, k$ *and* $\sigma_j = \sigma \,|\, \mathfrak{g}_j$ *for* $j = k + 1, \ldots, r$;

(3) *For* $i = 1, \ldots, k$, *each* $\mathfrak{g}_i$ *is isomorphic with* $\mathfrak{g}_i'$ *under* σ *and* $\mathfrak{h}_i = \{(X, \sigma(X)) \in \mathfrak{g}_i + \mathfrak{g}_i'; X \in \mathfrak{g}_i\}$. *For* $j = k + 1, \ldots, r$, $\mathfrak{h}_j = \mathfrak{g}_j \cap \mathfrak{h}$.

Proof. Let $\mathfrak{g}_1, \ldots, \mathfrak{g}_p$ be the simple ideals of $\mathfrak{g}$. Being an automorphism of $\mathfrak{g}$, σ permutes these simple ideals. Since σ is involutive, we have either

(1) $\sigma(\mathfrak{g}_i) = \mathfrak{g}_i$,

or

(2) $\sigma(\mathfrak{g}_i) = \mathfrak{g}_{i'}$ and $\sigma(\mathfrak{g}_{i'}) = \mathfrak{g}_i$.

Changing notations we may write

$$\mathfrak{g} = (\mathfrak{g}_1 + \mathfrak{g}_1') + \cdots + (\mathfrak{g}_k + \mathfrak{g}_k') + \mathfrak{g}_{k+1} + \cdots + \mathfrak{g}_r,$$

where $\sigma(\mathfrak{g}_i) = \mathfrak{g}_i'$ and $\sigma(\mathfrak{g}_i') = \mathfrak{g}_i$ for $i = 1, \ldots, k$ and $\sigma(\mathfrak{g}_j) = \mathfrak{g}_j$ for $j = k + 1, \ldots, r$. Theorem 5.2 follows now immediately.

QED.

A short exact sequence of symmetric Lie algebras may be considered as an infinitesimal version of the following situation.

PROPOSITION 5.3. *Let* (G, H, σ) *be a symmetric space and* (G', H', σ') *a closed normal symmetric subspace of* (G, H, σ). *If we set* $G^* = G/G'$, $H^* = H/H'$, *and define* σ^* *to be the involutive automorphism of* G^* *induced by* σ, *then* (G^*, H^*, σ^*) *is a symmetric space and the homogeneous space* G/H *is a fibre bundle over* G^*/H^* *with fibre* G'/H' *in a natural manner.*

The proof is trivial.

We may apply Proposition 5.3 to the case where G' is the radical of G, i.e., the connected Lie subgroup of G generated by the radical of 𝔤. As we shall see below, the radical G' is closed in G. Thus, G/H is a fibre bundle over G^*/H^* with fibre G'/H' where G^* is semi-simple and G' is solvable. By Theorem 4.1 the fibres are all totally geodesic with respect to the canonical connection of G/H. We shall now prove that the radical G' of G is closed in G. Let $\bar{G}'$ be the closure of G' in G. Since G' is a connected normal subgroup of G, so is $\bar{G}'$. It suffices therefore to prove that $\bar{G}'$ is solvable. But, in general, *a connected dense Lie subgroup G' of a connected Lie group $\bar{G}'$ is normal in $\bar{G}'$ and there is a connected abelian Lie subgroup A of $\bar{G}'$ such that $\bar{G}' = G'A$* (for the proof of this fact, see Hochschild [1; p. 190]). Hence, $\bar{G}'$ is solvable. Assume, in addition, that G is simply connected. Let $\mathfrak{g} = \mathfrak{g}' + \mathfrak{g}''$ be a semi-direct sum, where $\mathfrak{g}'$ is the radical of $\mathfrak{g}$ and $\mathfrak{g}''$ is a semi-simple subalgebra of $\mathfrak{g}$ invariant by σ (cf. Theorem 5.1). Let G'' be the connected Lie subgroup of G generated by $\mathfrak{g}''$. Since G is simply connected, G is written as the semi-direct product $G''G'$. Indeed, the projection $G \to G/G'$ induces a homomorphism $p: G'' \to G/G'$ which is obviously a local isomorphism. Hence, p is a covering map. On the other hand, since G is simply connected, so is G/G' (cf. Hochschild [1; pp. 135, 136]). It follows that p is an isomorphism of G'' onto G/G' and that G'' may be considered as (the image of) a global cross section of the fibring $G \to G/G'$. Our assertion follows now immediately. Setting $H'' = G'' \cap H$ and $\sigma'' = \sigma \mid G''$, we obtain a closed symmetric subspace (G'', H'', σ'') of (G, H, σ). In general, we have only $H''H' \subset H$. But $\mathfrak{h}$ is a semi-direct sum of $\mathfrak{h}'$ and $\mathfrak{h}''$. Hence, if H is connected, then $H = H''H'$. We may now conclude that *if G is simply connected and H is connected, then the fibre bundle G/H over G^*/H^* with fibre G'/H' admits a global cross section G''/H'' which is totally geodesic with respect to the canonical connection of G/H.*

Let (G, H, σ) be a symmetric space where G is simply connected and semi-simple and H is connected. Then corresponding to the direct sum decomposition of $(\mathfrak{g}, \mathfrak{h}, \sigma)$ in Theorem 5.2, we have the following direct product decomposition of (G, H, σ):

$$(G_1 \times G_1', H_1, \sigma_1) \times \cdots \times (G_k \times G_k', H_k, \sigma_k)$$
$$\times (G_{k+1}, H_{k+1}, \sigma_{k+1}) \times \cdots \times (G_r, H_r, \sigma_r),$$

and consequently the direct product decomposition of the affine symmetric space G/H:

$$(G_1 \times G_1')/H_1 \times \cdots \times (G_k \times G_k')/H_k$$
$$\times G_{k+1}/H_{k+1} \times \cdots \times G_r/H_r.$$

The proof is straightforward and is left to the reader.

Theorem 5.1 and Theorem 5.2 show that a symmetric Lie algebra is built of symmetric Lie algebras of the following three types:

(1) $(\mathfrak{g} + \mathfrak{g}, \Delta\mathfrak{g}, \sigma)$, where $\mathfrak{g}$ is simple, $\Delta\mathfrak{g} = \{(X, X); X \in \mathfrak{g}\}$, and $\sigma(X, Y) = (Y, X)$ for $X, Y \in \mathfrak{g}$;

(2) $(\mathfrak{g}, \mathfrak{h}, \sigma)$, where $\mathfrak{g}$ is simple;

(3) $(\mathfrak{g}, \mathfrak{h}, \sigma)$, where $\mathfrak{g}$ is solvable.

The symmetric Lie algebras of type (1) are in one-to-one correspondence with the simple Lie algebras in a natural manner. The symmetric Lie algebras of type (2) have been classified by Berger [2].

Example 5.1. Given a symmetric Lie algebra $(\mathfrak{g}, \mathfrak{h}, \sigma)$, we shall construct a new symmetric Lie algebra which will be denoted by $(T(\mathfrak{g}), T(\mathfrak{h}), T(\sigma))$. Let $T(\mathfrak{g})$ be the Lie algebra obtained by defining a new bracket operation $[\ ,\ ]'$ in $\mathfrak{g} + \mathfrak{g}$ as follows:

$$[(X, X'), (Y, Y')]' = ([X, Y], [X, Y'] + [X', Y])$$
$$\text{for } (X, Y) \in \mathfrak{g} + \mathfrak{g}.$$

Then $T(\mathfrak{g})$ is a semi-direct sum of an abelian ideal $\mathfrak{g}' = \{(0, X'); X' \in \mathfrak{g}\}$ and a subalgebra $\mathfrak{g}'' = \{(X, 0); X \in \mathfrak{g}\}$, which is naturally isomorphic with $\mathfrak{g}$. If $\mathfrak{g}$ is semi-simple, then $\mathfrak{g}'$ is the radical of $T(\mathfrak{g})$ and $\mathfrak{g}''$ is a semi-simple subalgebra of $T(\mathfrak{g})$. Let $T(\mathfrak{h})$ be the subalgebra of $T(\mathfrak{g})$ consisting of elements $(X, X') \in \mathfrak{h} + \mathfrak{h}$. We define an involutive automorphism $T(\sigma)$ of $T(\mathfrak{g})$ by setting

$$T(\sigma)(X, X') = (\sigma(X), \sigma(X')).$$

Setting
$$\mathfrak{h}' = \mathfrak{g}' \cap T(\mathfrak{h}), \qquad \sigma' = T(\sigma) \mid \mathfrak{g}',$$
$$\mathfrak{h}'' = \mathfrak{g}'' \cap T(\mathfrak{h}), \qquad \sigma'' = T(\sigma) \mid \mathfrak{g}'',$$

We obtain a semi-direct sum decomposition of a symmetric Lie algebra $(T(\mathfrak{g}), T(\mathfrak{h}), T(\sigma))$ into a symmetric ideal $(\mathfrak{g}', \mathfrak{h}', \sigma')$ and a symmetric subalgebra $(\mathfrak{g}'', \mathfrak{h}'', \sigma'')$. The corresponding construction for a symmetric space (G, H, σ) is more geometric. Let

$T(G)$ and $T(H)$ be the tangent bundles of G and H respectively; the differential of the multiplication $G \times G \to G$ defines a multiplication $T(G) \times T(G) \to T(G)$ to make $T(G)$ into a Lie group and $T(H)$ into a Lie subgroup of $T(G)$. Let $T(\sigma)$ be the differential of σ. Then $(T(G), T(H), T(\sigma))$ is a symmetric space and $T(G)/T(H) = T(G/H)$ in a natural manner. For differential geometric properties of this construction and of similar constructions under more general circumstances, see Yano and Kobayashi [1].

Example 5.2. We shall exhibit a symmetric Lie algebra $(\mathfrak{g}, \mathfrak{h}, \sigma)$ with the following properties:

(i) $\mathfrak{g}$ is simple and is a vector space direct sum $\mathfrak{g}_{-1} + \mathfrak{g}_0 + \mathfrak{g}_1$ with the relations

$$[\mathfrak{g}_0, \mathfrak{g}_0] \subset \mathfrak{g}_0, \quad [\mathfrak{g}_0, \mathfrak{g}_{-1}] \subset \mathfrak{g}_{-1}, \quad [\mathfrak{g}_0, \mathfrak{g}_1] \subset \mathfrak{g}_1,$$

$$[\mathfrak{g}_{-1}, \mathfrak{g}_1] \subset \mathfrak{g}_0, \quad [\mathfrak{g}_{-1}, \mathfrak{g}_{-1}] = 0, \quad [\mathfrak{g}_1, \mathfrak{g}_1] = 0;$$

(ii) The canonical decomposition $\mathfrak{g} = \mathfrak{h} + \mathfrak{m}$ is given by

$$\mathfrak{h} = \mathfrak{g}_0, \quad \mathfrak{m} = \mathfrak{g}_{-1} + \mathfrak{g}_1;$$

(iii) With respect to the Killing-Cartan form B of $\mathfrak{g}$, the subspaces $\mathfrak{g}_{-1}$ and $\mathfrak{g}_1$ are dual to each other and, moreover,

$$B(\mathfrak{g}_{-1}, \mathfrak{g}_{-1}) = 0 \quad \text{and} \quad B(\mathfrak{g}_1, \mathfrak{g}_1) = 0.$$

The following is an example of such a symmetric Lie algebra:

$$\mathfrak{g} = \mathfrak{sl}(p + q; \mathbf{R}),$$

$$\mathfrak{g}_0 = \left\{ \begin{pmatrix} X_{11} & 0 \\ 0 & X_{22} \end{pmatrix} \right\}, \quad \mathfrak{g}_{-1} = \left\{ \begin{pmatrix} 0 & X_{12} \\ 0 & 0 \end{pmatrix} \right\},$$

$$\mathfrak{g}_1 = \left\{ \begin{pmatrix} 0 & 0 \\ X_{21} & 0 \end{pmatrix} \right\},$$

where $X_{11}, X_{22}, X_{12}, X_{21}$ are matrices with p rows and p columns, q rows and q columns, p rows and q columns, q rows and p columns, respectively, and trace X_{11} + trace $X_{22} = 0$.

There are twelve classes of symmetric Lie algebras of classical type and six symmetric Lie algebras of exceptional type possessing the three properties above. They can be also characterized by the properties that $\mathfrak{g}$ is simple and that $\mathfrak{m}$ contains a proper subspace invariant by ad ($\mathfrak{h}$). For all these, see Berger [2], Kobayashi-Nagano [1], and Koh [1].

We consider now symmetric Lie algebras ($\mathfrak{g}$, $\mathfrak{h}$, σ) such that $\mathfrak{g}$ is simple and that ad ($\mathfrak{h}$) acts irreducibly on $\mathfrak{m}$. According to Berger's classification, there are 44 classes of such symmetric Lie algebras of classical type and 86 such symmetric Lie algebras of exceptional type (cf. Berger [2]). We give only one example.

Example 5.3. Let $\mathfrak{g} = \mathfrak{sl}(p + q; \mathbf{R})$ and

$$\mathfrak{h} = \left\{ \begin{pmatrix} X_{11} & X_{12} \\ X_{21} & X_{22} \end{pmatrix}; {}^tX_{11} = -X_{11}, {}^tX_{22} = -X_{22}, {}^tX_{12} = X_{21} \right\},$$

where $X_{11}, X_{12}, X_{21}, X_{22}$ are matrices of the sizes described in Example 5.2. We define an involutive automorphism σ of $\mathfrak{g}$ by setting

$$\sigma \begin{pmatrix} X_{11} & X_{12} \\ X_{21} & X_{22} \end{pmatrix} = \begin{pmatrix} -{}^tX_{11} & {}^tX_{21} \\ {}^tX_{12} & -{}^tX_{22} \end{pmatrix}.$$

To obtain a symmetric Lie algebra ($\mathfrak{g}$, $\mathfrak{h}$, σ) with $\mathfrak{g}$ solvable, it suffices to consider ($\mathfrak{g}_1 + \mathfrak{g}_1$, $\Delta\mathfrak{g}_1$, σ) with $\mathfrak{g}_1$ solvable (cf. Example 2.1). The following construction gives another example.

Example 5.4. Let $\mathfrak{h}$ be a Lie algebra and $\mathfrak{h}^c$ its complexification, i.e., $\mathfrak{h}^c = \mathfrak{h} + i\mathfrak{h}$. Choosing an arbitrary ideal $\mathfrak{h}_1$ of $\mathfrak{h}$, e.g., $\mathfrak{h}_1 = [\mathfrak{h}, \mathfrak{h}]$, we set $\mathfrak{g} = \mathfrak{h} + i\mathfrak{h}_1 \subset \mathfrak{h}^c$. Let $\sigma: \mathfrak{g} \to \mathfrak{g}$ be the complex conjugation, i.e., $\sigma(X + iY) = X - iY$ for $X \in \mathfrak{h}$ and $Y \in \mathfrak{h}_1$. Then ($\mathfrak{g}$, $\mathfrak{h}$, σ) is a symmetric Lie algebra. If $\mathfrak{h}$ is solvable, so are $\mathfrak{h}^c$ and $\mathfrak{g}$.

6. Riemannian symmetric spaces

A Riemannian manifold M is said to be *Riemannian locally symmetric* if it is affine locally symmetric with respect to the Riemannian connection. Similarly, a Riemannian manifold M is said to be *Riemannian (globally) symmetric* if it is affine symmetric with respect to the Riemannian connection.

PROPOSITION 6.1. *Let M be a Riemannian locally symmetric space. For each $x \in M$, the symmetry s_x is isometric.*

Proof. Since s_x is an affine transformation and its differential at x preserves the metric tensor at x, it suffices to prove the following.

LEMMA. *Let M and M' be Riemannian manifolds and $f: M \to M'$ an affine mapping. If $f_*: T_x(M) \to T_{f(x)}(M')$ is isometric for some point $x \in M$, then f is isometric.*

Proof of Lemma. Let g and g' be the metric tensors for M and M', respectively. Let $y \in M$ and let τ be a curve from y to x. Denote by the same letter τ the parallel displacement along τ. Set $\tau' = f(\tau)$. For $X, Y \in T_y(M)$ we have

$$g'_{f(y)}(f_*X, f_*Y) = g'_{f(x)}(\tau'(f_*X), \tau'(f_*Y))$$
$$= g'_{f(x)}(f_*(\tau X), f_*(\tau Y))$$
$$= g_x(\tau X, \tau Y) = g_y(X, Y).$$
QED.

From Theorem 1.1 we obtain

THEOREM 6.2. *A Riemannian manifold is Riemannian locally symmetric if and only if its curvature tensor field is parallel.*

From Theorem 1.2 we obtain

THEOREM 6.3. *A complete, simply connected, Riemannian locally symmetric space is Riemannian symmetric.*

From Theorem 1.3 we obtain

THEOREM 6.4. *A Riemannian symmetric space is complete.*

THEOREM 6.5. *Let M be a Riemannian symmetric space, G the largest connected group of isometries of M and H the isotropy subgroup of G at a point o of M. Let s_o be the symmetry of M at o and σ the involutive automorphism of G defined by*

$$\sigma(g) = s_o \circ g \circ s_o^{-1} \quad \text{for } g \in G.$$

Let G_σ be the closed subgroup of G consisting of elements fixed by σ. Then
(1) G is transitive on M so that $M = G/H$;
(2) H is compact and lies between G_σ and the identity component of G_σ.

Proof. (1) Although this may be obtained from Theorem 1.4, we shall prove it more directly. Let $x, y \in M$ and let $\tau = x_t$, $0 \leq t \leq 4a$, be a geodesic from x to y. (Since M is complete by Theorem 6.4,

x and y can be joined by a geodesic by Theorem 4.1 of Chapter IV.) If we set

$$f_t = s_{x_{2t}} \circ s_{x_t},$$

then f_t is a 1-parameter family of isometries by Proposition 6.1. Since f_0 is the identity transformation, this 1-parameter family of isometries f_t is contained in G. Clearly, f_a maps x into y.

(2) Let $\Im(M)$ be the group of isometries of M and $\Im_o(M)$ the isotropy subgroup of $\Im(M)$ at o. By Corollary 4.8 of Chapter I or by Theorem 3.4 of Chapter VI, $\Im_o(M)$ is compact. Being the identity component of $I(M)$, G is closed in $\Im(M)$. Hence $H = G \cap \Im_o(M)$ is compact. The fact that H lies between G_σ and the identity component of G_σ can be proved exactly in the same way as Theorem 1.5. QED.

For the study of Riemannian symmetric spaces, it is therefore natural to consider symmetric spaces (G, H, σ) satisfying the condition that

$$\mathrm{ad}_\mathfrak{g}(H) \quad \textit{is compact}.$$

This means that the image of H under the adjoint representation of G is a compact subgroup of the group of linear transformations of $\mathfrak{g}$. The condition is satisfied if H is compact. The converse holds if the adjoint representation of G restricted to H is faithful, that is, if H meets with the center of G only at the identity element. In particular, if (G, H, σ) is effective, then $\mathrm{ad}_\mathfrak{g}(H)$ is compact if and only if H is compact.

Assume that (G, H, σ) is a symmetric space with compact $\mathrm{ad}_\mathfrak{g}(H)$. Let $\mathfrak{g} = \mathfrak{h} + \mathfrak{m}$ be the canonical decomposition. Since $\mathfrak{h}$ and $\mathfrak{m}$ are invariant by $\mathrm{ad}_\mathfrak{g}(H)$ (cf. Proposition 2.2) and since $\mathrm{ad}_\mathfrak{g}(H)$ is compact, $\mathfrak{g}$ admits an $\mathrm{ad}_\mathfrak{g}(H)$-invariant inner product with respect to which $\mathfrak{h}$ and $\mathfrak{m}$ are perpendicular to each other. This inner product restricted to $\mathfrak{m}$ induces a G-invariant Riemannian metric on G/H. By Theorem 3.3, any G-invariant Riemannian metric on G/H defines the canonical connection of G/H. Hence G/H is a Riemannian symmetric space.

Remark. Proposition 6.1, Theorems 6.2, 6.3, and 6.4 are valid for the indefinite Riemannian case. Theorem 6.5, except the compactness of H, holds also in this case. But the proof of (1) has to be modified slightly; instead of a single geodesic from x to y, we need a broken geodesic from x to y. It is also possible to obtain (1) from Theorem 1.4.

THEOREM 6.6. *Let M be a simply connected Riemannian symmetric space and $M = M_0 \times M_1 \times \cdots \times M_k$ its de Rham decomposition, where M_0 is a Euclidean space and $M_1, \ldots, M_k$ are all irreducible. Then each factor M_i is a Riemannian symmetric space.*

Proof. We first note that M is complete by Theorem 6.4 and the de Rham decomposition theorem (cf. Theorem 6.2 of Chapter IV) can be applied to M. It suffices to prove the following lemma.

LEMMA. *Let M_1 and M_2 be Riemannian manifolds. If their Riemannian direct product $M_1 \times M_2$ is Riemannian symmetric, then both M_1 and M_2 are Riemannian symmetric.*

Proof of Lemma. Take arbitrary points $o_1 \in M_1$ and $o_2 \in M_2$. Let s be the symmetry of $M_1 \times M_2$ at (o_1, o_2). Let $X_1 \in T_{o_1}(M_1)$ and set $X = (X_1, 0) \in T_{(o_1, o_2)}(M_1 \times M_2)$. Then the symmetry s maps the geodesic $\exp tX = (\exp tX_1, o_2)$ upon the geodesic $\exp(-tX) = (\exp(-tX_1), o_2)$. It follows easily that s sends $M_1 \times \{o_2\}$ onto itself and induces a symmetry of M_1 at o_1. Hence, M_1 is Riemannian symmetric. Similarly for M_2. QED.

7. Structure of orthogonal symmetric Lie algebras

Let $(\mathfrak{g}, \mathfrak{h}, \sigma)$ be a symmetric Lie algebra. Consider the Lie algebra $\mathrm{ad}_\mathfrak{g}(\mathfrak{h})$ of linear endomorphisms of $\mathfrak{g}$ consisting of $\mathrm{ad}\, X$ where $X \in \mathfrak{h}$. If the connected Lie group of linear transformations of $\mathfrak{g}$ generated by $\mathrm{ad}_\mathfrak{g}(\mathfrak{h})$ is compact, then $(\mathfrak{g}, \mathfrak{h}, \sigma)$ is called an *orthogonal symmetric Lie algebra*. If (G, H, σ) is a symmetric space such that H has a finite number of connected components and if $(\mathfrak{g}, \mathfrak{h}, \sigma)$ is its symmetric Lie algebra, then $\mathrm{ad}_\mathfrak{g}(H)$ is compact if and only if $(\mathfrak{g}, \mathfrak{h}, \sigma)$ is an orthogonal symmetric Lie algebra. Let $\mathfrak{g} = \mathfrak{h} + \mathfrak{m}$ be the canonical decomposition of an orthogonal symmetric Lie algebra. Then $\mathfrak{g}$ admits an $\mathrm{ad}_\mathfrak{g}(\mathfrak{h})$-invariant inner product with respect to which $\mathfrak{h}$ and $\mathfrak{m}$ are perpendicular. By an $\mathrm{ad}_\mathfrak{g}(\mathfrak{h})$-invariant inner product, we mean an inner product $(\ ,\ )$ such that

$$([X, Y]), Z) + (Y, [X, Z]) = 0 \quad \text{for } X \in \mathfrak{h} \text{ and } Y, Z \in \mathfrak{g}.$$

Thus, $\mathrm{ad}\, X$ is skew-symmetric with respect to $(\ ,\ )$. The existence of such an inner product is obvious, since an inner product is $\mathrm{ad}_{\mathfrak{g}}(\mathfrak{h})$-invariant if and only if it is invariant by the connected compact Lie group of linear transformations of $\mathfrak{g}$ generated by $\mathrm{ad}_{\mathfrak{g}}(\mathfrak{h})$.

PROPOSITION 7.1. *Let* $(\mathfrak{g}, \mathfrak{h}, \sigma)$ *be an orthogonal symmetric Lie algebra and B the Killing-Cartan form of* $\mathfrak{g}$. *Let* $\mathfrak{c}$ *be the center of* $\mathfrak{g}$. *If* $\mathfrak{h} \cap \mathfrak{c} = 0$, *then B is negative-definite on* $\mathfrak{h}$.

Proof. Let (,) be an $\mathrm{ad}_{\mathfrak{g}}(\mathfrak{h})$-invariant inner product on $\mathfrak{g}$ and fix an orthonormal basis of $\mathfrak{g}$ with respect to (,). Then, for each $X \in \mathfrak{h}$, ad X is expressed by a skew-symmetric matrix $(a_{ij}(X))$. We have

$$B(X, X) = \sum_{i,j} a_{ij}(X) a_{ji}(X) = -\sum_{i,j} (a_{ij}(X))^2 \leq 0,$$

and the equality holds if and only if X is in the center $\mathfrak{c}$. QED.

Using Proposition 7.1 we shall prove

THEOREM 7.2. *Let* $(\mathfrak{g}, \mathfrak{h}, \sigma)$ *be an orthogonal symmetric Lie algebra such that* $\mathfrak{h}$ *and the center of* $\mathfrak{g}$ *have trivial intersection. Then* $(\mathfrak{g}, \mathfrak{h}, \sigma)$ *is a direct sum of orthogonal symmetric Lie algebras* $(\mathfrak{g}_0, \mathfrak{h}_0, \sigma_0)$ *and* $(\mathfrak{g}_1, \mathfrak{h}_1, \sigma_1)$ *such that*

(1) *If* $\mathfrak{g}_0 = \mathfrak{h}_0 + \mathfrak{m}_0$ *is the canonical decomposition, then* $[\mathfrak{m}_0, \mathfrak{m}_0] = 0$;

(2) $\mathfrak{g}_1$ *is semi-simple.*

Proof. We make a strong use of the following

LEMMA. *The radical* $\mathfrak{r}$ *of a Lie algebra* $\mathfrak{g}$ *is the orthogonal complement of* $[\mathfrak{g}, \mathfrak{g}]$ *with respect to the Killing-Cartan form B of* $\mathfrak{g}$.

For the proof, see Jacobson [1; p. 73] or Bourbaki [1; p. 69].

Let $\mathfrak{r}$ be the radical of $\mathfrak{g}$ and let $\mathfrak{g} = \mathfrak{h} + \mathfrak{m}$ be the canonical decomposition. Set

$$\mathfrak{m}_0 = \mathfrak{m} \cap \mathfrak{r}.$$

Let $\mathfrak{m}_1$ be an $\mathrm{ad}_{\mathfrak{g}}(\mathfrak{h})$-invariant subspace of $\mathfrak{m}$ such that

$$\mathfrak{m} = \mathfrak{m}_0 + \mathfrak{m}_1 \quad \text{(vector space direct sum)}.$$

(We have only to take the orthogonal complement of $\mathfrak{m}_0$ in $\mathfrak{m}$

with respect to an $\operatorname{ad}_{\mathfrak{g}}(\mathfrak{h})$-invariant inner product.) Set

$$\mathfrak{h}_1 = [\mathfrak{m}_1, \mathfrak{m}_1],$$
$$\mathfrak{h}_0 = \{X \in \mathfrak{h}; B(X, \mathfrak{h}_1) = 0\},$$
$$\mathfrak{g}_0 = \mathfrak{h}_0 + \mathfrak{m}_0,$$
$$\mathfrak{g}_1 = \mathfrak{h}_1 + \mathfrak{m}_1.$$

Since B is negative-definite on $\mathfrak{h}$ by Proposition 7.1, $\mathfrak{h}_0$ is a complement of $\mathfrak{h}_1$ in $\mathfrak{h}$. Hence $\mathfrak{g}$ is a vector space direct sum of $\mathfrak{g}_0$ and $\mathfrak{g}_1$. From the definition of $\mathfrak{m}_0$ and $\mathfrak{m}_1$ we obtain

(1) $[\mathfrak{h}, \mathfrak{m}_0] \subset \mathfrak{m}_0$,

(2) $[\mathfrak{h}, \mathfrak{m}_1] \subset \mathfrak{m}_1$.

Using $\mathfrak{h}_1 = [\mathfrak{m}_1, \mathfrak{m}_1]$, the Jacobi identity and (2), we have

(3) $[\mathfrak{h}, \mathfrak{h}_1] \subset \mathfrak{h}_1$.

From the invariance of B and (3) we obtain

$$B([\mathfrak{h}, \mathfrak{h}_0], \mathfrak{h}_1) \subset B(\mathfrak{h}_0, [\mathfrak{h}_1, \mathfrak{h}]) \subset B(\mathfrak{h}_0, \mathfrak{h}_1) = 0.$$

Hence,

(4) $[\mathfrak{h}, \mathfrak{h}_0] \subset \mathfrak{h}_0$.

From (cf. Lemma)

$$B([\mathfrak{m}_0, \mathfrak{m}_1], \mathfrak{h}) = B(\mathfrak{m}_0, [\mathfrak{m}_1, \mathfrak{h}]) \subset B(\mathfrak{r}, [\mathfrak{g}, \mathfrak{g}]) = 0$$

and from Proposition 7.1, we obtain

(5) $[\mathfrak{m}_0, \mathfrak{m}_1] = 0$.

From $\mathfrak{h}_1 = [\mathfrak{m}_1, \mathfrak{m}_1]$, the Jacobi identity and (5) we obtain

(6) $[\mathfrak{m}_0, \mathfrak{h}_1] = 0$.

We shall prove

(7) $[\mathfrak{h}_0, \mathfrak{m}_1] = 0$.

Since $[\mathfrak{h}_0, \mathfrak{m}_1] \subset \mathfrak{m}_1$ by (2) and $\mathfrak{m}_1 \cap \mathfrak{r} = \mathfrak{m}_1 \cap \mathfrak{r} \cap \mathfrak{m} = \mathfrak{m}_1 \cap \mathfrak{m}_0 = 0$, it suffices to show $[\mathfrak{h}_0, \mathfrak{m}_1] \subset \mathfrak{r}$. By the lemma, this will follow from $B([\mathfrak{h}_0, \mathfrak{m}_1], \mathfrak{g}) = 0$. Since $\mathfrak{h}$ and $\mathfrak{m}$ are perpendicular to each other with respect to B (cf. Lemma in the proof of Theorem 3.4), we have only to prove $B([\mathfrak{h}_0, \mathfrak{m}_1], \mathfrak{m}) = 0$.

This in turn follows from
$$B([\mathfrak{h}_0, \mathfrak{m}_1], \mathfrak{m}_0) \subset B([\mathfrak{g}, \mathfrak{g}], \mathfrak{r}) = 0$$
and
$$B([\mathfrak{h}_0, \mathfrak{m}_1], \mathfrak{m}_1) \subset B(\mathfrak{h}_0, [\mathfrak{m}_1, \mathfrak{m}_1]) = B(\mathfrak{h}_0, \mathfrak{h}_1) = 0.$$

From (1), (2), ..., (7) we see immediately that $\mathfrak{g}$ is a (Lie algebra) direct sum of $\mathfrak{g}_0$ and $\mathfrak{g}_1$.

From (cf. Lemma)
$$B([\mathfrak{m}_0, \mathfrak{m}_0], \mathfrak{h}) \subset B([\mathfrak{m}_0, \mathfrak{h}], \mathfrak{m}_0) \subset B([\mathfrak{g}, \mathfrak{g}], \mathfrak{r}) = 0$$
and from Proposition 7.1, we obtain

(8) $[\mathfrak{m}_0, \mathfrak{m}_0] = 0$.

It remains to prove that $\mathfrak{g}_1$ is semi-simple, i.e., $\mathfrak{g}_0$ contains the radical $\mathfrak{r}$ of $\mathfrak{g}$. Since $\mathfrak{r} = (\mathfrak{r} \cap \mathfrak{h}) + (\mathfrak{r} \cap \mathfrak{m}) = (\mathfrak{r} \cap \mathfrak{h}) + \mathfrak{m}_0$, it suffices to prove $\mathfrak{r} \cap \mathfrak{h} \subset \mathfrak{h}_0$. But this follows from
$$B(\mathfrak{r} \cap \mathfrak{h}, \mathfrak{h}_1) \subset B(\mathfrak{r}, \mathfrak{h}_1) = B(\mathfrak{r}, [\mathfrak{m}_1, \mathfrak{m}_1]) \subset B(\mathfrak{r}, [\mathfrak{g}, \mathfrak{g}]) = 0.$$
Hence $(\mathfrak{g}, \mathfrak{h}, \sigma)$ is a direct sum of $(\mathfrak{g}_0, \mathfrak{h}_0, \sigma_0)$ and $(\mathfrak{g}_1, \mathfrak{h}_1, \sigma_1)$. To complete the proof, we remark that the connected Lie group of linear transformations of $\mathfrak{g}$ generated by $\mathrm{ad}_\mathfrak{g}(\mathfrak{h})$ is isomorphic to the direct product of the two Lie groups generated by $\mathrm{ad}_{\mathfrak{g}_0}(\mathfrak{h}_0)$ and by $\mathrm{ad}_{\mathfrak{g}_1}(\mathfrak{h}_1)$ and hence both $(\mathfrak{g}_0, \mathfrak{h}_0, \sigma_0)$ and $(\mathfrak{g}_1, \mathfrak{h}_1, \sigma_1)$ are orthogonal symmetric Lie algebras. QED.

Remark. It is clear that in Theorem 7.2 if $(\mathfrak{g}, \mathfrak{h}, \sigma)$ is effective, so are $(\mathfrak{g}_0, \mathfrak{h}_0, \sigma_0)$ and $(\mathfrak{g}_1, \mathfrak{h}_1, \sigma_1)$.

Consider an orthogonal symmetric Lie algebra $(\mathfrak{g}, \mathfrak{h}, \sigma)$ with $\mathfrak{g}$ semi-simple. From Theorem 5.2 we see that $(\mathfrak{g}, \mathfrak{h}, \sigma)$ is a direct sum of orthogonal symmetric Lie algebras of the following two kinds:

(1) $(\mathfrak{g}' + \mathfrak{g}', \Delta\mathfrak{g}', \sigma)$, where $\mathfrak{g}'$ is simple, $\Delta\mathfrak{g}' = \{(X, X); X \in \mathfrak{g}'\}$ and $\sigma(X, Y) = (Y, X)$ for $X, Y \in \mathfrak{g}'$;

(2) $(\mathfrak{g}, \mathfrak{h}, \sigma)$, where $\mathfrak{g}$ is simple.

PROPOSITION 7.3. *Let $(\mathfrak{g}' + \mathfrak{g}', \Delta\mathfrak{g}', \sigma)$ be an orthogonal symmetric Lie algebra with $\mathfrak{g}'$ simple. Let $\mathfrak{g}' + \mathfrak{g}' = \Delta\mathfrak{g}' + \mathfrak{m}$ be the canonical decomposition. Then*

(1) $\mathrm{ad}\,(\Delta\mathfrak{g}')$ *is irreducible on* $\mathfrak{m}$;

(2) *The Killing-Cartan form B of $\mathfrak{g}' + \mathfrak{g}'$ is negative-definite.*

250   FOUNDATIONS OF DIFFERENTIAL GEOMETRY

Proof. Let $\mathfrak{m}'$ be a subspace of $\mathfrak{m}$ invariant by ad $(\Delta\mathfrak{g}')$. Then $\{X;\ (X,\ -X) \in \mathfrak{m}'\}$ is an ideal of $\mathfrak{g}'$. Since $\mathfrak{g}'$ is simple, it follows that either $\mathfrak{m}' = 0$ or $\mathfrak{m}' = \mathfrak{m}$, thus proving (1). Let B' be the Killing-Cartan form of $\mathfrak{g}'$. Then

$$B((X, Y), (X, Y)) = B'(X, X) + B'(Y, Y)$$

for $(X, Y) \in \mathfrak{g}' + \mathfrak{g}'$.

On the other hand, by Proposition 7.1 we have

$$0 > B((X, X), (X, X)) = 2B'(X, X)$$

for $(X, X) \in \Delta\mathfrak{g}',\ X \neq 0$.

It follows that B' and B are negative-definite.   QED.

PROPOSITION 7.4. *Let* $(\mathfrak{g},\ \mathfrak{h},\ \sigma)$ *be an orthogonal symmetric Lie algebra with* $\mathfrak{g}$ *simple. Let* $\mathfrak{g} = \mathfrak{h} + \mathfrak{m}$ *be the canonical decomposition. Then*

(1) ad $\mathfrak{h}$ *is irreducible on* $\mathfrak{m}$;

(2) *The Killing-Cartan form B of $\mathfrak{g}$ is (negative or positive) definite on* $\mathfrak{m}$.

Proof. Choose an inner product (,) on $\mathfrak{m}$ which is ad $\mathfrak{h}$-invariant. Let β be the linear transformation of $\mathfrak{m}$ defined by

$$(\beta X, Y) = B(X, Y) \quad \text{for } X, Y \in \mathfrak{m}.$$

Since B is symmetric and non-degenerate on $\mathfrak{m}$ (cf. Lemma in the proof of Theorem 3.4), the eigenvalues of β are all real and non-zero. Let $\mathfrak{m} = \mathfrak{m}_1 + \cdots + \mathfrak{m}_k$ be the eigenspace decomposition. Then $\mathfrak{m}_1, \ldots, \mathfrak{m}_k$ are mutually orthogonal with respect to (,). Hence, if $i \neq j$, then

$$0 = (\mathfrak{m}_i, \mathfrak{m}_j) = (\beta\mathfrak{m}_i, \mathfrak{m}_j) = B(\mathfrak{m}_i, \mathfrak{m}_j),$$

which implies that $\mathfrak{m}_1, \ldots, \mathfrak{m}_k$ are mutually orthogonal with respect to B also. On the other hand, since B and (,) are invariant by ad $\mathfrak{h}$, it follows that β commutes with ad $\mathfrak{h}$ and that $[\mathfrak{h}, \mathfrak{m}_i] \subset \mathfrak{m}_i$ for every i. Hence, if $i \neq j$, then

$$B([\mathfrak{m}_i, \mathfrak{m}_j], [\mathfrak{m}_i, \mathfrak{m}_j]) \subset B([\mathfrak{m}_i, \mathfrak{m}_j], \mathfrak{h})$$
$$= B(\mathfrak{m}_i, [\mathfrak{m}_j, \mathfrak{h}]) \subset B(\mathfrak{m}_i, \mathfrak{m}_j) = 0.$$

Since B is negative-definite on $\mathfrak{h}$, it follows that $[\mathfrak{m}_i, \mathfrak{m}_j] = 0$ for $i \neq j$. Using this we verify easily that each $[\mathfrak{m}_i, \mathfrak{m}_i] + \mathfrak{m}_i$ is an ideal of $\mathfrak{g}$. Since $\mathfrak{g}$ is simple, we must have $\mathfrak{m} = \mathfrak{m}_1$, i.e., that β has only one eigenvalue. This proves that B is proportional to $(\ ,\ )$. Hence B is definite on $\mathfrak{m}$, thus proving (2). To prove (1), let $\mathfrak{m}'$ be a subspace of $\mathfrak{m}$ invariant by ad $\mathfrak{h}$ and $\mathfrak{m}''$ the orthogonal complement of $\mathfrak{m}'$ in $\mathfrak{m}$ with respect to $(\ ,\ )$ (and hence with respect also to B, since B is proportional to $(\ ,\ )$). Then $\mathfrak{m}''$ is also invariant by ad $\mathfrak{h}$. Hence

$$B([\mathfrak{m}', \mathfrak{m}''], [\mathfrak{m}', \mathfrak{m}'']) \subset B([\mathfrak{m}', \mathfrak{m}''], \mathfrak{h})$$
$$= B(\mathfrak{m}', [\mathfrak{m}'', \mathfrak{h}]) \subset B(\mathfrak{m}', \mathfrak{m}'') = 0.$$

Since B is negative-definite on $\mathfrak{h}$, it follows that $[\mathfrak{m}', \mathfrak{m}''] = 0$. Using this we verify easily that $[\mathfrak{m}', \mathfrak{m}'] + \mathfrak{m}'$ is an ideal of $\mathfrak{g}$. Since $\mathfrak{g}$ is simple, it follows that either $\mathfrak{m}' = 0$ or $\mathfrak{m}' = \mathfrak{m}$.

QED.

As we can see from Example 5.2, it is essential that in Proposition 7.4 $(\mathfrak{g}, \mathfrak{h}, \sigma)$ is an *orthogonal* symmetric Lie algebra. The following proposition which holds for any symmetric Lie algebra may be considered as a converse to (1) of Propositions 7.3 and 7.4.

PROPOSITION 7.5. *Let* $(\mathfrak{g}, \mathfrak{h}, \sigma)$ *be an effective symmetric Lie algebra and* $\mathfrak{g} = \mathfrak{h} + \mathfrak{m}$ *the canonical decomposition. If* ad $\mathfrak{h}$ *is irreducible on* $\mathfrak{m}$, *then one of the following three holds:*

(1) $\mathfrak{g} = \mathfrak{g}' + \mathfrak{g}'$ *with* $\mathfrak{g}'$ *simple,* $\mathfrak{h} = \Delta \mathfrak{g}'$ *and* $\sigma(X, Y) = (Y, X)$ *for* $X, Y \in \mathfrak{g}'$;
(2) $\mathfrak{g}$ *is simple;*
(3) $[\mathfrak{m}, \mathfrak{m}] = 0$.

Proof. If $\mathfrak{g}$ is semi-simple, then we have either (1) or (2) by Theorem 5.2. It suffices therefore to prove that either $\mathfrak{g}$ is semi-simple or $[\mathfrak{m}, \mathfrak{m}] = 0$. Assuming that $\mathfrak{g}$ is not semi-simple, let $\mathfrak{r}$ be the radical of $\mathfrak{g}$ and consider the repeated commutators of $\mathfrak{r}$:

$$\mathfrak{r}^0 = \mathfrak{r}, \quad \mathfrak{r}^i = [\mathfrak{r}^{i-1}, \mathfrak{r}^{i-1}], \quad \text{for } i = 1, 2, \ldots.$$

There exists a non-negative integer k such that $\mathfrak{r}^k \neq 0$ but $\mathfrak{r}^{k+1} = 0$. Since $\mathfrak{r}$ is invariant by σ, so is $\mathfrak{r}^k$. Hence

$$\mathfrak{r}^k = \mathfrak{r}^k \cap \mathfrak{h} + \mathfrak{r}^k \cap \mathfrak{m}.$$

Since $\mathfrak{r}$ is an ideal of $\mathfrak{g}$, so is $\mathfrak{r}^k$ and $[\mathfrak{h}, \mathfrak{r}^k \cap \mathfrak{m}] \subset \mathfrak{r}^k \cap \mathfrak{m}$. Since ad $\mathfrak{h}$ is irreducible on $\mathfrak{m}$, either $\mathfrak{r}^k \cap \mathfrak{m} = 0$ or $\mathfrak{r}^k \cap \mathfrak{m} = \mathfrak{m}$. If

$\mathfrak{r}^k \cap \mathfrak{m} = 0$, then $\mathfrak{r}^k$ is contained in $\mathfrak{h}$, contradicting the assumption that $(\mathfrak{g}, \mathfrak{h}, \sigma)$ is effective. Hence, $\mathfrak{r}^k$ contains $\mathfrak{m}$. Since $[\mathfrak{r}^k, \mathfrak{r}^k] = 0$, we have $[\mathfrak{m}, \mathfrak{m}] = 0$. QED.

Remark. Since $[\mathfrak{m}, \mathfrak{m}] + \mathfrak{m}$ is an ideal of $\mathfrak{g}$, it follows that $\mathfrak{h} = [\mathfrak{m}, \mathfrak{m}]$ in (1) and (2) of Proposition 7.5. Hence we may conclude that, for an effective symmetric Lie algebra $(\mathfrak{g}, \mathfrak{h}, \sigma)$, ad $([\mathfrak{m}, \mathfrak{m}])$ is irreducible on $\mathfrak{m}$ if and only if (1) or (2) of Proposition 7.5 holds. For a symmetric space (G, H, σ), the irreducibility of ad $([\mathfrak{m}, \mathfrak{m}])$ acting on $\mathfrak{m}$ is precisely the irreducibility of the restricted linear holonomy group of the canonical connection on G/H (cf. Theorem 3.2).

We shall therefore say that an effective symmetric Lie algebra $(\mathfrak{g}, \mathfrak{h}, \sigma)$ is *irreducible* if ad $([\mathfrak{m}, \mathfrak{m}])$ is irreducible on $\mathfrak{m}$, i.e., if (1) or (2) of Proposition 7.5 holds.

In general, an orthogonal symmetric Lie algebra $(\mathfrak{g}, \mathfrak{h}, \sigma)$ with $\mathfrak{g}$ semi-simple is said to be of *compact type* or *non-compact type* according as the Killing-Cartan form B of $\mathfrak{g}$ is negative-definite or positive-definite on $\mathfrak{m}$. By Propositions 7.3, 7.4, and 7.5, every irreducible orthogonal symmetric Lie algebra is either of compact type or of non-compact type. From Theorem 5.2 we see also that every orthogonal symmetric Lie algebra $(\mathfrak{g}, \mathfrak{h}, \sigma)$ with $\mathfrak{g}$ semi-simple is a direct sum of an orthogonal symmetric Lie algebra of compact type and an orthogonal symmetric Lie algebra of non-compact type. (Of course, one of the factors might be trivial.) Since B is negative-definite on $\mathfrak{h}$ (cf. Proposition 7.1), $(\mathfrak{g}, \mathfrak{h}, \sigma)$ is of compact type if and only if B is negative-definite on $\mathfrak{g}$. A Lie algebra $\mathfrak{g}$ is said to be *of compact type* if its Killing-Cartan form B is negative-definite on $\mathfrak{g}$. Hence, an orthogonal symmetric Lie algebra $(\mathfrak{g}, \mathfrak{h}, \sigma)$ is of compact type if and only if $\mathfrak{g}$ is of compact type. It is known (cf. Hochschild [1; pp. 142–144]) that a connected, semi-simple Lie group G is compact if and only if its Lie algebra $\mathfrak{g}$ is of compact type (see also Example 3.2 of Chapter X). This justifies the term "compact type."

Example 7.1. We shall show that each point x of a Riemannian locally symmetric space M has a neighborhood which is isometric to a neighborhood of the origin of a certain Riemannian symmetric space G/H. Let $\mathfrak{m} = T_x(M)$ and let $\mathfrak{h}$ be the set of all linear endomorphisms U of $\mathfrak{m}$ which, when extended to a derivation, map g_x and R_x upon 0 (the condition $U \cdot g_x = 0$ means that

U is skew-symmetric). Then $\mathfrak{g} = \mathfrak{m} + \mathfrak{h}$ can be made into a symmetric Lie algebra with σ in the same manner as in Example 2.2. Let G be a simply connected Lie group with Lie algebra $\mathfrak{g}$ and let H be the connected Lie subgroup corresponding to $\mathfrak{h}$ so that (G, H, σ) is a symmetric space for $(\mathfrak{g}, \mathfrak{h}, \sigma)$. We show that $K = \mathrm{ad}\ (H)$ on $\mathfrak{m}$ is compact so that (G, H, σ) is a Riemannian symmetric space and $(\mathfrak{g}, \mathfrak{h}, \sigma)$ is an orthogonal symmetric Lie algebra. By construction of $\mathfrak{g}$, ad $(\mathfrak{h})$ on $\mathfrak{m}$ is nothing but the action of $\mathfrak{h}$ on $\mathfrak{m}$. Thus it is sufficient to show that the group K of orthogonal transformations generated by $\mathfrak{h}$ is compact. Consider the closure $\bar{K}$ of K in the orthogonal group on $\mathfrak{m}$ and let $\bar{\mathfrak{h}}$ be its Lie algebra. If $U \in \bar{\mathfrak{h}}$, then $u_t = \exp tU \in \bar{K}$ is the limit of a sequence of elements $u_n \in K$. Since u_n maps R_x into itself, so does u_t. This being the case for every t, we conclude that $U \cdot R_x = 0$, i.e. $U \in \mathfrak{h}$. Thus $\bar{K} = K$, proving our assertion that $K = \mathrm{ad}\ (H)$ on $\mathfrak{m}$ is compact. Now in the argument at the end of §3, the linear isomorphism F is isometric and hence the affine isomorphism f is an isometry by the lemma for Proposition 6.1.

8. Duality

We shall first consider an arbitrary symmetric Lie algebra $(\mathfrak{g}, \mathfrak{h}, \sigma)$ which is not necessarily orthogonal. Let $\mathfrak{g} = \mathfrak{h} + \mathfrak{m}$ be the canonical decomposition.

If we denote by $\mathfrak{g}^c$ and $\mathfrak{h}^c$ the complexifications of $\mathfrak{g}$ and $\mathfrak{h}$, respectively, and by σ^c the involutive automorphism of $\mathfrak{g}^c$ induced by σ, then $(\mathfrak{g}^c, \mathfrak{h}^c, \sigma^c)$ is a symmetric Lie algebra and $(\mathfrak{g}, \mathfrak{h}, \sigma)$ is a (real) symmetric subalgebra of $(\mathfrak{g}^c, \mathfrak{h}^c, \sigma^c)$. We set

$$\mathfrak{g}^* = \mathfrak{h} + i\mathfrak{m}.$$

It is then easy to verify that $\mathfrak{g}^*$ is a (real) subalgebra of $\mathfrak{g}^c$ and is invariant by σ^c. If we set $\sigma^* = \sigma^c \mid \mathfrak{g}^*$, then we obtain a symmetric subalgebra $(\mathfrak{g}^*, \mathfrak{h}, \sigma^*)$ of $(\mathfrak{g}^c, \mathfrak{h}^c, \sigma^c)$, which is called the *dual of* $(\mathfrak{g}, \mathfrak{h}, \sigma)$. It is evident that $(\mathfrak{g}, \mathfrak{h}, \sigma)$ is the dual of $(\mathfrak{g}^*, \mathfrak{h}, \sigma^*)$, which justifies the term "dual."

THEOREM 8.1. *Let $\mathfrak{g}$ be a Lie algebra and denote by σ the involutive automorphism of $\mathfrak{g} + \mathfrak{g}$ which sends (X, Y) into (Y, X). Then the dual of $(\mathfrak{g} + \mathfrak{g}, \Delta\mathfrak{g}, \sigma)$ is naturally isomorphic to $(\mathfrak{g}^c, \mathfrak{g}, \bar{\sigma})$, where $\bar{\sigma}$ denotes the complex conjugation in $\mathfrak{g}^c$ (with respect to $\mathfrak{g}$), and conversely.*

Proof. Consider the mapping of $(\mathfrak{g} + \mathfrak{g})^*$ into $\mathfrak{g}^c$ defined by
$$(X, X) + i(Y, -Y) \to X + iY, \quad X, Y \in \mathfrak{g}.$$
It is a simple matter to verify that this mapping gives an isomorphism of the dual of $(\mathfrak{g} + \mathfrak{g}, \Delta\mathfrak{g}, \sigma)$ onto $(\mathfrak{g}^c, \mathfrak{g}, \bar{\sigma})$. QED.

We shall now consider orthogonal symmetric Lie algebras.

PROPOSITION 8.2. *An orthogonal symmetric Lie algebra* $(\mathfrak{g} + \mathfrak{g}, \Delta\mathfrak{g}, \sigma)$ *with* $\mathfrak{g}$ *semi-simple is necessarily of compact type.*

Proof. Let B and $\bar{B}$ be the Killing-Cartan form of $\mathfrak{g}$ and $\mathfrak{g} + \mathfrak{g}$ respectively. Then
$$\bar{B}((X, X), (X, X)) = 2B(X, X) \quad \text{for } X \in \mathfrak{g}.$$
The left hand side is negative for every non-zero $X \in \mathfrak{g}$ by Proposition 7.1. Hence B is negative-definite on $\mathfrak{g}$. This implies that $\bar{B}$ is negative-definite on $\mathfrak{g} + \mathfrak{g}$. QED.

THEOREM 8.3. *An orthogonal symmetric Lie algebra* $(\mathfrak{g}, \mathfrak{h}, \sigma)$ *with* $\mathfrak{g}$ *simple is the dual of orthogonal symmetric Lie algebra of the form* $(\mathfrak{g}_1 + \mathfrak{g}_1, \Delta\mathfrak{g}_1, \sigma_1)$ *if and only if* $\mathfrak{g}$ *admits a complex structure J which is compatible with the Lie algebra structure of* $\mathfrak{g}$, *i.e., which satisfies*
$$\text{ad } (JX) = J \circ \text{ad } X = \text{ad } X \circ J \quad \text{for } X \in \mathfrak{g}.$$

Proof. Assume that $(\mathfrak{g}, \mathfrak{h}, \sigma)$ is the dual of $(\mathfrak{g}_1 + \mathfrak{g}_1, \Delta\mathfrak{g}_1, \sigma_1)$. By Theorem 8.1 $\mathfrak{g}$ is isomorphic with $\mathfrak{g}_1^c$. Multiplication by i in $\mathfrak{g}_1^c$ gives a complex structure J in $\mathfrak{g}$ compatible with the Lie algebra structure of $\mathfrak{g}$. Conversely, assume that $\mathfrak{g}$ admits a compatible complex structure J. If B is the Killing-Cartan form of $\mathfrak{g}$, then

$$B(JX, JY) = \text{trace } (\text{ad } JX \circ \text{ad } JY) = \text{trace } (J \circ \text{ad } X \circ J \circ \text{ad } Y)$$
$$= \text{trace } (J^2 \circ \text{ad } X \circ \text{ad } Y) = -\text{trace } (\text{ad } X \circ \text{ad } Y)$$
$$= -B(X, Y).$$

Using this we shall show that $\mathfrak{h} \cap J(\mathfrak{h}) = 0$. If $X \in \mathfrak{h} \cap J(\mathfrak{h})$, then $X = JY$ for some $Y \in \mathfrak{h}$. By the formula above, we have
$$B(X, X) = B(JY, JY) = -B(Y, Y).$$
On the other hand, by Proposition 7.1 both $B(X, X)$ and $B(Y, Y)$ are negative unless $X = 0$. Hence $\mathfrak{h} \cap J(\mathfrak{h}) = 0$. Since

ad $\mathfrak{h}$ is irreducible on $\mathfrak{m} = \mathfrak{g}/\mathfrak{h}$ by Proposition 7.4 and since $(\mathfrak{h} + J(\mathfrak{h}))/\mathfrak{h}$ is a non-zero subspace of $\mathfrak{g}/\mathfrak{h}$ which is invariant by ad $\mathfrak{h}$, we have $\mathfrak{g} = \mathfrak{h} + J(\mathfrak{h})$, i.e., $\mathfrak{g} \approx \mathfrak{h}^c$. Let $X, Y \in \mathfrak{h}$. Since (ad X) $\circ$ (ad JY) maps $\mathfrak{h}$ and $J(\mathfrak{h})$ into $J(\mathfrak{h})$ and $\mathfrak{h}$, respectively, we have $B(X, JY) = 0$, showing that $\mathfrak{h}$ is perpendicular to $J(\mathfrak{h})$ with respect to B. Hence $\mathfrak{g} = \mathfrak{h} + J(\mathfrak{h})$ is the canonical decomposition of $(\mathfrak{g}, \mathfrak{h}, \sigma)$ and σ is the complex conjugation in $\mathfrak{g} \approx \mathfrak{h}^c$ with respect to $\mathfrak{h}$. By Theorem 8.1 $(\mathfrak{g}, \mathfrak{h}, \sigma)$ is the dual of $(\mathfrak{h} + \mathfrak{h}, \Delta\mathfrak{h}, \sigma')$ where σ' maps $(X, Y) \in \mathfrak{h} + \mathfrak{h}$ into $(Y, X) \in \mathfrak{h} + \mathfrak{h}$. QED.

THEOREM 8.4. *The dual* $(\mathfrak{g}^*, \mathfrak{h}, \sigma^*)$ *of an orthogonal symmetric Lie algebra* $(\mathfrak{g}, \mathfrak{h}, \sigma)$ *is an orthogonal symmetric Lie algebra. If* $(\mathfrak{g}, \mathfrak{h}, \sigma)$ *is of compact type (resp. of non-compact type), then* $(\mathfrak{g}^*, \mathfrak{h}, \sigma^*)$ *is of non-compact type (resp. of compact type).*

Proof. Under the linear isomorphism $X \in \mathfrak{m} \to iX \subset i\mathfrak{m}$, ad $\mathfrak{h}$ acting on $\mathfrak{m}$ and ad $\mathfrak{h}$ acting on $i\mathfrak{m}$ are isomorphic. Hence, under the linear isomorphism $X + Y \in \mathfrak{h} + \mathfrak{m} \to X + iY \in \mathfrak{h} + i\mathfrak{m}$, ad $\mathfrak{h}$ acting on $\mathfrak{g}$ and ad $\mathfrak{h}$ acting on $\mathfrak{g}^*$ are isomorphic. If ad $\mathfrak{h}$ acting on $\mathfrak{g}$ generates a compact Lie group of linear transformations of $\mathfrak{g}$, then ad $\mathfrak{h}$ acting on $\mathfrak{g}^*$ also generates a compact Lie group of linear transformations of $\mathfrak{g}^*$. This proves our first assertion.

Let B and B^* be the Killing-Cartan forms of $\mathfrak{g}$ and $\mathfrak{g}^*$ respectively. Fix $X, Y \in \mathfrak{m}$. Then $B(X, Y)$ is the sum of the traces of the following two linear transformations:

$$U \in \mathfrak{h} \to [X, [Y, U]] \in \mathfrak{h} \quad \text{and} \quad V \in \mathfrak{m} \to [X, [Y, V]] \in \mathfrak{m}.$$

Similarly, $B^*(iX, iY)$ is the sum of the traces of the following two linear transformations:

$$U \in \mathfrak{h} \to [iX, [iY, U]] \in \mathfrak{h} \quad \text{and} \quad iV \in i\mathfrak{m} \to [iX, [iY, iV]] \in i\mathfrak{m}.$$

Hence, $B^*(iX, iY) = -B(X, Y)$. This shows that B^* is negative-definite (resp. positive-definite) on $i\mathfrak{m}$ if and only if B is positive-definite (resp. negative-definite) on $\mathfrak{m}$. QED.

From Proposition 7.5 and from the preceding results in this section, we obtain

THEOREM 8.5. *The irreducible orthogonal symmetric Lie algebras of compact type are divided into the following two classes:*

(I) $(\mathfrak{g}, \mathfrak{h}, \sigma)$ *where* $\mathfrak{g}$ *is a simple Lie algebra of compact type;*

(II) $(\mathfrak{g} + \mathfrak{g}, \Delta\mathfrak{g}, \sigma)$ where $\mathfrak{g}$ is a simple Lie algebra of compact type, σ maps $(X, Y) \in \mathfrak{g} + \mathfrak{g}$ into $(Y, X) \in \mathfrak{g} + \mathfrak{g}$ and $\Delta\mathfrak{g}$ is the diagonal of $\mathfrak{g} + \mathfrak{g}$.

The irreducible orthogonal symmetric Lie algebras of non-compact type are divided into the following two classes:

(III) $(\mathfrak{g}, \mathfrak{h}, \sigma)$ where $\mathfrak{g}$ is a simple Lie algebra of non-compact type which does not admit a compatible complex structure;

(IV) $(\mathfrak{g}^c, \mathfrak{g}, \sigma)$ where $\mathfrak{g}$ is a simple Lie algebra of compact type, $\mathfrak{g}^c$ denotes the complexification of $\mathfrak{g}$, and σ is the complex conjugation in $\mathfrak{g}^c$ with respect to $\mathfrak{g}$.

An orthogonal symmetric Lie algebras of type (I) is the dual of an orthogonal symmetric Lie algebra of type (III). An orthogonal symmetric Lie algebra of type (II) is the dual of an orthogonal symmetric Lie algebra of type (IV).

The types of symmetric spaces are related to their geometric properties as follows.

THEOREM 8.6. Let (G, H, σ) be a symmetric space with $\mathrm{ad}_\mathfrak{g}(H)$ compact and let $(\mathfrak{g}, \mathfrak{h}, \sigma)$ be its orthogonal symmetric Lie algebra. Take any G-invariant Riemannian metric on G/H. Then we have

(1) If $(\mathfrak{g}, \mathfrak{h}, \sigma)$ is of compact type, then G/H is a compact Riemannian symmetric space with non-negative sectional curvature and positive-definite Ricci tensor;

(2) If $(\mathfrak{g}, \mathfrak{h}, \sigma)$ is of non-compact type, then G/H is a simply connected non-compact Riemannian symmetric space with non-positive sectional curvature and negative-definite Ricci tensor and is diffeomorphic to a Euclidean space.

Proof. Let $\mathfrak{g} = \mathfrak{h} + \mathfrak{m}$ be the canonical decomposition and $(\,,\,)$ the inner product on $\mathfrak{m}$ corresponding to the given G-invariant Riemannian metric on G/H (cf. Proposition 3.1 of Chapter X). Since $(\mathfrak{g}, \mathfrak{h}, \sigma)$ is assumed to be either of compact or of non-compact type, $\mathfrak{g}$ is semi-simple. We decompose $(\mathfrak{g}, \mathfrak{h}, \sigma)$ into a direct sum of irreducible orthogonal symmetric Lie algebras:

$$(\mathfrak{g}, \mathfrak{h}, \sigma) = \sum_{i=1}^{r} (\mathfrak{g}_i, \mathfrak{h}_i, \sigma_i).$$

For each i, let $\mathfrak{g}_i = \mathfrak{h}_i + \mathfrak{m}_i$ be the canonical decomposition and B_i the Killing-Cartan form of $\mathfrak{g}_i$. Then every $\mathrm{ad}\,(\mathfrak{h})$-invariant symmetric bilinear form on $\mathfrak{m}$ can be expressed by means of $B_1, \ldots, B_r$ as follows:

LEMMA. *An* ad $(\mathfrak{h})$-*invariant symmetric bilinear form* (,) *on* $\mathfrak{m}$ *is a linear combination of* $B_1, \ldots, B_r$, *i.e., there exist constants* $a_1, \ldots, a_r$ *such that*

$$(X, Y) = \sum_{i=1}^{r} a_i B_i(X_i, Y_i) \quad \text{for } X = \Sigma_i X_i \text{ and } Y = \Sigma_i Y_i,$$

where $X_i, Y_i \in \mathfrak{m}_i$ *for* $i = 1, \ldots, r$.

Proof of Lemma. We first prove that, for $i \neq j$, $\mathfrak{m}_i$ and $\mathfrak{m}_j$ are perpendicular to each other with respect to (,). Let $X_i \in \mathfrak{m}_i$, $X_j \in \mathfrak{m}_j$ and $Z_i \in \mathfrak{h}_i$. Then $[Z_i, X_j] = 0$ implies

$$([Z_i, X_i], X_j) = -(X_i, [Z_i, X_j]) = 0.$$

Since $(\mathfrak{g}_i, \mathfrak{h}_i, \sigma_i)$ is irreducible, the set $\{[Z_i, X_i]; Z_i \in \mathfrak{h}_i, X_i \in \mathfrak{m}_i\}$ spans $\mathfrak{m}_i$. This shows that an arbitrary $X_j \in \mathfrak{m}_j$ is perpendicular to $\mathfrak{m}_i$ with respect to (,), thus proving our assertion. It remains to prove that the restriction of (,) to $\mathfrak{m}_i$ is proportional to B_i. But this follows easily from the fact that both (,) and B_i are invariant by ad $\mathfrak{h}_i$ acting irreducibly on $\mathfrak{m}_i$ (cf. Theorem 1 of Appendix 5). This completes the proof of Lemma.

We shall first prove the statements about the sectional curvature and the Ricci tensor. By virtue of the lemma, if suffices to consider the irreducible case. Then a G-invariant Riemannian metric on G/H is defined by an inner product of the form aB on $\mathfrak{m}$, where B is the Killing-Cartan form of $\mathfrak{g}$ and a is a constant which is positive or negative according as B is positive- or negative-definite on $\mathfrak{m}$, i.e., according as $(\mathfrak{g}, \mathfrak{h}, \sigma)$ is of non-compact or compact type. Let X and Y be elements of $\mathfrak{m}$ such that $aB(X, Y) = aB(Y, Y) = 1$ and $aB(X, Y) = 0$. Then the sectional curvature $R(X, Y, X, Y)_o$ of the plane spanned by X and Y at the origin o of G/H is given by (cf. (1) of Theorem 3.2)

$$R(X, Y, X, Y)_o = -aB([[X, Y], Y], X) = aB([X, Y], [X, Y]).$$

Since B is negative-definite on $\mathfrak{h}$ and $[X, Y] \in \mathfrak{h}$, it follows that $R(X, Y, X, Y)_o$ is non-negative or non-positive according as $(\mathfrak{g}, \mathfrak{h}, \sigma)$ is of compact or non-compact type. We note that $R(X, Y, X, Y)_o = 0$ if and only if $[X, Y] = 0$. Since the Ricci tensor of G/H is G-invariant, it must coincide with the metric tensor up to a constant factor by the lemma above. This shows that if $(\mathfrak{g}, \mathfrak{h}, \sigma)$ is irreducible, then G/H is an Einstein space.

Assume that $(\mathfrak{g}, \mathfrak{h}, \sigma)$ is of compact type. Since the sectional curvature is non-negative, the Ricci tensor of G/H is positive semi-definite. Since G/H is an Einstein space, the Ricci tensor is either positive-definite or zero. From the expression of the Ricci tensor by means of the sectional curvature (cf. p. 249 of Volume I), it follows that if the Ricci tensor is zero, then the sectional curvature is also zero. But this contradicts the irreducibility of $(\mathfrak{g}, \mathfrak{h}, \sigma)$. Hence the Ricci tensor is positive-definite. Similarly, if $(\mathfrak{g}, \mathfrak{h}, \sigma)$ is of non-compact type, then the Ricci tensor of G/H is negative-definite.

We consider now the general case where $(\mathfrak{g}, \mathfrak{h}, \sigma)$ need not be irreducible. From what we have just proved in the irreducible case, we see that if $(\mathfrak{g}, \mathfrak{h}, \sigma)$ is of compact type (resp. non-compact type), then the sectional curvature of G/H is non-negative (resp. non-positive) and the Ricci tensor of G/H is positive-definite (resp. negative-definite). If $(\mathfrak{g}, \mathfrak{h}, \sigma)$ is of compact type, then G/H is compact by Theorem 5.8 of Chapter VIII. If $(\mathfrak{g}, \mathfrak{h}, \sigma)$ is of non-compact type, then G/H is simply connected and is diffeomorphic to a Euclidean space by Theorem 8.3 of Chapter VIII. QED.

COROLLARY 8.7. *Let (G, H, σ) be a symmetric space with $\mathrm{ad}_\mathfrak{g}(H)$ compact such that its orthogonal symmetric Lie algebra $(\mathfrak{g}, \mathfrak{h}, \sigma)$ is irreducible. Take any G-invariant Riemannian metric on G/H (which is unique up to a constant factor). Then we have either*

(1) *$(\mathfrak{g}, \mathfrak{h}, \sigma)$ is of compact type and G/H is a compact Einstein space with non-negative sectional curvature and positive-definite Ricci tensor; or*

(2) *$(\mathfrak{g}, \mathfrak{h}, \sigma)$ is of non-compact type and G/H is a simply connected Einstein space with non-positive sectional curvature and negative-definite Ricci tensor and is diffeomorphic to a Euclidean space.*

Proof. At the end of §7 we explained that if an orthogonal symmetric Lie algebra $(\mathfrak{g}, \mathfrak{h}, \sigma)$ is irreducible, it is either of compact type or non-compact type. Corollary 8.7 follows now immediately from Theorem 8.6. The fact that G/H is an Einstein space was also established in the course of the proof of Theorem 8.6. QED.

Remark. In Theorem 8.6, G/H need not be an Einstein space. But it is not hard to see that there is a G-invariant metric on G/H for which G/H is an Einstein space.

9. Hermitian symmetric spaces

Let M be a manifold with an affine connection Γ and an almost complex structure J. We say that M is *complex affine locally symmetric* if for each point x of M the symmetry s_x (defined in §1) leaves Γ and J invariant. M is *complex affine (globally) symmetric* if each symmetry s_x can be extended to a global transformation of M which preserves Γ and J.

Let M be complex affine locally symmetric. Then the tensor ∇J is invariant by each symmetry s_x. Since the degree of ∇J is odd, we conclude that $\nabla J = 0$ by the lemma for Theorem 1.1. By Corollary 3.5 of Chapter IX we see that the torsion of J is 0. By Theorem 7.7 of Chapter VI we know that M and Γ are real analytic. The almost complex structure J being parallel, it is also real analytic. By the real analytic version of Theorem 2.5 of Chapter IX we conclude that J is integrable. Thus M is a complex manifold and Γ is complex.

Suppose that M is an almost complex manifold with a Hermitian metric g and let Γ be the Riemannian connection for g. If M is complex affine locally (or globally) symmetric with respect to Γ, then we say that M is *Hermitian locally* (or *globally*) *symmetric*. In this case, J is integrable and g is a Kaehler metric (since $\nabla J = 0$). Now suppose that a Kaehler manifold M is locally symmetric as a Riemannian manifold (see §6). Then each symmetry s_x maps J_x upon itself. For any point y in a normal neighborhood U of x, let τ be a geodesic from x to y in U. Since s_x is an affine transformation, we have $s_x \circ \tau = s_x(\tau) \circ s_x$, where τ and $s_x(\tau)$ denote the parallel displacement along τ and that along the image curve. Since J is parallel, we conclude that $s_x(J_y) = J_{s(x)y}$. Thus s_x preserves J and M is Hermitian locally symmetric.

Summing up we have

PROPOSITION 9.1. *Let M be a differentiable manifold with an almost complex structure J and a Hermitian metric g.*

(1) *If M is complex affine locally symmetric, then J is integrable and g is a Kaehler metric.*

(2) *If (J, g) is a Kaehler structure and if M is Riemannian locally symmetric, then M is Hermitian locally symmetric.*

Let M be a Hermitian globally symmetric space. Then M is a

Kaehler manifold such that each symmetry s_x can be extended to a global automorphism (i.e. holomorphic isometry) of M. Let $\mathfrak{K}(M)$ denote the group of all automorphisms of M. The same arguments as in the proof of Theorem 1.4 show that $\mathfrak{K}(M)$ is transitive on M. The identity component G of $\mathfrak{K}(M)$ is also transitive and M can be written as a homogeneous space G/H. We define an automorphism σ of G as in Theorem 1.5. We may conclude that $M = G/H$ is a symmetric space, where G is the largest connected group of automorphisms of M. We have proved

PROPOSITION 9.2. *A Hermitian globally symmetric space M can be represented as a symmetric space G/H, where G is the largest connected group of automorphisms of M.*

We shall now start with a symmetric space G/H with the canonical decomposition $\mathfrak{g} = \mathfrak{h} + \mathfrak{m}$ of the Lie algebra. Suppose that $\mathfrak{m}$ admits a complex structure I such that ad $(a) \circ I = I \circ$ ad (a) for every $a \in H$. By Proposition 6.5 of Chapter X we get an invariant almost complex structure J on G/H. The integrability condition in the same proposition is satisfied because $[\mathfrak{m}, \mathfrak{m}] \subset \mathfrak{h}$. Thus G/H is a complex homogeneous space. The canonical affine connection on G/H is then complex, since $\nabla J = 0$ by (4) of Theorem 3.2 (this fact also implies that J is integrable as we have already seen). If, in addition, $\mathfrak{m}$ admits an ad (H)-invariant inner product which is Hermitian with respect to I, then the corresponding invariant metric g on G/H is Hermitian with respect to J, and G/H is Hermitian globally symmetric. Summing up, we have

PROPOSITION 9.3. *Let G/H be a symmetric homogeneous space with the canonical decomposition $\mathfrak{g} = \mathfrak{h} + \mathfrak{m}$ of the Lie algebra.*

(1) If $\mathfrak{m}$ admits an ad (H)-*invariant complex structure I, then G/H admits an invariant complex structure such that the canonical affine connection is complex and G/H is complex affine symmetric.*

(2) If, in addition, $\mathfrak{m}$ admits an ad (H)-*invariant inner product which is Hermitian with respect to I, then G/H admits an invariant Kaehler metric and G/H is Hermitian symmetric.*

As an application of Proposition 9.3, we have

PROPOSITION 9.4. *Let G/H and G^*/H^* be symmetric homogeneous spaces with symmetric Lie algebras $(\mathfrak{g}, \mathfrak{h}, \sigma)$ and $(\mathfrak{g}^*, \mathfrak{h}^*, \sigma^*)$, respectively, such that $(\mathfrak{g}^*, \mathfrak{h}^*, \sigma^*)$ is the dual of $(\mathfrak{g}, \mathfrak{h}, \sigma)$. Then G/H is complex*

XI. SYMMETRIC SPACES   261

affine symmetric (resp. Hermitian symmetric) if and only if G^*/H^* is complex affine symmetric (resp. Hermitian symmetric).

This is immediate from the definition of the dual symmetric Lie algebra given in §8 and from Proposition 9.3.

PROPOSITION 9.5. *Let G/H be a complex affine symmetric space. Let $\mathfrak{m}'$ be a subspace of $\mathfrak{m}$ such that $[[\mathfrak{m}', \mathfrak{m}'], \mathfrak{m}'] \subset \mathfrak{m}'$ and let M' be the corresponding totally geodesic submanifold as in Theorem 4.3. Then M' is a complex submanifold of G/H if and only if $\mathfrak{m}'$ is invariant by the complex structure I on $\mathfrak{m}$.*

Proof. If M' is a complex submanifold, then $\mathfrak{m} = T'_o(M)$ is invariant by I. Conversely, suppose $\mathfrak{m}'$ is invariant by I. Since M' is totally geodesic and since J is parallel, it follows that, for each $x \in M'$, $T_x(M')$ is invariant by J_x, where J is the complex structure of G/H. Thus M' is a complex submanifold.   QED.

Examples of Hermitian symmetric spaces and their totally geodesic complex submanifolds are discussed in the next section.

The following theorem gives essentially a group-theoretic characterization of a Hermitian symmetric space.

THEOREM 9.6. (1) *Let G/H be a Hermitian symmetric space where H is compact and G is effective on G/H. If G is semi-simple, then there is an element Z_0 in the center of $\mathfrak{h}$ such that $I = \mathrm{ad}_\mathfrak{m}(Z_0)$ and $\mathfrak{h} = \{X \in \mathfrak{g}; [Z_0, X] = 0\}$;*

(2) *Let G/H be a symmetric space where H is compact and G is effective on G/H. If G is simple and if H has non-discrete center, then G/H admits an invariant Hermitian structure.*

Proof. (1) We extend the complex structure $I: \mathfrak{m} \to \mathfrak{m}$ to a linear endomorphism of $\mathfrak{g}$ by setting $I(X) = 0$ for $X \in \mathfrak{h}$. We show that I is a derivation of $\mathfrak{g}$, that is, $I[X, Y] = [IX, Y] + [X, IY]$ for all $X, Y \in \mathfrak{g}$. This identity is trivial for $X, Y \in \mathfrak{h}$. If $X \in \mathfrak{m}$ and $Y \in \mathfrak{h}$, then $IY = 0$. Also, I being invariant by ad (H), we have $[IX, Y] = I[X, Y] = 0$, proving the identity. It remains to consider the case $X, Y \in \mathfrak{m}$. By Proposition 4.5 of Chapter IX, we have $R(IX, IY) = R(X, Y)$ for the curvature tensor R. Since $R(X, Y) = -\mathrm{ad}_\mathfrak{m}([X, Y])$ by Theorem 3.2, we get $\mathrm{ad}_\mathfrak{m}([IX, IY]) = \mathrm{ad}_\mathfrak{m}([X, Y])$. The representation $\mathfrak{h} \to \mathrm{ad}_\mathfrak{m}(\mathfrak{h})$ being faithful (as we saw at the end of §1 of Chapter X), we get $[IX, IY] = [X, Y]$. Replacing Y by IY, we obtain

$[IX, Y] + [X, IY] = 0$. Since $I[X, Y] = 0$, we have thus proved the identity. Thus I is a derivation. Since G is semi-simple, every derivation of $\mathfrak{g}$ is inner (see Bourbaki [1], corollaire 3, p. 73). Thus there is an element $Z_0 \in \mathfrak{g}$ such that $I = \operatorname{ad}(Z_0)$. If we write $Z_0 = X_0 + Y_0$ with $X_0 \in \mathfrak{m}$ and $Y_0 \in \mathfrak{h}$, then $0 = \operatorname{ad}(Z_0) Z_0 = IZ_0 = IX_0$ so that $I^2 X_0 = -X_0 = 0$, that is, $Z_0 \in \mathfrak{h}$. Since $\operatorname{ad}(Z_0) Y = IY = 0$ for every $Y \in \mathfrak{h}$, we see that Z_0 is in the center of $\mathfrak{h}$. If $[Z_0, X] = 0$ for some $X \in \mathfrak{g}$, then $IX = 0$ and hence $X \in \mathfrak{h}$, proving that $\mathfrak{h}$ is the centralizer of Z_0 in $\mathfrak{g}$.

(2) Let H^0 denote the identity component of H. By Proposition 7.4, ad $\mathfrak{h}$ on $\mathfrak{m}$ is irreducible so that ad (H^0) on $\mathfrak{m}$ is irreducible. Choosing an ad (H)-invariant inner product on $\mathfrak{m}$, we consider ad (H^0) as a connected irreducible Lie subgroup of the orthogonal group on $\mathfrak{m}$. From the proof of Theorem 2 in Appendix 5 we see that the center of ad (H) is at most 1-dimensional. Since $H \to \operatorname{ad}_{\mathfrak{m}}(H)$ is faithful and since H has non-discrete center by assumption, we conclude that the identity component of the center of H is 1-dimensional and is isomorphic to the circle group. Thus there is an element $h \in H^0$ such that $\operatorname{ad}_{\mathfrak{m}}(h)^2 = -1$ on $\mathfrak{m}$. Thus $\operatorname{ad}_{\mathfrak{m}}(h)$ is an ad (H)-invariant complex structure on $\mathfrak{m}$. Now Proposition 9.3 completes the proof of our assertion. QED.

By Theorem 3.3 we know that the Ricci tensor of a symmetric space G/H is independent of the choice of a possible invariant Riemannian metric on G/H. The Ricci tensor of a Hermitian symmetric space G/H with semi-simple G has a simple expression.

PROPOSITION 9.7. *Let G/H be a Hermitian symmetric space where H is compact and G is effective on G/H. If G is semi-simple, then the Ricci tensor S of G/H is given by*

$$S(X, Y) = -\tfrac{1}{2} B(X, Y) \quad \text{for } X, Y \in \mathfrak{m},$$

where B is the Killing-Cartan form of $\mathfrak{g}$.

Proof. Let Z_0 be the element of the center of $\mathfrak{h}$ given in Theorem 9.6. By (2) of Proposition 4.5 of Chapter IX, we have

$$S(X, Y) = \tfrac{1}{2} \text{ trace of } I \circ R(X, IY) \quad \text{for } X, Y \in \mathfrak{m}.$$

Using $R(X, Y) = -\operatorname{ad}_{\mathfrak{m}}([X, Y])$ and $I = \operatorname{ad}_{\mathfrak{m}}(Z_0)$, we have

$$I \circ R(X, IY) = -\operatorname{ad}_{\mathfrak{m}}(Z_0) \circ \operatorname{ad}_{\mathfrak{m}}([X, [Z_0, Y]]).$$

Let $W = [X, [Z_0, Y]] \in \mathfrak{h}$. Since $[\mathfrak{m}, \mathfrak{h}] \subset \mathfrak{h}$ and $[\mathfrak{h}, \mathfrak{h}] \subset \mathfrak{h}$, we have

trace ad $(Z_0) \circ$ ad $(W) =$ trace ad$_\mathfrak{m}$ $(Z_0) \circ$ ad$_\mathfrak{m}$ (W)
$\qquad\qquad\qquad\qquad\qquad$ $+$ trace ad$_\mathfrak{h}$ $(Z_0) \circ$ ad$_\mathfrak{h}$ (W).

But ad$_\mathfrak{h}$ $(Z_0) \circ$ ad$_\mathfrak{h}$ $(W) = 0$ since Z_0 is in the center of $\mathfrak{h}$. Thus we get

trace $I \circ R(X, IY) = -$trace ad $(Z_0) \circ$ ad $(W) = -B(Z_0, W)$
$\qquad\qquad\qquad\quad = -B(Z_0, [X, [Z_0, Y]])$.

By invariance of B by ad $(\mathfrak{g})$, this is equal to

$B([X, Z_0], [Z_0, Y]) = -B([Z_0, X], [Z_0, Y])$
$\qquad\qquad\qquad = B([Z_0, [Z_0, X]], Y) = -B(X, Y)$,

since ad $(Z_0)^2 X = I^2 X = -X$. Therefore we have

$$S(X, Y) = -\tfrac{1}{2} B(X, Y) \quad \text{for } X, Y \in \mathfrak{m}. \qquad \text{QED.}$$

Let M be a simply connected Hermitian symmetric space and let $M = M_0 \times M_1 \times \cdots \times M_k$ be the de Rham decomposition of M as a Riemannian manifold, where M_0 is a Euclidean space and $M_1, \ldots, M_k$ are all irreducible. By Theorem 6.6, each factor is a Riemannian symmetric space. By Theorem 8.1 of Chapter IX, each factor is also a Kaehler manifold in a natural fashion. Hence each factor is a Hermitian symmetric space. In particular, M_0 is a complex Euclidean space and $M_1, \ldots, M_k$ are irreducible Hermitian symmetric spaces. Let $M = G/H$ be an irreducible Hermitian symmetric space where H is compact and G is effective on $M = G/H$. Let $(\mathfrak{g}, \mathfrak{h}, \sigma)$ be its orthogonal symmetric Lie algebra. Since $\mathfrak{h}$ has non-trivial center (see Theorem 9.6), $(\mathfrak{g}, \mathfrak{h}, \sigma)$ must be of type (I) or type (III) in Theorem 8.5 according as G/H is of compact type or non-compact type. We may now conclude that $(\mathfrak{g}, \mathfrak{h}, \sigma)$ *is the effective orthogonal symmetric Lie algebra of an irreducible Hermitian symmetric space G/H if and only if $\mathfrak{g}$ is simple and $\mathfrak{h}$ has non-trivial center.*

We consider a bounded domain D in $\mathbf{C}^n$ with the Bergman metric g (see Example 6.6 of Chapter IX). If D is Hermitian symmetric, it is called a *bounded symmetric domain*. For a bounded domain D every holomorphic transformation is necessarily an isometry (see Example 6.6 of Chapter IX). For a bounded symmetric domain D, the largest connected group G of holomorphic transformations is transitive on D and D is expressed as a Hermitian symmetric space G/H where H is compact. From Example 3.3 of Chapter IX we know that the Ricci tensor S of a bounded symmetric domain (or more generally, a homogeneous bounded domain) D is equal to $-g$ and hence is negative-definite. Let $\tilde{D}$ be the universal covering space of a bounded symmetric domain D. (We shall see in a moment that D is simply connected.) Since the Ricci tensor of $\tilde{D}$ is negative-definite, we see from Theorem 8.6 that all factors of $\tilde{D}$ in its de Rham decomposition are of non-compact type. (We may see this fact also as follows. The projection $p\colon \tilde{D} \to D \subset \mathbf{C}^n$ induces a holomorphic mapping from each factor of the de Rham decomposition of $\tilde{D}$ into $\mathbf{C}^n$. From the maximal modulus principle we see that $\tilde{D}$ has neither Euclidean nor compact factor.) Thus *a bounded symmetric domain D in $\mathbf{C}^n$ is of non-compact type and is simply connected* (see Theorem 8.6). It is known that, conversely, every Hermitian symmetric space of non-compact type can be realized as a bounded symmetric domain in $\mathbf{C}^n$; this result was obtained by E. Cartan [18] by case-by-case examination and the first *a priori* proof was given by Harish-Chandra (see Helgason [2]).

E. Cartan [18] proved that every homogeneous bounded domain D in $\mathbf{C}^2$ or $\mathbf{C}^3$ is symmetric. Hano [2] proved that a bounded domain D in $\mathbf{C}^n$ on which a unimodular Lie group of holomorphic transformations is transitive is necessarily symmetric. Pyatetzki-Shapiro [1], [2] discovered homogeneous bounded domains which are not symmetric.

10. Examples

In this section we discuss some examples of spaces from the point of view of symmetric spaces.

Example 10.1. $SO(n+1)/SO(n) = S^n$ (cf. Theorem 3.1 of Chapter V and Example in Note 7). Let $e_0, e_1, \ldots, e_n$ be the natural basis in $\mathbf{R}^{n+1}$ and let S^n be the unit sphere in $\mathbf{R}^{n+1}$. The rotation group $SO(n+1)$ acts on S^n transitively, and the isotropy group at $e_0 \in S^n$ consists of all $A \in SO(n+1)$ such that $Ae_0 = e_0$,

namely, such that
$$A = \begin{pmatrix} 1 & 0 \\ 0 & B \end{pmatrix}, \quad \text{where } B \in SO(n).$$

We denote this subgroup by $SO(n)$. We have thus a diffeomorphism f of $SO(n+1)/SO(n)$ onto S^n such that $f(\pi(A)) = Ae_0$ for every $A \in SO(n+1)$ where π denotes the natural projection of $SO(n+1)$ onto $SO(n+1)/SO(n)$. Let σ be an involutive automorphism of $SO(n+1)$ given by $\sigma(A) = SAS^{-1}$, where

$$S = \begin{pmatrix} -1 & 0 \\ 0 & I_n \end{pmatrix} \quad (I_n\text{: identity matrix of degree } n).$$

The subgroup $SO(n)$ coincides with the identity component of the subgroup of all fixed elements of σ. Thus $SO(n+1)/SO(n)$ is a symmetric space. The canonical decomposition of the Lie algebra is given by
$$\mathfrak{o}(n+1) = \mathfrak{o}(n) + \mathfrak{m},$$
where $\mathfrak{o}(n+1)$ is the Lie algebra of all $(n+1) \times (n+1)$ skew-symmetric matrices, $\mathfrak{o}(n)$ the subalgebra of $\mathfrak{o}(n+1)$ consisting of all matrices of the form

$$\begin{pmatrix} 0 & 0 \\ 0 & B \end{pmatrix} \quad (B\text{: skew-symmetric of degree } n),$$

and $\mathfrak{m}$ the subspace of all matrices of the form $\begin{pmatrix} 0 & -{}^t\xi \\ \xi & 0 \end{pmatrix}$, where ξ is a (column) vector in $\mathbf{R}^n$. The adjoint representation of $SO(n)$ in $\mathfrak{m}$ is given by

$$\operatorname{ad}\begin{pmatrix} 1 & 0 \\ 0 & B \end{pmatrix}\begin{pmatrix} 0 & -{}^t\xi \\ \xi & 0 \end{pmatrix} = \begin{pmatrix} 0 & -{}^t(B\xi) \\ B\xi & 0 \end{pmatrix};$$

in other words, it is essentially the action of $SO(n)$ on $\mathbf{R}^n$. Similarly, we have for $B \in \mathfrak{o}(n)$

$$\operatorname{ad}\begin{pmatrix} 0 & 0 \\ 0 & B \end{pmatrix}\begin{pmatrix} 0 & -{}^t\xi \\ \xi & 0 \end{pmatrix} = \begin{pmatrix} 0 & -{}^t(B\xi) \\ B\xi & 0 \end{pmatrix},$$

that is, the adjoint representation of $\mathfrak{o}(n)$ in $\mathfrak{m}$ is essentially the action of $\mathfrak{o}(n)$ on $\mathbf{R}^n$.

We transfer the Euclidean inner product (ξ, η) on $\mathbf{R}^n$ to $\mathfrak{m}$ under the identification

$$\xi \in \mathbf{R}^n \leftrightarrow \begin{pmatrix} 0 & -{}^t\xi \\ \xi & 0 \end{pmatrix} \in \mathfrak{m},$$

getting an inner product on $\mathfrak{m}$ which is invariant by ad $(SO(n))$. We note that if $\xi, \eta \in \mathbf{R}^n$ correspond to $X, Y \in \mathfrak{m}$, respectively, then $(\xi, \eta) = -\frac{1}{2}$ trace XY; thus our inner product on $\mathfrak{m}$ is the restriction of the inner product in $\mathfrak{o}(n+1)$ given by

$$(A, B) = -\tfrac{1}{2} \text{ trace } AB, \qquad A, B \in \mathfrak{o}(n+1),$$

which is invariant by ad $(SO(n+1))$ on $\mathfrak{o}(n+1)$. It is also known that the Killing-Cartan form φ of $\mathfrak{o}(n+1)$ is given by

$$\varphi(A, B) = (n-1) \text{ trace } AB.$$

Thus φ is negative-definite on $\mathfrak{o}(n+1)$, and $\varphi(X, Y) = -2(n-1)(X, Y)$ for $X, Y \in \mathfrak{m}$.

To find the differential f_* at the origin, let $\xi \in \mathbf{R}^n$ and let X be the corresponding element in $\mathfrak{m}$. Then

$$f_*(X) = [d(\exp tX \cdot e_0)/dt]_{t=0} = \begin{pmatrix} 0 \\ \xi \end{pmatrix} \in T_{e_0}(S^n).$$

It follows that f_* is isometric at o (with respect to the usual metric on S^n induced by the Euclidean metric on $\mathbf{R}^{n+1}$). Since the metrics on $SO(n+1)/SO(n)$ and S^n are invariant by $SO(n+1)$ and since $f \circ A = A \circ f$ for every $A \in SO(n+1)$, we see that f is an isometry of $SO(n+1)/SO(n)$ onto S^n.

The curvature tensor R of $SO(n+1)/SO(n)$ at the origin is given, under the identification of $\mathfrak{m}$ with $\mathbf{R}^n$, by $R(\xi, \eta) = -\text{ad}\,([\xi, \eta])$ (cf. Theorem 3.2). A simple computation of matrix brackets shows that $R(\xi, \eta) = \xi \wedge \eta$, where the endomorphism $\xi \wedge \eta$ maps ζ into $(\eta, \zeta)\xi - (\xi, \zeta)\eta$. This shows that $SO(n+1)/SO(n)$ has constant sectional curvature 1 (cf. §2 of Chapter V).

The geodesics of $SO(n+1)/SO(n)$ starting from the origin are of the form $\pi(\exp t\xi)$, where ξ is the element of $\mathfrak{m}$ corresponding

XI. SYMMETRIC SPACES 267

to $\xi \in \mathbf{R}^n$. For example, if $\xi = e_1$, then

$$\exp t\xi = \begin{pmatrix} \cos t & -\sin t & 0 \cdots 0 \\ \sin t & \cos t & 0 \cdots 0 \\ 0 & 0 & \\ \cdot & \cdot & \\ \cdot & \cdot & I_{n-1} \\ \cdot & \cdot & \\ 0 & 0 & \end{pmatrix}$$

so that

$$f(\pi(\exp t\xi)) = (\exp t\xi) \cdot e_0 = \begin{pmatrix} \cos t \\ \sin t \\ 0 \\ \cdot \\ \cdot \\ \cdot \\ 0 \end{pmatrix}$$

is a geodesic on S^n starting from e_0. All other geodesics on S^n starting from e_0 are obtained from the above geodesic by the rotations which fix e_0 (namely, $SO(n)$).

To find all m-dimensional totally geodesic submanifolds on S^n we look for subspaces $\mathfrak{m}'$ of $\mathfrak{m}$ such that $[[\mathfrak{m}', \mathfrak{m}'], \mathfrak{m}'] \subset \mathfrak{m}'$ (see Theorem 4.3). Consider the subspace $\mathfrak{m}'$ of $\mathfrak{m}$ consisting of all elements corresponding to the subspace $\mathbf{R}^m = \{\xi \in \mathbf{R}^n; \xi^k = 0 \text{ for } m + 1 \leq k \leq n\}$. A simple computation shows that $\mathfrak{h}' = [\mathfrak{m}', \mathfrak{m}']$ consists of all $B \in \mathfrak{o}(n)$ such that $Be_k = 0$ for $m + 1 \leq k \leq n$. We denote this subalgebra by $\mathfrak{o}(m)$. Thus we have $[\mathfrak{h}', \mathfrak{m}'] \subset \mathfrak{m}'$, and $\mathfrak{g}' = \mathfrak{m}' + \mathfrak{h}'$ coincides with $\{A \in \mathfrak{o}(n+1); Ae_k = 0 \text{ for } m + 1 \leq k \leq n\}$, which we denote by $\mathfrak{o}(m+1)$. The subalgebra $\mathfrak{g}' = \mathfrak{o}(m+1)$ generates $SO(m+1) = \{A \in SO(n+1); Ae_k = e_k \text{ for } m + 1 \leq k \leq n\}$ and $\mathfrak{h}' = \mathfrak{o}(m)$ generates $SO(m) = \{A \in SO(n); Ae_k = e_k \text{ for } m + 1 \leq k \leq n\}$. It follows that $SO(m+1)/SO(m)$ is a complete totally geodesic submanifold of $SO(n+1)/SO(n)$ corresponding to $\mathfrak{m}'$ (in the manner of Theorem 4.3), and

$$f(SO(m+1)/SO(m)) = \{x \in S^n; x^k = 0 \quad \text{for } m + 1 \leq k \leq n\}$$

268   FOUNDATIONS OF DIFFERENTIAL GEOMETRY

is a complete totally geodesic submanifold of S^n. All other m-dimensional complete totally geodesic submanifolds of S^n containing e_0 are obtained from the preceding one by the rotations belonging to $SO(n)$.

Example 10.2. *Hyperbolic space form* (cf. §3 of Chapter V). In $\mathbf{R}^{n+1}$ with the natural basis $e_0, e_1, \ldots, e_n$, $n \geq 2$, we consider a non-degenerate symmetric bilinear form

$$F(x,y) = -x^0 y^0 + \sum_{k=1}^{n} x^k y^k, \qquad x, y \in \mathbf{R}^{n+1}.$$

Let $O(1, n)$ be the orthogonal group for F, i.e.

$$O(1, n) = \{A \in GL(n+1; \mathbf{R}); F(Ax, Ay) = F(x,y), x, y \in \mathbf{R}^{n+1}\}$$
$$= \{A \in GL(n+1; \mathbf{R}); {}^t\!ASA = S\},$$

where

$$S = \begin{pmatrix} -1 & 0 \\ 0 & I_n \end{pmatrix}.$$

When $n = 3$, $O(1, 3)$ is called the *Lorentz group*. A matrix $A = (a_j^i)$, $0 \leq i, j \leq n$, belongs to $O(1, n)$ if and only if the $(n+1)$ column vectors, say, $\alpha_0, \alpha_1, \ldots, \alpha_n$, satisfy

$$F(\alpha_0, \alpha_0) = -1, \qquad F(\alpha_i, \alpha_i) = 1 \qquad \text{for } 1 \leq i \leq n,$$
$$F(\alpha_i, \alpha_j) = 0 \qquad \text{for } 0 \leq i < j \leq n.$$

In particular, $A \in O(1, n)$ satisfies $-(a_0^0)^2 + \sum_{k=1}^{n} (a_0^k)^2 = -1$ so that $a_0^0 \geq 1$ or $a_0^0 \leq -1$. Also, if $A \in O(1, n)$, then from ${}^t\!ASA = S$ we get $\det A = \pm 1$. Let $SO(1, n) = O(1, n) \cap SL(n+1; \mathbf{R})$. It is known that $O(1, n)$ has four connected components; a matrix $A \in O(1, n)$ belongs to the identity component if and only if $\det A = 1$ and $a_0^0 \geq 1$. In what follows we shall denote by G the identity component of $O(1, n)$, which is also the identity component of $SO(1, n)$. The Lie algebra $\mathfrak{g}$ of G is given by

$$\mathfrak{o}(1, n) = \{X \in \mathfrak{gl}(n+1; \mathbf{R}); {}^t\!XS + SX = 0\}.$$

Now the hypersurface M in $\mathbf{R}^{n+1}$ defined by $F(x, x) = -1$ is the disjoint union of two connected components:

$$M' = \{x \in M; x^0 \geq 1\} \quad \text{and} \quad M'' = \{x \in M; x^0 \leq -1\}.$$

XI. SYMMETRIC SPACES

By Witt's theorem (cf. §3 of Chapter V) or by elementary linear algebra we know that $O(1, n)$ acts transitively on M and that G acts transitively on M'. We shall confine our attention to the action of G on M'.

Let $x \in M'$, i.e. $F(x, x) = -1$ and $x^0 \geq 1$. The tangent space $T_x(M')$ is given, through the identification by parallel displacement in $\mathbf{R}^{n+1}$, by the subspace of all vectors $a \in \mathbf{R}^{n+1}$ such that $F(x, a) = 0$. The restriction of F to $T_x(M')$ is positive-definite. Thus the form F restricted to the tangent space at each point of M' gives rise to a Riemannian metric on M' which is obviously invariant by G.

For $e_0 \in M'$, the isotropy group is the subgroup H of G consisting of all matrices of the form

$$\begin{pmatrix} 1 & 0 \\ 0 & B \end{pmatrix}, \quad \text{where } B \subset SO(n).$$

We have thus a diffeomorphism f of G/H onto M' which is compatible with the actions of G on G/H and on M'. If we define an involutive automorphism σ of G by $\sigma(A) = SAS^{-1}$, then the set of fixed points coincides with H. Thus G/H is a symmetric space and the canonical decomposition of the Lie algebra is

$$\mathfrak{o}(1, n) = \mathfrak{h} + \mathfrak{m}^*,$$

where $\mathfrak{h}$ is $\mathfrak{o}(n)$ considered as a subalgebra of $\mathfrak{o}(1, n)$ and $\mathfrak{m}^*$ is the subspace of all matrices of the form

$$\begin{pmatrix} 0 & {}^t\xi \\ \xi & 0 \end{pmatrix}, \quad \text{where } \xi \text{ is a (column) vector in } \mathbf{R}^n.$$

Identifying $\mathfrak{m}^*$ with $\mathbf{R}^n$ we see that ad (H) on $\mathfrak{m}^*$ is nothing but the action of $SO(n)$ on $\mathbf{R}^n$, that is,

$$\text{ad} \begin{pmatrix} 1 & 0 \\ 0 & B \end{pmatrix} \cdot \begin{pmatrix} 0 & {}^t\xi \\ \xi & 0 \end{pmatrix} = \begin{pmatrix} 0 & {}^t(B\xi) \\ B\xi & 0 \end{pmatrix}.$$

Similarly, ad $(\mathfrak{h})$ on $\mathfrak{m}^*$ is expressed by

$$\text{ad} \begin{pmatrix} 0 & 0 \\ 0 & B \end{pmatrix} \cdot \begin{pmatrix} 0 & {}^t\xi \\ \xi & 0 \end{pmatrix} = \begin{pmatrix} 0 & {}^t(B\xi) \\ B\xi & 0 \end{pmatrix}.$$

The symmetric Lie algebra $\mathfrak{o}(1, n) = \mathfrak{o}(n) + \mathfrak{m}^*$ is isomorphic to the dual (cf. §8) of $\mathfrak{o}(n + 1) = \mathfrak{o}(n) + \mathfrak{m}$ in Example 10.1; in fact, we have the following isomorphism of the dual $\mathfrak{o}(n) + i\mathfrak{m}$ of $\mathfrak{o}(n + 1)$ onto $\mathfrak{o}(n) + \mathfrak{m}^*$:

$$\begin{pmatrix} 0 & 0 \\ 0 & B \end{pmatrix} \in \mathfrak{o}(n) \to \begin{pmatrix} 0 & 0 \\ 0 & B \end{pmatrix} \in \mathfrak{o}(n),$$

$$i\begin{pmatrix} 0 & -{}^t\xi \\ \xi & 0 \end{pmatrix} \in i\mathfrak{m} \to \begin{pmatrix} 0 & {}^t\xi \\ \xi & 0 \end{pmatrix} \in \mathfrak{m}^*.$$

The Killing-Cartan form φ^* of $\mathfrak{o}(1, n)$ satisfies

$$\varphi^*(\xi, \eta) = -\varphi(\xi, \eta),$$

where $\xi, \eta \in \mathbf{R}^n$ are identified with $\begin{pmatrix} 0 & {}^t\xi \\ \xi & 0 \end{pmatrix}, \begin{pmatrix} 0 & {}^t\eta \\ \eta & 0 \end{pmatrix} \in \mathfrak{m}^*$ on the left hand side and with $\begin{pmatrix} 0 & -{}^t\xi \\ \xi & 0 \end{pmatrix}, \begin{pmatrix} 0 & -{}^t\eta \\ \eta & 0 \end{pmatrix} \in \mathfrak{m}$ on the right hand side. (This is a special case of what we saw in the proof of Theorem 8.4.) We may also verify directly that

$$\varphi^*(X, Y) = 2(n - 1)(X, Y) \quad \text{for } X, Y \in \mathfrak{m}^*.$$

With the invariant metric arising from (X, Y) on $\mathfrak{m}$, G/H is a Riemannian symmetric space. The differential f^* of $f\colon G/H \to M'$ at the origin maps the element of $\mathfrak{m}^*$ corresponding to $\xi \in \mathbf{R}^n$ upon $\begin{pmatrix} 0 \\ \xi \end{pmatrix} \in T_{e_0}(M')$ and is thus isometric. Since f is compatible with the actions of G on G/H and on M' and since the metrics are invariant by G, we see that f is an isometry of G/H onto M'.

The curvature tensor R of G/H at the origin is expressed, under the identification of $\mathfrak{m}^*$ with $\mathbf{R}^n$, by $R(\xi, \eta) = -\xi \wedge \eta$; thus G/H has constant sectional curvature -1.

The geodesics of G/H starting from the origin are of the form

$\pi(\exp t\xi)$, where $\xi \in \mathfrak{m}^* = \mathbf{R}^n$. For example, if $\xi = e_1$, then

$$\exp t\xi = \begin{pmatrix} \cosh t & \sinh t & 0 \cdots 0 \\ \sinh t & \cosh t & 0 \cdots 0 \\ 0 & 0 & \\ \cdot & \cdot & \\ \cdot & \cdot & I_{n-1} \\ \cdot & \cdot & \\ 0 & 0 & \end{pmatrix}$$

so that $f(\exp t\xi) = (\cosh t, \sinh t, 0, \ldots, 0)$ is a geodesic in M' starting from e_0. All other geodesics starting from e_0 are obtained from this geodesic by the rotations in $SO(n)$.

Totally geodesic submanifolds in G/H can be found in the same manner as in Example 10.1. We can also proceed in the following manner (which is applicable to Example 10.1 as well). If $\mathfrak{m}'$ is an arbitrary m-dimensional subspace of $\mathfrak{m}^*$, then $[[\mathfrak{m}', \mathfrak{m}'], \mathfrak{m}'] \subset \mathfrak{m}'$. Indeed, identifying $\mathfrak{m}^*$ with $\mathbf{R}^n$, let $\alpha_1, \ldots, \alpha_m, \alpha_{m+1}, \ldots, \alpha_n$ be a basis of $\mathfrak{m}^*$ such that $\{\alpha_1, \ldots, \alpha_m\}$ is a basis of $\mathfrak{m}'$, then $[\alpha_i, \alpha_j] = \alpha_i \wedge \alpha_j \in \mathfrak{o}(n)$ and $[[\alpha_i, \alpha_j], \alpha_k] = (\alpha_i \wedge \alpha_j) \alpha_k \in \mathfrak{m}'$ for $1 \leq i, j, k \leq m$. Thus $[[\mathfrak{m}', \mathfrak{m}'], \mathfrak{m}'] \subset \mathfrak{m}'$. Hence $\mathfrak{m}'$ gives rise to an m-dimensional complete totally geodesic submanifold of G/H through the origin. In M', any complete totally geodesic submanifold of dimension m through e_0 can be transformed by a rotation in $SO(n)$ to the submanifold

$$\{x \in M'; x^{m+1} = \cdots = x^n = 0\}.$$

Recall that M' can be realized as the unit open ball D^n in $\mathbf{R}^{n+1}$ (see Theorem 3.2 of Chapter V). In this representation of M', the submanifold above is

$$D^m = \{y = (y^1, \ldots, y^n) \in D^n; y^{m+1} = \cdots = y^n = 0\}.$$

Example 10.3. *Oriented Grassmann manifold* $SO(p+q)/SO(p) \times SO(q)$. Let $G_{p,q}(\mathbf{R})$ be the Grassmann manifold of p-planes in $\mathbf{R}^{p+q}$ whose differentiable structure is defined in the manner of Example 2.4 of Chapter IX only by replacing $\mathbf{C}$ by $\mathbf{R}$ throughout. Let $\tilde{G}_{p,q}(\mathbf{R})$ denote the set of all *oriented* p-planes in $\mathbf{R}^{p+q}$. In order to define its differentiable structure we modify the preceding argument for $G_{p,q}(\mathbf{R})$ as follows: U_α is the set of all oriented p-planes S such that $x^{\alpha_1} \mid S, \ldots, x^{\alpha_p} \mid S$ form an ordered basis of S compatible with the given orientation of S, where $x^1, \ldots, x^{p+q}$ (instead of $z^1, \ldots, z^{p+q}$ in Example 2.4 of Chapter IX) is the natural coordinate system in $\mathbf{R}^{p+q}$. We call $\tilde{G}_{p,q}(\mathbf{R})$ the *oriented Grassmann manifold*. It is the two-fold covering of $G_{p,q}(\mathbf{R})$.

The rotation group $G = SO(p+q)$ acts transitively on $\tilde{G}_{p,q}(\mathbf{R})$. Let S_0 be the p-plane $x^{p+q} = \cdots = x^{p+q} = 0$ with the orientation given by the ordered coordinate system $x^1, \ldots, x^p$. The isotropy group H at S_0 consists of all matrices of the form

$$\begin{pmatrix} U & 0 \\ 0 & V \end{pmatrix}, \quad \text{where } U \in SO(p) \text{ and } V \in SO(q).$$

Denoting this subgroup by $SO(p) \times SO(q)$, we have $G/H = SO(p+q)/SO(p) \times SO(q) = \tilde{G}_{p,q}(\mathbf{R})$. We see that G/H is a symmetric space if we define an involutive automorphism σ of G by

$$\sigma(A) = SAS^{-1}, \quad \text{where } S = \begin{pmatrix} -I_p & 0 \\ 0 & I_q \end{pmatrix}.$$

The canonical decomposition of the Lie algebra: $\mathfrak{g} = \mathfrak{h} + \mathfrak{m}$ is given by

$$\mathfrak{g} = \mathfrak{o}(p+q),$$
$$\mathfrak{h} = \mathfrak{o}(p) + \mathfrak{o}(q) = \left\{ \begin{pmatrix} U & 0 \\ 0 & V \end{pmatrix}; U \in \mathfrak{o}(p), V \in \mathfrak{o}(q) \right\};$$

and

$$\mathfrak{m} = \left\{ \begin{pmatrix} 0 & -{}^t X \\ X & 0 \end{pmatrix}; X \in M(q, p; \mathbf{R}) \right\},$$

where $M(q, p; \mathbf{R})$ denotes the vector space of all real matrices with q rows and p columns. The adjoint representation of H on $\mathfrak{m}$ is given by

$$\operatorname{ad} \begin{pmatrix} U & 0 \\ 0 & V \end{pmatrix} \begin{pmatrix} 0 & -{}^t X \\ X & 0 \end{pmatrix} = \begin{pmatrix} 0 & -U {}^t X V^{-1} \\ VXU^{-1} & 0 \end{pmatrix}.$$

The restriction to $\mathfrak{m}$ of the inner product $(A, B) = -\frac{1}{2} \operatorname{trace} AB$

on $\mathfrak{o}(p+q)$ is, of course, invariant by ad (H) and induces an invariant metric on G/H.

Discussions on the curvature tensor, geodesics, etc. are similar to Example 10.1 (See also Leichtweiss [1], Wong [4].) We shall discuss another special case $p = 2$ and $q = n$ in Example 10.6.

Example 10.4. $SO^0(p, q)/SO(p) \times SO(q)$. Here $SO^0(p, q)$ denotes the identity component of the orthogonal group $O(p, q)$ for the bilinear form on $\mathbf{R}^{p+q}$:

$$F(x, y) = -\sum_{j=1}^{p} x^j y^j + \sum_{j=p+1}^{p+q} x^j y^j.$$

The involutive automorphism σ is defined in the same way as in Example 10.3. Discussions of this symmetric space are similar to the special case $p = 1$, $q = n$, which was treated in Example 10.2.

Example 10.5 *Complex projective space* $P_n(\mathbf{C})$. Before we discuss $P_n(\mathbf{C})$ from the point of view of Hermitian symmetric spaces, we shall reconsider the complex structure and the metric in $P_n(\mathbf{C})$ in Examples 2.4 and 6.3 of Chapter IX from a more geometric viewpoint. In $\mathbf{C}^{n+1}$ with the natural basis $e_0, e_1, \ldots, e_n$ we consider

$$h(z, w) = \sum_{k=0}^{n} z^k \bar{w}^k \quad \text{and} \quad g'(z, w) = \operatorname{Re} h(z, w),$$

which are related by $h(z, w) = g'(z, w) + ig'(z, iw)$. In fact, $g'(z, w)$ is nothing but the usual inner product when we identify $\mathbf{C}^{n+1}$ with $\mathbf{R}^{2n+2}$. For the unit sphere $S^{2n+1} = \{z \in \mathbf{C}^{n+1}; h(z, z) = 1\}$ the tangent space $T_z(S^{2n+1})$ at each point z can be identified (through parallel displacement in $\mathbf{C}^{n+1}$) with $\{w \in \mathbf{C}^{n+1}; g'(z, w) = 0\}$. Let T'_z be the orthogonal complement of the vector $iz \in T_z(S^{2n+1})$, that is,

$$T'_z = \{w \in \mathbf{C}^{n+1}; g'(z, w) = 0 \quad \text{and} \quad g'(iz, w) = 0\}.$$

When we consider S^{2n+1} as a principal fibre bundle over $P_n(\mathbf{C})$ with structure group S^1 as in Example 2.4 of Chapter IX, there is a connection such that T'_z is the horizontal subspace at z. The natural projection π of S^{2n+1} onto $P_n(\mathbf{C})$ induces a linear isomorphism of T'_z onto $T_p(P_n(\mathbf{C}))$, where $p = \pi(z)$. The Riemannian metric g' on S^{2n+1} (that is, the metric on S^{2n+1} induced by g') is invariant by the structure group. Thus we may define a

Riemannian metric $\tilde{g}$ on $P_n(\mathbf{C})$ by

$$\tilde{g}_p(X, Y) = \frac{4}{c} g'_z(X', Y'), \qquad X, Y \in T_p(P_n(\mathbf{C})), \qquad c > 0 \quad \text{fixed},$$

where z is any point of S^{2n+1} with $\pi(z) = p$ and X', Y' are the vectors in T'_z such that $\pi_*(X') = X$ and $\pi_*(Y') = Y$. On the other hand, the complex structures $J': w \to iw$ in the subspaces T'_z, $z \in S^{2n+1}$, are compatible with the action of S^1, that is, $J'(e^{i\theta}w) = e^{i\theta}(J'w)$. It follows that we may define a complex structure $\tilde{J}$ in each tangent space $T_p(P_n(\mathbf{C}))$ by

$$\tilde{J}(X) = \pi(J'X'), \qquad X \in T_p(P_n(\mathbf{C})),$$

where $\pi_*(X') = X$. Indeed, the almost complex structure so defined on $P_n(\mathbf{C})$ coincides with the usual complex structure of $P_n(\mathbf{C})$. Since each of them is invariant by the group $U(n + 1)$ acting transitively on $P_n(\mathbf{C})$, it is sufficient to show that $\tilde{J}$ at $p_0 = \pi(e_0)$ coincides with the usual complex structure at p_0 induced by a complex chart $\varphi: (z^0, z^1, \ldots, z^n) \in U_0 \to (t^1, \ldots, t^n) \in \mathbf{C}^n$, where $(z^0, \ldots, z^n)$ are homogeneous coordinates, $U_0 = \{z; z^0 \neq 0\}$, and $(t^1, \ldots, t^n)$ are inhomogeneous coordinates, i.e., $t^k = z^k/z^0$ for $1 \leq k \leq n$. We know that T'_{e_0} is spanned by $e_1, \ldots, e_n$, $ie_1, \ldots, ie_n$ as a real vector space. Now $\pi_*(e_k)$ is the tangent vector at p_0 of the curve

$$\pi((e_0 + se_k)/\sqrt{1 + s^2}) = (1, 0, \ldots, 0, s, 0, \ldots, 0),$$

where s is the $(k + 1)$st homogeneous coordinate; we see that $\varphi_*(\pi_*(e_k))$ is the k-th element of the natural basis of $\mathbf{C}^n$. Similarly, $\pi_*(ie_k)$ is the tangent vector at p_0 of the curve $(1, 0, \ldots, 0, is, 0, \ldots, 0)$, and $(\varphi_*(\pi_*(ie_k)) = i\varphi_*(\pi_*(e_k))$. Since $\tilde{J}\pi_*(e_k) = \pi_*(ie_k)$ by definition of $\tilde{J}$ at p_0 for $1 \leq k \leq n$, we conclude that $\varphi_* \tilde{J} = i\varphi_*$, that is, $\tilde{J}$ is nothing but the complex structure induced by the chart φ from the usual complex structure of $\mathbf{C}^n$.

The metric $\tilde{g}$ is Hermitian with respect to $\tilde{J}$, since J' is Hermitian with respect to g' on T'_z for each $z \in S^{2n+1}$. Observe also that both $\tilde{g}$ and $\tilde{J}$ are invariant by the action of $U(n + 1)$ on $P_n(\mathbf{C})$. At p_0 the expression of the metric in Theorem 7.8 (1) of Chapter IX reduces to $(4/c) \Sigma_{\alpha=1}^n dt^\alpha d\bar{t}^\alpha$ and coincides with our metric $\tilde{g}$. By invariance of both metrics by $U(n+1)$, we conclude that our metric $\tilde{g}$ coincides with that in Theorem 7.8 (1) of Chapter IX.

For the action of $U(n+1)$ on $P_n(\mathbf{C})$, the isotropy group at $p_0 = \pi(e_0)$ consists of all matrices of the form

$$\begin{pmatrix} e^{i\theta} & 0 \\ 0 & B \end{pmatrix}, \quad \text{where } B \in U(n).$$

Denoting this subgroup by $U(1) \times U(n)$ we have a diffeomorphism f of $U(n+1)/U(1) \times U(n)$ onto $P_n(\mathbf{C})$. The action of $U(n+1)$ is not effective; if we take the almost effective action of $SU(n+1)$ on $P_n(\mathbf{C})$, the isotropy group at p_0 is the subgroup $S(U(1) \times U(n))$ consisting of all matrices in $U(1) \times U(n)$ with determinant 1. We shall, however, proceed with the representation G/H with $G = U(n+1)$ and $H = U(1) \times U(n)$.

We define an involutive automorphism σ of $U(n+1)$ by

$$\sigma(A) = SAS^{-1}, \quad \text{where } S = \begin{pmatrix} -1 & 0 \\ 0 & I_n \end{pmatrix}.$$

The subgroup of all fixed elements is $U(1) \times U(n)$. Thus G/H is a symmetric space. The canonical decomposition of the Lie algebra is given by

$$\mathfrak{g} = \mathfrak{h} + \mathfrak{m},$$

where

$\mathfrak{g} = \mathfrak{u}(n+1)$: all skew-Hermitian matrices of degree $n+1$,

$$\mathfrak{h} = \mathfrak{u}(1) + \mathfrak{u}(n) = \left\{ \begin{pmatrix} \lambda & 0 \\ 0 & B \end{pmatrix}, \lambda + \bar{\lambda} = 0, B \in \mathfrak{u}(n) \right\},$$

and

$$\mathfrak{m} = \left\{ \begin{pmatrix} 0 & -{}^t\bar{\xi} \\ \xi & 0 \end{pmatrix}; \xi \in \mathbf{C}^n \right\}.$$

In the following discussion we shall identify $\mathfrak{m}$ with $\mathbf{C}^n$ in this obvious manner. The adjoint action of H on $\mathfrak{m}$ is expressed by

$$\operatorname{ad} \begin{pmatrix} e^{i\theta} & 0 \\ 0 & B \end{pmatrix} \cdot \xi = e^{-i\theta} B\xi, \quad \xi \in \mathfrak{m}.$$

The inner product

$$(A, B) = -\tfrac{1}{2} \text{ trace } AB = \tfrac{1}{2} \text{ trace } A^t\bar{B}, \qquad A, B \in \mathfrak{u}(n+1)$$

is ad $(U(n+1))$-invariant; its restriction to $\mathfrak{m}$ is given by

$$-\tfrac{1}{2} \text{ trace} \begin{pmatrix} 0 & -^t\bar{\xi} \\ \xi & 0 \end{pmatrix} \begin{pmatrix} 0 & -^t\bar{\eta} \\ \eta & 0 \end{pmatrix} = \tfrac{1}{2}\{h(\eta, \xi) + h(\xi, \eta)\}$$

$$= \text{Re } h(\xi, \eta)$$

where $h(\xi, \eta)$ is defined in $\mathbf{C}^n$ just as in the beginning. We shall take

$$g(\xi, \eta) = \frac{4}{c} \text{Re } h(\xi, \eta)$$

as an inner product on $\mathfrak{m}$ invariant by ad H and get an invariant Riemannian metric g on G/H.

We also consider an ad (H)-invariant complex structure on $\mathfrak{m}$ given by $\xi \to i\xi$, that is,

$$\begin{pmatrix} 0 & -^t\bar{\xi} \\ \xi & 0 \end{pmatrix} \to \begin{pmatrix} 0 & i\,^t\bar{\xi} \\ i\xi & 0 \end{pmatrix}$$

Since G/H is symmetric, we know that the induced G-invariant almost complex structure J on G/H is integrable (see Proposition 9.1). We also see that g is Hermitian with respect to J, since it is so on $\mathfrak{m}$.

We now wish to show that the invariant metric g and the invariant complex structure J on G/H correspond by f to $\tilde{g}$ and $\tilde{J}$ on $P_n(\mathbf{C})$. Since $\tilde{g}$ and $\tilde{J}$ are invariant by $U(n+1)$, it is sufficient to check our assertion for the differential f_* of f at the origin of G/H. We see that f_* maps $\xi \in \mathfrak{m} = \mathbf{C}^n$ upon

$$\left(\frac{d}{dt}\right)_0 \pi\left(\exp t\begin{pmatrix} 0 & -^t\bar{\xi} \\ \xi & 0 \end{pmatrix} \cdot e_0\right) = \pi_*\left(\begin{pmatrix} 0 & -^t\bar{\xi} \\ \xi & 0 \end{pmatrix} \cdot e_0\right) = \pi_*\begin{pmatrix} 0 \\ \xi \end{pmatrix},$$

where $\begin{pmatrix} 0 \\ \xi \end{pmatrix}$ belongs to $T'_{e_0}(S^{2n+1})$. Thus f_* preserves the metrics and the complex structures.

XI. SYMMETRIC SPACES 277

The curvature tensor R at the origin of G/H is computed as follows:

$$R(\xi, \eta)\zeta = \xi^t\bar{\eta}\zeta - \eta^t\bar{\xi}\zeta + \zeta^t\bar{\eta}\xi - \zeta^t\bar{\xi}\eta$$
$$= h(\zeta, \eta)\xi - h(\zeta, \xi)\eta + h(\xi, \eta)\zeta - h(\eta, \xi)\zeta$$
$$= \frac{c}{4}\{\xi \wedge \eta + J\xi \wedge J\eta + 2g(\xi, J\eta)J\}\zeta,$$

where $g = (4/c)\operatorname{Re} h$ and $\xi \wedge \eta$ is the endomorphism such that $(\xi \wedge \eta) \cdot \zeta = g(\zeta, \eta)\xi - g(\zeta, \xi)\eta$. It follows that the holomorphic sectional curvature of G/H is equal to the constant c (see §7 of Chapter IX).

To find geodesics in $P_n(\mathbf{C})$, let $\xi \in \mathfrak{m}$ be the element corresponding to $e_1 \in \mathbf{C}^n$. Then

$$\pi\left(\exp t\begin{pmatrix} 0 & -{}^t\bar{\xi} \\ \xi & 0 \end{pmatrix} \cdot e_0\right) = \pi \begin{pmatrix} \cos t \\ \sin t \\ 0 \\ \cdot \\ \cdot \\ \cdot \\ 0 \end{pmatrix}$$

so that the curve $(\cos t, \sin t, 0, \ldots, 0)$ in homogeneous coordinates is a geodesic starting from $p_0 = (1, 0, \ldots, 0)$. We note that it is contained in the complex projective line P_1 defined by $z^n = \cdots = z^n = 0$. All other geodesics starting from p_0 are obtained by transforming this geodesic by $U(1) \times U(n)$.

In order to find totally geodesic *complex* submanifolds of G/H through the origin, we have to find a J-invariant subspace $\mathfrak{m}'$ of $\mathfrak{m}$ such that $[[\mathfrak{m}', \mathfrak{m}'], \mathfrak{m}'] \subset \mathfrak{m}'$. But, actually, any J-invariant subspace $\mathfrak{m}'$ of $\mathfrak{m}$ satisfies this bracket condition. Indeed, if $\mathfrak{m}'$ corresponds to a complex subspace of $\mathbf{C}^n$, let $\alpha_1, \ldots, \alpha_m, \alpha_{m+1}, \ldots, \alpha_n$ be a basis of $\mathfrak{m}$ over $\mathbf{C}$ such that $\alpha_1, \ldots, \alpha_m$ is a basis of $\mathfrak{m}'$ over $\mathbf{C}$. Then $\alpha_1, \ldots, \alpha_m, J\alpha_1, \ldots, J\alpha_m$ form a basis of $\mathfrak{m}'$ over $\mathbf{R}$. We see that

$$[[\alpha_i, \alpha_j], \alpha_k] = -R(\alpha_i, \alpha_j) \cdot \alpha_k$$
$$= -\frac{c}{4}\{\alpha_i \wedge \alpha_j + J\alpha_i \wedge J\alpha_j + 2g(\alpha_i, J\alpha_j)J\} \cdot \alpha_k$$

belongs to $\mathfrak{m}'$ for $1 \leq i, j, k \leq m$. Similarly, the triple brackets among $\alpha_1, \ldots, \alpha_m, J\alpha_1, \ldots, J\alpha_m$ belong to $\mathfrak{m}'$, proving our assertion. If $\mathfrak{m}'$ is the complex subspace spanned by $e_1, \ldots, e_m$ of $\mathfrak{m} = \mathbf{C}^n$, then $\mathfrak{g}' = \mathfrak{m}' + [\mathfrak{m}', \mathfrak{m}']$ consists of all matrices of the form

$$\begin{pmatrix} A' & 0 \\ 0 & 0 \end{pmatrix}, \quad \text{where } A' \in \mathfrak{u}(m+1).$$

The subgroup G' generated by $\mathfrak{g}'$ is the subgroup of $U(n+1)$ consisting of all matrices of the form

$$\begin{pmatrix} A' & 0 \\ 0 & I_{n-m} \end{pmatrix}, \quad \text{where } A' \in U(m+1).$$

Hence the corresponding totally geodesic complex submanifold in $P_n(\mathbf{C})$ is $P_m(\mathbf{C}) = \{(z^0, \ldots, z^n) \in P_n(\mathbf{C}); z^{m+1} = \cdots = z^n = 0\}$. It follows that any complete totally geodesic complex submanifold is an m-dimensional complex projective subspace, where m is the complex dimension.

Example 10.6 *Complex quadric* $Q_{n-1}(\mathbf{C})$. In $P_n(\mathbf{C})$ with homogeneous coordinates $z^0, z^1, \ldots, z^n$ the complex quadric $Q_{n-1}(\mathbf{C})$ is a complex hypersurface defined by the equation

$$(z^0)^2 + (z^1)^2 + \cdots + (z^n)^2 = 0.$$

Let g be the Fubini-Study metric with holomorphic sectional curvature c for $P_n(\mathbf{C})$. Its restriction g to $Q_{n-1}(\mathbf{C})$ is a Kaehler metric (Example 6.7 and §9 of Chapter IX). We shall represent $Q_{n-1}(\mathbf{C})$ as a Hermitian symmetric space.

First we fix the following notations. In $\mathbf{C}^{n+1}$ with the natural basis $e_0, e_1, \ldots, e_n$ we denote by $H(z, w)$ the complex bilinear form defined by

$$H(z, w) = \sum_{k=0}^{n} z^k w^k, \quad z = (z^k), w = (w^k).$$

Then $Q_{n-1}(\mathbf{C}) = \{\pi(z); z \in \mathbf{C}^{n+1} - \{0\}, H(z, z) = 0\}$, where π is the natural projection of $\mathbf{C}^{n+1} - \{0\}$ onto $P_n(\mathbf{C})$. The unit vector

$$\beta_0 = (e_0 + ie_1)/\sqrt{2} \in S^{2n+1}$$

satisfies $H(\beta_0, \beta_0) = 0$. We let $q_0 = \pi(\beta_0) \in Q_{n-1}(\mathbf{C})$. The tangent space $T_{\beta_0}(S^{2n+1})$ has a basis consisting of $ie_0 - e_1, ie_0 + e_1,$

XI. SYMMETRIC SPACES   279

$-e_0 + ie_1, e_2, \ldots, e_n, ie_2, \ldots, ie_n$, where the first vector is $\sqrt{2}i\beta_0$. The subspace T'_{β_0} defined in Example 10.5 is spanned $ie_0 + e_1$, $-e_0 + ie_1, e_2, \ldots, e_n, ie_2, \ldots, ie_n$. Let T''_{β_0} be the subspace spanned by $e_2, \ldots, e_n, ie_2, \ldots, ie_n$. The natural projection π induces a linear isomorphism

$$\pi_*: T'_{\beta_0} \to T_{q_0}(P_n(\mathbf{C}))$$

and a linear isomorphism

$$\pi_*: T''_{\beta_0} \to T_{q_0}(Q_{n-1}(\mathbf{C})).$$

To prove the second assertion, it suffices to show that $\pi_*(T''_{\beta_0})$ is contained in $T_{q_0}(Q_{n-1}(\mathbf{C}))$, since both spaces have dimension $2n - 2$. For each k, $2 \leq k \leq n$, the curves in $P_n(\mathbf{C})$

$$a_t = (\cos t, i, 0, \ldots, 0, \sin t, 0, \ldots, 0)$$

and

$$b_t = (1, i \cos t, 0, \ldots, 0, i \sin t, 0, \ldots, 0),$$

where $\sin t$ and $i \sin t$ appear as the $(k + 1)$st homogeneous coordinates, lie in $Q_{n-1}(\mathbf{C})$. Thus their tangent vectors at $t = 0$

$$\dot{a}_0 = \pi_*(e_k) \quad \text{and} \quad \dot{b}_0 = \pi_*(ie_k)$$

are contained in $T_{q_0}(Q_{n-1}(\mathbf{C}))$, proving our assertion.

For an arbitrary point $q \in Q_{n-1}(\mathbf{C})$, we take a point $z \in S^{2n+1}$ such that $H(z, z) = 0$ and $\pi(z) = q$. We may write z uniquely in the form $z = x + iy$, where x and y are orthogonal real vectors in $\mathbf{C}^{n+1}$ of length $1/\sqrt{2}$. In the space T'_z defined in Example 10.5 let T''_z be the orthogonal complement (with respect to g') of the subspace spanned by $ix + y$ and $-x + iy$. We may then prove that π induces a linear isomorphism of T''_z onto $T_q(Q_{n-1}(\mathbf{C}))$; this can be done either directly as in the case where $z = (e_0 + ie_1)/\sqrt{2}$ or by using the transitive action of $SO(n + 1)$ on $Q_{n-1}(\mathbf{C})$, which we discuss in the following paragraph.

The group $SO(n + 1)$, considered as a subgroup of $U(n + 1)$, acts on $P_n(\mathbf{C})$ and leaves $Q_{n-1}(\mathbf{C})$ invariant, since $H(Az, Aw) = H(z, w)$ for all $A \in SO(n + 1)$ and $z, w \in \mathbf{C}^{n+1}$. Actually, $SO(n + 1)$ is transitive on $Q_{n-1}(\mathbf{C})$. In fact, for any $z \in S^{2n+1}$ such that $H(z, z) = 0$, we want to find $A \in SO(n + 1)$ such that $\pi(A\beta_0) = \pi(z)$. In the unique representation $z = x + iy$ we may normalize x and y and obtain two orthonormal real vectors α_0, α_1 such that

280   FOUNDATIONS OF DIFFERENTIAL GEOMETRY

$\pi(\alpha_0 + i\alpha_1) = \pi(z)$. By extending α_0, α_1 to an ordered orthonormal basis $\alpha_0, \alpha_1, \ldots, \alpha_n$ in $\mathbf{R}^{n+1}$ we get $A \in SO(n+1)$ with these column vectors. We have then $\pi(A(\beta_0)) = \pi(z)$.

For the action of $G = SO(n+1)$ on $Q_{n-1}(\mathbf{C})$, the isotropy group H at q_0 turns out to be the subgroup $SO(2) \times SO(n-1)$ of $SO(n+1)$ consisting of all matrices of the form

$$R(\theta) \times B = \begin{pmatrix} \cos\theta & -\sin\theta & 0 & \ldots & 0 \\ \sin\theta & \cos\theta & 0 & \ldots & 0 \\ 0 & 0 & & & \\ \cdot & \cdot & & B & \\ \cdot & \cdot & & & \\ 0 & 0 & & & \end{pmatrix},$$

where $B \in SO(n-1)$.

We have thus a diffeomorphism $f \colon G/H \to Q_{n-1}(\mathbf{C})$.

From Example 10.3 we know already that G/H is a symmetric space. The canonical decomposition of the Lie algebra is

$$\mathfrak{g} = \mathfrak{h} + \mathfrak{m},$$

where

$\mathfrak{g} = \mathfrak{o}(n+1)$

$\mathfrak{h} = \mathfrak{o}(2) + \mathfrak{o}(n-1) = \left\{ \begin{pmatrix} 0 & -\lambda & 0 \\ \lambda & 0 & 0 \\ 0 & 0 & B \end{pmatrix}, B \in \mathfrak{o}(n-1) \right\}$

$\mathfrak{m} = \left\{ \begin{pmatrix} 0 & 0 & -{}^t\xi \\ 0 & 0 & -{}^t\eta \\ \xi & \eta & 0 \end{pmatrix} ; \xi, \eta \text{ are column vectors in } \mathbf{R}^{n-1} \right\}.$

Identifying $(\xi, \eta) \in \mathbf{R}^{n-1} + \mathbf{R}^{n-1}$ with the matrix above in $\mathfrak{m}$, we see that ad (H) on $\mathfrak{m}$ is expressed by

$$\text{ad } (R(\theta) \times B) \cdot (\xi, \eta) = (B\xi, B\eta)R(-\theta),$$

where the right hand side is a matrix product. Similarly, ad $(\mathfrak{h})$ on

XI. SYMMETRIC SPACES 281

$\mathfrak{m}$ is expressed by

$$\operatorname{ad}\begin{pmatrix} 0 & -\lambda & 0 \\ \lambda & 0 & 0 \\ 0 & 0 & B \end{pmatrix} \cdot (\xi, \eta) = (B\xi, B\eta) + (\xi, \eta)\begin{pmatrix} 0 & +\lambda \\ -\lambda & 0 \end{pmatrix}.$$

The differential f_* at the origin of G/H maps $(\xi, \eta) \in \mathfrak{m}$ upon

$$\left(\frac{d}{dt}\right)_{t=0} (\pi(\exp t(\xi, \eta) \cdot \beta_0)) = \pi_*((\xi, \eta) \cdot \beta_0)$$

$$= \pi_* \begin{pmatrix} 0 \\ 0 \\ \xi + i\eta \end{pmatrix} \in T_{q_0}(Q_{n-1}(\mathbf{C})),$$

where the vector $\begin{pmatrix} 0 \\ 0 \\ \xi + i\eta \end{pmatrix}$ belongs to T''_{β_0}.

We define an inner product g on $\mathfrak{m} \times \mathfrak{m}$ by

$$g((\xi, \eta), (\xi', \eta')) = \frac{4}{c}\{\langle \xi, \xi' \rangle + \langle \eta, \eta' \rangle\},$$

where $\langle \xi, \xi' \rangle$ is the standard inner product in $\mathbf{R}^{n-1}$. We also define a complex structure J on $\mathfrak{m}$ by

$$J(\xi, \eta) = (-\eta, \xi).$$

Both g and J are invariant by ad (H) and give rise to a structure of a Hermitian symmetric space on G/H. It is now an easy matter to verify that f is an isomorphism of G/H onto $Q_{n-1}(\mathbf{C})$ as Kaehler manifolds.

The curvature tensor R at the origin is given as follows:

$$R((\xi, \eta), (\xi', \eta')) = \operatorname{ad}\begin{pmatrix} 0 & -\lambda & 0 \\ \lambda & 0 & 0 \\ 0 & 0 & B \end{pmatrix},$$

where $\lambda = \langle \xi', \eta \rangle - \langle \xi, \eta' \rangle$ and $B = \frac{c}{4}\{\xi \wedge \xi' + \eta \wedge \eta'\}$, the endomorphism $\xi \wedge \xi'$ being defined by using the standard inner product in $\mathbf{R}^{n-1}$. If we compute the Ricci tensor S by using the formula in Proposition 4.5 (2) of Chapter IX, we obtain

$$S = \frac{(n-1)c}{2} g.$$

Thus $Q_{n-1}(\mathbf{C})$ is an Einstein space.

For $n > 3$, it can be shown that ad (H) on $\mathfrak{m}$ is irreducible and therefore $Q_{n-1}(\mathbf{C})$ is irreducible. For $n = 3$ we see that

$$\mathfrak{m}_1 = \left\{ \begin{pmatrix} 0 & 0 & -a & b \\ 0 & 0 & -b & -a \\ a & b & 0 & 0 \\ -b & a & 0 & 0 \end{pmatrix} \right\}$$

and

$$\mathfrak{m}_2 = \left\{ \begin{pmatrix} 0 & 0 & -c & -d \\ 0 & 0 & -d & c \\ c & d & 0 & 0 \\ d & -c & 0 & 0 \end{pmatrix} \right\}$$

are invariant by ad (H). Indeed, $Q_2(\mathbf{C})$ can be shown to be isomorphic to $P_1(\mathbf{C}) \times P_1(\mathbf{C})$ as Kaehler manifolds.

Example 10.7. *Complex hyperbolic space.* In $\mathbf{C}^{n+1}$ with the standard basis $e_0, e_1, \ldots, e_n$ we consider a Hermitian form F defined by

$$F(z, w) = -z^0 \bar{w}^0 + \sum_{k=1}^{n} z^k \bar{w}^k.$$

Let

$$U(1, n) = \{A \in GL(n+1; \mathbf{C}); F(Az, Aw) = F(z, w), z, w \in \mathbf{C}^{n+1}\}$$
$$= \{A \in GL(n+1; \mathbf{C}); {}^t\bar{A}SA = S\},$$

where

$$S = \begin{pmatrix} -1 & 0 \\ 0 & I_n \end{pmatrix}.$$

XI. SYMMETRIC SPACES

It is known that $U(1, n)$ is connected. Its Lie algebra is

$$\mathfrak{u}(1, n) = \{X \in \mathfrak{gl}(n + 1; \mathbf{C}); {}^t\bar{X}S + SX = 0\}$$

$$= \left\{ \begin{pmatrix} i\lambda & {}^t\bar{\xi} \\ \xi & B \end{pmatrix}; \lambda \in \mathbf{R}, \xi \in \mathbf{C}^n, B \in \mathfrak{u}(n) \right\}.$$

Let M be a real hypersurface in $\mathbf{C}^{n+1}$ defined by $F(z, z) = -1$. We note that the group $U(1, n)$ acts transitively on M. On the other hand, the group $S^1 = \{e^{i\theta}\}$ acts freely on M by $z \to e^{i\theta}z$; let M' be the base manifold of the principal fibre bundle M with group S^1.

For $z \in M$, the tangent space $T_z(M)$ is represented by

$$\{w \in \mathbf{C}^{n+1}; \operatorname{Re} F(z, w) = 0\}.$$

In particular, $iz \in T_z(M)$. Let T'_z be the subspace of $T_z(M)$ defined by

$$T'_z = \{w \in T'_z(M); \operatorname{Re} F(iz, w) = 0\}$$
$$= \{w \in \mathbf{C}^{n+1}; F(z, w) = 0\}.$$

We observe that the restriction of F to T'_z is positive-definite.

We have a connection in M such that T'_z, $z \in M$, are the horizontal subspaces. The natural projection π of M onto M' induces a linear isomorphism of T'_z onto $T_{\pi(z)}(M')$. The complex structures $w \to iw$ on T'_z, $z \in M$, are compatible with the action of S^1 and induce an almost complex structure J' on M' such that $\pi_* i = J'\pi_*$. (We shall later see that J' is integrable so that M' is a complex manifold.) Fixing any negative constant c, we define in each tangent space $T_p(M')$, $p \in M'$, an inner product g' by

$$g'(X, Y) = -\frac{4}{c} \operatorname{Re} (F(X', Y')),$$

where $X', Y' \in T'_z$ and $\pi(z) = p$, $\pi_*(X') = X$, $\pi_*(Y') = Y$. The metric g' is Hermitian with respect to J'.

Now the group $G = U(1, n)$ acts transitively on M and hence on M'. Both g' and J' are invariant by G. The isotropy group at $p_0 = \pi(e_0)$ is the subgroup $H = U(1) \times U(n)$ of $U(1, n)$ consisting of all matrices of the form

$$\begin{pmatrix} e^{i\theta} & 0 \\ 0 & B \end{pmatrix}, \quad B \in U(n).$$

Thus we have a diffeomorphism f of G/H onto M'. The action of $U(1, n)$ is not effective; if we take $SU(1, n) = \{A \in U(1, n)\,; \det A = 1\}$, which acts almost effectively on M', then the isotropy group at p_0 is

$$S(U(1) \times U(n)) = \{A \in U(1) \times U(n)\,; \det A = 1\}.$$

If we take the conjugation by $S = \begin{pmatrix} -1 & 0 \\ 0 & I_n \end{pmatrix}$ as an involutive automorphism of G, then H is precisely the subgroup of all fixed elements. Thus G/H is a symmetric space. The canonical decomposition of the Lie algebra is

$$\mathfrak{g} = \mathfrak{h} + \mathfrak{m},$$

where
$$\mathfrak{g} = \mathfrak{u}(1, n)$$
$$\mathfrak{h} = \mathfrak{u}(1) + \mathfrak{u}(n) = \left\{ \begin{pmatrix} i\lambda & 0 \\ 0 & B \end{pmatrix}\,;\ \lambda \in \mathbf{R},\ B \in \mathfrak{u}(n) \right\},$$

and

$$\mathfrak{m} = \left\{ \begin{pmatrix} 0 & {}^t\bar{\xi} \\ \xi & 0 \end{pmatrix}\,;\ \xi \in \mathbf{C}^n \right\}.$$

Identifying $\xi \in \mathbf{C}^n$ with the above matrix in $\mathfrak{m}$, ad (H) on $\mathfrak{m}$ is expressed by

$$\mathrm{ad} \begin{pmatrix} e^{i\theta} & 0 \\ 0 & B \end{pmatrix} \cdot \xi = e^{-i\theta} B \xi.$$

The complex structure $\xi \to i\xi$ on $\mathfrak{m} = \mathbf{C}^n$ is invariant by ad (H) and induces an invariant complex structure on the symmetric space G/H. The inner product on $\mathfrak{m}$ given by

$$g(\xi, \eta) = -\frac{4}{c}\, \mathrm{Re}\, h(\xi, \eta),$$

where $h(\xi, \eta)$ is the standard Hermitian inner product on $\mathfrak{m} = \mathbf{C}^n$, is invariant by ad (H) and induces an invariant Hermitian metric on G/H. (The inner product g on $\mathfrak{m}$ is, of course, related to the restriction of the Killing-Cartan form of $\mathfrak{u}(1, n)$.)

The differential f_* at the origin of G/H maps $\xi \in \mathfrak{m}$ upon

$$\pi_* \begin{pmatrix} 0 \\ \xi \end{pmatrix} \in T_{p_0}(M'), \text{ where } \begin{pmatrix} 0 \\ \xi \end{pmatrix} \in T'_{e_0}.$$ Thus g and J on G/H correspond to g' and J' on M' by the diffeomorphism f. It follows that J' is actually integrable and f is an isomorphism of G/H onto M' as Kaehler manifolds.

The curvature tensor R at the origin is expressed on $\mathfrak{m}$ by

$$R(\xi, \eta) = \frac{c}{4}(\xi \wedge \eta + J\xi \wedge J\eta + 2g(\xi, J\eta)J),$$

which shows that G/H has constant holomorphic curvature equal to c.

If we take $\xi = e_1$ in $\mathbf{C}^n$, then

$$\pi(\exp t\xi \cdot e_0) = \pi \begin{pmatrix} \cosh t \\ \sinh t \\ 0 \\ \cdot \\ \cdot \\ \cdot \\ 0 \end{pmatrix}$$

is a geodesic in M'. All geodesics of M' are obtained from this geodesic by the transformations in $U(1, n)$. Any J-invariant subspace $\mathfrak{m}'$ of real dimension $2m$ satisfies the condition

$$[[\mathfrak{m}', \mathfrak{m}'], \mathfrak{m}'] \subset \mathfrak{m}'$$

and gives rise to a totally geodesic complex submanifold of complex dimension m in M'. (The situation is analogous to that of $P_n(\mathbf{C}) = SU(n+1)/S(U(1) \times U(n))$, which is the dual of our space $SU(1, n)/S(U(1) \times U(n))$.)

We may identify M' with the open unit ball in $\mathbf{C}^n$:

$$D_n(\mathbf{C}) = \left\{ w \in \mathbf{C}^n; \sum_{k=1}^n w^k \bar{w}^k < 1 \right\}$$

by the mapping

$$\pi(z^0, z^1, \ldots, z^n) \in M' \to (w^1, \ldots, w^n) \in D_n(\mathbf{C}),$$

where $F(z, z) = -1$ and $w^k = z^k/z^0$ (note that $F(z, z) = -1$ implies $z^0 \bar{z}^0 = 1 + \Sigma_{k=1}^n z^k \bar{z}^k > 0$).

Example 10.8. *Complex Grassmann manifold* $G_{p,q}(\mathbf{C})$. We take up Example 6.4 of Chapter IX from the point of view of symmetric spaces, generalizing Example 10.5. Let $M' = M'(p+q, p; \mathbf{C})$ be the space of complex matrices Z with $p+q$ rows and p columns such that ${}^t\bar{Z}Z = I_p$ (or, equivalently, the p column vectors are orthonormal with respect to the standard inner product in $\mathbf{C}^{p+q}$). The group $U(p)$ acts freely on M' on the right: $Z \to ZB$, where $B \in U(p)$. We may consider $G_{p,q}(\mathbf{C}) = M$ as the base space of the principal fibre bundle M' with group $U(p)$. On the other hand, the group $U(p+q)$ acts on M' to the left: $Z \to AZ$, where $A \in U(p+q)$. This action is transitive and induces a transitive action of $U(p+q)$ on M. Let $e_1, \ldots, e_{p+q}$ be the natural basis in $\mathbf{C}^{p+q}$ and let Z_0 be the matrix with $e_1, \ldots, e_p$ as the column vectors. Then $Z_0 \in M'$. The isotropy group at $p_0 = \pi(Z_0) \in M$ for the action of $U(p+q)$ is then

$$\{A \in U(p+q); AZ_0 = Z_0 B \quad \text{for some } B \in U(p)\}$$
$$= \left\{ A = \begin{pmatrix} B & 0 \\ 0 & C \end{pmatrix}, \quad \text{where } B \in U(p) \text{ and } C \in U(q) \right\},$$

which we shall denote by $U(p) \times U(q)$. Thus we have $M = U(p+q)/U(p) \times U(q)$.

For each point $Z \in M'$, the tangent space $T_Z(M')$ is the set of all matrices with $p+q$ rows and p columns such that ${}^t\bar{W}Z + {}^t\bar{Z}W = 0$. In $T_Z(M')$ we have an inner product $g(W_1, W_2) = \text{Re trace } {}^t\bar{W}_2 W_1$. Let $T''_Z = \{ZA; A \in \mathfrak{u}(p)\} \subset T_Z(M')$ and let T'_Z be the orthogonal complement of T''_Z in $T_Z(M')$ with respect to g. The subspace T'_Z admits a complex structure $W \to iW$. We see that $\pi: M' \to M$ induces a linear isomorphism of T'_Z onto $T_{\pi(Z)}(M)$. By transferring the complex structure i and the inner product g on T'_Z by π_* we get the usual complex structure J on M and the Hermitian metric on M which were defined in Examples 2.4 and 6.4 of Chapter IX, respectively.

On the other hand, $U(p+q)/U(p) \times U(q)$ is a symmetric space when we define an involutive automorphism σ by $\sigma(A) = SAS^{-1}$, where

$$S = \begin{pmatrix} -I_p & 0 \\ 0 & I_q \end{pmatrix}.$$

XI. SYMMETRIC SPACES

The Lie algebra decomposition is
$$\mathfrak{u}(p+q) = \mathfrak{u}(p) + \mathfrak{u}(q) + \mathfrak{m},$$
where
$$\mathfrak{u}(p) + \mathfrak{u}(q) = \left\{ \begin{pmatrix} B & 0 \\ 0 & C \end{pmatrix}; B \in \mathfrak{u}(p), C \in \mathfrak{u}(q) \right\}$$

$$\mathfrak{m} = \left\{ \begin{pmatrix} 0 & -{}^t\bar{X} \\ X & 0 \end{pmatrix}; \begin{array}{l} X \text{ a complex matrix with} \\ q \text{ columns and } p \text{ rows} \end{array} \right\}$$

The space $\mathfrak{m}$ admits a complex structure which is essentially $X \to iX$ and we obtain the corresponding invariant complex structure J on $U(p+q)/U(p) \times U(q)$. The restriction to $\mathfrak{m}$ of the inner product $g(A_1, A_2) = -\frac{1}{2}$ trace $A_1 A_2$ is given by

$$g\left(\begin{pmatrix} 0 & -{}^t\bar{X} \\ X & 0 \end{pmatrix}, \begin{pmatrix} 0 & -{}^t\bar{Y} \\ Y & 0 \end{pmatrix} \right) = \operatorname{Re}(\text{trace } {}^t\bar{Y}X).$$

This inner product on $\mathfrak{m}$ together with the complex structure J gives rise to a Hermitian structure of $U(p+q)/U(p) \times U(q)$. It is now an easy matter to verify that the Hermitian structure on $U(p+q)/U(p) \times U(q)$ is isomorphic to that on $M = G_{p,q}(\mathbf{C})$.

Example 10.9. $D_{p,q}$ in Example 6.5 of Chapter IX. This is a generalization of Example 10.7 and the dual of Example 10.8. Let M' be the space of complex matrices

$$Z' = \begin{pmatrix} Z_1 \\ Z_0 \end{pmatrix}, \quad \text{where } Z_0 \text{ is } p \times p \text{ and } Z_1 \text{ is } p \times q$$

such that $-{}^t\bar{Z}_0 Z_0 + {}^t\bar{Z}_1 Z_1 = -I_p$. If F is a Hermitian form on $\mathbf{C}^{p+q}$ defined by

$$F(z, w) = -\sum_{i=1}^{p} z^i \bar{w}^i + \sum_{j=p+1}^{p+q} z^j \bar{w}^j,$$

then a matrix Z' with column vectors $\gamma_1, \ldots, \gamma_p \in \mathbf{C}^{p+q}$ belongs to M' if and only if

$$F(\gamma_i, \gamma_i) = -1 \quad \text{for } 1 \leq i \leq p,$$
$$F(\gamma_i, \gamma_j) = 0 \quad \text{for } i \neq j.$$

288   FOUNDATIONS OF DIFFERENTIAL GEOMETRY

The group $U(p)$ acts freely on M' by $Z' \to Z'B$, where $B \in U(p)$. The quotient space M can then be viewed as the set of all p-dimensional subspaces V of $\mathbf{C}^{p+q}$ such that $F(x, x) \leq 0$ for every $x \in V$.

On the other hand, let

$$U(p, q) = \{A \in GL(p + q; \mathbf{C}); F(Ax, Ay) = F(x, y), x, y \in \mathbf{C}^{p+q}\}$$
$$= \{A \in GL(p + q; \mathbf{C}); {}^t\bar{A}SA = S\},$$

where $S = \begin{pmatrix} -I_p & 0 \\ 0 & I_q \end{pmatrix}$. It is known that $U(p, q)$ is connected.

This group acts on M' by $Z' \to AZ'$ and induces a transitive action on M. If

$$p_0 = \pi \begin{pmatrix} I_p \\ 0 \end{pmatrix}, \text{ where } \begin{pmatrix} I_p \\ 0 \end{pmatrix} \in M' \text{ and } \pi \text{ is the projection } M' \to M,$$

then the isotropy group at p_0 for the action of $U(p, q)$ on M turns out to be

$$U(p) \times U(q) = \left\{ \begin{pmatrix} B & 0 \\ 0 & C \end{pmatrix}; B \in U(p), C \in U(q) \right\}.$$

The involutive automorphism $\sigma\colon \sigma(A) = SAS^{-1}$ of $U(p, q)$ defines $U(p, q)/U(p) \times U(q) = M$ as a symmetric space. The natural Hermitian structure can be defined in the manner analogous to Example 10.8.

We shall indicate how we can identify our space M with the space $D_{p,q}$ defined in Example 6.5. If $Z' = \begin{pmatrix} Z_0 \\ Z_1 \end{pmatrix} \in M'$, then

$${}^t\bar{Z}_0 Z_0 = I_p + {}^t\bar{Z}_1 Z_1.$$

Hence ${}^t\bar{Z}_0 Z_0$ is positive-definite. This implies that if $Z' \in M'$, then Z_0 is non-singular. Let $Z = Z_1 Z_0^{-1}$. We show that the $p \times p$ Hermitian matrix $I_p - {}^t\bar{Z}Z$ is positive-definite so that $Z \in D_{p,q}$:

$$I_p - {}^t\bar{Z}Z = I_p - {}^t\bar{Z}_0^{-1}\,{}^t\bar{Z}_1 Z_1 Z_0^{-1}$$
$$= {}^t\bar{Z}_0^{-1}({}^t\bar{Z}_0 Z_0 - {}^t\bar{Z}_1 Z_1) Z_0^{-1}$$
$$= {}^t\bar{Z}_0^{-1} Z_0^{-1} > 0.$$

If $\begin{pmatrix} Z_0 \\ Z_1 \end{pmatrix} B = \begin{pmatrix} W_0 \\ W_1 \end{pmatrix}$ with $B \in U(p)$ in M', it follows that $Z_1 Z_0^{-1} = W_1 W_0^{-1}$. Thus the mapping $Z' = \begin{pmatrix} Z_0 \\ Z_1 \end{pmatrix} \to Z_1 Z_0^{-1}$ induces a mapping of M into $D_{p,q}$. This is one-to-one. To show that it is surjective, let $Z \in D_{p,q}$. Then there is a $p \times p$ positive-definite Hermitian matrix, say, P such that $I_p - {}^t\bar{Z}Z = P^2$. We let $Z_0 = P^{-1}$ and $Z_1 = ZP^{-1}$. Then $Z' = \begin{pmatrix} Z_0 \\ Z_1 \end{pmatrix} \in M'$ as can be easily verified, and $Z = Z_1 Z_0^{-1}$.

The action of $U(p+q)$ on M, when transferred to $D_{p,q}$ by this identification, is what we defined in Example 6.5.

11. An outline of the classification theory

By Theorem 6.6 the classification of the simply connected Riemannian symmetric spaces is reduced to that of the irreducible ones. Similarly, by Theorems 5.2 and 7.2 the classification of the effective orthogonal symmetric Lie algebras is reduced to that of the irreducible ones.

We shall say that two Riemannian manifolds M and M' with metric tensors g and g', respectively, are *homothetic* to each other if there exists a diffeomorphism φ of M onto M' such that $\varphi^*g' = c^2 g$, where c is a positive constant. Then we have

PROPOSITION 11.1. *The homothetic-equivalence classes of simply connected irreducible Riemannian symmetric spaces M are in one-to-one correspondence with the isomorphism classes of effective irreducible orthogonal symmetric Lie algebras $(\mathfrak{g}, \mathfrak{h}, \sigma)$, where the correspondence is given by Theorem 6.5.*

Proof. The proof is straightforward. Only the following remark will suffice. Given $(\mathfrak{g}, \mathfrak{h}, \sigma)$ we construct as in §2 an almost effective symmetric space (G, H, σ), where G is simply connected and H is connected so that $M = G/H$ is simply connected. Since $(\mathfrak{g}, \mathfrak{h}, \sigma)$ is irreducible, an invariant Riemannian metric on G/H is unique up to a constant factor. Let N be the discrete normal subgroup of G consisting of those elements which act trivially on $M = G/H$. Denoting G/N and H/N anew by G and H, respectively, we obtain an effective symmetric space (G, H, σ) where $M = G/H$

is simply connected. We claim that G is the largest connected group of isometries of M. Let $G_1 = \mathfrak{I}^0(M)$ be the largest connected group of isometries of M. The symmetry of M at the origin induces an involutive automorphism σ_1 of G_1 which extends the automorphism σ of G (cf. Theorem 1.5). In the canonical decompositions $\mathfrak{g} = \mathfrak{h} + \mathfrak{m}$ and $\mathfrak{g}_1 = \mathfrak{h}_1 + \mathfrak{m}_1$, we have $\mathfrak{m} = \mathfrak{m}_1$. Hence (cf. the remark following Proposition 7.5) we get

$$\mathfrak{h} = [\mathfrak{m}, \mathfrak{m}] = [\mathfrak{m}_1, \mathfrak{m}_1] = \mathfrak{h}_1.$$

This implies $\mathfrak{g} = \mathfrak{g}_1$ and $G = G_1$. QED.

By virtue of the duality theorem (cf. Theorem 8.5) we may further restrict our attention to the effective symmetric spaces (G, H, σ) with irreducible orthogonal symmetric Lie algebras $(\mathfrak{g}, \mathfrak{h}, \sigma)$ of non-compact type. If $(\mathfrak{g}, \mathfrak{h}, \sigma)$ is such a symmetric Lie algebra, then it belongs to one of the following two classes in Theorem 8.5:

(III) $(\mathfrak{g}, \mathfrak{h}, \sigma)$ where $\mathfrak{g}$ is a simple Lie algebra of non-compact type which does not admit a compatible complex structure;

(IV) $(\mathfrak{g}, \mathfrak{h}, \sigma)$ where $\mathfrak{h}$ is a simple Lie algebra of compact type, $\mathfrak{g} = \mathfrak{h}^c$ is the complexification of $\mathfrak{h}$ and σ is the complex conjugation in $\mathfrak{g}$ with respect to $\mathfrak{h}$.

In either of these cases, $\mathfrak{g}$ is simple and is of non-compact type (cf. Appendix 9). If (G, H, σ) is an effective symmetric space with irreducible orthogonal symmetric Lie algebra $(\mathfrak{g}, \mathfrak{h}, \sigma)$ of non-compact type, then G has trivial center (cf. Theorem 8.4 of Chapter VIII and Theorem 8.6) and H is a maximal compact subgroup of G (cf. Corollary 9.3 of Chapter VIII and Theorem 8.6). If (G_1, H_1, σ_1) is another effective symmetric space with irreducible orthogonal symmetric Lie algebra $(\mathfrak{g}_1, \mathfrak{h}_1, \sigma_1)$ of non-compact type and if $\mathfrak{g}_1 = \mathfrak{g}$, then $G_1 = G$ since both G and G_1 have trivial center, and H_1 is conjugate to H in G by Corollary 9.3 of Chapter VIII. In the canonical decompositions $\mathfrak{g} = \mathfrak{h} + \mathfrak{m}$ and $\mathfrak{g}_1 = \mathfrak{h}_1 + \mathfrak{m}_1$, $\mathfrak{m}$ and $\mathfrak{m}_1$ are the orthogonal complements of $\mathfrak{h}$ and $\mathfrak{h}_1$, respectively, in $\mathfrak{g} = \mathfrak{g}_1$ with respect to the Killing-Cartan form of $\mathfrak{g} = \mathfrak{g}_1$. We can now conclude that, for every simple Lie algebra $\mathfrak{g}$ of non-compact type, there is (up to an isomorphism) at most one irreducible orthogonal symmetric Lie algebra $(\mathfrak{g}, \mathfrak{h}, \sigma)$ of non-compact type. Assuming the result of Weyl on the existence of compact real form, we shall show that every simple Lie algebra $\mathfrak{g}$ of non-compact type gives rise to an irreducible orthogonal symmetric Lie algebra $(\mathfrak{g}, \mathfrak{h}, \sigma)$ of non-compact type.

A theorem of Weyl states that *if $\mathfrak{g}$ is a complex simple Lie algebra, there is a real simple Lie algebra $\mathfrak{u}$ of compact type such that $\mathfrak{g}$ is isomorphic to the complexification $\mathfrak{u}^c$ of $\mathfrak{u}$.* Such a $\mathfrak{u}$ is called a *compact real form* of $\mathfrak{g}$ (cf. Hochschild [1; p. 167]).

First we consider the case where $\mathfrak{g}$ is a simple Lie algebra which admits a compatible complex structure. Let $\mathfrak{h}$ be a compact real form of $\mathfrak{g}$. Then $(\mathfrak{g}, \mathfrak{h}, \sigma)$, where σ is the complex conjugation with respect to $\mathfrak{h}$, is an irreducible orthogonal symmetric Lie algebra of non-compact type.

We consider now the case where $\mathfrak{g}$ is a simple Lie algebra of non-compact type which does not admit a compatible complex structure. Let $\mathfrak{g}^c$ be the complexification of $\mathfrak{g}$; it is a simple Lie algebra of non-compact type. Let $\mathfrak{u}$ be a compact real form of $\mathfrak{g}^c$. Then $(\mathfrak{g}^c, \mathfrak{u}, \sigma_1)$, where σ_1 is the complex conjugation with respect to $\mathfrak{u}$, is an irreducible orthogonal symmetric Lie algebra of non-compact type. We denote by (G_1, U, σ_1) an effective symmetric space with symmetric Lie algebra $(\mathfrak{g}^c, \mathfrak{u}, \sigma_1)$. Denote by α the complex conjugation of $\mathfrak{g}^c$ with respect to $\mathfrak{g}$. Then α is an involutive automorphism of $\mathfrak{g}^c$ (as a real Lie algebra) and induces an involutive automorphism α of the simply connected Lie group $\tilde{G}_1$ with Lie algebra $\mathfrak{g}^c$. Since G_1 is the quotient group of $\tilde{G}_1$ by its center, α induces an involutive automorphism α of G_1. By Theorem 9.4 of Chapter VIII and Theorem 8.6, there exists a maximal compact subgroup K of G_1 which is invariant by α. By Corollary 9.3 of Chapter VIII and Theorem 8.6, K is conjugate to U in G_1. It follows that $\mathfrak{k}$ is also a compact real form of $\mathfrak{g}^c$. If we denote by σ the complex conjugation of $\mathfrak{g}^c$ with respect to $\mathfrak{k}$, then $(\mathfrak{g}^c, \mathfrak{k}, \sigma)$ is an irreducible orthogonal symmetric Lie algebra of non-compact type. Since K is invariant by α, in the canonical decomposition $\mathfrak{g}^c = \mathfrak{k} + i\mathfrak{k}$ associated with $(\mathfrak{g}^c, \mathfrak{k}, \sigma)$ both $\mathfrak{k}$ and $i\mathfrak{k}$ are invariant by α. Since $\mathfrak{g}$ consists of those elements of $\mathfrak{g}^c$ which are left fixed by α, we obtain

$$\mathfrak{g} = (\mathfrak{g} \cap \mathfrak{k}) + (\mathfrak{g} \cap i\mathfrak{k}).$$

If we set $\mathfrak{h} = \mathfrak{g} \cap \mathfrak{k}$ and denote also by σ the restriction of σ to $\mathfrak{g}$, then we obtain an irreducible orthogonal symmetric Lie algebra $(\mathfrak{g}, \mathfrak{h}, \sigma)$ of non-compact type.

We have thus established the fact that the irreducible orthogonal symmetric Lie algebras $(\mathfrak{g}, \mathfrak{h}, \sigma)$ of non-compact type are in one-to-one correspondence with the real simple Lie algebras $\mathfrak{g}$ of non-compact type.

The classification of real simple Lie algebras was first achieved by E. Cartan [17] in 1914. Using the theory of symmetric spaces, he gave a more systematic classification in 1929 (cf. E. Cartan [16]). A more algebraic and systematic proof was obtained by Gantmacher [1], [2]. A further simplification was achieved by Murakami [6] and Wallach [1], [2] independently along the same line. Using results of Satake [2], Araki [1] gave another systematic method of classification. The main difference between the method of Araki and that of Murakami-Wallach lies in different choices of Cartan subalgebras.

The results in Chapter XI on Riemannian symmetric spaces are largely due to E. Cartan [7], [15]. Generalizations to affine symmetric spaces are due to Nomizu [2]. For more details and references on Riemannian symmetric spaces, see Helgason [2]. For affine symmetric spaces, see Berger [2] and Koh [1].

CHAPTER XII

Characteristic Classes

1. Weil homomorphism

Let G be a Lie group with Lie algebra $\mathfrak{g}$. Let $I^k(G)$ be the set of symmetric multilinear mappings $f\colon \mathfrak{g} \times \cdots \times \mathfrak{g} \to \mathbf{R}$ such that $f((\operatorname{ad} a)t_1, \ldots, (\operatorname{ad} a)t_k) = f(t_1, \ldots, t_k)$ for $a \in G$ and $t_1, \ldots, t_k \in \mathfrak{g}$. A multilinear mapping f satisfying the condition above is said to be *invariant* (by G). Obviously, $I^k(G)$ is a vector space over $\mathbf{R}$. We set

$$I(G) = \sum_{k=0}^{\infty} I^k(G).$$

For $f \in I^k(G)$ and $g \in I^l(G)$, we define $fg \in I^{k+l}(G)$ by

$fg(t_1, \ldots, t_{k+l})$
$$= \frac{1}{(k+l)!} \sum_\sigma f(t_{\sigma(1)}, \ldots, t_{\sigma(k)}) g(t_{\sigma(k+1)}, \ldots, t_{\sigma(k+l)}),$$

where the summation is taken over all permutations σ of $(1, \ldots, k + l)$. Extending this multiplication to $I(G)$ in a natural manner, we make $I(G)$ into a commutative algebra over $\mathbf{R}$.

Let P be a principal fibre bundle over a manifold M with group G and projection p. Our immediate objective is to define a certain homomorphism of the algebra $I(G)$ into the cohomology algebra $H^*(M; \mathbf{R})$. We choose a connection in the bundle P. Let ω be its connection form and Ω its curvature form. For each $f \in I^k(G)$, let $f(\Omega)$ be the $2k$-form on P defined by

$f(\Omega)(X_1, \ldots, X_{2k})$
$$= \frac{1}{(2k)!} \sum_\sigma \varepsilon_\sigma f(\Omega(X_{\sigma(1)}, X_{\sigma(2)}), \ldots, \Omega(X_{\sigma(2k-1)}, X_{\sigma(2k)}))$$

for $X_1, \ldots, X_{2k} \in T_u(P)$

where the summation is taken over all permutations σ of $(1, 2, \ldots, 2k)$ and ε_σ denotes the sign of the permutation σ. The purpose of this section is to prove the following theorem.

THEOREM 1.1. *Let P be a principal fibre bundle over M with group G and projection π. Choosing a connection in P, let Ω be its curvature form on P.*

(1) For each $f \in I^k(G)$, the $2k$-form $f(\Omega)$ on P projects to a (unique) closed $2k$-form, say $\bar{f}(\Omega)$, on M, i.e., $f(\Omega) = \pi^*(\bar{f}(\Omega))$;

(2) If we denote by $w(f)$ the element of the de Rham cohomology group $H^{2k}(M; \mathbf{R})$ defined by the closed $2k$-form $\bar{f}(\Omega)$, then $w(f)$ is independent of the choice of a connection and $w \colon I(G) \to H^*(M; \mathbf{R})$ is an algebra homomorphism.

Proof. We first prove

LEMMA 1. *A q-form φ on P projects to a (unique) q-form, say $\bar{\varphi}$, on M if*

(a) $\varphi(X_1, \ldots, X_q) = 0$ *whenever at least one of the X_i's is vertical;*

(b) $\varphi(R_a X_1, \ldots, R_a X_q) = \varphi(X_1, \ldots, X_q)$ *for the right translation R_a defined by each element $a \in G$.*

Proof of Lemma 1. Let $V_1, \ldots, V_q$ be tangent vectors of M at a point of M. Define a q-form $\bar{\varphi}$ on M by

$$\bar{\varphi}(V_1, \ldots, V_q) = \varphi(X_1, \ldots, X_q),$$

where $X_1, \ldots, X_q$ are tangent vectors of P at a point $u \in P$ such that $\pi(X_i) = V_i$ for $i = 1, \ldots, q$. We have to show that $\bar{\varphi}(V_1, \ldots, V_q)$ is well defined (independent of the choice of $X_1, \ldots, X_q$). Let $Y_1, \ldots, Y_q \in T_v(P)$ with $\pi(Y_i) = V_i$ for $i = 1, \ldots, q$. To prove that $\varphi(X_1, \ldots, X_q) = \varphi(Y_1, \ldots, Y_q)$ we may assume by (b) that $u = v$. Since $X_i - Y_i$ is vertical for each i, it follows from (a) that

$$\begin{aligned} \varphi(X_1, \ldots, X_q) &= \varphi(Y_1, X_2, \ldots, X_q) \\ &= \varphi(Y_1, Y_2, X_3, \ldots, X_q) = \cdots \\ &= \varphi(Y_1, \ldots, Y_q). \end{aligned}$$

This completes the proof of Lemma 1.

We recall (p. 77 of Volume I) the definition of exterior covariant differentiation D: For any q-form φ on P, we set

$$(D\varphi)(X_1, \ldots, X_{q+1}) = (d\varphi)(hX_1, \ldots, hX_{q+1}),$$

where hX_i is the horizontal component of X_i.

LEMMA 2. *If a q-form φ on P projects to a q-form $\bar{\varphi}$ on M, i.e., $\varphi = \pi^*(\bar{\varphi})$, then $d\varphi = D\varphi$.*

Proof of Lemma 2. Let $X_1, \ldots, X_{q+1}$ be tangent vectors of P at a point of P. Then

$$(d\varphi)(X_1, \ldots, X_{q+1})$$
$$= (d\pi^*\bar{\varphi})(X_1, \ldots, X_{q+1}) = (\pi^* d\bar{\varphi})(X_1, \ldots, X_{q+1})$$
$$= (d\bar{\varphi})(\pi X_1, \ldots, \pi X_{q+1}) = (d\bar{\varphi})(\pi h X_1, \ldots, \pi h X_{q+1})$$
$$= (\pi^* d\bar{\varphi})(h X_1, \ldots, h X_{q+1}) = (d\pi^*\bar{\varphi})(h X_1, \ldots, h X_{q+1})$$
$$= (d\varphi)(h X_1, \ldots, h X_{q+1}) = (D\varphi)(X_1, \ldots, X_{q+1}).$$

This completes the proof of Lemma 2.

Applying Lemmas 1 and 2 to $f(\Omega)$ we shall prove (1) of Theorem 1.1. Since the curvature form Ω satisfies (a) of Lemma 1 by the very definition of Ω (cf. p. 77 of Volume I), so does $f(\Omega)$. Since Ω is a tensorial form of type ad G so that

$$R_a^*(\Omega) = (\text{ad } a^{-1})\Omega \quad \text{for } a \in G$$

and since f is invariant by G, $f(\Omega)$ satisfies (b) of Lemma 1. Hence $f(\Omega)$ projects to a unique $2k$-form, say $\bar{f}(\Omega)$, on M. To prove that $\bar{f}(\Omega)$ is closed, it suffices to prove that $f(\Omega)$ is closed. But, by Lemma 2, we have $d(f(\Omega)) = D(f(\Omega))$. Since D acts as a skew-derivation on the algebra of tensorial forms of P and since $D\Omega = 0$ (by Bianchi's identity, cf. Theorem 5.4 of Chapter II), it follows that $D(f(\Omega)) = 0$. (We recall that a form Ω on P is called a tensorial form if it is horizontal in the sense that $\Omega(X_1, \ldots, X_q) = 0$ whenever at least one of the X_i's is vertical, cf. p. 75 of Volume I.)

It is a routine matter to verify that $w: I(G) \to H^*(M; \mathbf{R})$ is an algebra homomorphism. It remains to show that w is independent of the choice of a connection. Consider two connection forms ω_0 and ω_1 on P and define

$$\alpha = \omega_1 - \omega_0,$$
$$\omega_t = \omega_0 + t\alpha \quad \text{for } 0 \leq t \leq 1.$$

The following lemma is immediate.

LEMMA 3. α *is a tensorial* 1-*form of type* ad G, *i.e.*,
(1) $\alpha(X) = 0$ *if X is vertical*;
(2) $R_a^*(\alpha) = \text{ad}\,(a^{-1})\alpha$ *for $a \in G$.*

$\{\omega_t; 0 \leq t \leq 1\}$ *is a* 1-*parameter family of connection forms on P*.

Let D_t be the exterior covariant differentiation and Ω_t the curvature form defined by the connection form ω_t. Then we have

LEMMA 4. $\dfrac{d}{dt}\Omega_t = D_t\alpha$.

Proof of Lemma 4. By the structure equation (cf. p. 78 of Volume I) we have

$$\Omega_t = d\omega_t + \tfrac{1}{2}[\omega_t, \omega_t] = d\omega_0 + t\,d\alpha + \tfrac{1}{2}[\omega_t, \omega_t].$$

Since $\dfrac{d}{dt}\omega_t = \alpha$, we obtain

$$\frac{d}{dt}\Omega_t = d\alpha + \tfrac{1}{2}[\alpha, \omega_t] + \tfrac{1}{2}[\omega_t, \alpha].$$

By Proposition 5.5 of Chapter II, the right hand side in the equality above is equal to $D_t\alpha$. This completes the proof of Lemma 4.

We are now in a position to show that w is independent of the choice of a connection. Given $f \in I^k(G)$ and $\mathfrak{g}$-valued forms $\varphi_1, \ldots, \varphi_k$ on P of degree, respectively, $q_1, \ldots, q_k$, we define a form $f(\varphi_1, \ldots, \varphi_k)$ of degree $q_1 + \cdots + q_k$ on P by

$$f(\varphi_1, \ldots, \varphi_k)(X_1, \ldots, X_{q_1+\cdots+q_k})$$
$$= A \sum \varepsilon_\sigma f(\varphi_1(X_{\sigma(1)}, \ldots, X_{\sigma(q_1)}), \ldots,$$
$$\varphi_k(X_{\sigma(q_1+\cdots+q_{k-1}+1)}, \ldots, X_{(q_1+\cdots+q_k)})),$$

where $A = 1/(q_1 + \cdots + q_k)!$, the summation is taken over all permutations σ of $(1, \ldots, q_1 + \cdots + q_k)$ and ε_σ is the sign of the permutation σ. The $2k$-form $f(\Omega)$ defined at the beginning of this section may be considered as an abbreviated notation for $f(\Omega, \ldots, \Omega)$. We may express $f(\varphi_1, \ldots, \varphi_k)$ as follows. Let $E_1, \ldots, E_r$ be a basis for $\mathfrak{g}$ and write

$$\varphi_i = \sum_{j=1}^{r} \varphi_i^j E_j \quad \text{for } i = 1, \ldots, k.$$

If we set
$$a_{j_1\cdots j_k} = f(E_{j_1}, \ldots, E_{j_k}),$$
then
$$f(\varphi_1, \ldots, \varphi_k) = \sum_{j_1,\ldots,j_k=1}^{r} a_{j_1\cdots j_k} \varphi_1^{j_1} \wedge \cdots \wedge \varphi_k^{j_k}.$$

We shall now complete the proof of Theorem 1.1 by showing the following

LEMMA 5. *Let $f \in I^k(G)$. If we set*
$$\Phi = k\int_0^1 f(\alpha, \Omega_t, \ldots, \Omega_t)\, dt,$$
then Φ projects to a (unique) $(2k-1)$-form on M and
$$d\Phi = f(\Omega_1, \ldots, \Omega_1) - f(\Omega_0, \ldots, \Omega_0).$$

Proof of Lemma 5. By Lemma 1, $f(\alpha, \Omega_t, \ldots, \Omega_t)$ projects to a $(2k-1)$-form on M. Hence Φ projects to a $(2k-1)$-form on M. By Lemmas 2 and 4 and by $D_t\Omega_t = 0$ (Bianchi's identity), we obtain

$$\begin{aligned} k\, d(f(\alpha, \Omega_t, \ldots, \Omega_t)) &= kD_t(f(\alpha, \Omega_t, \ldots, \Omega_t)) \\ &= kf(D_t\alpha, \Omega_t, \ldots, \Omega_t) \\ &= kf\left(\frac{d}{dt}\Omega_t, \Omega_t, \ldots, \Omega_t\right) \\ &= \frac{d}{dt}(f(\Omega_t, \Omega_t, \ldots, \Omega_t)). \end{aligned}$$

Hence,
$$\begin{aligned} d\Phi &= k\int_0^1 (d(f(\alpha, \Omega_t, \ldots, \Omega_t)))\, dt \\ &= \int_0^1 \frac{d}{dt}(f(\Omega_t, \ldots, \Omega_t))\, dt \\ &= f(\Omega_1, \ldots, \Omega_1) - f(\Omega_0, \ldots, \Omega_0). \end{aligned}$$

This completes the proof of Lemma 5 and hence that of Theorem 1.1. QED.

Theorem 1.1 is due to A. Weil, and $w: I(G) \to H^*(M; \mathbf{R})$ is called the *Weil homomorphism*. The reader will find a more algebraic proof in H. Cartan [2]. Our proof was taken from Chern [2]. Narasimhan and Ramanan [1] gave a different proof by making use of the existence of universal connections.

2. Invariant polynomials

Let V be a vector space over $\mathbf{R}$ and $S^k(V)$ the space of symmetric multilinear mappings f of $V \times \cdots \times V$ (k times) into $\mathbf{R}$. In the same way as we made $I(G)$ into a commutative algebra in §1, we define a multiplication in $S(V) = \Sigma_{k=0}^{\infty} S^k(V)$ to make it into a commutative algebra over $\mathbf{R}$.

Let $\xi^1, \ldots, \xi^r$ be a basis for the dual space of V. A mapping $p: V \to \mathbf{R}$ is called a *polynomial function* if it can be expressed as a polynomial of $\xi^1, \ldots, \xi^r$. The concept is evidently independent of the choice of $\xi^1, \ldots, \xi^r$. Let $P^k(V)$ denote the space of homogeneous polynomial functions of degree k on V. Then $P(V) = \Sigma_{k=0}^{\infty} P^k(V)$ is the algebra of polynomial functions on V.

PROPOSITION 2.1. *The mapping* $\varphi: S(V) \to P(V)$ *defined by*

$$(\varphi f)(t) = f(t, \ldots, t) \quad \text{for } f \in S^k(V) \text{ and } t \in V$$

is an isomorphism of $S(V)$ *onto* $P(V)$.

Proof. The mapping φ is easily seen to be an algebra homomorphism. We shall construct a mapping $\psi: P(V) \to S(V)$ and leave it to the reader to verify that ψ is the inverse mapping of φ. Using a basis $\xi^1, \ldots, \xi^r$ for the dual space of V we express each polynomial function $p \in P^k(V)$ as

$$\sum_{i_1,\ldots,i_k=1}^{r} a_{i_1 \cdots i_k} \xi^{i_1} \cdots \xi^{i_k},$$

where $a_{i_1 \cdots i_k}$ are symmetric in $i_1, \ldots, i_k$. Considering $\xi^1, \ldots, \xi^r$ as linear functionals on V, we set

$$(\psi p)(t_1, \ldots, t_k) = \sum a_{i_1 \cdots i_k} \xi^{i_1}(t_1) \cdots \xi^{i_k}(t_k) \quad \text{for } t_1, \ldots, t_k \in V.$$

QED.

PROPOSITION 2.2. *Given a group G of linear transformations of V, let $S_G(V)$ and $P_G(V)$ be the subalgebras of $S(V)$ and $P(V)$, respectively, consisting of G-invariant elements. Then the isomorphism $\varphi: S(V) \to P(V)$ defined in Proposition 2.1 induces an isomorphism of $S_G(V)$ onto $P_G(V)$.*

XII. CHARACTERISTIC CLASSES   299

The proof is straightforward and is left to the reader.

Applying Proposition 2.2 to the algebra $I(G)$ defined in §1, we obtain

COROLLARY 2.3. *Let G be a Lie group. Then the algebra $I(G)$ of (ad G)-invariant symmetric multilinear mappings of its Lie algebra $\mathfrak{g}$ into $\mathbf{R}$ may be identified with the algebra of (ad G)-invariant polynomial functions on $\mathfrak{g}$.*

The following theorem is useful in the actual determination of the algebra $I(G)$ defined in §1.

THEOREM 2.4. *Let G be a Lie group and $\mathfrak{g}$ its Lie algebra. Let G' be a Lie subgroup of G and $\mathfrak{g}'$ its Lie algebra. Let $I(G)$ (resp. $I(G')$) be the algebra of invariant symmetric multilinear mappings of $\mathfrak{g}$ (resp. $\mathfrak{g}'$) into $\mathbf{R}$. Set*

$$N = \{a \in G;\ (\operatorname{ad} a)\mathfrak{g}' \subset \mathfrak{g}'\}.$$

Considering N as a group of linear transformations acting on $\mathfrak{g}'$, let $I_N(G')$ be the subalgebra of $I(G')$ consisting of elements invariant by N. If

$$\mathfrak{g} = \{(\operatorname{ad} a)t';\ a \in G \quad \text{and} \quad t' \in \mathfrak{g}'\},$$

then the restriction map $I(G) \to I(G')$ maps $I(G)$ isomorphically into $I_N(G')$.

Proof. It is clear that an element of $I(G)$ restricted to $\mathfrak{g}'$ belongs to $I_N(G')$. To show that the mapping $I(G) \to I_N(G')$ is injective, let f be an element of $I^k(G)$ such that

$$f(t', \ldots, t') = 0 \quad \text{for } t' \in \mathfrak{g}'.$$

Our assumption says that every element t of $\mathfrak{g}$ is of the form $(\operatorname{ad} a)t'$ with $a \in G$ and $t' \in \mathfrak{g}'$. Then the invariance of f implies

$$f(t, \ldots, t) = f(t', \ldots, t') = 0.$$

By Corollary 2.3, we may conclude that $f = 0$.   QED.

Remark. In Theorem 2.4, *if G is compact and G' is a maximal torus T of G, then $I(G) \to I_N(T)$ is an isomorphism.* Since we do not use this fact, we give only an outline of the proof. Assume first that G is connected. Let $f' \in I_N(T)$ be a homogeneous polynomial of degree k. For $t \in \mathfrak{g}$, choose an element $a \in G$ such that $(\operatorname{ad} a)t$ is in the Lie algebra $\mathfrak{t}$ of T. We define $f(t)$ by $f(t) = f'((\operatorname{ad} a)t)$. If

b is another element of G such that $(\operatorname{ad} b)t$ is in $\mathfrak{t}$, then there is an element n of N such that $(\operatorname{ad} b)t = (\operatorname{ad} n)(\operatorname{ad} a)t$. (This follows from the following fact: if S is any subset of T and g is an element of G such that $gSg^{-1} \subset T$, there is an element $n \in N$ such that $nsn^{-1} = gsg^{-1}$ for all $s \in S$; cf Séminaire "Sophus Lie," No. 23.) Since f' is invariant by N, $f(t)$ is well defined. It is easy to see that f is a G-invariant differentiable function on $\mathfrak{g}$ which is homogeneous of degree k in the sense that $f(\lambda t) = \lambda^k f(t)$ for $\lambda \in \mathbf{R}$. Differentiate this identity k times with respect to λ and set $\lambda = 0$. The right hand side yields $k! f(t)$. Using a coordinate system $\xi^1, \ldots, \xi^r$ for $\mathfrak{g}$ and applying the chain rule, we obtain from the left hand side a polynomial $\Sigma\, a_{i_1 \cdots i_k} \xi^{i_1} \cdots \xi^{i_k}$, where $a_{i_1 \cdots i_k}$ denotes the value of $\partial^k f / \partial \xi^{i_1} \cdots \partial \xi^{i_k}$ at the origin. This shows that $f \in I(G)$. If G is not connected, the natural isomorphism $N/(N \cap G^0) \to G/G^0$ is used to reduce the problem to the case where G is connected.

As an application of Theorem 2.4 we shall determine $I(U(n))$.

THEOREM 2.5. *Define polynomial functions* $f_1, \ldots, f_n$ *on the Lie algebra* $\mathfrak{u}(n)$ *of* $U(n)$ *by*

$$\det(\lambda I_n + \sqrt{-1}\, X)$$
$$= \lambda^n - f_1(X)\lambda^{n-1} + f_2(X)\lambda^{n-2} - \cdots + (-1)^n f_n(X)$$
for $X \in \mathfrak{u}(n)$.

Then $f_1, \ldots, f_n$ *are algebraically independent and generate the algebra of polynomial functions on* $\mathfrak{u}(n)$ *invariant by* $\operatorname{ad} U(n)$.

Proof. Let T be the subgroup of $U(n)$ consisting of diagonal elements. Its Lie algebra $\mathfrak{t}$ consists of matrices of the form

$$\begin{pmatrix} \sqrt{-1}\,\xi^1 & & \\ & \ddots & \\ & & \sqrt{-1}\,\xi^n \end{pmatrix} \quad (\xi^i \text{ real}),$$

which will be denoted by $[\xi^1, \ldots, \xi^n]$ for the sake of simplicity. We shall use $\xi^1, \ldots, \xi^n$ as a coordinate system in $\mathfrak{t}$. If we set $N = \{A \in U(n)\,;\, AXA^{-1} \in \mathfrak{t} \text{ for } X \in \mathfrak{t}\}$, then Theorem 2.4 implies that $I(U(n)) \to I_N(T)$ is injective. It suffices now to establish the following three facts: (1) $f_1, \ldots, f_n$ are invariant by ad $(U(n))$; (2) Restricted to $\mathfrak{t}$, the polynomial functions $f_1, \ldots, f_n$ are the elementary symmetric functions of $\xi^1, \ldots, \xi^n$; (3) Every polynomial function on $\mathfrak{t}$ invariant by N is a symmetric function of $\xi^1, \ldots, \xi^n$. The first two facts are evident. To prove the third, we have only to show that, for every pair (i,j) with $i < j$, there exists an element of N which sends $[\xi^1, \ldots, \xi^i, \ldots, \xi^j, \ldots, \xi^n]$ into $[\xi^1, \ldots, \xi^j, \ldots, \xi^i, \ldots, \xi^n]$. Such an element is given by the matrix having 1 at the (i,j)-th place, the (j,i)-th place and the (k,k)-th places for $k \neq i,j$ and having 0 in all other places.

QED.

THEOREM 2.6. *Define polynomial functions* $f_1, \ldots, f_m$ *on the Lie algebra* $\mathfrak{o}(n)$ *of* $O(n)$ *(where* $n = 2m$ *or* $n = 2m + 1$*) by*

$$\det(\lambda I_n - X) = \lambda^n + f_1(X)\lambda^{n-2} + f_2(X)\lambda^{n-4} + \cdots$$

for $X \in \mathfrak{o}(n)$.

Then $f_1, \ldots, f_m$ *are algebraically independent and generate the algebra of polynomial functions on* $\mathfrak{o}(n)$ *invariant by* ad $(O(n))$.

Proof. The proof is similar to that of Theorem 2.5. We shall indicate only the necessary changes. According as $n = 2m$ or $n = 2m + 1$, let T be the connected Lie subgroup of $O(n)$ whose Lie algebra $\mathfrak{t}$ consists of matrices of the following form:

$$\begin{pmatrix} 0 & \xi^1 & & & & & \\ -\xi^1 & 0 & & & & & \\ & & \cdot & & & & \\ & & & \cdot & & & \\ & & & & \cdot & & \\ & & & & & 0 & \xi^m \\ & & & & & -\xi^m & 0 \end{pmatrix}$$

or

$$\begin{pmatrix} 0 & \xi^1 & & & & & & \\ -\xi^1 & 0 & & & & & & \\ & & \cdot & & & & & \\ & & & \cdot & & & & \\ & & & & \cdot & & & \\ & & & & & 0 & \xi^m & \\ & & & & & -\xi^m & 0 & \\ & & & & & & & 0 \end{pmatrix},$$

which will be denoted by $[\xi^1, \ldots, \xi^m]$ for the sake of simplicity. We first observe that the coefficient of λ^{n-i} in $\det(\lambda I_n - X)$ vanishes identically for each odd number i. This is due to the fact that the transpose of $\lambda I_n - X$ is equal to $\lambda I_n + X$ so that $\det(\lambda I_n - X) = \det(\lambda I_n + X)$. It suffices now to establish the following three facts: (1) $f_1, \ldots, f_m$ are invariant by $O(n)$; (2) Restricted to $\mathfrak{t}$, the polynomial functions $f_1, \ldots, f_m$ are the elementary symmetric functions of $(\xi^1)^2, \ldots, (\xi^m)^2$; (3) Every polynomial function on $\mathfrak{t}$ invariant by N is a symmetric function of $(\xi^1)^2, \ldots, (\xi^m)^2$. The first two facts being trivial, we shall show only the third. For each pair (i, j) with $i < j$ we shall construct an element A of N which sends $[\xi^1, \ldots, \xi^i, \ldots, \xi^j, \ldots, \xi^m]$ into $[\xi^1, \ldots, \xi^j, \ldots, \xi^i, \ldots, \xi^m]$ and for each i we shall construct an element B of N which sends $[\xi^1, \ldots, \xi^i, \ldots, \xi^m]$ into $[\xi^1, \ldots, -\xi^i, \ldots, \xi^m]$. In fact, A may be chosen from $N \cap SO(n)$; for example, for $i = 1$ and $j = 2$ we may take for A the following matrix:

$$\begin{pmatrix} 0 & I_2 & \\ I_2 & 0 & \\ & & I_{n-4} \end{pmatrix}.$$

For $i = 1$, for example, we may take for B the following matrix:

$$\begin{pmatrix} 0 & 1 & \\ 1 & 0 & \\ & & I_{n-2} \end{pmatrix}.$$

If n is odd, we may change the (n, n)-th element in the matrix above from 1 to -1 and thus choose B from $SO(n) \cap N$. The fact that A and B for n odd may be chosen not only from N but also from $N \cap SO(n)$ will be used in the proof of the next theorem.
QED.

THEOREM 2.7. *Define polynomial functions* $f_1, \ldots, f_m$ *on the Lie algebra* $\mathfrak{o}(n)$ *of* $SO(n)$ *as in Theorem* 2.6.

(1) *If* $n = 2m + 1$, *then* $f_1, \ldots, f_m$ *are algebraically independent and generate the algebra of polynomial functions on* $\mathfrak{o}(n)$ *invariant by* $\mathrm{ad}\,(SO(n))$;

(2) *If* $n = 2m$, *then there exists a polynomial function* g *(unique up to a sign) such that* $f_m = g^2$ *and the functions* $f_1, \ldots, f_{m-1}, g$ *are*

algebraically independent and generate the algebra of polynomial functions on $\mathfrak{o}(n)$ *invariant by* ad $(SO(n))$.

Proof. Set

$$N = \{A \in SO(n) ; AXA^{-1} \in \mathfrak{t} \quad \text{for } X \in \mathfrak{t}\}.$$

This subgroup N of $SO(n)$ is the intersection of $SO(n)$ with the subgroup N of $O(n)$ defined in the proof of Theorem 2.6.

(1) In this case, the proof of Theorem 2.6 is valid since the matrices A and B in the proof of Theorem 2.6 may be chosen from the present subgroup N of $SO(n)$.

(2) In this case, the matrix A in the proof of Theorem 2.6 may be chosen from N. For each pair (i,j) with $i < j$, we shall construct an element C of N which sends $[\xi^1, \ldots, \xi^i, \ldots, \xi^j, \ldots, \xi^m]$ into $[\xi^1, \ldots, -\xi^i, \ldots, -\xi^j, \ldots, \xi^m]$. For $i = 1$ and $j = 2$, for example, we set

$$C = \begin{pmatrix} 0 & 1 & & & \\ 1 & 0 & & & \\ & & 0 & 1 & \\ & & 1 & 0 & \\ & & & & I_{n-4} \end{pmatrix}.$$

From the fact that a polynomial function belonging to $I_N(T)$ is invariant by A, it follows that every element of $I_N(T)$ is a symmetric function of $\xi^1, \ldots, \xi^m$. On the other hand, for a polynomial of $\xi^1, \ldots, \xi^m$ invariant by C, a term containing an odd power of ξ^i must contain also an odd power of ξ^j. Consequently, every polynomial function belonging to $I_N(T)$ has the property that each term is of either even degree in every one of $\xi^1, \ldots, \xi^m$ or odd degree in every one of $\xi^1, \ldots, \xi^m$. In other words, every element $f \in I_N(T)$ may be expressed in the following form:

$$f = p + (\xi^1 \cdots \xi^m) q,$$

where both p and q are polynomials in $(\xi^1)^2, \ldots, (\xi^m)^2$. Since f is a symmetric function of $\xi^1, \ldots, \xi^m$, it follows that both p and q are symmetric functions of $(\xi^1)^2, \ldots, (\xi^m)^2$. To complete the proof we shall construct a polynomial function g on $\mathfrak{o}(n)$ invariant

by ad $(SO(n))$ such that $f_m = g^2$ and that $g = \xi^1 \cdots \xi^m$ on $\mathfrak{t}$. Let
$$X = (x^{ij}) \in \mathfrak{o}(2m) \quad \text{with } x_{ij} = -x_{ji}.$$
Set
$$g(X) = \frac{1}{2^m m!} \sum \varepsilon_{i_1 i_2 \cdots i_{2m-1} i_{2m}} x_{i_1 i_2} \cdots x_{i_{2m-1} i_{2m}},$$
where the summation is taken over all permutations of $(1, \ldots, 2m)$ and $\varepsilon_{i_1 \cdots i_{2m}}$ is 1 or -1 according as $(i_1, \ldots, i_{2m})$ is an even or odd permutation of $(1, \ldots, 2m)$. From the usual definition of the determinant it follows that g is invariant by ad $(SO(n))$. The restriction of g to $\mathfrak{t}$ is clearly equal to $\xi^1 \cdots \xi^m$. Hence $f_m = g^2$ on $\mathfrak{t}$. Since $I(SO(n)) \to I(T)$ is injective by Theorem 2.4, we may conclude that $f_m = g^2$ on $\mathfrak{o}(n)$. QED.

We shall now consider the group $Sp(m)$ which consists of unitary matrices $X \in U(2m)$ such that
$$^t X J X = J,$$
where
$$J = \begin{pmatrix} 0 & I_m \\ -I_m & 0 \end{pmatrix}.$$

THEOREM 2.8. *Define polynomial functions* $f_1, \ldots, f_m$ *on the Lie algebra* $\mathfrak{sp}(m)$ *of* $Sp(m)$ *by*
$$\det(\lambda I_{2m} + \sqrt{-1}\, X)$$
$$= \lambda^{2m} - f_1(X) \lambda^{2m-2} + f_2(X) \lambda^{2m-4} - \cdots + (-1)^m f_m(X)$$
$$\text{for } X \in \mathfrak{sp}(m).$$

Then $f_1, \cdots, f_m$ *are algebraically independent and generate the algebra of polynomial functions on* $\mathfrak{sp}(m)$ *invariant by* ad $(Sp(m))$.

Proof. Let T be the subgroup of $Sp(m)$ consisting of diagonal matrices. Its Lie algebra $\mathfrak{t}$ consists of matrices of the form

$$\begin{pmatrix} \sqrt{-1}\,\xi^1 & & & & & \\ & \ddots & & & & \\ & & \sqrt{-1}\,\xi^m & & & \\ & & & -\sqrt{-1}\,\xi^1 & & \\ & & & & \ddots & \\ & & & & & -\sqrt{-1}\,\xi^m \end{pmatrix}$$

(ξ^i real)

which will be denoted by $[\xi^1, \ldots, \xi^m]$ for the sake of simplicity. Set $N = \{A \in Sp(m); AXA^{-1} \in \mathfrak{t} \text{ for } X \in \mathfrak{t}\}$. Then $I(Sp(m)) \to I_N(T)$ is injective by Theorem 2.4. Since $f_1, \ldots, f_m$ are invariant by ad $(Sp(m))$ and since, restricted to $\mathfrak{t}, f_1, \ldots, f_m$ are the elementary symmetric functions of $(\xi^1)^2, \ldots, (\xi^m)^2$, it suffices to prove that every polynomial function on $\mathfrak{t}$ invariant by N is a symmetric function of $(\xi^1)^2, \ldots, (\xi^m)^2$. But this can be accomplished as in the proof of Theorem 2.6 by constructing certain matrices A and B belonging to N. More explicitly, for every pair (i, j) with $i < j$, consider the matrix $A \in Sp(m)$ having 1 at the (i, j)-th place, the (j, i)-th place, the $(m+i, m+j)$-th place, the $(m+j, m+i)$-th place and the (k, k)-th places for $k \neq i, j, m+i, m+j$ and having 0 in all other places. Then A is an element of N which sends $[\xi^1, \ldots, \xi^i, \ldots, \xi^j, \ldots, \xi^m]$ into $[\xi^1, \ldots, \xi^j, \ldots, \xi^i, \ldots, \xi^m]$. For each i with $1 \leq i \leq m$, let $B \in Sp(m)$ be the matrix having $\sqrt{-1}$ at the $(i, m+i)$-th place and the $(m+i, i)$-th place, having 1 at the (k, k)-th places for $1 \leq k \leq 2m$ and $k \neq i, m+i$ and having 0 in all other places. Then B is an element of N which sends $[\xi^1, \ldots, \xi^i, \ldots, \xi^m]$ into $[\xi^1, \ldots, -\xi^i, \ldots, \xi^m]$. QED.

Remark. In the proofs of Theorems 2.5, 2.6, 2.7, and 2.8, we used only the fact that the restriction map $I(G) \to I_N(T)$ is injective although it is actually an isomorphism; see the remark following Theorem 2.4. But, from the proofs of these theorems we see directly that $I(G) \to I_N(T)$ is an isomorphism when G is $U(n)$, $O(n)$, $SO(n)$, or $Sp(m)$. For further results on $I(G)$, see H. Cartan [3]. The quotient group N/T is called the *Weyl group* of G. Hence, $I_N(T)$ is isomorphic to the algebra of polynomial functions on $\mathfrak{t}$ which are invariant by the Weyl group N/T. By a general result of Chevalley [2], $I_N(T)$ is finitely generated. We exhibited generators when G is one of the compact groups cited above.

3. Chern classes

We recall the axiomatic definition of Chern classes (Hirzebruch [3] and Husemoller [1]). We consider the category of differentiable complex vector bundles over differentiable manifolds.

Axiom 1. For each complex vector bundle E over M and for each integer $i \geq 0$, the *i-th Chern class* $c_i(E) \in H^{2i}(M; \mathbf{R})$ is given, and $c_0(E) = 1$.

We set $c(E) = \Sigma_{i=0}^{\infty} c_i(E)$ and call $c(E)$ the *total Chern class* of E.

Axiom 2 (Naturality). Let E be a complex vector bundle over M and $f: M' \to M$ a differentiable map. Then

$$c(f^{-1}E) = f^*(c(E)) \in H^*(M'; \mathbf{R}),$$

where $f^{-1}E$ denotes the complex vector bundle over M' induced by f from E.

Axiom 3 (Whitney sum formula). Let $E_1, \ldots, E_q$ be complex line bundles over M, i.e., complex vector bundles with fibre $\mathbf{C}$. Let $E_1 \oplus \cdots \oplus E_q$ be their Whitney sum, i.e., $E_1 \oplus \cdots \oplus E_q = d^{-1}(E_1 \times \cdots \times E_q)$, where $d: M \to M \times \cdots \times M$ maps each point $x \in M$ into the diagonal element $(x, \ldots, x) \in M \times \cdots \times M$. Then

$$c(E_1 \oplus \cdots \oplus E_q) = c(E_1) \cdots c(E_q).$$

To state Axiom 4, we need to define a certain natural complex line bundle over the n-dimensional complex projective space $P_n(\mathbf{C})$. A point x of $P_n(\mathbf{C})$ is a 1-dimensional complex subspace, denoted by F_x, of $\mathbf{C}^{n+1}$. To each $x \in P_n(\mathbf{C})$ we assign the corresponding F_x as the fibre over x, thus obtaining a complex line bundle over $P_n(\mathbf{C})$ which will be denoted by E_n. Instead of describing the complex structure of E_n in detail, we exhibit its associated principal bundle. Let $\mathbf{C}^*$ be the multiplicative group of non-zero complex numbers. Then $\mathbf{C}^*$ acts on the space $\mathbf{C}^{n+1} - \{0\}$ of non-zero vectors in $\mathbf{C}^{n+1}$ by

$$((z^0, z^1, \ldots, z^n), w) \in (\mathbf{C}^{n+1} - \{0\}) \times \mathbf{C}^*$$
$$\to (z^0 w, z^1 w, \ldots, z^n w) \in \mathbf{C}^{n+1} - \{0\}.$$

Under this action of $\mathbf{C}^*$, the space $\mathbf{C}^{n+1} - \{0\}$ is the principal fibre bundle over $P_n(\mathbf{C})$ with group $\mathbf{C}^*$ associated with the natural line bundle E_n. If we denote by p the projection of this principal bundle, and by U_i the open subset of $P_n(\mathbf{C})$ defined by $z^i \neq 0$, then

$$p^{-1}(U_i) = \{(z^0, \ldots, z^n) \in \mathbf{C}^{n+1}; z^i \neq 0\}.$$

If we denote by φ_i the mapping $p^{-1}(U_i) \to \mathbf{C}^*$ defined by

$\varphi_i(z^0, \ldots, z^n) = z^i$, then the transition function ψ_{ji} is given by
$$\psi_{ji}(p(z^0, \ldots, z^n)) = z^j/z^i \quad \text{on } U_i \cap U_j.$$
For the normalization axiom we need to consider only E_1.

Axiom 4 (Normalization). $-c_1(E_1)$ is the generator of $H^2(P_1(\mathbf{C}); \mathbf{Z})$; in other words, $c_1(E_1)$ evaluated (or integrated) on the fundamental 2-cycle $P_1(\mathbf{C})$ is equal to -1.

Although we considered the category of differentiable complex vector bundles and defined $c_i(E)$ as an element of $H^{2i}(M; \mathbf{R})$ instead of as an element of $H^{2i}(M; \mathbf{Z})$, the usual proof of the existence and uniqueness of the Chern classes holds in our case without any modification (cf. for instance, Husemoller [1]).

Let E be a complex vector bundle over M with fibre $\mathbf{C}^r$ and group $GL(r; \mathbf{C})$. Let P be its associated principal fibre bundle. We shall now give a formula which expresses the k th Chern class $c_k(E)$ by a closed differential form γ_k of degree $2k$ on M. We define first polynomial functions $f_0, f_1, \ldots, f_r$ on the Lie algebra $\mathfrak{gl}(r; \mathbf{C})$ by

$$\det\left(\lambda I_r - \frac{1}{2\pi\sqrt{-1}} X\right) = \sum_{k=0}^{r} f_k(X) \lambda^{r-k} \quad \text{for } X \in \mathfrak{gl}(r; \mathbf{C}).$$

Then they are invariant by ad $(GL(r; \mathbf{C}))$. Let ω be a connection form on P and Ω its curvature form. By Theorem 1.1 there exists a unique closed $2k$-form γ_k on M such that

$$p^*(\gamma_k) = f_k(\Omega),$$

where $p: P \to M$ is the projection. The cohomology class determined by γ_k is independent of the choice of connection. From the definition of the γ_k's we may write

$$\det\left(I_r - \frac{1}{2\pi\sqrt{-1}} \Omega\right) = p^*(1 + \gamma_1 + \cdots + \gamma_r).$$

THEOREM 3.1. *The k-th Chern class $c_k(E)$ of a complex vector bundle E over M is represented by the closed $2k$-form γ_k defined above.*

Proof. We shall show that the real cohomology classes represented by the γ_k's satisfy the four axioms.

(1) Evidently γ_0 represents $1 \in H^0(M; \mathbf{R})$.

(2) Let P be the principal bundle associated with a complex

vector bundle E over M. Given a map $f: M' \to M$, it is clear that the induced bundle $f^{-1}P$ is the principal bundle associated with the induced vector bundle $f^{-1}E$. Denoting also by f the natural bundle map $f^{-1}P \to P$ and by ω a connection form on P, we set

$$\omega' = f^*(\omega).$$

Then ω' is a connection form on $f^{-1}P$ and its curvature form Ω' is related to the curvature form Ω of ω by $\Omega' = f^*(\Omega)$ (cf. Proposition 6.2 of Chapter II). If we define a closed $2k$-form γ'_k on M using Ω' in the same way as we define γ_k using Ω, then it is clear that $f^*(\gamma_k) = \gamma'_k$.

(3) Let $E_1, \ldots, E_q$ be complex line bundles over M and $P_1, \ldots, P_q$ their associated principal bundles. For each i, let ω_i be a connection form on P_i and Ω_i its curvature form. Since $P_1 \times \cdots \times P_q$ is a principal fibre bundle over $M \times \cdots \times M$ with group $\mathbf{C}^* \times \cdots \times \mathbf{C}^*$, where $\mathbf{C}^* = GL(1; \mathbf{C})$, the diagonal map $d: M \to M \times \cdots \times M$ induces a principal fibre bundle $P = d^{-1}(P_1 \times \cdots \times P_q)$ on M with group $\mathbf{C}^* \times \cdots \times \mathbf{C}^*$. The group $\mathbf{C}^* \times \cdots \times \mathbf{C}^*$ may be considered as the subgroup of $GL(q; \mathbf{C})$ consisting of diagonal matrices. The Whitney sum $E = E_1 \oplus \cdots \oplus E_q$ is a vector bundle with fibre $\mathbf{C}^q$. Its associated principal fibre bundle Q with group $GL(q; \mathbf{C})$ contains P as a subbundle. Let $p_i: P \to P_i$ be the restriction of the projection $P_1 \times \cdots \times P_q \to P_i$ to P and set

$$\omega = \omega_1^* + \cdots + \omega_q^* \quad \text{where } \omega_i^* = p_i^*(\omega_i).$$

Then ω is a connection form on P and its curvature form Ω is given by

$$\Omega = \Omega_1^* + \cdots + \Omega_q^* \quad \text{where } \Omega_i^* = p_i^*(\Omega_i).$$

Let $\tilde{\omega}$ be the connection form on Q which extends ω. Let $\tilde{\Omega}$ be its curvature form on Q. Then the restriction of

$$\det\left(I_q - \frac{1}{2\pi\sqrt{-1}}\tilde{\Omega}\right)$$

to P is equal to

$$\left(1 - \frac{1}{2\pi\sqrt{-1}}\Omega_1^*\right) \wedge \cdots \wedge \left(1 - \frac{1}{2\pi\sqrt{-1}}\Omega_q^*\right).$$

This establishes the Whitney sum formula.

XII. CHARACTERISTIC CLASSES

(4) Let $P = \mathbf{C}^2 - \{0\}$; P is the principal bundle over $P_1(\mathbf{C})$ with group $\mathbf{C}^*$ associated with the natural line bundle E_1. We define a 1-form ω on P by

$$\omega = (\bar{z}, dz)/(\bar{z}, z),$$

where $(\bar{z}, dz) = \bar{z}^0 dz^0 + \bar{z}^1 dz^1$ and $(\bar{z}, z) = \bar{z}^0 z^0 + \bar{z}^1 z^1$. Then ω is a connection form and its curvature form Ω is given by

$$\Omega = d\omega = \{(\bar{z}, z)(d\bar{z}, dz) - (z, d\bar{z}) \wedge (\bar{z}, dz)\}/(\bar{z}, z)^2,$$

where

$$(d\bar{z}, dz) = d\bar{z}^0 \wedge dz^0 + d\bar{z}^1 \wedge dz^1.$$

Let U be the open subset of $P_1(\mathbf{C})$ defined by $z^0 \neq 0$. If we set $w = z^1/z^0$, then w may be used as a local coordinate system in U. Substituting $z^1 = z^0 w$ in the formula above for Ω, we obtain

$$\Omega = (d\bar{w} \wedge dw)/(1 + w\bar{w})^2.$$

Then $\gamma_1 = \gamma_1(E_1)$ can be written as follows:

$$-\gamma_1 = (d\bar{w} \wedge dw)/2\pi\sqrt{-1}(1 + w\bar{w})^2 \quad \text{on } U.$$

If we set

$$w = re^{2\pi\sqrt{-1}t},$$

then

$$-\gamma_1 = (2r\, dr \wedge dt)/(1 + r^2)^2 \quad \text{on } U.$$

Since $P_1(\mathbf{C}) - U$ is just a point, the integral $-\int_{P_1(\mathbf{C})} \gamma_1$ is equal to the integral $-\int_U \gamma_1$. We wish to show that the latter is equal to 1. From the formula above for γ_1 in terms of r and t, we obtain

$$-\int_U \gamma_1 = \int_0^1 \left(\int_0^\infty \frac{2r\, dr}{(1 + r^2)^2} \right) dt = 1.$$

QED.

If we express the curvature form Ω by a matrix-valued 2-form (Ω^i_j), then the $2k$-form γ_k representing the k-th Chern class $c_k(E)$ can be written as follows:

$$p^*(\gamma_k) = \frac{(-1)^k}{(2\pi\sqrt{-1})^k k!} \sum \delta^{j_1 \cdots j_k}_{i_1 \cdots i_k} \Omega^{i_1}_{j_1} \wedge \cdots \wedge \Omega^{i_k}_{j_k},$$

where the summation is taken over all ordered subsets $(i_1, \ldots, i_k)$ of k elements from $(1, \ldots, r)$ and all permutations $(j_1, \ldots, j_k)$ of $(i_1, \ldots, i_k)$ and the symbol $\delta_{i_1 \cdots i_k}^{j_1 \cdots j_k}$ denotes the sign of the permutation $(i_1, \ldots, i_k) \to (j_1, \ldots, j_k)$. The verification is easy and is left to the reader.

Let P be the principal bundle with group $GL(r; \mathbf{C})$ associated with a complex vector bundle E over M. We shall show that the algebra of characteristic classes of P defined in §1 is generated by the Chern classes of E. Reducing the structure group $GL(r; \mathbf{C})$ to $U(r)$ we consider a subbundle P' of P and choose a connection form ω' on P' with curvature form Ω'. Let ω be the connection form on P which extends ω' and Ω its curvature form. Let f be an ad $(GL(r; \mathbf{C}))$-invariant polynomial function on $\mathfrak{gl}(r; \mathbf{C})$ and f' its restriction to $\mathfrak{u}(r)$. Then f' is invariant by $U(r)$. Since the restriction of $f(\Omega)$ to P' is equal to $f'(\Omega')$, the characteristic class of P defined by f coincides with the characteristic class of P' defined by f'. In §2 we determined all ad $(U(r))$-invariant polynomial functions on $\mathfrak{u}(r)$ and our assertion now follows from the definition of the γ_k's and from Theorem 2.5.

If we restrict the polynomial function $\det\left(I_r - \dfrac{1}{2\pi\sqrt{-1}} X\right)$ used to define the Chern classes to the subalgebra $\mathfrak{t}$ of $\mathfrak{gl}(r; \mathbf{C})$ consisting of diagonal matrices $[\xi_1, \ldots, \xi_r]$ with diagonal entries $\sqrt{-1}\,\xi_1, \ldots, \sqrt{-1}\,\xi_r$, then

$$(*) \quad \det\left(I_r - \frac{1}{2\pi\sqrt{-1}} X\right)$$

$$= \left(1 - \frac{\xi_1}{2\pi}\right) \cdots \left(1 - \frac{\xi_r}{2\pi}\right) \quad \text{for } X = [\xi_1, \ldots, \xi_r].$$

Similarly, if we restrict the function defined by the power series

$$\operatorname{trace}\left(\exp\left(\frac{-1}{2\pi\sqrt{-1}} X\right)\right)$$

$$= \operatorname{trace}\left(\sum_{k=0}^{\infty} \frac{(-1)^k}{k!\,(2\pi\sqrt{-1})^k} X^k\right) \quad \text{for } X \in \mathfrak{gl}(r; \mathbf{C})$$

XII. CHARACTERISTIC CLASSES

to t, then we obtain

$$(**) \quad \sum_{j=1}^{r} \exp\left(\frac{\xi_j}{2\pi}\right)$$
$$= \sum_{k=0}^{\infty} \left(\frac{1}{k!\,(2\pi)^k} \sum_{j=1}^{r} \xi_j^k\right) \quad \text{for } X = [\xi_1, \ldots, \xi_r].$$

It is trivial to see that this power series defined on $\mathfrak{gl}(r; \mathbf{C})$ is invariant by ad $(GL(r; \mathbf{C}))$. We shall substitute the curvature form Ω into X. Since trace (X^k) is invariant by ad $(GL(r; \mathbf{C}))$, trace (Ω^k) projects to a closed $2k$-form on M by Theorem 1.1. It follows that trace $(\Omega^k) = 0$ if $2k > \dim M$. The form

$$\text{trace}\left(\exp\left(\frac{-1}{2\pi\sqrt{-1}}\,\Omega\right)\right)$$

defined on the bundle P projects upon a closed form on M. The cohomology class defined by this closed form on M is called the *Chern character* of the vector bundle E and is denoted by $ch(E)$. The Chern character $ch(E)$ is usually defined as follows by means of a formal factorization (cf. Chapter III of Hirzebruch [3]):

If $\sum_{i=0}^{r} c_i(E)x_i = \prod_{j=1}^{r} (1 + \eta_j x)$, then $ch(E) = \sum_{j=1}^{r} \exp \eta_j$.

Comparing the formulas (∗) and (∗∗) above, we see easily that our definition of the Chern character coincides with the usual one. The Chern character satisfies (cf. Hirzebruch [3])

$$ch(E \oplus E') = ch(E) + ch(E'),$$
$$ch(E \otimes E') = ch(E)ch(E').$$

The Chern class was first introduced in Chern [10].

Example 3.1. Let E be a Hermitian vector bundle over a complex manifold M with fibre $\mathbf{C}^r$ and fibre metric h (cf. §10 of Chapter IX). We may use the curvature form of the Hermitian connection to express the Chern classes of E. Since the first Chern class $c_1(E)$ can be represented by a closed 2-form γ_1 on M such that $p^*\gamma_1 = \dfrac{-1}{2\pi\sqrt{-1}}\,\text{trace }\Omega$, we see from formulas (5) and (10) of §10 of Chapter IX that $c_1(E)$ can be represented by the closed

2-form

$$\frac{-1}{2\pi\sqrt{-1}} \sum_{i,j=1}^{n} K_{i\bar{j}} \, dz^i \wedge d\bar{z}^j$$

where $K_{i\bar{j}}$ may be calculated by formula (14) of §10 of Chapter IX. Hence $c_1(E)$ can be represented by the closed 2-form

$$\frac{1}{2\pi\sqrt{-1}} \bar{\partial} \, \partial \log H, \quad \text{where } H = \det(h_{\alpha\bar{\beta}}).$$

4. Pontrjagin classes

Let E be a real vector bundle over a manifold M with fibre $\mathbf{R}^q$. The complexification E^c of E is a complex vector bundle over M with fibre $\mathbf{C}^q$ obtained by complexifying each fibre of E. Let P (resp. P^c) be the principal fibre bundle with group $GL(q; \mathbf{R})$ (resp. $GL(q; \mathbf{C})$) associated with E (resp. E^c). Then P is a subbundle of P^c.

The k-th Pontrjagin class $p_k(E)$ of E is defined to be

$$(-1)^k c_{2k}(E^c) \in H^{4k}(M; \mathbf{R}),$$

where $c_{2k}(E^c)$ denotes the $2k$-th Chern class of the complex vector bundle E^c. The total Pontrjagin class $p(E)$ is defined to be $1 + p_1(E) + p_2(E) + \cdots \in H^*(M; \mathbf{R})$.

We define ad $(GL(q; \mathbf{R}))$-invariant polynomial functions $g_0, g_1, \ldots, g_q$ on $\mathfrak{gl}(q; \mathbf{R})$ by

$$\det\left(\lambda I_q - \frac{1}{2\pi} X\right) = \sum_{k=0}^{q} g_k(X) \lambda^{q-k} \quad \text{for } X \in \mathfrak{gl}(q; \mathbf{R}).$$

Let ω be a connection form on P and Ω its curvature form. By Theorem 1.1 there exists a unique closed $4k$-form β_k on M such that

$$p^*(\beta_k) = g_{2k}(\Omega),$$

where $p \colon P \to M$ is the projection. (The reason for not considering $g_k(\Omega)$ for odd k will be given later.)

THEOREM 4.1. *The closed $4k$-form β_k on M defined above represents the k-th Pontrjagin class of a vector bundle E.*

Proof. Let φ be the connection form on P^c which extends ω.

Let Φ be the curvature form of φ. As in §3 we define polynomial functions $f_0, f_1, \ldots, f_q$ on $\mathfrak{gl}(q; \mathbf{C})$ by

$$\det\left(\lambda I_q - \frac{1}{2\pi\sqrt{-1}} X\right) = \sum_{k=0}^{q} f_k(X) \lambda^{q-k} \quad \text{for } X \in \mathfrak{gl}(q; \mathbf{C}).$$

We see easily that

$$g_{2k}(X) = (-1)^k f_{2k}(X) \quad \text{for } X \in \mathfrak{gl}(q; \mathbf{R}),$$

and hence that the restriction of $(-1)^k f_{2k}(\Phi)$ to P coincides with $g_{2k}(\Omega)$. Our assertion now follows from Theorem 3.1 and the definition of γ_{2k} in §3. QED.

If we write Ω as a matrix-valued form (Ω_j^i), then we may express $g_{2k}(\Omega)$ as follows:

$$g_{2k}(\Omega) = \frac{1}{(2\pi)^{2k}(2k)!} \sum \delta_{i_1 \cdots i_{2k}}^{j_1 \cdots j_{2k}} \Omega_{j_1}^{i_1} \wedge \cdots \wedge \Omega_{j_{2k}}^{i_{2k}},$$

where the summation runs over all ordered subsets $(i_1, \ldots, i_{2k})$ of $2k$ elements of $(1, \ldots, q)$ and all permutations $(j_1, \ldots, j_{2k})$ of $(i_1, \ldots, i_{2k})$ and where $\delta_{j_1 \cdots j_{2k}}^{i_1 \cdots i_{2k}}$ denotes the sign of the permutation.

Let E be a real vector bundle with fibre $\mathbf{R}^q$ and P its associated principal fibre bundle with group $GL(q; \mathbf{R})$. Then the algebra of characteristic classes of P defined in §1 is generated by Pontrjagin classes of E. The proof is analogous to the case of complex vector bundles; in §3 replace $U(r)$ by $O(q)$ and use Theorem 2.6 in place of Theorem 2.5.

We may consider in the proof of Theorem 4.1 the closed $2k$-form α_k on M defined by $p^*(\alpha_k) = g_k(\Omega)$ for odd k. But α_k is always cohomologous to zero. In fact, reducing the structure group $GL(q; \mathbf{R})$ to $O(q)$ we consider a subbundle Q of P and make use of a connection form on Q and its curvature form which take their values in $\mathfrak{o}(q)$. It is then clear that we have only to prove $g_k(X) = 0$ for $X \subset \mathfrak{o}(q)$ and for odd k. Since $X \subset \mathfrak{o}(q)$ is a skew symmetric matrix, we have

$$\det\left(\lambda I_q - \frac{1}{2\pi} X\right) = \det{}^t\!\left(\lambda I_q - \frac{1}{2\pi} X\right) = \det\left(\lambda I_q + \frac{1}{2\pi} X\right)$$

and hence

$$\sum_{k=0}^{q} g_k(X) \lambda^{q-k} = \sum_{k=0}^{q} (-1)^k g_k(X) \lambda^{q-k} \quad \text{for } X \in \mathfrak{o}(q),$$

which implies our assertion. It follows that the total Pontrjagin class may be represented by

$$\det \left(I_q - \frac{1}{2\pi} \Omega \right) \quad \text{(projected onto the base } M\text{)}.$$

The Pontrjagin class was first introduced in Pontrjagin [2].

5. Euler classes

Let E be a real vector bundle over a manifold M with fibre $\mathbf{R}^q$. Let P be its associated principal fibre bundle with group $GL(q; \mathbf{R})$. Let $GL^+(q; \mathbf{R})$ be the subgroup of $GL(q; \mathbf{R})$ consisting of matrices with positive determinants; it is a subgroup of index 2. A vector bundle E is said to be *orientable* if the structure group of P can be reduced to $GL^+(q; \mathbf{R})$. If E is orientable and if such a reduction is chosen, E is said to be *oriented*.

Let f be a mapping of another manifold M' into M and $f^{-1}E$ the induced vector bundle over M'. If E is orientable, so is $f^{-1}E$. If E is oriented, so is $f^{-1}E$ in a natural manner.

Let E and E' be two real vector bundles over M with fibre $\mathbf{R}^q$ and $\mathbf{R}^{q'}$ respectively. Since

$$GL(q; \mathbf{R}) \times GL(q'; \mathbf{R}) \subset GL(q + q'; \mathbf{R})$$

and

$$GL^+(q; \mathbf{R}) \times GL^+(q'; \mathbf{R}) \subset GL^+(q + q'; \mathbf{R})$$

in a natural manner, it follows that if E and E' are orientable so is their Whitney sum $E \oplus E'$ and that if E and E' are oriented so is $E \oplus E'$ in a natural manner.

Let E be a complex vector bundle over M with fibre $\mathbf{C}^r$. It may be considered as a real vector bundle with fibre $\mathbf{R}^{2r}$. Since the associated principal fibre bundle of E as a complex vector bundle has a structure group $GL(r; \mathbf{C}) \subset GL^+(2r; \mathbf{R})$, E is oriented in a natural manner as a real vector bundle.

We shall now give an axiomatic definition of Euler classes. We consider the category of differentiable oriented real vector bundles over differentiable manifolds.

Axiom 1. For each oriented real vector bundle E over M with fibre $\mathbf{R}^q$, the *Euler class* $\chi(E) \in H^q(M; \mathbf{R})$ is given and $\chi(E) = 0$ for q odd.

Axiom 2 (Naturality). If E is an oriented real vector bundle over M and if f is a mapping of another manifold M' into M, then

$$\chi(f^{-1}E) = f^*(\chi(E)) \in H^*(M'; \mathbf{R}),$$

where $f^{-1}E$ is the vector bundle over M' induced by f from E.

Axiom 3 (Whitney sum formula). Let $E_1, \ldots, E_r$ be oriented real vector bundles over M with fibre $\mathbf{R}^2$. Then

$$\chi(E_1 \oplus \cdots \oplus E_r) = \chi(E_1) \cdots \chi(E_r).$$

Axiom 4 (Normalization). Let E_1 be the natural complex line bundle over the 1-dimensional complex projective space $P_1(\mathbf{C})$ (cf. §3). Then its Euler class $\chi(E_1)$ coincides with the first Chern class $c_1(E_1)$.

By a *Riemannian vector bundle* we shall mean a pair (E, g) of a real vector bundle E and a fibre metric g in E. By definition g defines an inner product g_x in the fibre at $x \in M$ and the family of inner products g_x depends differentiably on x (cf. p. 116 of Volume I).

Let E be a vector bundle over M with fibre $\mathbf{R}^q$ and P its associated principal fibre bundle with group $GL(q; \mathbf{R})$. Then the set of fibre metrics on E are in a natural one-to-one correspondence with the set of reductions of P to subbundles with structure group $O(q)$ (cf. Proposition 5.6 and Example 5.5 of Chapter I, as well as p. 117 of Volume I). If E is oriented, then each fibre metric g defines a reduction of P to a subbundle with group $SO(q)$, which will be called the principal fibre bundle associated with the oriented Riemannian vector bundle (E, g).

Given a Riemannian vector bundle (E, g) over M and a mapping $f: M' \to M$, we denote by $f^{-1}(E, g)$ the Riemannian vector bundle over M' consisting of the induced vector bundle $f^{-1}E$ over M' and the fibre metric naturally induced by f from g. Given two Riemannian vector bundles (E, g) and (E', g') on M, we denote by $(E, g) \oplus (E', g')$ the Riemannian vector bundle over M consisting of $E \oplus E'$ and the naturally defined fibre metric $g + g'$ and we call it the Whitney sum of (E, g) and (E', g').

Let E_n be the natural complex line bundle over $P_n(\mathbf{C})$ defined in §3. A point of $P_n(\mathbf{C})$ is a 1-dimensional complex subspace of $\mathbf{C}^{n+1}$ and the fibre of E_n at that point is precisely the corresponding subspace of $\mathbf{C}^{n+1}$. Hence the natural inner product in $\mathbf{C}^{n+1}$ induces an inner product in each fibre of E_n and defines what we call the natural fibre metric in E_n.

We now consider the following cohomology class $\chi(E, g)$ defined axiomatically:

Axiom 1'. For each oriented Riemannian vector bundle (E, g) over M with fibre $\mathbf{R}^q$, the class $\chi(E, g) \in H^q(M; \mathbf{R})$ is given and $\chi(E, g) = 0$ for q odd.

Axiom 2' (Naturality). If (E, g) is an oriented Riemannian vector bundle over M and if f is a mapping of M' into M, then

$$\chi(f^{-1}(E, g)) = f^*(\chi(E, g)) \in H^*(M; \mathbf{R}).$$

Axiom 3' (Whitney sum formula). Let $(E_1, g_1), \ldots, (E_r, g_r)$ be oriented Riemannian vector bundles over M with fibre $\mathbf{R}^2$. Then

$$\chi((E_1, g_1) + \cdots + (E_r, g_r)) = \chi(E_1, g_1) \cdots \chi(E_r, g_r).$$

Axiom 4' (Normalization). Let E_1 be the natural complex line bundle over $P_1(\mathbf{C})$ and g_1 the natural fibre metric in E_1. Then $\chi(E_1, g_1)$ coincides with the first Chern class $c_1(E_1)$.

In contrast to the Chern class, the Euler class is usually defined in a constructive manner and not axiomatically (see, for example, Husemoller [1]). This is due to the fact that in algebraic topology the Euler class is defined to be an element of $H^*(M; \mathbf{Z})$, not of $H^*(M; \mathbf{R})$. But we are interested in the Euler class as an element of $H^*(M; \mathbf{R})$. Since the Euler class defined in the usual manner in algebraic topology satisfies Axioms 1, 2, 3, and 4, the existence of $\chi(E)$ satisfying Axioms 1, 2, 3, and 4 is assured. It is clear that $\chi(E)$ satisfying Axioms 1, 2, 3, and 4 satisfies Axioms 1', 2', 3', and 4'. If we prove the uniqueness of $\chi(E, g)$ satisfying Axioms 1', 2', 3', and 4', then we will have the uniqueness of $\chi(E)$ satisfying Axioms 1, 2, 3, and 4. Assuming certain facts from algebraic topology we shall show the uniqueness of $\chi(E, g)$.

Let (E, g) be an oriented Riemannian vector bundle with fibre

$\mathbf{R}^{2p}$. Let $\tilde{G}_{2p,2k}$ denote the Grassmann manifold of oriented $2p$-planes in $\mathbf{R}^{2p+2k}$ so that

$$\tilde{G}_{2p,2k} = SO(2p + 2k)/SO(2p) \times SO(2k).$$

Attaching to each point of $\tilde{G}_{2p,2k}$ (which is a $2p$-plane in $\mathbf{R}^{2p+2k}$) the corresponding $2p$-dimensional oriented subspace of $\mathbf{R}^{2p+2k}$, we obtain an oriented vector bundle $E_{2p,2k}$ over $\tilde{G}_{2p,2k}$ with fibre $\mathbf{R}^{2p}$. The natural inner product of $\mathbf{R}^{2p+2k}$ induces a fibre metric g_0 in $E_{2p,2k}$. The principal fibre bundle

$$\tilde{V}_{2p,2k} = SO(2p + 2k)/\{1\} \times SO(2k)$$

over $\tilde{G}_{2p,2k}$ with group $SO(2p)$ is the bundle associated with the vector bundle $E_{2p,2k}$. The classification theorem implies that, for a sufficient large k, there exists a mapping $f\colon M \to \tilde{G}_{2p,2k}$ such that $(E, g) = f^{-1}(E_{2p,2k}, g_0)$. Let T denote the maximal torus $SO(2) \times \cdots \times SO(2)$ ($p + k$ times) lying in a natural manner in $SO(2p) \times SO(2k)$. Let $h\colon SO(2p + 2k)/T \to \tilde{G}_{2p,2k}$ be the natural projection. Then the structure group of $h^{-1}(E_{2p,2k})$ can be reduced to T in a natural manner. In other words,

$$h^{-1}(E_{2p,2k}, g_0) = (E_1, g_1) + \cdots + (E_p, g_p),$$

where $(E_1, g_1), \ldots, (E_p, g_p)$ are oriented Riemannian vector bundles over $SO(2p + 2k)/T$ with fibre $\mathbf{R}^2$. Assume that both $\chi(E, g)$ and $\bar{\chi}(E, g)$ satisfy Axioms 1', 2', and 3'. Then

$$\chi(E, g) = f^*\chi(E_{2p+2k}, g_0), \quad \bar{\chi}(E, g) = f^*\bar{\chi}(E_{2p,2k}, g_0),$$
$$\chi(E_1, g_1) \cdots \chi(E_p, g_p) = h^*\chi(E_{2p+2k}, g_0),$$
$$\bar{\chi}(E_1, g_1) \cdots \bar{\chi}(E_p, g_p) = h^*\bar{\chi}(E_{2p,2k}, g_0).$$

Assume that $\chi(E_i, g_i) = \bar{\chi}(E_i, g_i)$ for $i = 1, \ldots, p$. Since h^* is an isomorphism of $H^*(\tilde{G}_{2p,2k}; \mathbf{R})$ into $H^*(SO(2p + 2k)/T; \mathbf{R})$, we obtain $\chi(E_{2p,2k}, g_0) = \bar{\chi}(E_{2p,2k}, g_0)$ and hence $\chi(E, g) = \bar{\chi}(E, g)$. It remains to be shown that $\chi(E_i, g_i) = \bar{\chi}(E_i, g_i)$. Assuming now that (E, g) is an oriented Riemannian vector bundle with fibre $\mathbf{R}^2$, we shall show that $\chi(E, g) = \bar{\chi}(E, g)$. By the classification theorem, there is a mapping $f\colon M \to P_1(C)$ such that $(E, g) = f^{-1}(E_1, g_1)$, where E_1 denotes the natural complex line bundle over $P_1(C)$ and g_1 denotes the natural fibre metric in E_1. Hence $\chi(E, g) = f^*\chi(E_1, g_1) = f^*\bar{\chi}(E_1, g_1) = \bar{\chi}(E, g)$ by Axioms 2' and 4'. This completes the proof of the fact that $\chi(E, g)$ is unique. The proof given here is essentially the same as the proof of the uniqueness for the Chern class (see for instance Husemoller [1]). The main difference in the proof lies in the fact that

$$H^*(\tilde{G}_{2p,2k}; \mathbf{R}) \to H^*(SO(2p + 2k)/T; \mathbf{R})$$

is injective but

$$H^*(\tilde{G}_{2p,2k}; \mathbf{Z}) \to H^*(SO(2p + 2k)/T; \mathbf{Z})$$

is not, whereas

$$H^*(U(p + k)/U(p) \times U(k); \mathbf{Z}) \to H^*(U(p + k)/T; \mathbf{Z})$$

is injective (where T denotes also the maximal torus $U(1) \times \cdots \times U(1)$ of $U(p) \times U(k)$). This is precisely the reason why the Euler class as an element of the integral cohomology group is not usually defined axiomatically.

We shall now express the Euler class $\chi(E)$ of an oriented real vector bundle E over M with fibre $\mathbf{R}^{2p}$ by a closed $2p$-form on M. We choose a fibre metric g in E and let Q be the principal fibre bundle with group $SO(2p)$ associated with the Riemannian vector bundle (E, g). Let $\omega = (\omega_j^i)$ be a connection form on Q and $\Omega = (\Omega_j^i)$ its curvature form. From Theorems 1.1 and 2.7 (cf. the expression of the polynomial function g in the proof of Theorem 2.7) it follows that there exists a unique closed $2p$-form γ on M such that

$$\pi^*(\gamma) = \frac{(-1)^p}{2^{2p}\pi^p \cdot p!} \sum \varepsilon_{i_1 \cdots i_{2p}} \Omega_{i_2}^{i_1} \wedge \cdots \wedge \Omega_{i_{2p}}^{i_{2p-1}}.$$

THEOREM 5.1. *The Euler class of an oriented real vector bundle E over M with fibre $\mathbf{R}^{2p}$ is represented by the closed $2p$-form γ on M defined above.*

Proof. Once a fibre metric g on E is chosen and the associated principal fibre bundle Q with group $SO(2p)$ is constructed, the cohomology class represented by γ is independent of the connection chosen in Q by Theorem 1.1. It is, however, not clear *a priori* that the class represented by γ is independent of the fibre metric chosen. We shall therefore prove that the cohomology class represented by the closed form γ satisfies Axioms 1', 2', 3', and 4' for $\chi(E, g)$. It then follows that γ represents the class $\chi(E, g)$

and hence $\chi(E)$. We set $\gamma = 0$ if the dimension of the fibre is odd. Then Axiom 1' is trivially satisfied. Axioms 2' and 3' can be verified in the same way as Axioms 2 and 3 in the proof of Theorem 3.1. To verify Axiom 4', let P, ω, and Ω be as in the proof of Theorem 3.1. Let Q be the unit sphere in $\mathbf{C}^2$, i.e., $Q = \{(z^0, z^1); z^0 \bar{z}^0 + z^1 \bar{z}^1 = 1\}$. Then Q is a subbundle of P with group $U(1)$ and is the principal fibre bundle associated with the natural complex line bundle E_1 over $P_1(\mathbf{C})$ with natural fibre metric g_1. We identify $U(1)$ with $SO(2)$ by

$$a + b\sqrt{-1} \in U(1) \to \begin{pmatrix} a & -b \\ b & a \end{pmatrix} \in SO(2) \quad \text{where } a, b \in \mathbf{R}.$$

Let ω be the connection form on P defined in the proof (4) of Theorem 3.1, i.e., $\omega = (\bar{z}, dz)/(\bar{z}, z)$. Let ω' be its restriction to Q. Then ω' is purely imaginary, i.e., takes its values in the Lie algebra $\mathfrak{u}(1)$. This means that ω' is a connection form on Q. Its curvature form Ω' is the restriction to Q of the curvature form Ω of ω. Under the identification of $U(1)$ with $SO(2)$ described above, $\mathfrak{u}(1)$ is identified with $\mathfrak{so}(2)$ by

$$a\sqrt{-1} \in \mathfrak{u}(1) \to \begin{pmatrix} 0 & -a \\ a & 0 \end{pmatrix} \in \mathfrak{so}(2) \quad \text{where } a \in \mathbf{R}.$$

Accordingly, the connection form ω' and its curvature form Ω' are identified respectively with the following forms with values in $\mathfrak{so}(2)$:

$$\begin{pmatrix} 0 & \sqrt{-1}\,\omega' \\ -\sqrt{-1}\,\omega' & 0 \end{pmatrix} \quad \text{and} \quad \begin{pmatrix} 0 & \sqrt{-1}\,\Omega' \\ -\sqrt{-1}\,\Omega' & 0 \end{pmatrix}.$$

It is now clear that γ coincides with the form $\gamma_1(E_n)$ defined in §3. QED.

If M is an oriented compact Riemannian manifold of dimension $2p$ and if E is the tangent bundle of M, then the closed $2p$-form γ integrated over M gives the Euler number of M. This is the so-called Gauss-Bonnet theorem. Our axiomatic approach minimizes calculation. For the history of the Gauss-Bonnet theorem, see Note 20.

In expressing the Chern classes and the Pontrjagin classes in Theorems 3.1 and 4.1, we can use any connection in the vector bundle E. But in expressing the Euler class in Theorem 5.1, we have to use a connection compatible with a fibre metric of E. It follows from Theorems 3.1 and 4.1 that if E admits a connection with vanishing curvature, then its Chern and Pontrjagin classes are trivial. But in order to apply Theorem 5.1 to prove the vanishing of the Euler class of E we need a connection with vanishing curvature which is compatible with a fibre metric, that is, a connection with vanishing curvature whose holonomy group is contained in $SO(2p)$.

APPENDIX 8

Integrable real analytic almost complex structures

In Theorem 2.5 of Chapter IX, we stated the following theorem of Newlander and Nirenberg [1]:

An almost complex structure is a complex structure if it is integrable, i.e., if it has no torsion.

It is beyond the scope of this book to give a complete proof of this theorem. We shall prove here the theorem under the stronger assumption that the given almost complex structure is real analytic and integrable.

Let M be a $2n$-dimensional manifold with almost complex structure J.

LEMMA 1. *If every point of M has a neighborhood U and n complex-valued functions $f^1, \ldots, f^n$ on U such that $df^1, \ldots, df^n$ are of type $(1, 0)$ and linearly independent everywhere on U, then the almost complex structure J of M is a complex structure.*

Proof of Lemma 1. Taking U small if necessary we may assume that $f^1, \ldots, f^n$ define a one-to-one mapping of U into $\mathbf{C}^n$, which will be denoted by f for the sake of simplicity. Let V be a small neighborhood of another point with similar functions $g^1, \ldots, g^n$ and the corresponding one-to-one mapping $g: V \to \mathbf{C}^n$. Assuming that $U \cap V$ is non-empty, we shall show that $g \circ f^{-1}: f(U \cap V) \to g(U \cap V)$ is holomorphic. By Proposition 2.9 of Chapter IX, both f and g are almost complex mappings. Hence, $g \circ f^{-1}$ is also an almost complex mapping. By Proposition 2.3 of Chapter IX, $g \circ f^{-1}$ is holomorphic. This shows that by taking (U, f) as a chart we obtain a complex structure on M which induces the given almost complex structure J. QED.

Lemma 1 shows that the theorem to be proved is of local nature. Let U be a small neighborhood and $\omega^1, \ldots, \omega^n$ real analytic complex 1-forms of type $(1, 0)$ on U which are linearly independent everywhere on U. If $x^1, \ldots, x^{2n}$ is a local coordinate system in U, we may choose $\omega^1, \ldots, \omega^n$ from the $2n$ 1-forms $dx^1 + iJ\,dx^1, \ldots, dx^{2n} + iJ\,dx^{2n}$ of type $(1, 0)$. We set

$$\omega^{\bar{j}} = \bar{\omega}^j \quad \text{for } j = 1, \ldots, n.$$

By Theorem 2.8 of Chapter IX, we have

LEMMA 2.

$$d\omega^j = \sum_{h<k} a^j_{hk} \omega^h \wedge \omega^k + \sum_{h,k} b^j_{h\bar{k}} \omega^h \wedge \omega^{\bar{k}},$$

$$d\omega^{\bar{j}} = \sum_{h<k} \bar{a}^j_{hk} \omega^{\bar{h}} \wedge \omega^{\bar{k}} + \sum_{h,k} \bar{b}^j_{h\bar{k}} \omega^{\bar{h}} \wedge \omega^k.$$

Let us write now

$$\omega^j = \sum_{A=1}^{2n} f^j_A(x)\, dx^A \quad \text{and} \quad \omega^{\bar{j}} = \sum_{A=1}^{2n} \bar{f}^j_A(x)\, dx^A,$$

where each $f^j_A(x)$ is complex-valued and real analytic in $x^1, \ldots, x^{2n}$ and $\bar{f}^j_A(x)$ is the complex conjugate of $f^j_A(x)$ and hence is also real analytic in $x^1, \ldots, x^{2n}$.

We may consider U as a neighborhood of the origin in $\mathbf{R}^{2n}$ with coordinate system $x^1, \ldots, x^{2n}$. Consider $\mathbf{C}^{2n}$ with coordinate system $z^1, \ldots, z^{2n}$, where $z^A = x^A + iy^A$ for $A = 1, \ldots, 2n$. In this way we consider $\mathbf{R}^{2n}$ as a subspace in $\mathbf{C}^{2n}$. Expanding $f^j_A(x)$ and $\bar{f}^j_A(x)$ in Taylor series of $x^1, \ldots, x^{2n}$, we replace $x^1, \ldots, x^{2n}$ formally by $z^1, \ldots, z^{2n}$. Then the resulting power series of $z^1, \ldots, z^{2n}$ are convergent in a neighborhood, say U^*, of the origin in $\mathbf{C}^{2n}$ and define holomorphic functions, say $F^j_A(z)$ and $F^{\bar{j}}_A(z)$, on U^*. (It should be emphasized that $F^{\bar{j}}_A(z)$ is holomorphic and is *not* the complex conjugate of $F^j_A(z)$.) We set

$$\Omega^j = \sum_{A=1}^{2n} F^j_A(z)\, dz^A \quad \text{and} \quad \Omega^{\bar{j}} = \sum_{A=1}^{2n} F^{\bar{j}}_A(z)\, dz^A$$

for $j = 1, \ldots, n$. Then Ω^j and $\Omega^{\bar{j}}$ are holomorphic 1-forms on U^*. (Again we point out that $\Omega^{\bar{j}}$ is *not* the complex conjugate of Ω^j.)

Since $\omega^1, \ldots, \omega^n$ are 1-forms of type $(1, 0)$ and are linearly independent, $\omega^1, \ldots, \omega^n, \omega^{\bar{1}}, \ldots, \omega^{\bar{n}}$ are also linearly independent. The $2n \times 2n$ matrix formed by the coefficients $f^j_A(x)$ and $\bar{f}^j_A(x)$ is therefore non-singular. Since $F^j_A(x) = f^j_A(x)$ and $F^{\bar{j}}_A(x) = \bar{f}^j_A(x)$ for

x in a neighborhood of the origin in $\mathbf{R}^{2n}$, the $2n \times 2n$ matrix formed by $F_A^j(z)$ and $F_A^{\bar{j}}(z)$ is non-singular for z in a small neighborhood of the origin in $\mathbf{C}^{2n}$. It follows that the $2n$ holomorphic 1-forms $\Omega^1, \ldots, \Omega^n, \Omega^{\bar{1}}, \ldots, \Omega^{\bar{n}}$ are linearly independent in a neighborhood of the origin in $\mathbf{C}^{2n}$. Taking U^* small, we may assume that they are linearly independent on U^*. We write

$$d\Omega^j = \sum_{h<k} A^j_{hk}\Omega^h \wedge \Omega^k + \sum_{h,k} B^j_{hk}\Omega^h \wedge \Omega^{\bar{k}} + \sum_{h<k} C^j_{hk}\Omega^{\bar{h}} \wedge \Omega^{\bar{k}},$$

where the coefficients A^j_{hk}, B^j_{hk}, C^j_{hk} are all holomorphic functions of $z^1, \ldots, z^{2n}$. Since the restriction of Ω^j to U coincides with ω^j and the restriction of $\Omega^{\bar{j}}$ to U coincides with $\bar{\omega}^j$, restricting the equation above to U we see from Lemma 2 that

$$A^j_{hk}(x) = a^j_{hk}(x), \qquad B^j_{hk}(x) = b^j_{hk}(x), \qquad C^j_{hk}(x) = 0$$

for x in U. Since C^j_{hk} is holomorphic, it vanishes identically on U^*. Hence,

LEMMA 3. $d\Omega^j = \Sigma_{h<k} A^j_{hk}\Omega^h \wedge \Omega^k + \Sigma_{h,k} B^j_{hk}\Omega^h \wedge \Omega^{\bar{k}}$.

Now we need the following theorem of Frobenius:

THEOREM (Frobenius). *Let $\varphi^1, \ldots, \varphi^r$ be holomorphic 1-forms defined in a neighborhood V of the origin in $\mathbf{C}^m$. If $\varphi^1, \ldots, \varphi^r$ are linearly independent everywhere on V and if*

$$d\varphi^j = \sum_{k=1}^{r} \psi^j_k \wedge \varphi^k \qquad \text{for } j = 1, \ldots, r,$$

where each ψ^j_k is a holomorphic 1-form on U, then there exist a smaller neighborhood W of the origin and holomorphic functions $g^1, \ldots, g^r$ on W such that

$$\varphi^j = \sum_{k=1}^{r} p^j_k \, dg^k \qquad \text{for } j = 1, \ldots, r,$$

where each p^j_k is a holomorphic function on W.

The proof of this theorem is essentially the same as the proof of the ordinary Frobenius theorem in the real case and hence is omitted.

We apply the theorem of Frobenius to the holomorphic 1-forms

$\Omega^1, \ldots, \Omega^n$ of Lemma 3. We see then that there exist holomorphic functions $F^1, \ldots, F^n$ in a small neighborhood (which will be still denoted by U^*) of the origin of $\mathbf{C}^{2n}$ such that

$$\Omega^j = \sum_{k=1}^{n} P_k^j \, dF^k \quad \text{for } j = 1, \ldots, n,$$

where each P_k^j is a holomorphic function in U^*. If we denote by f_k and p_k^j the restrictions of F_k and P_k^j to U, then we have

$$\omega^j = \sum_{k=1}^{n} p_k^j \, df^k.$$

Since $\omega^1, \ldots, \omega^n$ are linearly independent 1-forms of type $(1, 0)$ on U, $df^1, \ldots, df^n$ are also linearly independent 1-forms of type $(1, 0)$ on U. Hence $f^1, \ldots, f^n$ satisfy the requirement stated in Lemma 1. This completes the proof of the theorem that *an integrable real analytic almost complex structure is a complex structure.*

As we have already stated, Newlander and Nirenberg [1] proved that an integrable almost complex structure of differentiability class $C^{2n+\lambda}$, $0 < \lambda < 1$, is a complex structure. Nijenhuis and Woolf [1] extended the result to the case of class $C^{1+\lambda}$. An alternative proof was given by Kohn [1]; it makes use of an analogue of the Hodge-Kodaira decomposition theorem for the so-called $\bar{\partial}$-Neumann problem (see also Hörmander [1]). The theorem in the real analytic case proved in this appendix is due to Ehresmann [5] and also to Eckmann and Frölicher [1] (see also Frölicher [1]). Recently, Malgrange obtained a proof which reduces the differentiable case to the real analytic case.

For $n = 1$, the theorem of Newlander-Nirenberg means essentially the existence of isothermal coordinates (see Chern [16]).

APPENDIX 9

Some definitions and facts on Lie algebras

We summarize here basic facts on Lie algebras which are needed in this volume, particularly in Chapter XI. For the proofs which are omitted, we refer the reader to Bourbaki [1], Hochschild [1], and Jacobson [1].

Throughout this appendix we consider only finite-dimensional Lie algebras over the field **R** of real numbers or the field **C** of complex numbers.

1. Let $\mathfrak{g}$ be a Lie algebra. The *derived series* of $\mathfrak{g}$ is the decreasing sequence of ideals $D^0\mathfrak{g}, D^1\mathfrak{g}, \ldots$ of $\mathfrak{g}$ defined inductively by

$$D^0\mathfrak{g} = \mathfrak{g}, \qquad D^{p+1}\mathfrak{g} = [D^p\mathfrak{g}, D^p\mathfrak{g}].$$

The *descending central series* of $\mathfrak{g}$ is the decreasing sequence of ideals $C^0\mathfrak{g}, C^1\mathfrak{g}, \ldots$ of $\mathfrak{g}$ defined inductively by

$$C^0\mathfrak{g} = \mathfrak{g}, \qquad C^{p+1}\mathfrak{g} = [\mathfrak{g}, C^p\mathfrak{g}].$$

Evidently,

$$D^p\mathfrak{g} \subset C^p\mathfrak{g}.$$

We say that $\mathfrak{g}$ is *abelian* if $D^1\mathfrak{g} = 0$, *nilpotent* if $C^p\mathfrak{g} = 0$ for some p, and *solvable* if $D^p\mathfrak{g} = 0$ for some p. Every Lie algebra $\mathfrak{g}$ has a unique *maximum nilpotent ideal* $\mathfrak{n}$, i.e., a nilpotent ideal which contains every nilpotent ideal of $\mathfrak{g}$. Every Lie algebra $\mathfrak{g}$ has a unique maximum solvable ideal $\mathfrak{r}$, called the *radical* of $\mathfrak{g}$. A Lie algebra $\mathfrak{g}$ is said to be *semi-simple* if its radical $\mathfrak{r}$ reduces to zero. A Lie algebra $\mathfrak{g}$ is said to be *simple* if it is not abelian and has no non-zero ideal other than $\mathfrak{g}$ itself.

2. The *Killing-Cartan form* B of a Lie algebra $\mathfrak{g}$ is the symmetric bilinear form on $\mathfrak{g}$ defined by

$$B(X, Y) = \text{trace ad } (X)\text{ad } (Y) \qquad \text{for } X, Y \in \mathfrak{g},$$

where ad (X) is an inner derivation of $\mathfrak{g}$ defined by ad $(X)Z = [X, Z]$ for all $Z \in \mathfrak{g}$. For any automorphism α of $\mathfrak{g}$, we have

$$B(\alpha X, \alpha Y) = B(X, Y) \quad \text{for } X, Y \in \mathfrak{g}.$$

Similarly, for any derivation D of $\mathfrak{g}$, we have

$$B(DX, Y) + B(X, DY) = 0 \quad \text{for } X, Y \in \mathfrak{g}.$$

(A derivation D of $\mathfrak{g}$ is, by definition, a linear endomorphism of $\mathfrak{g}$ satisfying $D[X, Y] = [DX, Y] + [X, DY]$.) In particular, we have

$$B([Z, X], Y) + B(X, [Z, Y]) = 0 \quad \text{for } X, Y, Z \in \mathfrak{g}.$$

For a subspace $\mathfrak{a}$ of $\mathfrak{g}$, we denote by $\mathfrak{a}^\perp$ the subspace of $\mathfrak{g}$ defined by

$$\mathfrak{a}^\perp = \{X \in \mathfrak{g}; B(X, Y) = 0 \text{ for all } Y \in \mathfrak{a}\}.$$

If $\mathfrak{a}$ is an ideal of $\mathfrak{g}$, then $\mathfrak{a}^\perp$ is also an ideal of $\mathfrak{g}$. If we denote by $\mathfrak{n}$ and $\mathfrak{r}$ the maximum nilpotent ideal of $\mathfrak{g}$ and the radical of $\mathfrak{g}$, respectively, then we have $\mathfrak{g} \supset \mathfrak{r} \supset \mathfrak{g}^\perp \supset \mathfrak{n}$.

If $\mathfrak{a}$ is an ideal of $\mathfrak{g}$, the Killing-Cartan form of $\mathfrak{a}$ is the restriction of the Killing-Cartan form B of $\mathfrak{g}$ to $\mathfrak{a}$.

3. Any one of the following properties of $\mathfrak{g}$ is equivalent to semi-simplicity of $\mathfrak{g}$:
 (1) Its radical $\mathfrak{r} = 0$ (by definition);
 (2) Its maximum nilpotent ideal $\mathfrak{n} = 0$;
 (3) $\mathfrak{g}^\perp = 0$, i.e., its Killing-Cartan form B is non-degenerate;
 (4) Every abelian ideal of $\mathfrak{g}$ is 0;
 (5) $\mathfrak{g}$ is isomorphic to a direct sum of simple Lie algebras;
 (6) Every finite-dimensional representation of $\mathfrak{g}$ is semi-simple, i.e., completely reducible (H. Weyl).

A Lie algebra $\mathfrak{g}$ is said to be *reductive* if its adjoint representation is semi-simple. $\mathfrak{g}$ is reductive if and only if it is isomorphic to a direct sum of a semi-simple Lie algebra and an abelian Lie algebra. A subalgebra $\mathfrak{h}$ of $\mathfrak{g}$ is said to be *reductive in* $\mathfrak{g}$ if the restriction of the adjoint representation of $\mathfrak{g}$ to $\mathfrak{h}$ is semi-simple. If $\mathfrak{h}$ is reductive in $\mathfrak{g}$ and if ρ is a finite-dimensional semi-simple representation of $\mathfrak{g}$, then the restriction of ρ to $\mathfrak{h}$ is semi-simple. If $\mathfrak{h}$ is a subalgebra of $\mathfrak{g}$ reductive in $\mathfrak{g}$ and if $\mathfrak{k}$ is a subalgebra of $\mathfrak{h}$ reductive in $\mathfrak{h}$, then $\mathfrak{k}$ is reductive in $\mathfrak{g}$. If a subalgebra $\mathfrak{h}$ of $\mathfrak{g}$ is semi-simple, it is reductive in $\mathfrak{g}$ by property (6) of a semi-simple Lie algebra.

If H is a compact subgroup of a Lie group G, then the Lie algebra $\mathfrak{h}$ of H is reductive in the Lie algebra $\mathfrak{g}$ of G. In fact, if Ad_G denotes the adjoint representation of G, then $\mathrm{Ad}_G(H)$ is compact, and there is a (positive-definite) inner product in $\mathfrak{g}$ which is invariant by $\mathrm{Ad}_G(H)$. If $\mathfrak{a}$ is a subspace of $\mathfrak{g}$ which is invariant by $\mathrm{ad}_\mathfrak{g}(\mathfrak{h})$, then its orthogonal complement with respect to this inner product is also invariant by $\mathrm{ad}_\mathfrak{g}(\mathfrak{h})$. In particular, *the Lie algebra $\mathfrak{g}$ of a compact Lie group is reductive.*

4. A subalgebra $\mathfrak{s}$ of a Lie algebra $\mathfrak{g}$ is called a *Levi subalgebra* if $\mathfrak{g} = \mathfrak{r} + \mathfrak{s}$ (vector space direct sum), where $\mathfrak{r}$ is the radical of $\mathfrak{g}$. Since $\mathfrak{s}$ is isomorphic to $\mathfrak{g}/\mathfrak{r}$, it is semi-simple. The theorem of Levi states that every Lie algebra $\mathfrak{g}$ has a Levi subalgebra. We prove the following result which is needed in §5 of Chapter XI:

If A is a compact group of automorphisms of a Lie algebra $\mathfrak{g}$, then $\mathfrak{g}$ has a Levi subalgebra which is invariant by A.

The proof is by induction on the dimension of the radical $\mathfrak{r}$ of $\mathfrak{g}$.

(i) Case $[\mathfrak{r}, \mathfrak{r}] \neq 0$. We set $\bar{\mathfrak{g}} = \mathfrak{g}/[\mathfrak{r}, \mathfrak{r}]$ and $\bar{\mathfrak{r}} = \mathfrak{r}/[\mathfrak{r}, \mathfrak{r}]$. Then $\bar{\mathfrak{r}}$ is the radical of $\bar{\mathfrak{g}}$. Since A leaves $\mathfrak{r}$ invariant, it acts on $\bar{\mathfrak{g}}$ as an automorphism group. By inductive hypothesis, there is a Levi subalgebra $\bar{\mathfrak{s}}$ of $\bar{\mathfrak{g}}$ which is invariant by A. If we denote by p the natural homomorphism $\mathfrak{g} \to \bar{\mathfrak{g}} = \mathfrak{g}/[\mathfrak{r}, \mathfrak{r}]$, then $[\mathfrak{r}, \mathfrak{r}]$ is the radical of $p^{-1}(\bar{\mathfrak{s}})$. Since A leaves $p^{-1}(\bar{\mathfrak{s}})$ invariant, A may be considered as an automorphism group of $p^{-1}(\bar{\mathfrak{s}})$. Again by inductive hypothesis, there is a Levi subalgebra $\mathfrak{s}$ of $p^{-1}(\bar{\mathfrak{s}})$ which is invariant by A. Then $\mathfrak{s}$ is a Levi subalgebra of $\mathfrak{g}$ which is invariant by A.

(ii) Case $[\mathfrak{r}, \mathfrak{r}] = 0$. We fix one Levi subalgebra $\mathfrak{s}_0$ of $\mathfrak{g}$. Let $\mathfrak{s}$ be an arbitrary Levi subalgebra of $\mathfrak{g}$. For each $X \in \mathfrak{s}_0$, we write

$$X = f(X) + X_\mathfrak{s}, \quad \text{where } f(X) \in \mathfrak{r} \text{ and } X_\mathfrak{s} \in \mathfrak{s}.$$

Then $f: \mathfrak{s}_0 \to \mathfrak{r}$ is a linear mapping. If $X, Y \in \mathfrak{s}_0$, then we have

$$[X_\mathfrak{s}, Y_\mathfrak{s}] = [X - f(X), Y - f(Y)]$$
$$= [X, Y] - [X, f(Y)] - [f(X), Y].$$

Hence,

(*) $f([X, Y]) = [X, f(Y)] + [f(X), Y] \quad \text{for } X, Y \in \mathfrak{s}_0.$

Conversely, let $f: \mathfrak{s}_0 \to \mathfrak{r}$ be a linear mapping satisfying (∗). Then the mapping $X \in \mathfrak{s}_0 \to X - f(X) \in \mathfrak{g}$ is an isomorphism of $\mathfrak{s}_0$ into $\mathfrak{g}$, and its image is a Levi subalgebra of $\mathfrak{g}$. Let V be the space of all linear mappings $f: \mathfrak{s}_0 \to \mathfrak{r}$ satisfying (∗). It is now easy to see that the construction of f given above gives a one-to-one correspondence between the set of Levi subalgebras $\mathfrak{s}$ and the vector space V of linear mappings $f: \mathfrak{s}_0 \to \mathfrak{r}$ satisfying (∗). (The reader who is familiar with the Lie algebra cohomology theory will recognize that V is the space $Z^1(\mathfrak{s}_0, \mathfrak{r})$ of 1-cocycles of $\mathfrak{s}_0$ with respect to the representation ρ of $\mathfrak{s}_0$ defined by $\rho(X)U = [X, U]$ for $X \in \mathfrak{s}_0$ and $U \in \mathfrak{r}$. But this fact will not be used here.) Every automorphism of $\mathfrak{g}$ sends each Levi subalgebra of $\mathfrak{g}$ into a Levi subalgebra of $\mathfrak{g}$. Hence, the group A acts on the set of Levi subalgebras of $\mathfrak{g}$ and, consequently, on V. The problem now is to find a fixed point of A acting on V. We shall show that A is a group of affine transformations acting on V. Let α be an element of A. Let $f_0, f_0^\alpha, f,$ and f^α be the elements of V corresponding to Levi subalgebras $\mathfrak{s}_0$, $\alpha(\mathfrak{s}_0)$, $\mathfrak{s}$ and $\alpha(\mathfrak{s})$, respectively. Let $X \in \mathfrak{s}_0$ and denote by $\bar{\alpha}(X)$ the $\mathfrak{s}_0$-component of $\alpha(X)$ with respect to the decomposition $\mathfrak{g} = \mathfrak{r} + \mathfrak{s}_0$ so that $\alpha(X) = \beta(X) + \bar{\alpha}(X)$ where $\beta(X) \in \mathfrak{r}$. If we rewrite this as $\bar{\alpha}(X) = -\beta(X) + \alpha(X)$, we may consider $-\beta(X)$ as the $\mathfrak{r}$-component of $\bar{\alpha}(X)$ with respect to the decomposition $\mathfrak{g} = \mathfrak{r} + \alpha(\mathfrak{s}_0)$. Since f_0^α corresponds to $\alpha(\mathfrak{s}_0)$ by definition, $-\beta(X)$ is equal to $f_0^\alpha(\bar{\alpha}(X))$. Hence, we have

$$\bar{\alpha}(X) = f_0^\alpha(\bar{\alpha}(X)) + \alpha(X).$$

The right hand side may be written as $\{f_0^\alpha(\bar{\alpha}(X)) + \alpha(f(X))\} + \{\alpha(X) - \alpha(f(X))\}$. Since $X - f(X)$ belongs to $\mathfrak{s}$ from the definition of f, $\{\alpha(X) - \alpha(f(X))\}$ belongs to $\alpha(\mathfrak{s})$, and $\{f_0^\alpha((\mathfrak{s}X)) + \alpha(f(X))\}$ is the $\mathfrak{r}$-component of $\bar{\alpha}(X)$ with respect to the decomposition $\mathfrak{g} = \mathfrak{r} + \alpha(\mathfrak{s})$. Hence

$$f^\alpha(\bar{\alpha}(X)) = f_0^\alpha(\bar{\alpha}(X)) + \alpha(f(X)).$$

Since X is an arbitrary element of $\mathfrak{s}_0$, we have

$$f^\alpha = f_0^\alpha + \alpha \circ f \circ \bar{\alpha}^{-1}.$$

This shows that $f \in V \to f^\alpha \in V$ is an affine transformation of V with translation part f_0^α and linear part $f \to \alpha \circ f \circ \bar{\alpha}^{-1}$. In general, if A is a compact group of affine transformations of a vector space

V, then there is a fixed point of A in V. Although this is a special case of Theorem 9.2 of Chapter VIII since V admits an inner product with respect to which A is a group of Euclidean motions, we can give a more explicit expression for a fixed point. Let μ be the Haar measure on A with total measure $\mu(A) = 1$. Then the point of V given by

$$\int_A f_0^\alpha \, d\mu(\alpha)$$

is left fixed by A.

In our application in §5 of Chapter XI, A is a finite group of order 2 generated by an involutive automorphism α of $\mathfrak{g}$. In this case, the formula above for a fixed point reduces to

$$\tfrac{1}{2}(f_0 + f_0^\alpha).$$

The result we obtained here has been proved by Taft [1], [2] for a wider class of algebras with finite automorphism groups.

5. We shall prove the following fact quoted in §11 of Chapter XI. *If $\mathfrak{g}$ is a real simple Lie algebra, then either $\mathfrak{g}$ admits a compatible complex structure (i.e., a complex structure J such that $J([X, Y]) = [X, JY]$ for $X, Y \in \mathfrak{g}$) or its complexification $\mathfrak{g}^c$ is simple over $\mathbf{R}$* (and hence over $\mathbf{C}$ also). Assuming that $\mathfrak{g}^c$ is not simple, let $\mathfrak{a}$ be any non-trivial ideal of $\mathfrak{g}^c$. Then $\mathfrak{a} \cap \mathfrak{g}$ is an ideal of $\mathfrak{g}$ and hence is equal to either $\mathfrak{g}$ or 0. If $\mathfrak{a} \cap \mathfrak{g} = \mathfrak{g}$, then

$$\mathfrak{a} \supset [\mathfrak{a}, \mathfrak{g}^c] \supset [\mathfrak{g}, \mathfrak{g} + i\mathfrak{g}] = [\mathfrak{g}, \mathfrak{g}] + i[\mathfrak{g}, \mathfrak{g}] = \mathfrak{g} + i\mathfrak{g} = \mathfrak{g}^c,$$

which is a contradiction. Hence, $\mathfrak{a} \cap \mathfrak{g} = 0$. It is an easy matter to verify that $\{Y \in \mathfrak{g}; iY \in \mathfrak{a} \cap i\mathfrak{g}\}$ is an ideal of $\mathfrak{g}$. If this ideal is equal to $\mathfrak{g}$, then $\mathfrak{a} \supset [\mathfrak{a}, \mathfrak{g}^c] \supset [i\mathfrak{g}, g + i\mathfrak{g}] = i\mathfrak{g} + \mathfrak{g} = \mathfrak{g}^c$, which is a contradiction. Hence, $\mathfrak{a} \cap i\mathfrak{g} = 0$. We define two linear mappings p and q of $\mathfrak{a}$ into $\mathfrak{g}$ by

$$p(X + iY) = X \quad \text{and} \quad q(X + iY) = Y \quad \text{for } X + iY \in \mathfrak{a}.$$

The image of p is an ideal of $\mathfrak{g}$ and the kernel of p is $\mathfrak{a} \cap i\mathfrak{g}$. Hence, p is bijective. Similarly, q is also bijective. We define a linear isomorphism $J\colon \mathfrak{g} \to \mathfrak{g}$ by $J = q \circ p^{-1}$ so that every element of $\mathfrak{a}$ is of the form $X + iJX$, where $X \in \mathfrak{g}$. It is straightforward to verify that J is a compatible complex structure in $\mathfrak{g}$.

6. A real semi-simple Lie algebra $\mathfrak{g}$ is said to be of *compact type* if its Killing-Cartan form B is negative-definite. We shall show

that *if $\mathfrak{g}$ is the Lie algebra of a compact Lie group G, then the Killing-Cartan form B of $\mathfrak{g}$ is negative semi-definite*. Since Ad (G) is compact, there is a positive-definite inner product $(\,,\,)$ in $\mathfrak{g}$ which is invariant by Ad (G). For any $X \in \mathfrak{g}$, ad (X) is then skew-symmetric. Since $(\text{ad }(X) \text{ ad }(X)Y, Y) = -(\text{ad }(X)Y, \text{ad }(X)Y) \leq 0$ for all $Y \in \mathfrak{g}$, it follows that (ad (X)ad (X)) is negative semi-definite. Its trace $B(X, X)$ is therefore non-positive, which proves our assertion. Since B is non-degenerate for a semi-simple Lie algebra $\mathfrak{g}$, we can conclude that *if G is a compact Lie group with semi-simple Lie algebra $\mathfrak{g}$, then $\mathfrak{g}$ is a compact type. Conversely, if G is a connected Lie group with semi-simple Lie algebra $\mathfrak{g}$ of compact type, then G is compact* (cf. Example 3.2 of Chapter X).

If a Lie algebra $\mathfrak{g}$ admits a compatible complex structure J, then ad (JX) ad $(JY) = -\text{ad }(X)$ ad (Y) and hence $B(JX, JY) = -B(X, Y)$, and B cannot be negative-definite.

7. A real subalgebra $\mathfrak{h}$ of a complex Lie algebra $\mathfrak{g}$ is called a *real form* of $\mathfrak{g}$ if $\mathfrak{g} = \mathfrak{h} + i\mathfrak{h}$ and $\mathfrak{h} \cap i\mathfrak{h} = 0$ so that $\mathfrak{g} = \mathfrak{h}^c$. Every complex semi-simple Lie algebra $\mathfrak{g}$ has a real form which is of compact type (Theorem of Weyl). The proof requires the knowledge of root system, but is otherwise simple (cf. Hochschild [1; p. 167]).

NOTES

Note 12. Connections and holonomy groups (Supplement to Note 1)

1. In Theorem 7.2 of Chapter IV we proved that if M is a connected, simply connected, and complete Riemannian manifold such that its (restricted) affine holonomy group Φ^0 at a point $x \in M$ leaves a point of the Euclidean tangent space $T_x(M)$ fixed, then M is isometric to a Euclidean space. Nomizu and Yano [1] gave a simpler proof to this theorem by making use of the fact that if M is a complete, irreducible Riemannian manifold other than a line, then every affine transformation of M is necessarily an isometry (cf. Theorem 3.6 of Chapter VI).

2. The decomposition theorem of de Rham (Theorem 6.1 of Chapter IV) has been extended by H. Wu [1], [2], [3] to the case of an indefinite Riemannian metric. To explain his result, let V be a vector space with non-degenerate symmetric bilinear form g. A subspace V_1 of V is said to be *non-degenerate* if the restriction of g to V_1 is non-degenerate. If V_1 is non-degenerate, its orthogonal complement V_2 is also non-degenerate and $V = V_1 + V_2$ (direct sum). Let H be a group of linear transformations of V which leave g invariant. We say that H is *weakly irreducible* if V and (0) are the only non-degenerate subspaces of V which are invariant by H. Let M be a simply connected manifold with an indefinite Riemannian metric g. Let $\Psi(x)$ be the (restricted) linear holonomy group of M with reference point $x \in M$. We apply the concepts introduced above to $T_x(M)$ and $\Psi(x)$. Let $T_x^{(0)}$ denote the subspace of $T_x(M)$ consisting of those vectors which are left fixed by $\Psi(x)$. *Assume that $T_x^{(0)}$ is non-degenerate*. Then its orthogonal complement in $T_x(M)$ can be decomposed into a direct sum $\Sigma_{i=1}^k T_x^{(i)}$ of mutually orthogonal, $\Psi(x)$-invariant and weakly irreducible subspaces. We call $T_x(M) = \Sigma_{i=0}^k T_x^{(i)}$ a *canonical decomposition* (or *de Rham decomposition*) of $T_x(M)$. Then a canonical decomposition is unique up to an order and $\Psi(x)$ is a direct product $\Psi_0(x) \times \Psi_1(x) \times \cdots \times \Psi_k(x)$ of normal subgroups, where $\Psi_i(x)$ is trivial on $T_x^{(j)}$ if

$i \neq j$ and is weakly irreducible on $T_x^{(i)}$ for each $i = 1, \ldots, k$, and $\Psi_0(x)$ consists of the identity only. If M is moreover complete, then M is isometric to a direct product $M_0 \times M_1 \times \cdots \times M_k$ of manifolds with indefinite metrics, where M_0 is flat and $M_1, \ldots, M_k$ have weakly irreducible holonomy groups. The affine symmetric spaces described in Example 5.2 of Chapter XI admit invariant indefinite metrics and have reducible, but weakly irreducible holonomy groups.

3. In their second paper, Narasimhan and Ramanan [1] proved that *given a Lie group G with a finite number of connected components and a positive integer n, there exist a principal bundle $E(N, G)$ and a connection Γ_0 on E such that any connection Γ in any principal bundle $P(M, G)$, $\dim M \leq n$, can be obtained as the inverse of Γ_0 by a certain homomorphism of P into E.* This extends their result in their first paper (cf. also p. 289 of Volume I) in which G was assumed to be compact. (See also Lehmann [2].)

4. The results of Weyl and Cartan mentioned on pages 288–289 of Volume I have been clarified by Kobayashi and Nagano [2] as follows. *Let G be a Lie subgroup of $GL(n; \mathbf{R})$ and $\mathfrak{g} \subset \mathfrak{gl}(n; \mathbf{R})$ its Lie algebra. Then every G-structure P on an n-dimensional manifold M admits a torsion-free connection if and only if $H^{0,2}(\mathfrak{g}) = 0$*, where $H^{0,2}(\mathfrak{g})$ is a certain cohomology group defined by Spencer which depends not only on $\mathfrak{g}$ but on the representation $\mathfrak{g} \subset \mathfrak{gl}(n; \mathbf{R})$. The classification of linear Lie algebras $\mathfrak{g}$ with $H^{0,2}(\mathfrak{g}) = 0$ was given by Kobayashi-Nagano completing the result of E. Cartan.

5. On flat connections, their holonomy groups (§4 of Chapter V and Note 5) and their characteristic classes, see also Auslander [1]–[10], Auslander-Auslander [1], Auslander-Kuranishi [1], Auslander-Marcus [1], [2], Auslander-Szczarba [1], Benzécri [1], [2], Charlap [1], Charlap-Vasquez [1], Kuiper [1]–[4], Thorpe [2], Milnor [6], Kamber-Tondeur [1], Lehmann [3]. It is still an open question whether a compact and complete flat affinely connected manifold has vanishing Euler number.

Note 13. The automorphism group of a geometric structure
(Supplement to Note 9)

In Note 9, we enumerated geometric structures whose automorphism groups are Lie groups. The content of Note 9 has been expanded in an expository article by Chu and Kobayashi [1]. We shall make a report here on recent progress on the subject.

Let G be a Lie subgroup of $GL(n; \mathbf{R})$ and let M be a manifold of dimension n. Let $P = P(M, G)$ be a G-structure on M, i.e., a subbundle of the bundle $L(M)$ of linear frames with group $G \subset GL(n; \mathbf{R})$ (cf. p. 288 of Volume I). A transformation f of M is called an *automorphism* of the G-structure P if the transformation $\tilde{f}$ of $L(M)$ induced by f(cf. p. 226 of Volume I) maps P into itself. For example, an $O(n)$-structure is nothing but a Riemannian metric and its automorphisms are isometries. To describe the results on the automorphism group of a G-structure, we briefly define the following notions.

Let $\mathfrak{gl}(V)$ be the Lie algebra of linear endomorphisms of an n-dimensional vector space V (say, over $\mathbf{R}$). Let $\mathfrak{g}$ be a Lie subalgebra of $\mathfrak{gl}(V)$. The k-*th prolongation* $\mathfrak{g}^{(k)}$ of $\mathfrak{g}$ is, by definition, the vector space consisting of all symmetric $(k+1)$-linear mappings

$$a: (x_1, \ldots, x_{k+1}) \in V \times \cdots \times V \to a(x_1, \ldots, x_{k+1}) \in V$$

such that, for arbitrarily fixed $x_1, \ldots, x_k \in V$, the linear endomorphism of V which sends x_{k+1} into $a(x_1, \ldots, x_k, x_{k+1})$ is an element of $\mathfrak{g}$. We say that $\mathfrak{g}$ is of *finite type* if $\mathfrak{g}^{(k)} = 0$ for some k. Otherwise, $\mathfrak{g}$ is said to be of *infinite type*. For $\mathfrak{g} = \mathfrak{o}(n)$, Lie algebra of all skew-symmetric matrices of degree n, we have $\mathfrak{g}^{(1)} = 0$. For $\mathfrak{g} = \mathfrak{gl}(m, \mathbf{C})$ considered as a real Lie subalgebra of $\mathfrak{gl}(2m; \mathbf{R})$ in a natural manner, $\mathfrak{g}$ is of infinite type.

The following theorem was first announced by Ehresmann [7].

THEOREM 1. *Let G be a Lie subgroup of $GL(n; \mathbf{R})$ whose Lie algebra $\mathfrak{g} \subset \mathfrak{gl}(n; \mathbf{R})$ is of finite type with $\mathfrak{g}^{(k)} = 0$. Then for any G-structure on a manifold M of dimension n, the group of automorphisms of the G-structure is a Lie group of dimension at most $n + \dim \mathfrak{g} + \Sigma_{i=1}^{k-1} \dim \mathfrak{g}^{(i)}$.*

Leaving the detail of the proof to Sternberg [1], Corollary 4.2, p. 348 or Ruh [1], we shall explain only the main idea. Given a G-structure P on M, we construct, following Chern [3], a tower of principal bundles $P = P^{(0)}, P^{(1)}, P^{(2)}, \ldots$ such that $P^{(i)}$ is a $\mathfrak{g}^{(i)}$ structure on $P^{(i-1)}$. If $\mathfrak{g}^{(k)} = 0$, then $P^{(k)}$ is an $\{e\}$-structure on $P^{(k-1)}$, where $\{e\}$ is the trivial group consisting of the identity element alone. We lift every automorphism of P to an automorphism of $P^{(i)}$ for each i in a natural manner. In this way the proof is reduced to the special case of $\{e\}$-structure which has been treated by Kobayashi [1] for the purpose of giving a unified proof to the theorem of Myers-Steenrod on the group of isometries and the theorem of Nomizu-Hano-Morimoto on the group of affine transformations (cf. Note 9), as well as the theorem that the group of conformal transformations of a Riemannian manifold and the group of projective transformations of a projectively connected manifold. All these theorems are now essentially contained in Theorem 1.

Boothby-Kobayashi-Wang [1] proved that the automorphism group of an almost complex structure on a compact manifold is a Lie group by showing that the automorphisms satisfy a system of elliptic partial differential equations. This generalizes the theorem of Bochner-Montgomery on the automorphism group of a compact complex manifold (cf. Note 9). Ruh [1] generalized the result of Boothby-Kobayashi-Wang to a certain class of G-structures. To describe the most general result in this direction, we make one more definition. We say that $\mathfrak{g} \subset \mathfrak{gl}(V)$ is *elliptic* if it does not contain any linear endomorphism of rank 1. If $\mathfrak{g}$ is of finite type, it is elliptic. This can be seen as follows. If $A \in \mathfrak{g}$ is an element of rank 1, then $A = e \otimes \omega$, where $e \in V$ and ω is in the dual space of V, that is, $A(x) = \omega(x)e$ for all $x \in V$. For any positive integer k, the symmetric $(k+1)$-linear mapping $a: V \times \cdots \times V \to V$ defined by

$$a(x_1, \ldots, x_{k+1}) = \omega(x_1) \cdots \omega(x_{k+1})e$$

is a non-zero element of $\mathfrak{g}^{(k)}$. This proves our assertion. Considered as a real subalgebra of $\mathfrak{gl}(2m; \mathbf{R})$ in a natural manner, $\mathfrak{gl}(m; \mathbf{C})$ is elliptic and of infinite type. The following result is due to Ochiai [1]. (It was pointed out by Tandai that the theorem of Ochiai and that of Ruh are equivalent (cf. Fujimoto [4]).)

THEOREM 2. *Let G be a Lie subgroup of $GL(n; \mathbf{R})$ whose Lie algebra $\mathfrak{g} \subset \mathfrak{gl}(n; \mathbf{R})$ is elliptic. For any G-structure on a compact manifold M of dimension n, the automorphism group is a Lie group.*

In this case no estimate on the dimension of the automorphism group is known. As in the case of an almost complex structure, the proof is achieved by the construction of a system of elliptic partial differential equations satisfied by the automorphisms.

We shall mention two results which are not contained in Theorems 1 and 2. Let M be a real hypersurface of dimension $2n - 1$ in $\mathbf{C}^n$. A diffeomorphism of M onto itself is called a *pseudo-conformal transformation* of M if it can be extended to a holomorphic diffeomorphism of a neighborhood of M onto a neighborhood of M. We say that M is non-degenerate at $x \in M$ if the so-called Levi-Krzoska form is non-degenerate at x. Tanaka [3] proved:

THEOREM 3. *If M is a real hypersurface of $\mathbf{C}^n$ which is non-degenerate at one point, then the group of pseudo-conformal transformations is a Lie group of dimension at most $n^2 + 2n$.*

If M is a compact real hypersurface of $\mathbf{C}^n$, there is a point where M is non-degenerate and Theorem 3 applies to M. The proof of Theorem 3 is closely related to that of Theorem 1.

Kobayashi [24], [26] introduced a class of complex manifolds, called hyperbolic manifolds, which include all Kaehler manifolds whose holomorphic sectional curvatures are bounded above by a negative constant. Then we have (Kobayashi [24], [26], and Wu [5])

THEOREM 4. *The group of holomorphic transformations of a hyperbolic manifold M of complex dimension n is a Lie group of dimension at most $n^2 + 2n$ with compact isotropy subgroups.*

The proof relies on the theorem of van Dantzig and van der Waerden on the group of isometries of a locally compact metric space (cf. Theorem 4.7 of Chapter I). (In Wu's paper where the notion of tight manifold is used, the proof makes use of properties of normal families of holomorphic mappings.) See also Kaup [1] on the automorphisms of a complex space.

Brickell [1] gives another geometric structure (called an area measure) whose automorphism group is a Lie group.

Beyond the fact that the automorphism groups are Lie groups in the cases listed above and in Note 9, very little is known about the structures of these Lie groups. We first note that *the group $\mathfrak{I}(M)$ of isometries of a connected compact homogeneous Riemannian manifold M is semi-simple if the Euler number $\chi(M)$ is non-zero*. In fact, since M is compact, so is $\mathfrak{I}(M)$ by Theorem 3.4 of Chapter VI. The Lie algebra $\mathfrak{i}(M)$ of $\mathfrak{I}(M)$ is therefore reductive (cf. Appendix 9). If

X is a vector field on M belonging to the center of $\mathfrak{i}(M)$, the transitivity of $\mathfrak{I}(M)$ implies that X has constant length. If $\chi(M) \neq 0$, X must vanish identically. The proof shows that any Lie subgroup of $\mathfrak{I}(M)$ which is transitive on M is reductive if M is compact and that it is semi-simple if, moreover, $\chi(M) \neq 0$. In the proof of Theorem 8.4 of Chapter VIII, we established that *if M is a connected homogeneous Riemannian manifold with negative-definite Ricci tensor and if G is a Lie subgroup of $\mathfrak{I}(M)$ which is transitive on M, then the center of G is discrete.* It follows then that *if M is a connected homogeneous Riemannian manifold with (positive or negative) definite Ricci tensor, then no nilpotent Lie subgroup of $\mathfrak{I}(M)$ is transitive on M*; this fact, which was pointed out to us by G. R. Jensen, generalizes partially the result of Wolf [15] to the effect that if M is a connected, non-flat homogeneous Riemannian manifold with everywhere non-negative or everywhere non-positive sectional curvature, then no nilpotent Lie subgroup of $\mathfrak{I}(M)$ is transitive on M. See also Hermann [14] and Wolf [16].

On the structure of the group of holomorphic transformations of a Kaehler manifold, we have the following result of Matsushima [3]. *For a compact Einstein-Kaehler manifold M, the Lie algebra of holomorphic vector fields is naturally isomorphic to* $\mathfrak{i}(M) + \sqrt{-1}\,\mathfrak{i}(M)$. Since $\mathfrak{i}(M)$ is reductive, its complexification is also reductive. This result follows from the Hodge decomposition theorem applied to the 1-form corresponding to a holomorphic vector field. This theorem of Matsushima has been generalized to the case of compact Kaehler manifold with constant scalar curvature by Lichnerowicz [8]. See also Yano [8; pp. 92–98].

A theorem of Bochner states that if M is a compact Riemannian manifold with negative-definite Ricci tensor, then the group $\mathfrak{I}(M)$ of isometries of M is finite (cf. Corollary 5.4 of Chapter VI). As a variation of this result of Bochner, Frankel [5] proved that *if M is a compact Riemannian manifold with sectional curvature ≤ 0 and with negative-definite Ricci tensor, then any isometry of M that is homotopic to the identity (in the group of all diffeomorphisms of M) is the identity*. Its proof is related to that of Theorem 8.3 of Chapter VIII. Another variation is the following theorem of Kobayashi [15]. *If M is a compact Kaehler manifold with negative-definite Ricci tensor (more generally, with negative first Chern class), then the group of holomorphic transformations of M is finite.* This generalizes similar results by

Bochner [3], Hawley [2], and Sampson [1] on compact complex manifolds which can be uniformized by bounded domains in $\mathbf{C}^n$. It strengthens the theorem of Bochner [1] that the group of holomorphic transformations of a compact Kaehler manifold with negative-definite Ricci tensor is discrete. For related results, see Peters [1].

For the group $\Im(M)$ of isometries of a Riemannian symmetric space M, see Takeuchi [2]; while the determination of the identity component $\Im^0(M)$ is simple, that of $\Im(M)$ itself is very delicate.

For many other results on automorphism groups of Riemannian and Kaehlerian structures and other G-structures, we refer the reader to Libermann [1], [3], Lichnerowicz [3], [10], [11], Yano [2], [8], and references therein.

Note 14. The Laplacian

Let M be a Riemannian manifold. For any differentiable function f, we define the *gradient* of f, denoted by grad f, to be a unique vector field X such that

$$g(X, Y) = df(Y) = Yf \quad \text{for all vector fields } Y,$$

where g denotes the (positive-definite) metric of M.

For any vector field X on M, we define the *divergence* of X, denoted by div X, to be the function such that at each point x of M

$(\operatorname{div} X)_x = $ trace of the endomorphism $V \to \nabla_V X$ of $T_x(M)$.

In the case where M is oriented, we have defined div X by using the volume element of M (cf. Appendix 6). Of course, these definitions coincide.

In terms of local coordinates $(x^1, \ldots, x^n)$, we have

$$df = \sum_{i=1}^{n} f_i \, dx^i, \quad \text{where} \quad f_i = \partial f/\partial x^i$$

and

$$\operatorname{grad} f = \sum_{i=1}^{n} f^i \, \partial/\partial x^i, \quad \text{where} \quad f^i = \sum_{j=1}^{n} g^{ij} f_j,$$

where (g^{ij}) has been defined on p. 156 of Volume I. If $X = \sum_{i=1}^{n} \xi^i \, \partial/\partial x^i$, then

$$\text{div } X = \sum_{i=1}^{n} \xi^i{}_{;i}.$$

The *Laplacian* (or *Laplace-Beltrami operator*) Δ on M is a mapping which associates to any differentiable function f the function $\Delta f = \text{div} (\text{grad} f)$. In terms of local coordinates we have

$$\Delta f = \sum_{i=1}^{n} f^i{}_{;i} = \sum_{i,j=1}^{n} g^{ij} f_{j;i} = \sum_{i,j=1}^{n} g^{ij} f_{i;j}.$$

In particular, if $x^1, \ldots, x^n$ are normal coordinates with origin p, then

$$(\Delta f)_p = \sum_{i=1}^{n} (\partial^2 f/\partial x^{i^2})_p$$

because $g_{ij} = \delta_{ij}$, $g^{ij} = \delta^{ij}$, and $f_{i;j} = \partial f_i/\partial x^j$ at p.

It is easy to verify that
(i) $\text{div } (fX) = Xf + f \text{ div } X$
for any function f and any vector field X;
(ii) $\Delta(f^2) = 2f \Delta f + (\text{grad} f) \cdot f$
$= 2f \Delta f + g(df, df)$
for any function f, where $g(df, df)$ denotes the inner product in the cotangent space $T^*_x(M)$ for each x (cf. p. 156, Volume I). We shall prove

LEMMA (E. Hopf). *Let M be a compact Riemannian manifold. If f is a differentiable function on M such that $\Delta f \geq 0$ everywhere, then f is a constant function (and $\Delta f = 0$).*

Proof. We may assume that M is oriented by taking the two-fold covering of M if necessary. Let $X = \text{grad} f$. By Green's theorem (Appendix 6) we have

$$\int_M \Delta f \, dv = \int_M \text{div } X \, dv = 0,$$

where dv is the volume element of M. Since $\Delta f \geq 0$ by assumption, we must have $\Delta f = 0$ everywhere on M. By using Green's theorem for $f^2/2$ and by using (ii) above, we obtain

$$\int_M \Delta(f^2/2) \, dv = \int_M f \Delta f \, dv + \int_M g(df, df) \, dv = 0.$$

Since $\Delta f = 0$ as we have shown, we obtain

$$\int_M g(df, df)\, dv = 0,$$

which implies $df = 0$ everywhere, that is, f is a constant function.

A function f is said to be *harmonic* if $\Delta f = 0$. The lemma shows that a compact Riemannian manifold has no harmonic function except for constant functions.

The lemma of Hopf was used successfully in the work of Bochner, Lichnerowicz, Yano, and others (see Yano and Bochner [1] and Yano [2]). Lichnerowicz [3, p. 4] gave a generalization to the case of a tensor field: *if K is a tensor field with components $K^{s\cdots}_{t\cdots}$ and if*

$$\sum_{i,j=1}^{n} g^{ij} K^{s\cdots}_{t\cdots;i;j} = 0,$$

then

$$K^{s\cdots}_{t\cdots;i} = 0 \quad (\text{that is, } \nabla K = 0).$$

We shall give some examples of computation for Δ. Let M be an n-dimensional Riemannian manifold isometrically immersed in $\mathbf{R}^{n+p}$. Since the computation is done locally, we identify a point x of M with the position vector x of the corresponding point in $\mathbf{R}^{n+p}$. Let a be any constant vector in $\mathbf{R}^{n+p}$ and consider the function $f(x) = (x, a)$ on M. (If a is the i-th vector of the standard basis $e_1, \ldots, e_{n+p}$ of $\mathbf{R}^{n+p}$, then f is the i-th coordinate function of the immersion.) We compute Δf.

For any vector field Y on M, we have

$$Yf = (Y, a),$$

where Y on the right hand side is the $\mathbf{R}^{n+p}$-valued vector function which represents the vector field Y on M. For any vector field X on M, we have

$$XYf = (\nabla'_X Y, a) = (\nabla_X Y, a) + (\alpha(X, Y), a)$$

by using the formula of Gauss (§3, Chapter VII)

$$\nabla'_X Y = \nabla_X Y + \alpha(X, Y),$$

where $\alpha(X, Y)$ is the second fundamental form of M.

Since

$$(\nabla_X Y)f = (\nabla_X Y, a),$$

we obtain
$$XYf - (\nabla_X Y)f = (\alpha(X, Y), a).$$
The left hand side is equal to $(\nabla^2 f)(; Y; X)$. If $\{X_1, \ldots, X_n\}$ is an orthonormal basis in $T_x(M)$, then

$$\Delta f = \sum_{i,j=1}^n g^{ij}(\nabla^2 f)(; X_i; X_j)$$
$$= \sum_{i=1}^n (\nabla^2 f)(; X_i; X_i) = (\sum_{i=1}^n \alpha(X_i, X_i), a)$$

If $\xi_1, \ldots, \xi_p$ is an orthonormal basis in the normal space at x and if we set
$$\alpha(X, Y) = \sum_{k=1}^p h^k(X, Y)\xi_k,$$
then
$$\sum_{i=1}^n h^k(X_i, X_i) = \text{trace } A_k,$$
where A_k is the operator corresponding to h^k. Thus
$$\frac{1}{n}\sum_{i=1}^n \alpha(X_i, X_i)$$
is nothing but the mean curvature normal η at x (cf. the remark following Example 5.3 of Chapter VII). Thus
$$\Delta f = n(\eta, a) \qquad \text{for } f(x) = (x, a).$$

Example 1. We may obtain another proof of the result that there is no compact minimal submanifold in a Euclidean space (cf. the remark cited above). Indeed, *the mean curvature normal η of M in $\mathbf{R}^{n+p}$ is zero if and only if $\Delta f = 0$ for every function f of the form (x, a)*. If M is compact, the harmonic function f must be a constant. By taking $a = e_i$, where $\{e_1, \ldots, e_{n+p}\}$ is an orthonormal basis of the Euclidean space, we see that the i-th coordinate function of the immersion is a constant. Thus M reduces to a single point.

Example 2. Let $S^n(r)$ be the hypersphere of radius $r > 0$ in $\mathbf{R}^{n+1}$: $(x^0)^2 + (x^1)^2 + \cdots + (x^n)^2 = r^2$. As in Example 4.2 of Chapter VII, we take the outward unit normal $\xi = x/r$ and get

$A_\xi = (-1/r)I$. Thus the mean curvature normal η is equal to

$$\frac{1}{n}(\text{trace } A_\xi)\xi = -x/r^2.$$

We have for a function f of the form $f(x) = (x, a)$

(1) $\Delta f = (-n/r^2)f$.

In particular, each coordinate function x^i, as a function on $S^n(r)$, is an eigenfunction for the eigenvalue $-n/r^2$ of the Laplacian Δ. We have also

(2) $\nabla^2 f + (f/r^2)g = 0$,

that is, $(\nabla^2 f)(; Y; X) + (f/r^2)g(X, Y) = 0$ for all vector fields X and Y. This follows from

$$(\nabla^2 f)(; Y; X) = (\alpha(X, Y), a),$$

which appeared in the computation above and from

$$(\alpha(X, Y), a) = (h(X, Y)\xi, a) = (-g(X, Y)x/r^2, a)$$
$$= (-f/r^2)g(X, Y)$$

for $S^n(r)$. If we set $Z = \text{grad } f$, then we may also prove

(3) $L_Z g = 2(f/r^2)g$,

which shows that Z is an infinitesimal conformal transformation (which is not homothetic because f is not a constant).

It is known that *a connected complete Riemannian manifold M of dimension $n > 2$ admits a non-zero solution f of the differential equation (2) if and only if M is isometric to $S^n(r)$. If M is a connected, compact, orientable Einstein space of scalar curvature 1, then M admits an eigenfunction f with $\Delta f = -nf$ if and only if M is isometric to $S^n(1)$.* For these results and the corresponding results for the complex projective space $P_n(\mathbf{C})$, see Obata [1] and the references cited there (also see Note 11). We note that the Laplacian in Obata [1] is the negative of our Laplacian.

Example 3. Let M be a k-dimensional submanifold of $S^n(r)$ in $\mathbf{R}^{n+1}$. In addition to the unit normal $\xi = x/r$ to $S^n(r)$, choose $n - k$ vector fields ξ_j which are normal to M and tangent to $S^n(r)$ such that $\xi, \xi_1, \ldots, \xi_{n-k}$ are orthonormal at every point of M.

We denote by D, ∇^0, and ∇ the Riemannian connections of $\mathbf{R}^{n+1}$, $S^n(r)$, and M, respectively. For vector fields X and Y tangent to M we have
$$D_X Y = \nabla^0_X Y + h(X, Y)\xi$$
and
$$\nabla^0_X Y = \nabla_X Y + \sum_{j=1}^{n-k} h^j(X, Y)\xi_j$$
so that
$$D_X Y = \nabla_X Y + h(X, Y)\xi + \sum_{j=1}^{n-k} h^j(X, Y)\xi_j,$$
where h is the second fundamental form for $S^n(r)$ in $\mathbf{R}^{n+1}$ and $h^1, \ldots, h^{n-k}$ are the second fundamental forms of M as a submanifold of $S^n(r)$.

For any constant vector a in $\mathbf{R}^{n+1}$, consider $f(x) = (x, a)$ as a function on M. By the same sort of computation as in Example 1 we get for the Laplacian Δ for M
$$\Delta f = (\sum_{i=1}^{k} h(X_i, X_i)\xi + \sum_{i=1}^{k}\sum_{j=1}^{n-k} h^j(X_i, X_i)\xi_j, a),$$
where $X_1, \ldots, X_k$ is an orthonormal basis of the tangent space of M at a point. Of course, we have
$$\sum_{i=1}^{k} h(X_i, X_i) = -k/r^2,$$
and $\sum_{i=1}^{k}\sum_{j=1}^{n-k} h^j(X_i, X_i)\xi_j$ is equal to $k\zeta$, where ζ is the mean curvature vector of M in $S^n(r)$. Thus we obtain
$$\Delta f = -(k/r^2)f + (k\zeta, a).$$
In particular, for each coordinate function x^i considered as a function on M, we have
$$\Delta x^i = -(k/r^2)x^i + k\zeta^i, \quad 0 \leq i \leq n,$$
where $(\zeta^0, \ldots, \zeta^n)$ are the components of ζ as a vector in $\mathbf{R}^{n+1}$. This formula implies that M is minimal as a submanifold of $S^n(r)$ if and only if
$$\Delta x^i = -(k/r^2)x^i, \quad 0 \leq i \leq n.$$
See Takahashi [2], Nagano [9], Takeuchi-Kobayashi [1], Simons

[2], [3], and Calabi [7] for further results in this direction. For eigenvalues of the Laplacian, see also Berger [22], McKean-Singer [1], Milnor [7].

For the theory of invariant differential operators (generalized Laplacians) and spherical functions on Riemannian symmetric spaces, we refer to Helgason [2, Chapter X].

The Laplacian can be defined for differential forms on a Riemannian manifold. For the theory of harmonic integrals, see Hodge [1], Weil [2], and de Rham [2]. For its applications to various problems in differential geometry, see Yano and Bochner [1], and Lichnerowicz [3].

Note 15. Surfaces of constant curvature in $\mathbf{R}^3$

In Chapter VII we proved that a connected complete Riemannian manifold of dimension $n \geq 3$ isometrically immersed in $\mathbf{R}^{n+1}$ is actually imbedded as a hypersphere if it has non-zero constant sectional curvature. For the case $n = 2$, we have

THEOREM 1. *Let M be a connected complete 2-dimensional Riemannian manifold isometrically immersed in $\mathbf{R}^3$. If M has constant curvature $k \neq 0$, then*
(a) $k > 0$;
(b) *M is imbedded as a sphere.*

Part (a) is due to Hilbert [2]; we omit the proof. We shall prove here part (b) assuming that $k > 0$. Under this assumption, it follows that M is compact (in fact, a sphere or a projective plane) by virtue of Theorem 1 of Note 4. By replacing M by its universal covering space if necessary, we may assume that M is actually a sphere with the usual metric; our assertion amounts to saying that *an isometric immersion of a sphere into $\mathbf{R}^3$ is nothing but an ordinary imbedding (rigidity of a sphere)*. Thus it is a special case of the theorem of Cohn-Vossen [1], [2] (see Chern [18]).

We shall prove the rigidity of a sphere in the following form:

THEOREM 2. *Let M be a 2-dimensional connected compact Riemannian manifold of constant curvature k. If M is isometrically immersed in $\mathbf{R}^3$, then M is imbedded as a sphere (and $k > 0$).*

We prepare a few lemmas, for which we assume M to be orientable and choose a field of unit normals so that the second fundamental form h and the corresponding symmetric operator A are defined globally on M.

LEMMA 1. *For each $x \in M$, let $\lambda(x)$ and $\mu(x)$ be the eigenvalues of A such that $\lambda(x) \geq \mu(x)$. The functions λ and μ are continuous on M and differentiable on $W = \{x \in M; \lambda(x) > \mu(x)\}$.*

Proof. If (x^1, x^2) is a local coordinate system around a point x, then A can be expressed by a matrix

$$\begin{pmatrix} a(x^1, x^2) & b(x^1, x^2) \\ c(x^1, x^2) & d(x^1, x^2) \end{pmatrix}.$$

The eigenvalues λ and μ are the roots of the quadratic equation

$$f(t) = \det\begin{pmatrix} t - a & -b \\ -c & t - d \end{pmatrix} = t^2 - \varphi t + \psi,$$

where $\varphi(x^1, x^2) = a(x^1, x^2) + d(x^1, x^2)$ and $\psi(x^1, x^2) = a(x^1, x^2) d(x^1, x^2) - b(x^1, x^2) c(x^1, x^2)$ are differentiable functions together with a, b, c, and d. Our assertion easily follows from the formula for the roots of the quadratic equation.

LEMMA 2. *Assume $\lambda(x_0) > \mu(x_0)$. Then there exist differentiable vector fields X and Y on a neighborhood U of x_0 such that*

$$g(X, X) = g(Y, Y) = 1, \quad g(X, Y) = 0,$$
$$AX = \lambda X, \quad \text{and} \quad AY = \mu Y$$

at each point of U.

Proof. Let V be a neighborhood of x_0 in which $\lambda > \mu$. Choose a tangent vector Z_0 at x_0 such that $AZ_0 \neq \lambda Z_0$ and $AZ_0 \neq \mu Z_0$ (this is possible because $\lambda \neq \mu$ at x_0). Extend Z_0 to a differentiable vector field Z on V. Then

$$X = AZ - \mu Z \quad \text{and} \quad Y = AZ - \lambda Z$$

are differentiable vector fields on V such that $AX = \lambda X$ and $AY = \mu Y$ at every point. By our choice of Z_0, X and Y are not 0 at x_0, and therefore they are not 0 in a neighborhood U of x_0. Since X and Y are eigenvectors of A for the distinct eigenvalues, X and Y are orthogonal at each point. By normalizing X and Y we get the vector fields we want.

LEMMA 3. *In the notation of Lemma 2 we have*

(1) $\nabla_X X = aY$, $\nabla_Y X = bY$, $\nabla_X Y = -aX$, $\nabla_Y Y = -bY$,

where a and b are given by

$$a = Y\lambda/(\lambda - \mu) \quad \text{and} \quad b = X\mu/(\lambda - \mu);$$

(2) *the curvature k is given by*

$$k = -(a^2 + b^2) - Xb + Ya.$$

Proof.
(1) From $g(X, X) = 1$, we obtain $g(\nabla_X X, X) = 0$ so that $\nabla_X X = aY$ for a certain function a. Similarly, $g(\nabla_Y X, X) = 0$ so that $\nabla_Y X = bY$ for a certain function b. From $g(Y, Y) = 1$, we get $\nabla_X Y = cX$ and $\nabla_Y Y = dX$ in the same fashion. From $g(X, Y) = 0$, we get $g(\nabla_X X, Y) + g(X, \nabla_X Y) = 0$ so that $a + c = 0$. Similarly, $g(\nabla_Y X, Y) + g(X, \nabla_Y Y) = 0$ so that $b + d = 0$. The functions a and b can be determined by using the equation of Codazzi $(\nabla_X A)(Y) = (\nabla_Y A)(X)$.

(2) The results in (1) give

$$R(X, Y)Y = -(a^2 + b^2)X - (Xb)X + (Ya)X$$

and hence

$$k = g(X, R(X, Y)Y) = -a^2 - b^2 - Xb + Ya.$$

LEMMA 4. *Assume $\lambda(x_0) > \mu(x_0)$. If λ has a relative maximum at x_0 and μ has a relative minimum at x_0, then the curvature k is ≤ 0 at x_0.*

Proof. For the vector fields X and Y in Lemma 1 we have $X\lambda = X\mu = Y\lambda = Y\mu = 0$ at x_0. By computation based on Lemma 3, we have at x_0

$$Ya = Y^2\lambda/(\lambda - \mu), \quad Xb = X^2\mu/(\lambda - \mu)$$

and

$$k = Y^2\lambda/(\lambda - \mu) - X^2\mu/(\lambda - \mu).$$

Since λ has a relative maximum at x_0, we have $Y^2\lambda \leq 0$ at x_0. Since μ has a relative minimum at x_0, we have $X^2\mu \geq 0$ at x_0. Thus $k(x_0) \leq 0$ from the formula above.

Still assuming that M is orientable, we now prove Theorem 2. Since M is compact, there is a point where the second fundamental form is (positive or negative) definite (cf. Proposition 4.6 of Chapter VII). Thus $k = \lambda\mu$ is positive. Hence the constant curvature k is positive. Since M is compact, the function λ has a maximum at x_0, and at x_0 the function μ has a minimum. If $\lambda(x_0) > \mu(x_0)$, then Lemma 4 gives a contradiction to $k > 0$. Hence $\lambda(x_0) = \mu(x_0)$, which implies that $\lambda(x) = \mu(x)$ for every $x \in M$. Every point being umbilical, we see that M is a sphere by Theorem 5.1 of Chapter VII. If M is not orientable, let $\tilde{M}$ be the orientable double covering space of M. Applying the reasoning above to $\tilde{M}$, we see that $\tilde{M}$, immersed in $\mathbf{R}^{n+1}$ as a double covering of M, is umbilical at every point. Hence M is umbilical at every point and is a sphere by Theorem 5.1 of Chapter VII.

We can make use of Lemma 4 to prove another classical theorem. A connected compact surface in $\mathbf{R}^3$ is called an *ovaloid* if the curvature is positive everywhere. *An ovaloid whose mean curvature is a constant is a sphere.* (Note that an ovaloid is necessarily orientable; the condition $H =$ constant is independent of the choice of the field of unit normals, i.e., just a change of sign.) This result is due to Liebmann [1]. We shall here prove a more general result due to Aleksandrov [2] and Chern [9].

THEOREM 3. *If an ovaloid M is isometrically immersed in $\mathbf{R}^3$ in such a way that $\mu = f(\lambda)$, where λ and μ are the functions giving the eigenvalues of A as in Lemma 1 and f is a non-increasing function (i.e. $t \leq s$ implies $f(t) \geq f(s)$), then M is imbedded as a sphere.*

Proof. Let x_0 be a point where λ has a maximum. Then $\lambda(x) \leq \lambda(x_0)$ and hence $\mu(x) \geq \mu(x_0)$ for every x, that is, μ has a minimum at x_0. If $\lambda(x_0) > \mu(x_0)$, then Lemma 4 gives $k(x_0) \leq 0$, which is a contradiction to the assumption $k > 0$. Thus $\lambda(x_0) = \mu(x_0)$ and $\lambda(x) = \mu(x)$ for every x as in the proof of Theorem 2.

Theorem 3 includes the case of constant mean curvature H ($\lambda + \mu =$ constant) as well as Theorem 2 ($\mu = k/\lambda$ where k is a positive constant).

Hopf [7] proved that a connected, compact orientable surface in $\mathbf{R}^3$ such that $H =$ constant is a sphere provided that its genus is 0. Aleksandrov [3] proved this result without the assumption on the genus.

We now consider the case where the constant curvature k is 0 in Theorem 1. Pogorelov [3], [4] and Hartman-Nirenberg [1] proved that M is immersed as a plane or a cylinder. Massey [1] gave a more elementary proof. Hartmann-Nirenberg actually proved a general theorem concerning hypersurfaces in a Euclidean space whose Gaussian mapping has Jacobian of fixed sign ≥ 0 (or ≤ 0) and derived the following result: *If f is an isometric immersion of an n-dimensional connected complete flat Riemannian manifold M into $\mathbf{R}^{n+1}$, then $f(M)$ is of the form $C \times \mathbf{R}^{n-1}$, where $\mathbf{R}^{n-1}$ is a Euclidean subspace of dimension $n - 1$ and C is a curve lying on a plane $\mathbf{R}^2$ perpendicular to $\mathbf{R}^{n-1}$.* A more direct proof generalizing the method of Massey can be found in Nomizu [12] in which some related problem was treated.

Sacksteder [1] and O'Neill [5] studied a generalization of the above theorem to the case where M has non-negative sectional curvature everywhere or M is of larger codimension. Hartman [1] proved the following general result in this direction. *Let f be an isometric immersion of an n-dimensional complete Riemannian manifold M into $\mathbf{R}^{n+p}$. Assume that*

(1) *the sectional curvature of $M \geq 0$ for all planes;*
(2) *the index of relative nullity $\nu(x)$ is a positive constant, say, ν.*

Then there exist an $(n - \nu)$-dimensional complete Riemannian manifold M', an isometric immersion f' of M' into $\mathbf{R}^{n+p-\nu}$, and an isometry φ of M onto $M' \times \mathbf{R}^\nu$ such that the diagram

$$\begin{array}{ccc} M & \xrightarrow{f} & \mathbf{R}^{n+p} \\ \downarrow \varphi & & \downarrow \mathrm{id} \\ M' \times \mathbf{R}^\nu & \xrightarrow{f' \times \mathrm{id}} & \mathbf{R}^{n+p-\nu} \times \mathbf{R}^\nu \end{array}$$

is commutative, where id *denotes the identity mapping.* (For the notion of index of relative nullity, see Note 16.)

For rigidity of hypersurfaces, see also Chern [18], [22]; Chern-Hano-Hsiung [1]; Hsiung [1]–[4]; Katsurada [1]–[4]; Aeppli [1].

Note 16. Index of nullity

For a Riemannian manifold M the *index of nullity* at x is the dimension $\mu(x)$ of the subspace

$$T_0(x) = \{X \in T_x(M); R(X, Y) = 0 \text{ for all } Y \in T_x(M)\}.$$

If M is isometrically immersed in a Euclidean space $\mathbf{R}^{n+p}$, where

$n = \dim M$, the *index of relative nullity* $\nu(x)$ at x is the dimension of the subspace

$$T''_x(M) = \{X \in T_x(M) \, ; A_\xi(X) = 0 \text{ for all } \xi \in T_x^\perp(M)\},$$

where $T_x(M)^\perp$ is the normal space to M at x. These notions were defined in Chern-Kuiper [1].

When the codimension p is 1, take a unit normal ξ at x. The subspace $T''(x)$ is just the null space of $A = A_\xi$ so that $n - \nu(x)$ is equal to the rank of A, namely, the type number at x. In Theorem 6.1 of Chapter VII, we showed that if the type number at x is ≥ 2, then $\nu(x) = \mu(x)$.

We prove the following result of Chern and Kuiper.

PROPOSITION 1. *For an arbitrary codimension p, we have*

$$\nu(x) \leq \mu(x) \leq \nu(x) + p.$$

Proof. As in the case of codimension 1 we have $T'' \subset T_0$ by the equation of Gauss. Hence $\nu(x) \leq \mu(x)$. Let S be the orthogonal complement of T'' in the space T_0 so that $T_0 = T'' + S$. If we set $m = \dim S$, then $m = \mu(x) - \nu(x)$ and we wish to prove $m \leq p$.

We observe that T'' can also be defined as

$$\{X \in T_x(M) \, ; \alpha(X, Y) = 0 \text{ for all } Y \in T_x(M)\};$$

this follows easily from Proposition 3.3, (2) of Chapter VII. Consider the mapping $\alpha \colon S \times S \to T_x^\perp$ (normal space). If $p < m$, then the proof of the lemma for Theorem 4.7 of Chapter VII gives a pair of vectors $X, Y \in S$, not both 0, such that

$$\alpha(X, X) = \alpha(Y, Y) \quad \text{and} \quad \alpha(X, Y) = 0.$$

Since $X \in S \subset T_0$, we have $R(X, Y) = 0$. From the formula

$$g(R(X, Y)Y, X) = g(\alpha(X, X), \alpha(Y, Y)) - g(\alpha(X, Y), \alpha(X, Y))$$

it follows that $(\alpha(X, X), \alpha(Y, Y)) = 0$, that is, $\alpha(X, X) = \alpha(Y, Y) = 0$. Since $R(X, Z) = 0$ for an arbitrary Z in $T_x(M)$, we have

$$0 = g(R(X, Z)Z, X) = (\alpha(X, X), \alpha(Z, Z)) - (\alpha(X, Z), \alpha(X, Z))$$
$$= -(\alpha(X, Z), \alpha(X, Z)).$$

Thus $\alpha(X, Z) = 0$ for every $Z \in T_x(M)$, showing $X \in T''$. Thus $X \in T'' \cap S$ and $X = 0$. Similarly, $Y = 0$, contradicting the fact that X and Y are not both 0. We have thus proved $p \geq m$.

The nullity spaces T_0 on a Riemannian manifold were studied by Maltz [1], Rosenthal [1]. As in A. Gray [4] the index of nullity can be defined for any curvature-like tensor field L of type $(1, 3)$ on a Riemannian manifold M. A tensor field L of type $(1, 3)$ is curvature-like if it has the same formal properties as the curvature tensor field R, namely,

(1) $L(X, Y)$ is a skew-symmetric endomorphism;
(2) $L(X, Y) = -L(Y, X)$;
(3) $\mathfrak{S} L(X, Y)Z = 0$;
(4) $\mathfrak{S} (\nabla_X L)(Y, Z) = 0$.

Here $\mathfrak{S}$ denotes the cyclic sum over X, Y, and Z.

For L, define the nullity at x as the dimension of the nullity space

$$T_0(x) = \{X \in T_x(M); L(X, Y) = 0 \text{ for all } Y \in T_x(M)\}.$$

We have then

PROPOSITION 2. *For a curvature-like tensor field L,*
(1) *the orthogonal complement $T_1(x)$ of $T_0(x)$ in $T_x(M)$ is spanned by $L(X, Y)Z$, where X, Y, and Z are arbitrary vectors at x;*
(2) *if $\dim T_0(x) = $ constant on M, then the distribution T_0 is involutive and totally geodesic (that is, $\nabla_X Y \in T_0$ for any vector fields X, Y belonging to T_0).*

Note 17. *Type number and rigidity of imbedding*

Let V be a real vector space of n dimensions with inner product. We identify $\wedge^2 V$ with the space $E(V)$ of skew-symmetric endomorphisms of V: $(x \wedge y)z = (y, z)x - (x, z)y$.

Given a system $\{A_1, \ldots, A_k\}$ of symmetric endomorphisms that are linearly independent, we define the *type number* of the system to be the largest integer r for which there are r vectors $x_1, \ldots, x_r$ in V such that the kr vectors $A_\alpha(x_i)$, $1 \leq \alpha \leq k$, $1 \leq i \leq r$, are linearly independent.

When $k = 1$, the type number of the system $\{A\}$ of one single endomorphism A is just equal to the rank of A.

We have

THEOREM 1. *Let* $\{A_1, \ldots, A_k\}$ *and* $\{\bar{A}_1, \ldots, \bar{A}_k\}$ *be two systems of symmetric endomorphisms and assume that* $\{A_1, \ldots, A_k\}$ *is linearly independent with type number* ≥ 3. *If*

$$\sum_{\alpha=1}^{k} A_\alpha \wedge A_\alpha = \sum_{\alpha=1}^{k} \bar{A}_\alpha \wedge \bar{A}_\alpha$$

that is, if

$$\sum_{\alpha=1}^{k} A_\alpha(x) \wedge A_\alpha(y) = \sum_{\alpha=1}^{k} \bar{A}_\alpha(x) \wedge \bar{A}_\alpha(y) \quad \text{for all } x, y \in V,$$

then there exists an orthogonal matrix $S = (s_{\beta\alpha})$ *of degree k such that*

$$\bar{A}_\alpha = \sum_{\beta=1}^{k} s_{\beta\alpha} A_\beta, \quad 1 \leq \alpha \leq k.$$

This theorem as well as the following proof are reformulations of Chern [5].

LEMMA 1. *Suppose* $y_1, \ldots, y_k, z_1, \ldots, z_k$ *are linearly independent and let* $f = \sum_{\alpha=1}^{k} y_\alpha \wedge z_\alpha$. *Considering f as an element in $E(V)$, we have*

(i) *range of* f = *subspace spanned by* $\{y_1, \ldots, y_k, z_1, \ldots, z_k\}$.

(ii) *If f is equal to* $\sum_{\alpha=1}^{k} \bar{y}_\alpha \wedge \bar{z}_\alpha$, *where* $\bar{y}_1, \ldots, \bar{y}_k, \bar{z}_1, \ldots, \bar{z}_k$ *are in V, then* $\bar{y}_1, \ldots, \bar{y}_k, \bar{z}_1, \ldots, \bar{z}_k$ *are linearly independent and*

$$Sp\{\bar{y}_1, \ldots, \bar{y}_k, \bar{z}_1, \ldots, \bar{z}_k\} = Sp\{y_1, \ldots, y_k, z_1, \ldots, z_k\},$$

where $Sp\{\cdots\}$ *denotes the subspace spanned by the elements in* $\{\cdots\}$.

Proof. Since $f(x) = \sum_{\alpha=1}^{k}\{(x, z_\alpha)y_\alpha - (x, y_\alpha)z_\alpha\}$, it is clear that range $f \subset Sp\{y_1, \ldots, y_k, z_1, \ldots, z_k\}$. To prove the inclusion in the other direction, take any β and choose $x \in V$ such that

$$(x, y_\beta) = 1$$
$$(x, y_\alpha) = 0 \quad \text{for all } \alpha \neq \beta$$

and

$$(x, z_\alpha) = 0 \quad \text{for all } \alpha.$$

This is possible since $\{y_\alpha, z_\alpha\}$, $1 \leq \alpha \leq k$, are linearly independent. Then we have $f(x) = y_\beta$, showing that $y_\beta \in \text{range } f$. Similarly, $z_\beta \in \text{range } f$ for each β. Thus

$$Sp\{y_1, \ldots, y_k, z_1, \ldots, z_k\} \subset \text{range } f.$$

To prove (ii), we first observe that range $f \subset Sp\{\bar{y}_1, \ldots, \bar{y}_k, \bar{z}_1, \ldots, \bar{z}_k\}$ so that by (i) we have

$$Sp\{y_1, \ldots, y_k, z_1, \ldots, z_k\} \subset Sp\{\bar{y}_1, \ldots, \bar{y}_k, \bar{z}_1, \ldots, \bar{z}_k\}.$$

Since $\{y_\alpha, z_\alpha\}$, $1 \leq \alpha \leq k$, are linearly independent,

$$\dim Sp\{y_1, \ldots, y_k, z_1, \ldots, z_k\} = 2k,$$

which implies that $\dim Sp\{\bar{y}_1, \ldots, \bar{y}_k, \bar{z}_1, \ldots, \bar{z}_k\} = 2k$, proving that $\{\bar{y}_\alpha, \bar{z}_\alpha\}$, $1 \leq \alpha \leq k$, are also linearly independent. The second assertion in (ii) is also clear.

LEMMA 2. *Suppose that $x_1, \ldots, x_k$ are linearly independent. If $y_1, \ldots, y_k \in V$ and if $\sum_{\alpha=1}^{k} x_\alpha \wedge y_\alpha = 0$, then each y_α is linearly dependent on $x_1, \ldots, x_k$.*

Proof. Choose $x_{k+1}, \ldots, x_n$ so that $\{x_1, \ldots, x_k, x_{k+1}, \ldots, x_n\}$ is a basis of V and let

$$y_i = \sum_{j=1}^{k} a_{ij} x_j, \quad 1 \leq i \leq k.$$

We also set

$$a_{ij} = 0 \quad \text{for } i > k \text{ and } 1 \leq j \leq n.$$

Then

$$\sum_{i=1}^{k} x_i \wedge y_i = \sum_{i=1}^{k} x_i \wedge \left(\sum_{j=1}^{n} a_{ij} x_j\right) = \sum_{i=1}^{k} \sum_{j=1}^{n} a_{ij} x_i \wedge x_j$$

$$= \sum_{i,j=1}^{n} a_{ij} x_i \wedge x_j = \sum_{i<j} (a_{ij} - a_{ji}) x_i \wedge x_j.$$

Since $\sum_{i=1}^{k} x_i \wedge y_i = 0$ and since $x_i \wedge x_j$ are linearly independent, we have $a_{ij} - a_{ji} = 0$ for all $i < j$. Thus

$$a_{ij} = a_{ji} = 0 \quad \text{for } 1 \leq i \leq k \text{ and } j > k.$$

This shows that y_i, $1 \leq i \leq k$, are linearly dependent on $x_1, \ldots, x_k$.

Proof of the theorem. Let $x_1, x_2, x_3 \in V$ be such that $A_\alpha(x_i)$,

$1 \leq \alpha \leq k$, $1 \leq i \leq 3$, are linearly independent. From

$$\sum_{\alpha=1}^{k} A_\alpha(x_1) \wedge A_\alpha(x_2) = \sum_{\alpha=1}^{k} \bar{A}_\alpha(x_1) \wedge \bar{A}_\alpha(x_2)$$

we conclude by Lemma 1 that each $\bar{A}_\alpha(x_1)$ is a linear combination of $A_\beta(x_1)$ and $A_\beta(x_2)$, $1 \leq \beta \leq k$. From

$$\sum_{\alpha=1}^{k} A_\alpha(x_1) \wedge A_\alpha(x_3) = \sum_{\alpha=1}^{k} \bar{A}_\alpha(x_1) \wedge \bar{A}_\alpha(x_3)$$

we see that each $\bar{A}_\alpha(x_1)$ is a linear combination of $A_\beta(x_1)$ and $A_\beta(x_3)$, $1 \leq \beta \leq k$. Since $A_\beta(x_1)$, $A_\beta(x_2)$, $A_\beta(x_3)$, $1 \leq \beta \leq k$, are linearly independent, it follows that each $\bar{A}_\alpha(x_1)$ is a linear combination of $A_\beta(x_1)$, $1 \leq \beta \leq k$. Similarly, each $\bar{A}_\alpha(x_2)$ is a linear combination of $A_\beta(x_2)$, $1 \leq \beta \leq k$ and each $\bar{A}_\alpha(x_3)$ is a linear combination of $A_\beta(x_3)$, $1 \leq \beta \leq k$. We set

$$\bar{A}_\alpha(x_1) = \sum_{\beta=1}^{k} s_{\beta\alpha} A_\beta(x_1)$$

$$\bar{A}_\alpha(x_2) = \sum_{\beta=1}^{k} t_{\beta\alpha} A_\beta(x_2)$$

$$\bar{A}_\alpha(x_3) = \sum_{\beta=1}^{k} u_{\beta\alpha} A_\beta(x_3).$$

From

$$\sum_{\alpha=1}^{k} \bar{A}_\alpha(x_i) \wedge \bar{A}_\alpha(x_j) = \sum_{\alpha=1}^{k} A_\alpha(x_i) \wedge A_\alpha(x_j), \quad 1 \leq i, j \leq 3,$$

we get for the matrices $S = (s_{\beta\alpha})$, $T = (t_{\beta\alpha})$, and $U = (u_{\beta\alpha})$ the relations

$$S\,^tT = I = T\,^tS, \qquad T\,^tU = I = U\,^tT, \qquad S\,^tU = I = \,^tUS,$$

where t denotes the transpose and I denotes the identity matrix. They imply $S = T = U$ and $S\,^tS = I$, that is, S is an orthogonal matrix. If we set

$$\bar{\bar{A}}_\alpha = \sum_{\beta=1}^{k} s_{\beta\alpha} A_\beta,$$

we know that

$$\bar{\bar{A}}_\alpha(x_i) = \bar{A}_\alpha(x_i), \quad 1 \leq i \leq 3.$$

We wish to show that $\bar{\bar{A}}_\alpha(x_j) = \bar{A}_\alpha(x_j)$ for $j \geq 4$.

For $j \geq 4$, we have
$$\sum_{\alpha=1}^{k} \bar{A}_\alpha(x_j) \wedge \bar{A}_\alpha(x_i) = \sum_{\alpha=1}^{k} \bar{\bar{A}}_\alpha(x_j) \wedge \bar{\bar{A}}_\alpha(x_i)$$
i.e.,
$$\sum_{\alpha=1}^{k} (\bar{A}_\alpha(x_j) - \bar{\bar{A}}_\alpha(x_j)) \wedge \bar{A}_\alpha(x_i) = 0, \quad 1 \leq i \leq 3.$$

We know that $\bar{A}_\alpha(x_1)$, $1 \leq \alpha \leq k$, are linearly independent. From the above equation for $i = 1$, we conclude by Lemma 2 that $\bar{A}_\alpha(x_j) - \bar{\bar{A}}_\alpha(x_j)$ is a linear combination of $\bar{A}_\beta(x_1)$, $1 \leq \beta \leq k$. Similarly, we see that $\bar{A}_\alpha(x_j) - \bar{\bar{A}}_\alpha(x_j)$ is a linear combination of $\bar{A}_\beta(x_2)$, $1 \leq \beta \leq k$. Since $\bar{A}_\beta(x_1)$, $\bar{A}_\beta(x_2)$, $1 \leq \beta \leq k$, are linearly independent (as are $A_\beta(x_1)$, $A_\beta(x_2)$, $1 \leq \beta \leq k$), it follows that $\bar{A}_\alpha(x_j) - \bar{\bar{A}}_\alpha(x_j) = 0$, $1 \leq \alpha \leq k, j \geq 4$, proving our assertion. This completes the proof of Theorem 1.

Let $f: M \to \mathbf{R}^{n+p}$ be an isometric imbedding of an n-dimensional Riemannian manifold M into a Euclidean space $\mathbf{R}^{n+p}$. We shall deal only with a local theory and use the notations developed in §3 of Chapter VII.

The mapping $\xi \to A_\xi$ of $T_x(M)^\perp$ into the vector space of all symmetric endomorphisms of $T_x(M)$ was defined in Proposition 3.3 of Chapter VII. Let $k(x)$ be the rank of this linear mapping. If we denote by $N_0(x)$ the null space of the linear mapping $\xi \to A_\xi$, then the orthogonal complement of $N_0(x)$ in $T_x(M)^\perp$ is called the *first normal space* at x. Clearly, $k(x)$ is equal to the dimension of the first normal space at x. If we choose a basis $\xi_1, \ldots, \xi_k$, where $k = k(x)$, in the first normal space at x, then the corresponding endomorphisms $A_1, \ldots, A_k$ are linearly independent.

Now the local rigidity theorem by Allendoerfer [1], which is an extension of the result on a hypersurface (Corollary 6.5 of Chapter VII), can be stated as follows.

THEOREM 2. *Let f and $\bar{f}$ be two isometric imbeddings of an n-dimensional Riemannian manifold M into a Euclidean space $\mathbf{R}^{n+p}$. Assume that, for a neighborhood U of a point $x_0 \in M$, we have*

(1) *the dimensions of the first normal spaces at $x \in U$ for both f and $\bar{f}$ are equal to a constant, say, k;*

(2) *the type number of f at each $x \in U$ is at least 3.*

Then there is an isometry τ of $\mathbf{R}^{n+p}$ such that $\tilde{f} = \tau \circ f$ on a neighborhood of x_0.

We note that Theorem 1 can be utilized for the proof of Theorem 2. See also Chern [15].

Note 18. Isometric imbeddings

The theory of surfaces in 3-dimensional Euclidean space differs considerably from the general theory of submanifolds in N-dimensional Euclidean space in its method and results (see Note 15). The problem of isometric imbedding of a 2-dimensional Riemannian manifold into the 3-dimensional Euclidean space, in particular the so-called Weyl problem, will not be discussed here.

The first result in the general isometric imbedding is the theorem of Janet [1] and E. Cartan [14] which states that a real analytic Riemannian manifold M of dimension n can be locally isometrically imbedded into any real analytic Riemannian manifold V of dimension $\frac{1}{2}n(n + 1)$. (See Burstin [1] for comments on Janet's proof.) The generalization to the C^∞ case is open even when V is a Euclidean space. A global isometric imbedding theorem was obtained by Nash [2]:

THEOREM 1. *Every compact n-dimensional Riemannian manifold M of class C^k ($3 \leq k \leq \infty$) can be C^k-isometrically imbedded in any small portion of a Euclidean space $\mathbf{R}^N$, where $N = \frac{1}{2}n(3n + 11)$. Every non-compact n-dimensional Riemannian manifold M of class C^k ($3 \leq k \leq \infty$) can be C^k-isometrically imbedded in any small portion of a Euclidean space $\mathbf{R}^N$, where $N = \frac{1}{2}n(n + 1)(3n + 11)$.*

For a C^1-isometric imbedding, the dimension of the receiving space can be very small. The result of Nash [1] and Kuiper [5] states:

THEOREM 2. *Let M be a compact n-dimensional Riemannian manifold of class C^1 with boundary (which can be empty). If M can be C^1-imbedded in a Euclidean space $\mathbf{R}^N$, $N \geq n + 1$, then it can be C^1-isometrically imbedded in $\mathbf{R}^N$.*

In particular, M can be locally C^1-isometrically imbedded in $\mathbf{R}^{n+1}$. A C^1-isometric imbedding can be quite pathological. An n-dimensional torus can be C^1-isometrically imbedded in $\mathbf{R}^{n+1}$ but it cannot be C^4-isometrically imbedded in $\mathbf{R}^{2n-1}$ (Tompkins [1], cf. Corollary 5.3 of Chapter VII).

Once we know that every Riemannian manifold can be isometrically imbedded in a Euclidean space of sufficiently large dimension, we naturally seek for a Euclidean space of smallest possible dimension in which a Riemannian manifold can be isometrically imbedded. Hilbert [2] proved that a complete surface of constant negative curvature cannot be C^4-isometrically immersed in $\mathbf{R}^3$, as is mentioned in Note 15 as well. Hilbert's result has been generalized by Efimov [1] to a complete surface of bounded negative curvature.

The result of Chern and Kuiper [1] generalizing the theorem of Tompkins mentioned above says that a compact n-dimensional Riemannian manifold with non-positive sectional curvature can not be isometrically immersed in $\mathbf{R}^{2n-1}$ (cf. Corollary 5.2 of Chapter VII). A further generalization by O'Neill [2] states that if N is a complete simply connected Riemannian manifold of dimension $2n - 1$ with sectional curvature $K_N \leq 0$, then a compact n-dimensional Riemannian manifold M with sectional curvature $K_M \leq K_N$ cannot be isometrically immersed in N.

Otsuki [2] proved that an n-dimensional Riemannian manifold M of constant negative curvature cannot be (even locally) isometrically immersed in $\mathbf{R}^{2n-2}$. For more results in this direction, see Otsuki [2], [3].

The holonomy group gives also some restrictions on isometric immersions. If a compact n-dimensional Riemannian manifold M can be isometrically immersed in $\mathbf{R}^{n+1}$, then its restricted linear holonomy group must be $SO(n)$ (see Kobayashi [8]). If a compact n-dimensional Riemannian manifold M can be isometrically immersed in $\mathbf{R}^{n+2}$, then its restricted linear holonomy group must be $SO(k) \times SO(n-k)$ except when $n = 4$ (see R. L. Bishop [1]). If a compact n-dimensional Riemannian manifold M with restricted holonomy group $SO(k) \times SO(n-k)$ is isometrically imbedded in $\mathbf{R}^{n+2}$, then the imbedding is a product of two imbeddings of hypersurfaces, with an exception occurring in the case $k = 1$ or $n - k = 1$ (see S. B. Alexander [1]).

For a non-compact n-dimensional Riemannian manifold M isometrically immersed in $\mathbf{R}^{n+1}$, there are certain restrictions on the holonomy group (see Dolbeault-Lemoine [1]). According to R. L. Bishop [1], when a non-compact n-dimensional Riemannian manifold M is isometrically immersed $\mathbf{R}^{n+2}$, its restricted linear holonomy group is either of the form $SO(k) \times SO(n-k)$ or $U(m)$, where $n = 2m$ (unless M is flat).

The isometric imbedding problem is easier if the receiving space is allowed to be infinite-dimensional. Bieberbach [4] proved that a simply connected complete surface H_2 of constant negative curvature can be isometrically imbedded in a Hilbert space. This was generalized by Blanuša [1] to an isometric imbedding of a simply connected complete n-dimensional space H_n of constant negative curvature into a Hilbert space. Later Blanuša [2] improved his own result by showing that H_n can be isometrically imbedded into $\mathbf{R}^{6n-5}$ and H_2 into $\mathbf{R}^6$. A holomorphic isometric imbedding of a Kaehler manifold into a complex Euclidean space is not always possible. A compact complex manifold of positive dimension can never be immersed in a complex Euclidean space. According to Propositions 8.3 and 8.4 of Chapter IX, a Kaehler manifold which can be holomorphically and isometrically immersed into a complex Euclidean space must have non-positive holomorphic sectional curvature and negative semi-definite Ricci tensor. Bochner [2] proved that a certain class of Kaehler manifolds (including at least all classical bounded symmetric domains with Bergman metric) cannot be holomorphically and isometrically immersed into a finite-dimensional complex Euclidean space. A systematic study of imbedding a Kaehler manifold into a complex Hilbert space and other infinite-dimensional spaces was done by Calabi [1]. It is interesting to note (cf. Bochner [2] and Kobayashi [14]) that the Bergman metric of a complex manifold M is so defined that M is holomorphically and isometrically imbedded in a complex projective space (generally of infinite dimensions) in a natural manner.

Consider now the following generalization of the isometric imbedding problem. Let M be a Riemannian manifold and G a group of isometries of M. The problem is to find not only an isometric imbedding of M into a Euclidean space $\mathbf{R}^N$ but also a compatible imbedding of G into the group $\Im(\mathbf{R}^N)$ of Euclidean motions of $\mathbf{R}^N$. Such an imbedding (rather a pair of imbeddings) is called an *equivariant isometric imbedding* of (M, G). If G is transitive on M so that $M = G/H$ and if the linear isotropy representation of H is irreducible, then a differentiable imbedding of M into $\mathbf{R}^N$

with a compatible imbedding of G into $\Im(\mathbf{R}^N)$ is isometric since there is, up to a constant factor, only one Riemannian metric on G/H invariant by G. It is known (Lichnerowicz [3]; pp. 158–167) that a Hermitian symmetric space G/H of compact type can be equivariantly and isometrically imbedded into $\mathbf{R}^N$, where $N = \dim G$. This has been extended to almost all symmetric spaces of compact type (Nagano [10] and Kobayashi [25]).

Finally we mention a paper by Friedman [1] in which a generalization of the Janet-Cartan theorem to an indefinite Riemannian metric is obtained.

Note 19. Equivalence problems for Riemannian manifolds

A transformation f between two Riemannian manifolds M and M' is said to be *strongly curvature-preserving* if f maps $\nabla^m R$ upon $\nabla'^m R'$ for every $m = 0, 1, 2, \ldots$, where $\nabla^m R$ and $\nabla'^m R'$ denote the m-th covariant differentials of the curvature tensor fields R of M and R' of M', respectively. Nomizu and Yano [3], [4] proved

THEOREM 1. *A strongly curvature-preserving diffeomorphism between irreducible and analytic Riemannian manifolds M and M' of dimension ≥ 2 is a homothetic transformation.*

This combined with Lemma 2 on p. 242 of Volume I yields

COROLLARY. *A strongly curvature-preserving diffeomorphism of a complete irreducible analytic Riemannian manifold M of dimension ≥ 2 onto itself is an isometry.*

The corresponding result for infinitesimal transformations was proved in Nomizu and Yano [2]. See Tanno [1] for generalizations to the case of an indefinite metric.

We say that a Riemannian manifold M is *strongly curvature-homogeneous* if, for any points x and y of M, there is a linear isomorphism of $T_x(M)$ onto $T_y(M)$ which maps g_x (the metric at x) and $(\nabla^m R)_x$, $m = 0, 1, 2, \ldots$ upon g_y and $(\nabla^m R)_y$, $m = 0, 1, 2, \ldots$ Then

THEOREM 2. *If a connected Riemannian manifold M is strongly curvature-homogeneous, then it is locally homogeneous. If, moreover, M is complete and simply connected, then it is homogeneous.*

Theorem 2 was originally proved by Singer [2] when M is complete and simply connected. The local version as stated in Theorem 2 was later obtained by Nomizu [9].

Kulkarni [1] studied the following question related to Theorem 1 above and Schur's theorem (Theorem 2.2 of Chapter V): When there exists a diffeomorphism which preserves sectional curvature between two Riemannian manifolds M and M', are M and M' isometric?

Note 20. Gauss-Bonnet theorem

1. The Gauss-Bonnet theorem for a compact orientable 2-dimensional Riemannian manifold M states that

$$\frac{1}{2\pi} \int_M K \, dA = \chi(M),$$

where K is the Gaussian curvature of the surface M, dA denotes the area element of M, and $\chi(M)$ is the Euler number of M. This is usually derived from the Gauss-Bonnet formula for a piece of surface. Let D be a simply connected region on M bounded by a piecewise differentiable curve C consisting of m differentiable curves. Then the Gauss-Bonnet formula for D states

$$\int_C k_g \, ds + \sum_{i=1}^m (\pi - \alpha_i) + \int_D K \, dA = 2\pi,$$

where k_g is the geodesic curvature of C and $\alpha_1, \ldots, \alpha_m$ denote the inner angles at the points where C is not differentiable. Triangulating M and applying the Gauss-Bonnet formula to each triangle we obtain the Gauss-Bonnet theorem for M.

2. The first step toward a generalization of the Gauss-Bonnet theorem to higher dimensional manifolds was taken by H. Hopf [2]. The following lemma of Hopf is basic.

LEMMA. *Let N be an n-dimensional compact manifold in $\mathbf{R}^n$ with boundary ∂N. Then the degree of the spherical map of Gauss $\partial N \to S^{n-1}$ is equal to the Euler number $\chi(N)$ of N.*

For the proof, see, for instance, Milnor [5].

Now consider an n-dimensional compact submanifold M (without boundary) in $\mathbf{R}^{n+k}$. Let N_ε denote the closed ε-neighborhood of M. For ε sufficiently small N_ε is a differentiable manifold with boundary. Since $\chi(N_\varepsilon) = \chi(M)$, the lemma above implies that the degree of the spherical map of Gauss $\partial N_\varepsilon \to S^{n+k-1}$ is equal to the Euler number $\chi(M)$ of M.

Consider the special case where $k = 1$ and M is orientable. Then ∂N_ε consists of two components. Let d be the degree of the spherical map $M \to S^n$. Then the degree of the spherical map $\partial N_\varepsilon \to S^n$ is given by $d + (-1)^n d$. Hence

$$d + (-1)^n d = \chi(M).$$

We shall now express the degree d by integrals. Let $y^0, y^1, \ldots, y^n$ be the natural coordinate system in $\mathbf{R}^{n+1}$. The volume element of the unit sphere S^n is given by

$$\sum_{i=0}^{n} (-1)^i y^i \, dy^0 \wedge \cdots \wedge \widehat{dy^i} \wedge \cdots \wedge dy^n.$$

Let ξ be the spherical map $M \to S^n$; it can be given by $n + 1$ functions $\xi^0, \xi^1, \ldots, \xi^n$ on M with $(\xi^0)^2 + (\xi^1)^2 + \cdots + (\xi^n)^2 = 1$. Then

$$d = \int_M \sum_{i=0}^{n} \xi^i \, d\xi^0 \wedge \cdots \wedge \widehat{d\xi^i} \wedge \cdots \wedge d\xi^n.$$

The formula of Weingarten states (cf. §3 of Chapter VII)

$$\xi_*(X) = -A(X) \qquad \text{for every vector field } X \text{ on } M,$$

where $A = A_\xi$ is the symmetric transformation of each tangent space $T_x(M)$ defined by the second fundamental form. By a simple calculation we obtain

$$d = \frac{1}{\omega_n} \int_M K_n \, dv \qquad (\omega_n = \text{volume of the unit } n\text{-sphere}),$$

where K_n is the Gaussian curvature of M and dv denotes the volume element of M.

For n even, we can replace d by $\tfrac{1}{2}\chi(M)$ and express K_n as a polynomial of the Riemannian curvature. The resulting formula may be considered a generalization of the Gauss-Bonnet formula. We note that $2K_n \, dv$ is equal to the n-form γ defined in §5 of Chapter XII for the vector bundle $T(M)$ and the Riemannian connection of M.

360   FOUNDATIONS OF DIFFERENTIAL GEOMETRY

3. A further generalization was obtained by Allendoerfer [1] and Fenchel [1] independently. Let M be an n-dimensional compact orientable submanifold of $\mathbf{R}^{n+k}$ and N_ε the closed ε-neighborhood of M for a sufficiently small ε. Let d be the degree of the spherical map $\partial N_\varepsilon \to S^{n+k-1}$ and K_{n+k-1} the Gaussian curvature of the hypersurface ∂N_ε. Then by the result above, we have

$$\tfrac{1}{2}\chi(\partial N_\varepsilon) = d = \frac{1}{\omega_{n+k-1}} \int_{\partial N_\varepsilon} K_{n+1-k}\, dv,$$

where dv denotes the volume element of ∂N_ε. Assume that n and $k-1$ are both even. Since ∂N_ε is a sphere bundle over M, we have

$$\chi(\partial N_\varepsilon) = \chi(M) \cdot \chi(S^{k-1}) = 2\chi(M).$$

On the other hand, integrating K_{n+k-1} along the fibres we obtain (by a non-trivial calculation)

$$\frac{1}{\omega_{n+k-1}} \int_{\partial N_\varepsilon} K_{n+k-1}\, dv = \int_M \gamma,$$

where γ is the n-form defined in §5 of Chapter XII. Hence,

$$\chi(M) = \int_M \gamma.$$

The assumption that $k-1$ is even is not restrictive. If $k-1$ is odd, it suffices to imbed M into $\mathbf{R}^{n+k+1}$ in a natural manner.

4. As we have mentioned in Note 18, every Riemannian manifold can be isometrically imbedded in a Euclidean space of sufficiently high dimension. But this imbedding theorem was not established at the time when Allendoerfer and Fenchel obtained the Gauss-Bonnet theorem for submanifolds of Euclidean spaces. But the local imbedding theorem of Janet and Cartan (cf. Note 18) was then available. In 1943, Allendoerfer and Weil [1] obtained the Gauss-Bonnet theorem for arbitrary Riemannian manifolds by proving a generalized Gauss-Bonnet formula for a piece of a Riemannian manifold isometrically imbedded in a Euclidean space.

5. The so-called intrinsic proof was obtained by Chern [4] in 1944. Let M be a $2m$-dimensional orientable compact Riemannian manifold and $S(M)$ the tangent sphere bundle over M, i.e., the

bundle of unit tangent vectors of M. Let $p: S(M) \to M$ be the projection. Chern constructs a $(2m-1)$-form π on $S(M)$ such that $d\pi = p^*(\gamma)$ and that the integral of π along each fibre of $S(M)$ is 1. Let X be a unit vector field on M with isolated singularities at $x_1, \ldots, x_k$. Let $a_1, \ldots, a_k$ be the index of X at $x_1, \ldots, x_n$. By a theorem of Hopf, $a_1 + \cdots + a_k$ is equal to the Euler number $\chi(M)$ (see for instance, Milnor [5]). The cross section X of $S(M)$ may be considered as a submanifold of $S(M)$ with boundary, and its boundary ∂X is given by $a_1 S_1 + \cdots + a_k S_k$ where $S_1, \ldots, S_k$ are the fibres of $S(M)$ at $x_1, \ldots, x_k$. Hence

$$\int_M \gamma = \int_X p^*(\gamma) = \int_X d\pi = \int_{\partial X} \pi = \sum_{i=1}^k a_i \int_{S_i} \pi = \sum_{i=1}^k a_i$$
$$= \chi(M).$$

We see in Chern's proof the birth of the concept of transgression. Chern [23] extended the Gauss-Bonnet theorem to the case of indefinite Riemannian metric. Our axiomatic proof may be easily generalized to include also the case of indefinite Riemannian metric. See also Eells [1], Bishop-Goldberg [2], Avez [1] on the Gauss-Bonnet theorem. On some estimates of the Euler number with the aid of the Gauss-Bonnet theorem, see Chern [14], Berger [14], Berger-Bott [1].

6. Cohn-Vossen [3] investigated global properties of non-compact Riemannian manifolds of dimension 2. Among other things he proved that for a complete orientable Riemannian manifold M of dimension 2 the inequality

$$\int_M \gamma \leq \chi(M)$$

holds provided that the integral exists as an improper integral. No generalization to higher dimensions is known.

7. We also mention a result in Milnor [6] that a compact orientable surface of genus $g \geq 2$ does not admit any affine connection with zero curvature. See also Benzécri [2].

Note 21. Total curvature

As in Note 20, let M be an n-dimensional manifold immersed in a Euclidean space $\mathbf{R}^N$ and B the bundle of unit normal spheres over M. The degree of the canonical map $\nu\colon B \to S^{N-1}$ was defined by

$$\frac{1}{\omega_{N-1}} \int_B G \, dv_B,$$

where ω_{N-1} is the volume of the unit $(N-1)$-sphere S^{N-1} and $G(\xi)$ is the determinant of the symmetric transformation A_ξ defined by the second fundamental form of M. The *total curvature* of M immersed in $\mathbf{R}^N$ by an immersion map $f\colon M \to \mathbf{R}^N$ is defined by

$$\tau(M, f, \mathbf{R}^N) = \frac{1}{\omega_{N-1}} \int_B |G| \, dv_B.$$

We shall relate the total curvature of M with the topology of M.

Given a function φ on M, a point x on M is a *critical point* of φ if $(d\varphi)_x = 0$. A critical point x of φ is said to be *non-degenerate* if the Hessian form of φ at x is non-degenerate. In this case the dimension of a maximal-dimensional subspace of $T_x(M)$ on which the Hessian form is negative-definite is called the *index* of the critical point x. We introduce the following notations.

$\Phi(M) = $ the set of functions φ on M with only non-degenerate critical points;

$\beta_k(\varphi) = $ the number of the critical points of index k of φ;

$\beta(\varphi) = \Sigma_{k=0}^n \beta_k(\varphi) = $ the number of the critical points of φ;

$\beta_k(M) = \min \{\beta_k(\varphi); \varphi \in \Phi(M)\}$;

$\beta(M) = \min \{\beta(\varphi); \varphi \in \Phi(M)\}$.

For an arbitrary field K, set

$$b_k(M, K) = \dim_K H_k(M, K);$$

$$b(M, K) = \Sigma_{k=0}^n b_k(M, K);$$

$$b(M) = \max \{b(M, K); K \text{ field}\}.$$

Then the basic relations between the various numbers introduced above may be stated as follows:

THEOREM 1. *For a compact n-dimensional manifold M, we have*

$$\inf_f \tau(M, f, \mathbf{R}^N) = \beta(M) \geq \sum_{k=0}^{n} \beta_k(M) \geq b(M),$$

where the infimum is taken for all immersions f and for variable N.

Chern and Lashof [1], [2] proved the inequality $\tau(M, f, \mathbf{R}^N) \geq \beta(M)$. The equality $\inf_f \tau(M, f, \mathbf{R}^N) = \beta(M)$ is due to Kuiper [7]. The inequality $\beta(M) \geq \Sigma \beta_k(M)$ is evident and the inequality $\Sigma \beta_k(M) \geq b(M)$ is a Morse inequality (see, for instance, Milnor [3]). An immersion f is said to be *minimal* if $\tau(M, f, \mathbf{R}^N) = \beta(M)$. This is not to be confused with the term used on p. 34.

We list other known facts on total curvature.

THEOREM 2. *Let $f: M \to \mathbf{R}^N$ and $f': M' \to \mathbf{R}^{N'}$ be immersions. Then*

$$\tau(M \times M, f \times f', \mathbf{R}^N \times \mathbf{R}^{N'}) = \tau(M, f, \mathbf{R}^N)\tau(M', f', \mathbf{R}^{N'}).$$

In particular, letting M' be a point, we obtain

COROLLARY. *Let $f: M \to \mathbf{R}^N$ be an immersion and $i: \mathbf{R}^N \to \mathbf{R}^{N+k}$ an imbedding of $\mathbf{R}^N$ as a linear subspace of $\mathbf{R}^{N+k}$. Then*

$$\tau(M, i \circ f, \mathbf{R}^{N+k}) = \tau(M, f, \mathbf{R}^N).$$

The corollary above is due to Chern-Lashof [2] and its generalization (Theorem 2) is due to Kuiper [7]. The following result is also due to Chern-Lashof [2].

THEOREM 3. *Let M be an n-dimensional compact manifold and $f: M \to \mathbf{R}^N$ an immersion. Then $\tau(M, f, \mathbf{R}^N) = 2$ if and only if the immersion f is an imbedding and $f(M)$ is a convex hypersurface in a linear subspace $\mathbf{R}^{n+1}$ of $\mathbf{R}^N$.*

As pointed out by Kuiper [7], Theorem 3 implies that if a manifold M homeomorphic to a sphere can be minimally immersed into $\mathbf{R}^N$, then M is diffeomorphic to an ordinary sphere. (For a manifold M homeomorphic to a sphere admits a function with only two critical points, that is, M satisfies $\beta(M) = 2$.)

We know of no general method to characterize those compact manifolds which can be minimally immersed into a Euclidean space. Here are some manifolds which admit minimal immersions.

THEOREM 4. (Kuiper [9]). *Every orientable closed surface and also every non-orientable closed surface with Euler number ≤ -2 can be minimally immersed in $\mathbf{R}^3$. The real projective plane and the Klein bottle cannot be minimally immersed in $\mathbf{R}^3$.*

In an earlier paper, Kuiper [7] exhibited a minimal immersion of the real projective plane in $\mathbf{R}^4$.

THEOREM 5. (Kobayashi [23]). *Every compact homogeneous Kaehler manifold can be minimally imbedded into a Euclidean space of sufficiently high dimension.*

For further properties of the minimal imbedding in Theorem 5 such as the dimension of the receiving Euclidean space and the equivariance, see Kobayashi [23]. Tai [1] constructed minimal imbeddings of all projective spaces and the Cayley planes. Takeuchi and Kobayashi [1] obtained minimal imbeddings of all R-spaces, generalizing Theorem 5 and the result of Tai.

For results in the theory of knots in connection with total curvature, see Ferus [1].

An article by Kuiper [11] lists a number of unsolved problems on total curvature. See also Kuiper [10], Otsuki [7], Wilson [1].

Note 22. Topology of Riemannian manifolds with positive curvature

Let δ be a positive number with $0 < \delta \leq 1$. An n-dimensional Riemannian manifold M is said to be δ-*pinched* if its sectional curvature K satisfies

$$A\delta \leq K \leq A$$

for some positive number A. The constant A in the inequalities above is not essential. In general, the sectional curvature K of a Riemannian metric g and the sectional curvature $\bar{K}$ of the Riemannian metric $\bar{g} = cg$ (where c is a positive constant) are related by $\bar{K} = K/c$. We can therefore "normalize" the metric so that if M is δ-pinched then $\delta \leq K \leq 1$.

In Volume I, we proved the following theorem of Hopf [1] (cf. Theorem 3.1 of Chapter V and Theorem 7.10 of Chapter VI).

THEOREM 1. *A complete, simply connected 1-pinched Riemannian manifold M is isometric to an ordinary sphere.*

Recently, J. Wolf [7] obtained a complete classification of the complete 1-pinched Riemannian manifolds.

We know very little about δ-pinched Riemannian manifolds in general. The following list seems to exhaust all that we know at present.

THEOREM 2. *A complete δ-pinched Riemannian manifold M with $\delta > 0$ is compact and has a finite fundamental group.*

Theorem 2 is due to Myers [1] and is proved in Theorem 5.8 of Chapter VIII. Actually, the theorem of Myers is stronger than the one stated above since it assumes only that the eigenvalues of the Ricci tensor are bounded below by a positive constant. The following result is due to Synge [2]:

THEOREM 3. *A complete δ-pinched Riemannian manifold M (with $\delta > 0$) of even dimension is either (1) simply connected or (2) non-orientable with $\pi_1(M) = Z_2$.*

Theorem 1 suggests that if a Riemannian manifold M is complete and δ-pinched with δ sufficiently close to 1, then M is similar to a sphere in one sense or another. The first result in this direction was obtained by Rauch [1] who proved that a complete simply connected Riemannian manifold which is 0.75-pinched is homeomorphic to a sphere. For an even-dimensional manifold, this pinching number was improved to $\delta = 0.54\ldots$ by Klingenberg [2] who made the first systematic study of cut loci since Myers and J. H. C. Whitehead. Improving Klingenberg's method, Berger [6, 7] finally obtained

THEOREM 4. *Let M be a complete simply connected Riemannian manifold of even dimensions which is δ-pinched. If $\delta > \frac{1}{4}$, then M is homeomorphic to a sphere. If $\delta = \frac{1}{4}$, then M is either homeomorphic to a sphere or isometric to a compact symmetric space of rank 1.*

The first part of Theorem 4 was obtained independently also by Toponogov [3]. Refining his results on cut loci and using Berger's proof of Theorem 4, Klingenberg [5, 6] obtained

THEOREM 5. *A complete simply connected Riemannian manifold of odd dimensions which is $\frac{1}{4}$-pinched is homeomorphic to a sphere.*

The first parts of Theorem 4 and Theorem 5 are known as "*Sphere Theorem*."

Since all complex and quaternionic projective spaces and the Cayley plane are known to be $\frac{1}{4}$-pinched, Theorem 4 is the best possible result. On the other hand, in the odd-dimensional case it is not known whether Theorem 5 can be improved. The proof of the Sphere Theorem has been partly simplified since then by Tsukamoto [3].

Klingenberg [8] gave a new approach to the Sphere Theorem. Let M be a complete simply connected Riemannian manifold. He proved that if the conjugate locus of one particular point of M is similar to that of a compact simply connected symmetric space M', then $H^*(M; \mathbf{Z})$ is isomorphic to $H^*(M'; \mathbf{Z})$. In particular, he showed that if the sectional curvature along the geodesics emanating from one particular point of M is δ-pinched with $\delta > \frac{1}{4}$, then M has the homotopy type of a sphere and hence is homeomorphic to a sphere when $\dim M \geq 5$ (by Smale's solution of the generalized Poincaré conjecture, Smale [1]).

In the Sphere Theorem, it is not known if M is diffeomorphic with an ordinary sphere. But Gromoll [1] proved the following

THEOREM 6. *There exists a sequence of numbers* $\frac{1}{4} = \delta_1 < \delta_2 < \delta_3 < \cdots$, $\lim_{\lambda \to \infty} \delta_\lambda = 1$, *such that if M is a complete simply connected n-dimensional Riemannian manifold which is δ_{n-2}-pinched, then M is diffeomorphic to an ordinary sphere.*

In the hope of the eventual classification of δ-pinched Riemannian manifolds, it is important to obtain as many properties as possible of a δ-pinched Riemannian manifold. An interesting conjecture by Chern is that a complete δ-pinched Riemannian manifold M of even dimensions has positive Euler number $\chi(M)$. For $\dim M = 2$, the conjecture is trivially true. For $\dim M = 4$, we have $\chi(M) = 2 + \dim H^2(M; \mathbf{R}) > 0$. In fact, if M is compact with positive Ricci tensor, then $H^1(M; \mathbf{R}) = 0$ by Theorem 5.8 of Chapter VIII and $H^3(M; \mathbf{R}) = 0$ by the Poincaré duality. For $\dim M = 4$, $\chi(M) > 0$ may be also verified by the Gauss-Bonnet formula (cf. Chern [14]). Also using the Gauss-Bonnet formula Berger [14] proved that if M is a complete δ-pinched Riemannian manifold of dimension $2m$, then $|\chi(M)| \leq 2^{-m}(2m)! \cdot \delta^{-m}$. Using harmonic forms, Berger [6] proved that, for a complete Riemannian manifold of odd dimension $2m + 1$ which is $2(m - 1)/(8m - 1)$-pinched, its second Betti number vanishes.

There are very few examples of Riemannian manifolds with positive sectional curvature. The only compact simply connected manifolds which are known to carry Riemannian metrics of positive sectional curvature are the ordinary spheres, the complex projective spaces, the quaternionic projective spaces, the Cayley plane and the two homogeneous spaces (of dimension 7 and 13) discovered by Berger [8] (cf. also Eliasson [1]). In particular, one does not know if there is any compact product manifold $M = M' \times M''$ which can carry a Riemannian metric of positive sectional curvature. This question is open even for the product of two 2-spheres $S^2 \times S^2$.

Closely connected with the concept of manifold of positive sectional curvature is that of manifold of positive curvature operator. At each point x of M the Riemannian curvature tensor R defines a linear transformation of $\wedge^2 T_x(M)$ into $\wedge^2 T_x^*(M)$ which sends $X \wedge Y$ into $R(\cdot, \cdot, X, Y)$. By the duality between $\wedge^2 T_x(M)$ and $\wedge^2 T_x^*(M)$ defined by the Riemannian metric, this linear transformation can be identified with a linear endomorphism of the space of 2-forms $\wedge^2 T_x^*(M)$, called the *curvature operator* at x. We shall denote by ρ_x the curvature operator at x. Since $R(X, Y, U, V) = R(U, V, X, Y)$, ρ_x is symmetric (i.e., self-adjoint) with respect to the inner product defined on $\wedge^2 T_x^*(M)$ by the Riemannian metric. Hence the eigenvalues of ρ_x are all real. Let λ and Λ be two real numbers. By the notation $\lambda \leq \rho_x \leq \Lambda$ we shall mean that the eigenvalues of ρ_x are contained in the closed interval $[\lambda, \Lambda]$. Similarly, by $\lambda \leq \rho(M) \leq \Lambda$ we shall mean that the eigenvalues of ρ_x for all $x \in M$ are contained in $[\lambda, \Lambda]$. If $\{X_i\}$ is an orthonormal basis for $T_x(M)$ and $\{X_i^*\}$ is the dual basis for $T_x^*(M)$, then $g(\rho_x(X_1^* \wedge X_2^*), X_1^* \wedge X_2^*) = R(X_1, X_2, X_1, X_2)$. Hence, *if* $0 < \lambda \leq \rho(M) \leq \Lambda$, *then* M *is* (λ/Λ)-*pinched* (cf. Berger [11]). The following result is due to Berger [11].

THEOREM 7. *Let M be a compact Riemannian manifold with positive curature operator, i.e., $0 < \lambda \leq \rho(M) \leq \Lambda$. Then $H^2(M; \mathbf{R}) = 0$. If the curvature operator is non-negative, i.e., $0 \leq \rho(M) \leq \Lambda$, then the harmonic 2-forms of M are parallel tensor fields.*

Using the Gauss-Bonnet formula, K. Johnson proved the following

THEOREM 8. *If M is a compact Riemannian manifold of even dimension with positive curvature operator, then its Euler number $\chi(M)$ is positive.*

The following result is due to Bochner and Yano (Bochner-Yano [1] and Yano-Bochner [1; p. 83]).

THEOREM 9. *If M is a compact n-dimensional Riemannian manifold with*

$$0 < \tfrac{1}{2}\Lambda \leq \rho(M) \leq \Lambda,$$

then all the Betti numbers $b_1, \ldots, b_{n-1}$ of M vanish.

An elliptic paraboloid is a complete simply connected 2-dimensional Riemannian manifold with positive sectional curvature. Since it is non-compact, the sectional curvature is not bounded by a positive constant from below. Recently, Gromoll and Meyer [1] have shown the following

THEOREM 10. *A complete non-compact Riemannian manifold of dimension ≥ 5 with positive curvature is diffeomorphic with $\mathbf{R}^n$.*

Note 23. Topology of Kaehler manifolds with positive curvature

Let M be a Kaehler manifold of complex dimension n. In addition to the sectional curvature and the holomorphic sectional curvature we introduced in §7 of Chapter IX, we define the Kaehlerian sectional curvature of M as follows. Let p be a plane in $T_x(M)$, i.e., a real 2-dimensional subspace of $T_x(M)$. Let X and Y be an orthonormal basis. As in §7 of Chapter IX we define the angle $\alpha(p)$ between p and $J(p)$ by $\cos \alpha(p) = |g(X, JY)|$. We know (cf. Proposition 7.4 of Chapter IX) that the sectional curvature of a space of constant holomorphic sectional curvature 1 is given by $\tfrac{1}{4}(1 + 3 \cos^2 \alpha(p))$. Denote by $K(p)$ the sectional curvature of M. Then it is quite natural to set

$$K^*(p) = 4K(p)/(1 + 3 \cos^2 \alpha(p)).$$

We shall call $K^*(p)$ the *Kaehlerian sectional curvature* of the plane section p.

For a Kaehler manifold M we consider three kinds of pinchings. We say that M is δ-*pinched* if it is so as a Riemannian manifold (cf. Note 22). We say that M is δ-*Kaehler pinched* if there is a constant A such that

$$\delta A \leq K^*(p) \leq A \quad \text{for all planes } p.$$

We say that M is δ-*holomorphically pinched* if there is a constant A such that

$$\delta A \leq K^*(p) \leq A \quad \text{for all planes } p \text{ invariant by } J.$$

Now let X and Y be arbitrary tangent vectors of M at x. If we define $Q(X)$ by

$$Q(X) = R(X, JX, X, JX),$$

then by polarization we obtain (cf. Bishop and Goldberg [2])

$$R(X, Y, X, Y) = \tfrac{1}{32}[3Q(X + JY) + 3Q(X - JY)$$
$$- Q(X + Y) - Q(X - Y)$$
$$- 4Q(X) - 4Q(Y)].$$

From this identity we can deduce certain relations between the three pinchings introduced above. In particular (cf. Berger [5], Bishop and Goldberg [2]), if M is δ-holomorphically pinched, then $\tfrac{1}{4}(3\delta - 2)A \leq K(p) \leq A$ for all planes p.

We shall now list implications of various pinching assumptions.

THEOREM 1. *A complete Kaehler manifold M of complex dimension n is holomorphically isometric with the complex projective space $P_n(\mathbf{C})$ with a canonical metric if any one of the following three conditions is satisfied:*

(1) M is $\tfrac{1}{4}$-*pinched;*
(2) M is 1-*holomorphically pinched;*
(3) M is 1-*Kaehler pinched.*

As stated at the end of §7 of Chapter IX, case (2) follows from Theorems 7.8 and 7.9 (cf. Igusa [1]). Condition (3) clearly implies condition (2). By a simple algebraic manipulation, we obtain condition (2) from condition (1) (cf. Berger [5]).

THEOREM 2. *A complete δ-holomorphically pinched Kaehler manifold with* $\delta > 0$ *is compact and simply connected.*

The proof of Theorem 2 is essentially the same as that for the theorem of Myers (Theorem 5.8 of Chapter VIII) as pointed out by Tsukamoto [1]. (In the proof of Theorem 3.3 of Chapter VIII, we set $Y = J(\dot{\tau})$ and then argue in the same way as in the proof of Theorem 5.8 of Chapter VIII.)

The theorem of Myers can be sharpened in the Kaehlerian case as follows (cf. Kobayashi [17]):

THEOREM 3. *A compact Kaehler manifold M with positive-definite Ricci tensor is simply connected.*

The proof makes use of Myers' theorem, the Riemann-Roch Theorem of Hirzebruch [3], and the following result of Bochner [1] (cf. Lichnerowicz [3]):

THEOREM 4. *A compact Kaehler manifold M with positive-definite Ricci tensor admits no non-zero holomorphic p-forms for $p \geq 1$. In other words,* $H^{p,0}(M; \mathbf{C}) = 0$ *for* $p \geq 1$.

An analogous theorem for a compact Kaehler manifold with positive holomorphic sectional curvature is not known.

The following result is due to Bishop and Goldberg [3].

THEOREM 5. *Let M be a complete Kaehler manifold which is either δ-pinched with $\delta > 0$ or δ-holomorphically pinched with $\delta > \frac{1}{2}$. Then*

$$\dim H^2(M; \mathbf{R}) = 1.$$

Since the direct product of two copies of complex projective space is $\frac{1}{2}$-holomorphically pinched, $\frac{1}{2}$ in Theorem 5 cannot be lowered. It would be of some interest to note that a $\frac{1}{2}$-holomorphically pinched Kaehler manifold has positive-definite Ricci tensor (Berger [5]). Theorem 5 implies that if M is a product of two compact complex manifolds, then M cannot admit a Kaehler metric of positive sectional curvature or of holomorphic pinching $> \frac{1}{2}$.

A Kaehlerian analogue of the Sphere Theorem would be that if M is a compact Kaehler manifold of dimension n with positive sectional curvature, then M is holomorphically homeomorphic to $P_n(\mathbf{C})$; it is, however, yet to be proved. Rauch [3] obtained some preparatory theorems in this direction. Do Carmo [1] obtained a Kaehlerian analogue of Rauch's comparison theorem and used it to prove that a compact n-dimensional Kaehler manifold M with Kaehlerian pinching $\delta > 0.8$ has the same $\mathbf{Z}_2$ cohomology ring as the complex projective space $P_n(\mathbf{C})$. This Kaehlerian pinching number has been improved by different methods. Klingenberg [8] proved that if M has Kaehlerian pinching $\delta > 0.64$, then M has the same homotopy type as $P_n(\mathbf{C})$. Kobayashi [20] gave the Kaehlerian pinching number $\frac{4}{7} = 0.571\ldots$. To date, the following result of Klingenberg [8] gives the best estimate for Kaehlerian pinching.

THEOREM 6. *A compact n-dimensional Kaehler manifold M with Kaehlerian pinching $> \frac{9}{16} = 0.562\ldots$ has the same homotopy type as the complex projective space $P_n(\mathbf{C})$.*

While Klingenberg makes use of Morse theory, Kobayashi reduces the problem to the Sphere Theorem by establishing the following theorem.

THEOREM 7. *If M is a compact Kaehler manifold with Kaehlerian pinching $> \delta$, then there exists a principal circle bundle P over M with a Riemannian metric whose (Riemannian) pinching is greater than $\delta/(4-3\delta)$.*

Thanks to Theorem 5, the proof of Theorem 7 (Kobayashi [20]) can be simplified considerably.

The proof of Theorem 7 gives also the following result.

THEOREM 8. *If M is a compact Kaehler manifold with holomorphic pinching $> \delta$, then there exists a principal circle bundle P over M with a Riemannian metric whose (Riemannian) pinching is greater than $(3\delta - 2)/(4 - 3\delta)$.*

Theorem 8 is an improvement by Bishop and Goldberg [1] of a similar result obtained by Kobayashi [20]. From Theorem 8 we obtain

THEOREM 9. *A compact n-dimensional Kaehler manifold M with holomorphic pinching $> \frac{4}{5}$ has the same homotopy type as the complex projective space $P_n(\mathbf{C})$.*

Berger [19] proved that an n-dimensional compact Kaehler-Einstein manifold M with positive sectional curvature is isometric to $P_n(\mathbf{C})$ with a canonical metric. For a compact Kaehler manifold with constant scalar curvature, the so-called Ricci form is harmonic. From Theorem 5, it follows (Bishop and Goldberg [4]) that a compact Kaehler manifold with constant scalar curvature and with positive sectional curvature is an Einstein manifold. Hence,

THEOREM 10. *An n-dimensional compact Kaehler manifold with constant scalar curvature and with positive sectional curvature is holomorphically isometric to $P_n(\mathbf{C})$ with a canonical metric.*

Analyzing the proof of Berger [19], Bishop and Goldberg [4] obtained also the following result.

THEOREM 11. *An n-dimensional compact Kaehler manifold with constant scalar curvature and with holomorphic pinching $>\frac{1}{2}$ is holomorphically isometric to $P_n(\mathbf{C})$ with a canonical metric.*

For a Kaehler manifold of dimension 2 we have the best possible result due to Andreotti and Frankel (see Frankel [3]):

THEOREM 12. *A 2-dimensional compact Kaehler manifold with positive sectional curvature is holomorphically homeomorphic to the complex projective space $P_2(\mathbf{C})$.*

The proof relies on the known classification of algebraic surfaces.

Goldberg and Kobayashi [3] introduced the notion of holomorphic bisectional curvature. Given two J-invariant planes p and p' in $T_x(M)$, the *holomorphic bisectional curvature* $H(p, p')$ is defined by

$$H(p, p') = R(X, JX, Y, JY),$$

where X (resp. Y) is a unit vector in p (resp. p'). It is a simple matter to verify that $R(X, JX, Y, JY)$ depends only on p and p'. It is clear that $H(p, p)$ is the holomorphic sectional curvature determined by p. By Bianchi's identity (Theorem 5.3 of Chapter III) we have

$$H(p, p') = R(X, Y, X, Y) + R(X, JY, X, JY).$$

The right hand side is a sum of two sectional curvatures (up to constant factors). We may therefore consider the concept of holomorphic bisectional curvature as an intermediate concept between those of sectional curvature and holomorphic sectional curvature. In Theorems 5, 10, and 12 above, the assumption of positive sectional curvature may be replaced by that of positive holomorphic bisectional curvature (cf. Goldberg-Kobayashi [3]).

Note 24. Structure theorems on homogeneous complex manifolds

The first systematic study of compact homogeneous complex manifolds was done by H. C. Wang [4] who classified completely the C-spaces, i.e., the simply connected compact homogeneous complex manifolds. One of his main results is stated in the following theorem.

THEOREM 1. *Let K be a connected compact semi-simple Lie group, T a toral subgroup of K, and $C(T)$ the centralizer of T in K. Let U be a closed connected subgroup of K such that $(C(T))_s \subset U \subset C(T)$, where $(C(T))_s$ denotes the semi-simple part of $C(T)$. Then the coset space K/U, if even-dimensional, has an invariant complex structure. Conversely, every C-space can be thus obtained.*

A C-space K/U is a holomorphic fibre bundle over a homogeneous Kaehler manifold $K/C(T)$ with complex toral fibre $U/(C(T))_s$. According to Goto [2], the base space $K/C(T)$ is projective algebraic and rational. We may say that a C-space K/U lies between $K/(C(T))_s$ (called an M-space in Wang [4]) and a homogeneous Kaehler manifold $K/C(T)$.

This result was later generalized by Grauert-Remmert [1]:

THEOREM 2. *Every compact homogeneous complex manifold M is a holomorphic fibre bundle over a homogeneous projective algebraic manifold V with a complex parallelizable fibre F.*

A complex manifold is said to be *complex parallelizable* if its tangent bundle is complex analytically trivial. According to H. C. Wang [5], a compact complex parallelizable manifold is of the form G/D, where G is a complex Lie group and D is a discrete subgroup of G (see Example 2.3 of Chapter IX).

Wang has shown also that the following three conditions on a C-space K/U are equivalent: (1) $U = C(T)$; (2) the second Betti number of K/U is non-zero; (3) the Euler number of K/U is non-zero. In particular, a Kaehlerian C-space is necessarily of the form $K/C(T)$. The following result is due to Matsushima [4]:

THEOREM 3. *Every compact homogeneous Kaehler manifold is a Kaehlerian direct product of a Kaehlerian C-space and a flat complex torus.*

By a homogeneous Kaehler manifold we mean of course a Kaehler manifold on which the group of holomorphic isometric transformations is transitive. Borel-Remmert [1] generalized the result of Matsushima to a compact Kaehler manifold on which the group of holomorphic transformations is transitive. (See also Aeppli [2].)

In the non-compact case, results are not so complete. Let $M = G/H$ be a homogeneous complex manifold with an invariant volume element v. In terms of a local coordinate system $z^1, \ldots, z^n$ of M, v may be written in the form $v = V\, dz^1 \wedge \cdots \wedge d\bar{z}^1 \wedge \cdots \wedge d\bar{z}^n$. Then we define a Hermitian form $h = 2 \Sigma\, h_{i\bar{j}}\, dz^i\, d\bar{z}^j$ by

$$h_{i\bar{j}} = \partial^2 \log V / \partial z^i\, \partial \bar{z}^j.$$

The Hermitian form h is well-defined, independently of the local coordinate system chosen. (Note that the same construction was used in obtaining the Bergman metric from the Bergman kernel form in §6 of Chapter IX). Koszul [2] considered the case where the canonical Hermitian form h is non-degenerate. If H is compact, then a homogeneous complex manifold $M = G/H$ admits always an invariant volume element v and hence the canonical Hermitian form h. Among other things Koszul proved:

THEOREM 4. *Let $M = G/H$ be a homogeneous complex manifold with G connected and semi-simple and H compact. If the canonical Hermitian form h is non-degenerate, then*

(1) *H is connected;*
(2) *the center of G is finite;*
(3) *the number of the negative squares in the canonical Hermitian form h is equal to the difference between the dimension of a maximal compact subgroup of G and the dimension of H.*

If h is positive-definite in Theorem 4, then H is a maximal compact subgroup of G. Hence

COROLLARY. *Let $M = G/H$ be a homogeneous complex manifold with G connected and semi-simple and H compact. If the canonical Hermitian form h is positive-definite, then G/H is a Hermitian symmetric space of non-compact type.*

This corollary was also obtained by Borel [1] and was later generalized by Hano [2] to the case where G is unimodular. If M is a homogeneous bounded domain in $\mathbf{C}^n$, then M is of the form G/H with H compact and its canonical Hermitian form h is nothing but the Bergman metric of M and hence is positive-definite. Pyatetzki-Shapiro [1, 2] discovered homogeneous bounded domains in $\mathbf{C}^n$ which are not symmetric.

Hano-Kobayashi [1] considered the case where h is degenerate and obtained the following

THEOREM 5. *Let $M = G/H$ be a homogeneous complex manifold with an invariant volume element v. Then there is a closed subgroup L of G containing H such that*

(1) G/L is a homogeneous symplectic manifold, i.e., admits a closed 2-form of maximum rank invariant by G;

(2) L/H is a connected complex submanifold of G/H and is complex parallelizable.

In the fibration of G/H over G/L with fibre L/H, the fibres are the maximal integral submanifolds of G/H defined by the distribution $\{X \in T(G/H); h(X, \cdot) = 0\}$. It is not known whether the base space G/L is homogeneous complex and the fibration is holomorphic. In the special case where G is semi-simple and G/H is compact, G/L is a homogeneous Kaehler manifold and the fibration is holomorphic (see Matsushima [7]).

In Theorems 4 and 5, the group G is real and usually far from being complex as in the case where $M = G/H$ is a bounded domain. Matsushima [6] considered homogeneous complex manifolds G/H where G is a complex Lie group and H is a closed complex subgroup, and he characterized those which are Stein manifolds.

Note 25. Invariant connections on homogeneous spaces

1. In §2 of Chapter X, we discussed the existence of invariant affine connections on reductive homogeneous spaces. Conversely,

given a differentiable manifold M with an affine connection we may ask under what conditions M admits a transitive group of affine transformations. Theorem 2.8 in Chapter X provides one answer. Generalizing this theorem, Kostant [5] gave the following result. Let ∇ and ∇' be two affine connections on a differentiable manifold M, and let S be the tensor field of type (1, 2) which is the difference of ∇' and ∇, that is, $S(X, Y) = \nabla'_X Y - \nabla_X Y$, as in Proposition 7.10 of Chapter III. We say that ∇ is *rigid* with respect to ∇' if S is parallel with respect to ∇' (i.e., $\nabla'S = 0$). A result of Kostant is: *Let ∇ be an affine connection on a simply connected manifold M. Then M is a reductive homogeneous space of a connected Lie group with ∇ as an invariant affine connection if and only if there is an affine connection ∇' on M such that*

(1) ∇' *is invariant under parallelism* (see p. 262 of Volume I);
(2) ∇ *is rigid with respect to ∇';*
(3) ∇' *is complete.*

A similar result was obtained by Molino [1]. We say that an affine connection ∇ is locally invariant with respect to another affine connection ∇' if each of the tensor fields T, R, $\nabla^m T$, $\nabla^m R$, $1 \leq m < \infty$ (namely, the torsion and curvature tensor fields and their successive covariant differentials) is parallel with respect to ∇'. A result of Molino states: *If an analytic affine connection ∇ on a real analytic manifold M is locally invariant with respect to another affine connection, then M with ∇ is locally isomorphic to a homogeneous space with a certain invariant affine connection.*

2. Nguyen-Van Hai [2], [3], [4] studied conditions for the existence of an invariant affine connection on a (not necessarily reductive) homogeneous space. His result in [5] generalizes some results in Nomizu [8], [9] and is related to the problem of characterizing an affine connection which admits a transitive group of affine transformations. The problem of characterizing a Riemannian manifold which admits a transitive group of isometries was studied by Ambrose and Singer [2], Singer [2], Nomizu [8], [9] (see Note 19).

3. In Corollary 5.4 of Chapter X, we proved that a simply connected naturally reductive homogeneous space G/H with an invariant Riemannian metric is irreducible (as a Riemannian manifold) if G is simple. *Let G/H be a Riemannian homogeneous space such that G is compact and the Euler number $\chi(G/H)$ is non-zero (and*

hence positive). *If G is simple, G/H is an irreducible Riemannian manifold* (Hano and Matsushima [1]). *Conversely, if G/H is an irreducible Riemannian manifold, G is simple provided that G is effective on G/H* (Kostant [4]). Kostant gave an example which shows that the condition on $\chi(G/H)$ cannot be dropped. For detailed analysis of the holonomy groups and their reducibility for naturally reductive homogeneous spaces, and for arbitrary Riemannian homogeneous spaces, see Kostant [2] and [4], respectively.

4. As part of the general problem of determining all complete Riemannian manifolds with strictly positive curvature (see Note 22), Berger [8] studied a compact homogeneous space G/H with an invariant Riemannian metric which is naturally reductive in the manner of Corollary 3.6 of Chapter X. In the notation there, the condition that the sectional curvature is positive implies that $[X, Y] \neq 0$ for any linearly independent vectors in the subspace $\mathfrak{m}$. Studying this algebraic condition for the Lie algebra, he obtained all such spaces and showed, among other things, that with two exceptions such spaces G/H are homeomorphic, but not necessarily isometric, to compact Riemannian symmetric spaces of rank 1. For example, S^{2n+1} admits a Riemannian metric of the type considered which is not of constant curvature. For the exceptional cases, he showed that they are not homeomorphic to compact Riemannian symmetric spaces of rank 1.

5. It is known that a Kaehlerian homogeneous space G/H of a reductive Lie group G is the direct product of Kaehlerian homogeneous spaces $W_0, W_1, \ldots, W_m$, where W_0 is the center of G with an invariant Kaehlerian structure (hence a complex torus) and $W_1, \ldots, W_m$ are simply connected Kaehlerian homogeneous spaces of simple Lie groups (Borel [1], Lichnerowicz [1], Matsushima [4]). Hano and Matsushima [1] showed that this decomposition is precisely the de Rham decomposition given in Theorem 8.1 of Chapter IX; they showed that if G is simple, then G/H is irreducible—a result to be contrasted with the Riemannian case described in 3 of this note. They also proved that if G/H is a Kaehlerian homogeneous space with semi-simple G, then the decomposition of the Lie algebra $\mathfrak{g} = \mathfrak{m} + \mathfrak{h}$ such that $\text{ad}(H)\mathfrak{m} = \mathfrak{m}$ is unique and that if the corresponding canonical connection coincides with the Kaehlerian connection, then G/H is Hermitian symmetric.

6. Let G/H be a Riemannian homogeneous space where G is connected and effective on G/H. If the connected linear isotropy group $\tilde{H}^0$ is irreducible, then G/H is simply connected and Riemannian symmetric provided that G/H is non-compact (a result of Matsushima; for the proof, see the appendix in Nagano [8]).

Note 26. Complex submanifolds

1. The results of Thomas-E. Cartan and Fialkow on Einstein hypersurfaces M of $\mathbf{R}^{n+1}$, $n \geq 3$, in Theorem 5.3 of Chapter VII were extended by Fialkow [1] to the case of hypersurfaces in a real space form (i.e. a Riemannian manifold of constant curvature c). The analogous problem for complex hypersurfaces in a complex space form (i.e. a Kaehler manifold of constant holomorphic sectional curvature) was studied by B. Smyth [1]. The standard models of complex space forms are the complex Euclidean space $\mathbf{C}^n$ with flat metric, the complex projective space $P_n(\mathbf{C})$ with Fubini-Study metric (see Example 6.3 of Chapter IX and Example 10.5 of Chapter XI), and the unit ball $D_n(\mathbf{C})$ with Bergman metric (see Example 6.5 of Chapter IX and Example 10.7 of Chapter XI). Smyth showed that an Einstein complex hypersurface in a complex space form is locally symmetric, and he proved the following classification theorem:

THEOREM. *The only (simply connected) complete Einstein complex hypersurfaces M in $\mathbf{C}^{n+1}$ [resp. $D_{n+1}(\mathbf{C})$], $n \geq 2$, are $\mathbf{C}^n$ [resp. $D_n(\mathbf{C})$]. The only complete Einstein complex hypersurfaces M in $P_{n+1}(\mathbf{C})$, $n \geq 2$, are $P_n(\mathbf{C})$ or complex quadrics (see Example 10.6 of Chapter XI).*

The corresponding local theorem was proved by Chern [27]. Takahashi [3] showed that the condition that M is Einstein can be relaxed to the condition that the Ricci tensor S of M is parallel. Kobayashi [28] obtained the following partial generalization: *If M is a complete complex hypersurface with constant scalar curvature in $P_{n+1}(\mathbf{C})$, $n \geq 2$, then M is either a projective hyperplane $P_n(\mathbf{C})$ or a complex quadric.*

2. Continuing Smyth [1], Nomizu and Smyth [1] extended the above results covering the case $n = 1$ and also removing the assumption of simply-connectedness for M in the classification theorem. They also proved that the linear holonomy group of any

complex hypersurface in $D_{n+1}(\mathbf{C})$ [resp. $P_{n+1}(\mathbf{C})$] is isomorphic to $U(n)$ [resp. $U(n)$ or $SO(n) \times S^1$, the second case arising only when M is locally a complex quadric]. The linear holonomy group of a complex hypersurface in $\mathbf{C}^{n+1}$ was also studied by Kerbrat [1]. Nomizu and Smyth [1] also studied local rigidity and other related problems for complex hypersurfaces.

3. Let f be a Kaehlerian (i.e. holomorphic and isometric) immersion of a Kaehler manifold M into a Kaehler manifold $\tilde{M}$. As we saw in Proposition 9.2 of Chapter IX, the difference $\Delta(p)$ of the holomorphic sectional curvatures of M and $\tilde{M}$ for a holomorphic plane spanned by unit vectors X and JX of M is equal to $-\|\alpha(X, X)\|^2$, where α is the second fundamental form. Thus $\Delta(p) \leq 0$. O'Neill [4] studied Kaehlerian immersions for which $\Delta(p)$ is a constant for all holomorphic planes p and proved that such an immersion is totally geodesic (i.e., $\alpha = 0$) if $m < n(n + 3)/2$, where n and m are the complex dimensions of M and $\tilde{M}$. He also showed that this result no longer holds if $m = n(n + 3)/2$.

4. For more remarks and results on Kaehlerian immersions, see Note 18. For other results and problems on complex submanifolds of a complex projective space, see Chern [19] and Pohl [2].

Note 27. Minimal submanifolds

An isometric immersion f of a Riemannian manifold M into another Riemannian manifold $\tilde{M}$ is said to be *minimal* if the mean curvature normal (or its length, which is the mean curvature) defined in §5 of Chapter VII is 0. We also say that M is a *minimal submanifold* immersed in $\tilde{M}$. It is locally obtained as an extremal for n-dimensional volume for deformations with a fixed boundary, where $n = \dim M$. (See Eisenhart [1].)

A totally geodesic submanifold is, of course, minimal. We showed that there is no compact minimal submanifold in a Euclidean space (§5 of Chapter VII). See also Example 1 of Note 14 for another proof based on the Laplacian. The same result holds when the Euclidean space is replaced by a simply connected Riemannian manifold with non-positive sectional curvature (see O'Neill [2]). See also Myers [5].

Frankel [4] proved the following. Let M be a *compact minimal submanifold immersed in a complete Riemannian manifold $\tilde{M}$.* Then

(1) *If $\tilde{M}$ has non-positive sectional curvature, the fundamental group $\pi_1(M)$ is infinite.*

(2) *If $\tilde{M}$ has positive definite Ricci tensor, the natural homomorphism $\pi_1(M) \to \pi_1(\tilde{M})$ is surjective.*

In Note 14 we showed that a k-dimensional submanifold M of $S^n(r) \subset \mathbf{R}^{n+1}$ is minimal if and only if

$$\Delta x^i = -\frac{k}{r^2} x^i, \quad 0 \leq i \leq n,$$

where $x^0, x^1, \ldots, x^n$ is a rectangular coordinate system in $\mathbf{R}^{n+1}$. Takahashi [2] proved that a compact homogeneous Riemannian manifold with irreducible linear isotropy group admits a minimal immersion into a Euclidean sphere. Wu-yi Hsiang [1] has shown that every Riemannian homogeneous space of a compact Lie group G/H admits a minimal immersion in a sphere in such a way that the image is homogeneous with respect to a certain subgroup of the rotation group of the sphere. For more results on minimal submanifolds of a sphere, see Takeuchi-Kobayashi [1], Simons [2], [3], and Calabi [7].

In the complex case it is known that any complex submanifold M of a Kaehler manifold $\tilde{M}$ is minimal. We may prove this fact as follows. For any vector ξ normal to M at a point x of M we have a symmetric endormorphism A_ξ of the tangent space $T_x(M)$ such that $g(A_\xi X, Y) = g(\alpha(X, Y), \xi)$, where g is the metric and α the second fundamental form (see §3 of Chapter VII and §9 of Chapter IX). Thus we get

$$g(A_\xi JX, Y) = g(\alpha(JX, Y), \xi),$$

where J denotes the complex structure. On the other hand, we have

$$g(JA_\xi X, Y) = -g(A_\xi X, JY) = -g(\alpha(X, JY), \xi).$$

Since $\alpha(JX, Y) = \alpha(X, JY)$ by Proposition 9.1 of Chapter IX, we get

$$g(A_\xi JX, Y) = -g(JA_\xi X, Y),$$

that is, $A_\xi J = -JA_\xi$. A simple argument in linear algebra then shows that trace $A_\xi = 0$. Thus M is minimal in $\tilde{M}$.

The results for 2-dimensional minimal surfaces are extensive; see Nitsche [1], also Chern [24], [28], Pohl [2], Almgren [1], [2].

Note 28. Contact structure and related structures

1. On a differentiable manifold M of dimension $2n + 1$, a 1-form ω is said to be a *contact form* if $\omega \wedge (d\omega)^n \neq 0$ at each point of M. In particular, on $\mathbf{R}^{2n+1}$ with coordinates $(z, x^1, \ldots, x^n, y^1, \ldots, y^n)$ we have the canonical contact form

$$\omega_0 = dz - \sum_{i=1}^{n} y^i \, dx^i.$$

A local transformation f of $\mathbf{R}^{2n+1}$ is called a local contact transformation if $f^*\omega_0 = \alpha\omega_0$ on the domain of f, where α is a non-vanishing differentiable function. The set of all local contact transformations forms a pseudogroup, which we denote by Γ.

A *contact structure* on a $(2n + 1)$-dimensional differentiable manifold M is defined by a complete atlas compatible with Γ in the sense of §1, Chapter I. A chart (U, φ) belonging to the atlas defines a contact form $\omega_U = \varphi^*\omega_0$ on U, and for two charts (U_i, φ_i), $i = 1, 2$, with non-empty $U_1 \cap U_2$, we have $\omega_{U_1} = \alpha_{12}\omega_{U_2}$ on $U_1 \cap U_2$ with a certain non-vanishing function α_{12}. A 1-form ω on M is called a *contact form belonging to the given contact structure* if, for any chart (U, φ), we have $\omega = \alpha_U \omega_U$ on U, where α_U is a non-vanishing function on U. An automorphism f of the contact structure is called a *contact transformation*; it may be defined as a differentiable transformation f of M such that if ω is a contact form defined on an open subset of M belonging to the contact structure then $f^*\omega$ is a contact form belonging to the contact structure.

If a differentiable manifold M of dimension $2n + 1$ admits a contact form ω globally defined on M, then ω determines a contact structure. This is based on a result of E. Cartan [19]: each point x of M admits a coordinate system $z, x^1, \ldots, x^n, y^1, \ldots, y^n$ in a neighborhood U such that $\omega = dz - \Sigma_{i=1}^{n} y^i \, dx^i$ on U. The set of such coordinate neighborhoods forms an atlas compatible with the pseudogroup Γ and defines a contact structure on M.

Conversely, a contact structure on a manifold of dimension $2n + 1$ with n even admits a globally defined contact form belonging to the contact structure if and only if the manifold is orientable (see J. W. Gray [1] and a remark following Theorem 4.3 in Sasaki [3]). On the other hand, a $(2n + 1)$-dimensional manifold with a contact structure is always orientable if n is odd.

A typical model for a contact manifold is S^{2n+1}. In fact, if $\{x^1, \ldots, x^{2n+2}\}$ is the natural coordinate system in $\mathbf{R}^{2n+2}$, the restriction ω to S^{2n+2} of the form

$$\sum_{i=1}^{n+1} (x^{2i-1}\, dx^{2i} - x^{2i}\, dx^{2i-1})$$

is a contact form on S^{2n+1}. Another example is the unit cotangent sphere bundle $S^*(M)$, namely, the set of all unit covectors of a Riemannian manifold M. In fact, if we take the form ω on the cotangent bundle $T^*(M)$ defined in Example 6.10 of Chapter IX, its restriction to $S^*(M)$ is a contact form.

The study of contact transformations goes back to S. Lie and F. Engel [1]. The pseudogroup of local contact transformations is an infinite continuous simple pseudogroup in the sense of E. Cartan [19] (see Chern [3]). Contact structures, their automorphisms, and deformations were studied in J. W. Gray [1], Wang, and Boothby [1]. The latter have shown that a compact contact manifold is, under a certain condition, a principal fibre bundle over a symplectic manifold with a circle group as the structure group. The fibring of S^{2n+1} over $P_n(\mathbf{C})$ in Example 10.5 is a typical model for this result.

2. It is known (see Chern [3]) that a contact structure on M defined by a global contact form admits a G-structure (namely, a reduction of the structure group of the tangent bundle to a subgroup G of $GL(2n + 1; \mathbf{R})$) with $G = 1 \times U(n)$ consisting of all matrices of the form

$$\begin{pmatrix} 1 & 0 & 0 \\ 0 & A & -B \\ 0 & B & A \end{pmatrix},$$
where A, B are real matrices of degree n such that $A + iB$ is unitary.

A $(1 \times U(1))$-structure was called an almost contact manifold in J. W. Gray [1]. Forgetting about any metric condition, one can consider, as in Chern [3], a G-structure on a $(2n + 1)$-dimensional manifold M where G consists of all matrices of the form

$$\begin{pmatrix} c & 0 & 0 \\ 0 & A & -B \\ 0 & B & A \end{pmatrix},$$ where $c > 0$ and A, B are real matrices of degree n.

The structure is then analogous to an almost complex structure (see Proposition 3.1 of Chapter IX). Let $\mathbf{R}^{2n+1} = \mathbf{R} + \mathbf{R}^{2n}$ with coordinate z for $\mathbf{R}$ and coordinates $(x^1, \ldots, x^n, y^1, \ldots, y^n)$ for $\mathbf{R}^{2n}$; the group G above acts on $\mathbf{R}^{2n+1}$ accordingly. For each $x \in M$, take an element u of the G-subbundle P of $L(M)$ such that $\pi(u) = x$ and define, considering u as a linear isomorphism $\mathbf{R}^{2n+1} \to T_x(M)$,

$$T_k^0 = u(\mathbf{R}), \quad S_x = u(\mathbf{R}^{2n}), \quad \text{and} \quad J_x = u \circ J_0 \circ u^{-1},$$

where J_0 is the linear transformation of $\mathbf{R}^{2n+1}$ given by the matrix

$$\begin{pmatrix} 0 & 0 \\ 0 & J \end{pmatrix} \quad \text{with } J = \begin{pmatrix} 0 & -I_n \\ I_n & 0 \end{pmatrix}.$$

Then T^0 is a 1-dimensional distribution with an orientation, S is a $2n$-dimensional distribution, and J is a tensor field of type $(1, 1)$ such that $J = 0$ on T^0 and $J^2 = -1$ on S. This structure is named an *almost cocomplex structure* in Bouzon [1]. Sasaki [5] defined the essentially same structure in a slightly different way; he used a triple (φ, ξ, η) where φ is a tensor field of type $(1, 1)$, ξ is a vector field, and η is a 1-form satisfying

$$\eta(\xi) = 1 \quad \text{and} \quad \varphi^2 = -1 + \xi \otimes \eta.$$

At any rate, such a structure has close analogy with an almost complex structure, and the analogy to a Hermitian metric can be defined as a Riemannian metric g which makes T^0 and S orthogonal and which is Hermitian on each S_x with respect to the complex structure given by the restriction of J_x to S_x. If M is a real hypersurface (of dimension $2n + 1$) imbedded in an almost Hermitian manifold M of dimension $2n + 2$, then there is a natural almost cocomplex structure together with such a metric. On the other hand, if M is an almost cocomplex structure, then we can define an almost complex structure in $\tilde{M} = M \times \mathbf{R}$ by

$$\tilde{J} = J \text{ on } S, \quad \tilde{J}(Z) = W, \quad \tilde{J}(W) = -Z,$$

where Z is a non-vanishing vector field in the direction of T^0 and W denotes the standard vector field on the real line $\mathbf{R}$. More generally, if M_1 and M_2 are manifolds with almost cocomplex structures, then $M_1 \times M_2$ admits an almost complex structure. From this viewpoint Morimoto [4] obtained a generalization of the result of Calabi-Eckmann that $S^{2p+1} \times S^{2q+1}$ admits a complex structure (see Example 2.5 of Chapter IX).

For many results on almost cocomplex structures (or almost contact structures), see Bouzon [1] and Sasaki [3], [4], and the references contained therein.

3. Yano [11] has unified the notions of almost complex structure and almost cocomplex structure by considering a tensor field f of type $(1, 1)$ on an n-dimensional manifold M such that $f^3 + f = 0$. and such that the rank of f is equal to a constant k everywhere. Obviously, k has to be even. In the case where $k = n$ (even), f is an almost complex structure. In the case where $k = 2m$ and $n = 2m + 1$, we have essentially an almost cocomplex structure. In the general case, each tangent space T_x is decomposed as the direct sum $T_x^0 + S_x$, where

$$T_x^0 = \{X \in T_x(M) ; fX = 0\}$$

and

$$S_x = \{X \in T_x(M) ; (f^2 + 1)X = 0\},$$

as follows easily from the consideration of the minimal polynomial. Thus we get two complementary distributions T^0 and S of dimensions $n - k$ and k (even), respectively. For the study of this structure, see Yano and Ishihara [1] and the references contained therein.

More generally, a structure defined by a number of complementary distributions T_i, $1 \leq i \leq k$, is called an *almost product structure*. A. G. Walker [7] studied the existence of affine connections related to such a structure. See also Walker [3], [4], [5], Willmore [1] to [6], E. T. Kobayashi [1], [2], [3], Lehmann-Lejeune [1], [2], Clark [1], Clark and Bruckheimer [1], Kurita [4].

4. Contact forms and contact transformations can be defined for a complex manifold M of complex dimension $2n + 1$. A *complex contact form* on M is a holomorphic 1-form ω such that $\omega \wedge (d\omega)^n \neq 0$ everywhere. On $\mathbf{C}^{2n+1}$ with complex coordinates, we have the standard contact form

$$\omega^0 = dz^0 - \sum_{k=1}^{n} z^{n+k} \, dz^k.$$

A local contact transformation of $\mathbf{C}^{2n+1}$ is a local holomorphic transformation f such that $f^*\omega_0 = \alpha\omega_0$ where α is a non-vanishing holomorphic function. The definition of a complex contact structure is now analogous to the real case. For basic results, see Kobayashi [13]. A complex contact structure admits a globally defined complex contact form if and only if the first Chern class $c_1(M)$ is 0. Boothby [3], [4] studied homogeneous complex contact manifolds using the results of Wang on C-spaces (see Note 24). (See also Wolf [8].)

Bibliography for Volumes I and II

The following Bibliography includes all those references that are quoted throughout Volumes I and II; those in the Bibliography for Volume I are reproduced in the same numbering. The Bibliography also contains books and papers on the topics closely related to the content of our treatises. References marked with asterisk have themselves an extensive list of publications which may have been omitted in our Bibliography. We note that the order of listing for each author is not necessarily chronological.

We give reference to *Mathematical Reviews* whenever possible. For example, MR 7, 412 means p. 412 of Volume 7 and MR 27 #2945 means #2945 of Volume 27.

ABRAHAM, R.
[1] Piecewise differentiable manifolds and the space-time of general relativity, J. Math. Mech. 11 (1962), 553–592. (MR 25 #2895.)

ADAMS, J. F.
[1] On the non-existence of elements of Hopf invariant one, Ann. of Math. 72 (1960), 20–104. (MR 25 #4530.)

ADLER, A. W.
[1] Characteristic classes of homogeneous spaces, Trans. Amer. Math. Soc. 86 (1957), 348–365. (MR 19, 1181.)
[2] A characteristic relation in fibre bundles, Amer. J. Math. 79 (1957), 713–724. (MR 21 #902.)
[3] Classifying spaces for Kaehler metrics, I. Math. Ann. 152 (1963), 164–184 (MR 28 #1569); II. *ibid.* 154 (1964), 257–266 (MR 28 #5408); III. *ibid.* 156 (1964), 378–392 (MR 29 #6458); IV. *ibid.* 160 (1965), 41–58 (MR 34 #5039.)
[4] Integrability conditions for almost complex structures, Michigan Math. J. 13 (1966), 499–505. (MR 36 #3298.)

AEPPLI, A.
[1] Einige Ähnlichkeits und Symmetriesätze für differenzierbare Flächen in Raum, Comment. Math. Helv. 33 (1959), 174–195. (MR 22 #5945.)
[2] Some differential geometric remarks on complex homogeneous manifolds, Arch. Math. 16 (1965), 60–68. (MR 32 #464.)

ALEKSANDROV, A. D.
[1] Die Innere Geometrie der Konvexen Flächen, Akademie-Verlag, Berlin, 1955. (MR 17, 74.)
[2] Ein allgemeiner Eindeutigkeitssatz für geschlossenen Flächen, C. R. Acad. Sci URSS (Doklady) 19 (1938), 227–229.
[3] Uniqueness theorems for surfaces in the large I, Vestnik Leningrad Univ. 11 (1956), No. 19, 5–7. (MR 19, 167; also, 27 #698.)
[4] A characteristic property of spheres, Ann. Math. Pura Appl. 58 (1962), 303–315.

ALEKSEEVSKII, D. V.
[1] Holonomy groups of Riemannian spaces, Ukrain. Math. Z. 19 (1967), 100–104. (MR 35 #934.)

ALEXANDER, S. B.
[1] Reducibility of Euclidean immersions of low codimension, Thesis, Univ. of Illinois, 1967. (J. Differential Geometry, to appear.)

ALLAMIGEON, A. C.
[1] Espaces homogènes symétriques harmoniques à groupe semi-simple, C. R. Acad. Sci Paris 243 (1956), 121–123. (MR 18, 496.)
[2] Espaces harmoniques décomposables, C. R. Acad. Sci. Paris 245 (1957), 1498–1500 (MR 19, 879.)
[3] Espaces homogènes symétriques harmoniques, C. R. Acad. Sci. Paris 246 (1958), 1004–1005 (MR 20 #2757.)
[4] Quelques propérités des espaces homogènes réductifs à groupe nilpotent, C. R. Acad. Sci. Paris 247 (1958), 628–631. (MR 20 #7075.)
[5] Propriétés globales des espaces de Riemann harmoniques, Ann. Inst. Fourier (Grenoble) 15 (1965), 91–132. (MR 33 #6549.)

ALLENDOERFER, C. B.
[1] Rigidity for spaces of class greater than one, Amer. J. Math. 61 (1939), 633–644. (MR 1, 28.)
[2] The Euler number of a Riemannian manifold, Amer. J. Math. 62 (1940), 243–248. (MR 2, 20.)
[3] Global theorems in Riemannian geometry, Bull. Amer. Math. Soc. 54 (1948), 249–259. (MR 10, 266.)
[4] Einstein spaces of class one, Bull. Amer. Math. Soc. 43 (1937), 265–270.

ALLENDOERFER, C. B. and EELLS, J., Jr.
[1] On the cohomology of smooth manifolds, Comment. Math. Helv. 32 (1958), 165–179. (MR 21 #868.)

ALLENDOERFER, C. B. and WEIL, A.
[1] The Gauss-Bonnet theorem for Riemannian polyhedra, Trans. Amer. Math. Soc. 53 (1943), 101–129. (MR 4, 169.)

ALMGREN, F. J. Jr.
[1] Three theorems on manifolds with bounded mean curvature, Bull. Amer. Math. Soc. 71 (1965), 755–756. (MR 31 #5182.)
[2] Some interior regularity theorems for minimal surfaces and extension of Bernstein's theorem, Ann. of Math. 84 (1966), 277–292. (MR 34 #702.)

AMBROSE, W.
[1] Parallel translation of Riemannian curvature, Ann. of Math. 64 (1956), 337–363. (MR 21 #1627.)
[2] The Cartan structural equations in classical Riemannian geometry, J. Indian Math. Soc. 24 (1960), 23–76. (MR 23 #A1317.)
[3] Higher order Grassmann bundles, Topology 3 (1964), suppl 2, 199–238. (MR 30 #1518.)
[4] The index theorem in Riemannian geometry, Ann. of Math. 73 (1961), 49–86. (MR 24 #A3608.) (See Takahashi [1], MR 30 #545, for correction.)

AMBROSE, W. and SINGER, I. M.
[1] A theorem on holonomy, Trans. Amer. Math. Soc. 75 (1953), 428–443. (MR 16, 172.)
[2] On homogeneous Riemannian manifolds, Duke Math. J. 25 (1958), 647–669. (MR 21 #1628.)

AOMOTO, K.
[1] L'analyse harmonique sur les espaces riemanniens à courbure riemannienne négative I, J. Fac. Sci. Univ. Tokyo, 13 (1966), 85–105. (MR 34 #3496.)

APTE, M.
[1] Sur certaines classes caractéristiques des variétés kählériennes compactes, C. R. Acad. Sci. Paris 240 (1955), 149–151. (MR 16, 625.)

APTE, M. and LICHNEROWICZ, A.
[1] Sur les transformations affines d'une variété presque hermitienne compacte, C. R. Acad. Sci. Paris 242 (1956), 337–339. (MR 17, 787.)

ARAGNOL, A.
[1] Sur la géométrie différentielle des espaces fibrés, Ann. Sci. École Norm. Sup. 75 (1958), 257–407. (MR 23 #A1322b.)

ARAKI, S.
[1] On root systems and an infinitesimal classification of irreducible symmetric spaces, J. Math. Osaka City Univ. 13 (1962), 1–34. (MR 27 #3743.)

ARENS, R
[1] Topologies for homeomorphism groups, Ann. of Math. 68 (1946), 593–610. (MR 8, 479.)

ARNOLD, V.
[1] Sur la courbure de Riemann des groupes de difféomorphismes, C. R. Acad. Sci. Paris 260 (1965), 5668–5671. (MR 31 #2692.)

ARTIN, E.
[1] Geometric Algebra, Interscience Tracts # 3, Interscience, New York, 1957. (MR 18, 553.)

ATIYAH, M. F.
[1] Complex analytic connections in fibre bundles, Trans. Amer. Math. Soc. 85 (1957), 181–207 (MR 19, 172); same title, Intern. Symp. Algebraic Topology, Mexico City, 1958, 77–82 (MR 20 #4656).

ATIYAH, M. F. and BOTT, R.
[1] A Lefschetz fixed point formula for elliptic differential operators, Bull. Amer. Math. Soc. 72 (1966), 245–250. (MR 32 #8360.)

AUBIN, T.
[1] Sur la courbure scalaire des variétés riemanniennes compactes, C. R. Acad. Sci. Paris 262 (1966), 130–133. (MR 33 #3232.)
[2] Sur le groupe fondamental de certaines variétés riemanniennes compactes, C. R. Acad. Sci. Paris 261 (1965), 3032–3034. (MR 32 #4632.)
[3] Sur la courbure conforme des variétés riemanniennes, C. R. Acad. Sci. Paris, Ser. A-B 262 (1966), A 391–393. (MR 33 #670.)
[4] Variétés hermitiennes compactes localement conformément kählériennes, C. R. Acad. Sci. Paris 261 (1965), 2427–2430. (MR 32 #3021.)

AUSLANDER, L.
[1] Examples of locally affine spaces, Ann. of Math. 64 (1956), 255–259. (MR 18, 332.)
[2] On holonomy covering spaces, Proc. Amer. Math. Soc. 7 (1956), 685–689. (MR 18, 507.)
[3] On the group of affinities of locally affine spaces, Proc. Amer. Math. Soc. 9 (1958), 471–473. (MR 20 #2014.)
[4] Some compact solvmanifolds and locally affine spaces, J. Math. Mech. 7 (1958), 963–975. (MR 21 #93.)
[5] On the sheeted structure of compact locally affine spaces, Michigan Math. J. 5 (1958), 163–168. (MR 21 #2268.)
[6] Bieberbach's theorems on space groups and discrete uniform subgroups of Lie groups, Ann. of Math. 71 (1960), 579–590 (MR 22 #12161); II, Amer. J. Math. 83 (1961), 276–280 (MR 23 #A962.)
[7] On the theory of solvmanifolds and generalizations with applications to differential geometry, Proc. Symp. Pure Math. (Tucson), Vol. III Amer. Math. Soc. 1961, 138–143. (MR 23 #A2483.)
[8] On the Euler characteristic of compact locally affine spaces, Comment Math. Helv. 35 (1961), 25–27 (MR 19, 172); II, Bull. Amer. Math. Soc. 67 (1961), 405–406 (MR 24 #A1680).

[9] The structure of complete locally affine manifolds, Topology 3 (1964), Sup. 1, 131–139. (MR 28 #4463.)
[10] Four dimensional compact locally Hermitian manifolds, Trans. Amer. Math. Soc. 84 (1957), 379–391. (MR 18, 762.)

AUSLANDER, L. and AUSLANDER, M.
[1] Solvable Lie groups and locally Euclidean Riemann spaces, Proc. Amer. Math. Soc. 9 (1958), 933–941. (MR 21 #2021.)

AUSLANDER, L. and KURANISHI, M.
[1] On the holonomy group of locally Euclidean spaces, Ann. of Math. 65 (1957), 411–415. (MR 19, 168.)

AUSLANDER, L. and MARKUS, L.
[1] Holonomy of flat affinely connected manifolds, Ann. of Math. 62 (1955), 139–151. (MR 17, 298.)
[2] Flat Lorentz 3-manifold, Memoir Amer. Math. Soc. #30, 1959. (MR 24 #A1689.)

AUSLANDER, L. and SZCZARBA, R. H.
[1] Characteristic classes of compact solvmanifolds, Ann. of Math. 76 (1962), 1–8. (MR 25 #4547.)

AUSLANDER, L., GREEN, L., HAHN, F.
[1] Flows on Homogeneous Spaces, Ann. of Math. Studies, No. 53, Princeton Univ. Press, 1963. (MR 29 #4841.)

AVEZ, A.
[1] Formule de Gauss-Bonnet-Chern en métrique de signature quelconque, C. R. Acad. Sci. Paris 255 (1962), 2049–2051 (MR 26 #2993); same title, De Revista de la Unión Mat. Argentina, 21 (1963), 191–197 (MR 29 #6452).
[2] Applications de la formule de Gauss-Bonnet-Chern aux variétés à quatre dimensions, C. R. Acad. Sci. Paris 256 (1963), 5488–5490. (MR 28 #555.)
[3] Espaces harmoniques compactes, C. R. Acad. Sci. Paris 258 (1964), 2727–2729. (MR 28 #5404.)
[4] Essais de géométrie riemannienne hyperbolique globale, Ann. Inst. Fourier (Grenoble) 13 (1963), fasc. 2, 105–190. (MR 29 #5205.)
[5] Index des variétés de dimension 4, C. R. Acad. Sci. Paris 259 (1964), 1934–1937. (MR 30 #1477.)
[6] Valeur moyenne du scalaire de courbure sur une variété compacte. Applications relativistes, C. R. Acad. Sci. Paris 256 (1963), 5271–5273. (MR 27 #1909.)
[7] Remarques sur les variétés de dimension 4, C. R. Acad. Sci. Paris 264 (1967), 738–740.
[8] Conditions nécessaires et suffisantes pour qu'une variété soit un espace d'Einstein, C. R. Acad. Sci. Paris 24 (1959), 1113–1115. (MR 20 #7301.)

BA, B.
[1] Structures presque complexes, structures conformes et dérivations, Cahiers de Topologie et Géométrie Différentielle, VIII (1966), 1-74.

BARBANCE, C.
[1] Transformations conforme d'une variété riemannienne compacte, C. R. Acad. Sci. Paris 260 (1965), 1547–1549. (MR 31 #1634.)
[2] Decomposition d'un tenseur symétrique sur un espace d'Einstein, C. R. Acad. Sci. Paris 258 (1964), 5336–5338. (MR 29 #1602.)

DE BARROS, C. M.
[1] Espaces infinitésimaux, Cahiers de Topologie et Géométrie Différéntielle 7 (1965). (MR 34 #728.)
[2] Variétés hor-symplectiques, C. R. Acad. Sci. Paris 259 (1964), 1291–1294. (MR 30 #2440.)
[3] Variétés presque hor-complexes, C. R. Acad. Sci. Paris 260 (1965), 1543–1546. (MR 30 #5241.)
[4] Opérateurs infinitésimaux sur l'algèbre des formes différentielles extérieures, C. R. Acad. Sci. Paris 261 (1965), 4594–4597. (MR 32 #6354.)
[5] Sur la construction d'opérateurs infinitésimaux gradués, C. R. Acad. Sci. Paris 263 (1966), 72–75. (MR 33 #7957.)

BEEZ, R.
[1] Zur Theorie der Krümmungmasses von Mannigfaltigkeiten höherer Ordunung, Z. Math. Physik 21 (1876), 373–401.

BENZÉCRI, J. P.
[1] Sur les variétés localement affines et localement projectives, Bull. Soc. Math. France 88 (1960), 229–332. (MR 23 #A1325.)
[2] Sur la classe d'Euler de fibrés affins plats, C. R. Acad. Sci. Paris 260 (1965), 5442–5444. (MR 31 #2734.)

BERGER, M.
[1] Sur les groupes d'holonomie des variétés à connexion affine et des variétés riemanniennes, Bull. Soc. Math. France 83 (1955), 279–330. (MR 18, 149.)
[2] Les espaces symétriques noncompacts, Ann. Sci. École Norm. Sup. 74 (1957), 85–177. (MR 21 #3516.)
[3] Les variétés riemanniennes à courbure positive, Bull. Soc. Math. Belg. 10 (1958), 89–104.
[4] Variétés riemanniennes à courbure positive, Bull. Soc. Math. France 87 (1959), 285–292. (MR 23 #A1338.)
[5] Pincement riemannien et pincement holomorphe, Ann. Scuola Norm. Sup. Pisa 14 (1960), 151–159 (MR 25 #3477); (correction) ibid. 16 (1962), 297 (MR 28 #558.)

[6] Sur quelques variétés riemanniennes suffisament pincées, Bull. Soc. Math. France 88 (1960), 57–71. (MR 24 #A3606.)

[7] Les variétés riemanniennes $\frac{1}{4}$-pincées, Ann. Scuola Norm. Sup. Pisa 14 (1960), 161–170. (MR 25 #3478.)

[8] Les variétés riemanniennes homogènes normales simplement connexes à courbure strictement positive, Ann. Scuola Norm. Sup. Pisa 15 (1961), 179–246. (MR 24 #A2919.)

[9] Sur quelques variétés d'Einstein compactes, Ann. Mat. Pura Appl. 53 (1961), 89–95. (MR 24 #A519.)

[10] Sur les variétés à courbure positive de diamètre minimum, Comment Math. Helv. 35 (1961), 28–34. (MR 25 #2560.)

[11] Sur les variétés à opérateur de courbure positif, C. R. Acad. Sci. Paris 253 (1961), 2832–2834. (MR 25 #3479.)

[12] Les sphères parmi les variétés d'Einstein, C. R. Acad. Sci. Paris 254 (1962), 1564–1566. (MR 24 #A3607.)

[13] An extension of Rauch's metric comparison theorem and some applications, Illinois J. Math. 6 (1962), 700–712. (MR 26 #719.)

[14] On the characteristic of positively pinched Riemannian manifolds, Proc. Nat. Acad. Sci. USA 48 (1962), 1915–1917. (MR 26 #720.)

[15] Les variétés riemanniennes dont la courbure satisfait certaines conditions, Proc. Internat. Congr. Math. Stockholm (1962), 447–456. (MR 31 #695.)

[16] Sur les variétés (4/23)-pincées de dimension 5, C. R. Acad. Sci. Paris 257 (1963), 4122–4125. (MR 28 #1557.)

[17] Les variétés kählériennes compactes d'Einstein de dimension quatre à courbure positive, Tensor 13 (1963), 71–74. (MR 27 #5209.)

[18] Sur quelques variétés riemanniennes compactes d'Einstein, C. R. Acad. Sci. Paris 260 (1965), 1554–1557. (MR 31 #696.)

[19] Sur les variétés d'Einstein compactes, C. R. IIIe Reunion Math. Expression latine, Namur (1965), 35–55.

[20] Trois remarques sur les variétés riemanniennes à courbure positive, C. R. Acad. Sci. Paris 263 (1966), 76–78. (MR 33 #7966.)

[21] Remarques sur les groupes d'holonomie des variétés riemanniennes, C. R. Acad. Sci. Paris 262 (1966), 1316–1318. (MR 34 #746.)

[22] Sur le spectre d'une variété riemannienne, C. R. Acad. Sci. Paris 263 (1966), 13–16. (MR 34 #1964.)

BERGER, M. and BOTT, R.

[1] Sur les variétés à courbure strictement positive, Topology 1 (1962), 301–311. (MR 26 #4296.)

BERGMAN, S.

[1] Über die Kernfunktion eines Bereiches und ihr Verhalten am Rande, J. Reine Angew. Math. 169 (1933), 1–42 and 172 (1935), 89–128.

BERNARD, D.
[1] Sur la géométrie différentielle des G-structures, Ann. Inst. Fourier (Grenoble) 10 (1960), 151–270. (MR 23 #A4094.)

BERWALD, L.
[1] Una forma invariante della seconda variazione, Rendiconti dei Lincei 7 (1928), 301–306.

BIEBERBACH, L.
[1] Über die Bewegungsgruppen der Euklidischen Räume I, Math. Ann. 70 (1911), 297–336; II, 72 (1912), 400–412.
[2] Über die Minkowskische Reduktion der positiven quadratschen Formen und die endliche Gruppen linearer ganzzählige Substitutionen, Göttingen Nachr. (1912), 207–216.
[3] Über topologischen Typen der offenen Euklidischen Raumformen, Sitz. Ber. d. Pr. Akad. Wiss., Phys. Math. Kl. (1929), 612.
[4] Eine singularitätfreie Fläche konstanter negativer Krümmung im Hilbertschen Raum, Comment. Math. Helv. 4 (1932), 248–255.

BIEBERBACH, L. and SCHUR, I.
[1] Über die Minkowskische Reduktionstheorie der positiven quadratischen Formen, Sitz. Ber. Preuss. Akad. Wiss. Berlin 33 (1928), 510–535.

BISHOP, E.
[1] Partially analytic spaces, Amer. J. Math. 83 (1961), 669–692. (MR 25 #5191.)

BISHOP, R. L.
[1] The holonomy algebra of immersed manifolds of codimension two, J. Differential Geometry, to appear.

BISHOP, R. L. and CRITTENDEN, R. J.
[1] Geometry of Manifolds, Academic Press, New York, 1964. (MR 29 #6401.)

BISHOP, R. L. and GOLDBERG, S. I.
[1] On the topology of positively curved Kaehler manifolds, Tôhoku Math. J. 15 (1963), 359–364. (MR 28 #2511.)
[2] Some implications of the generalized Gauss-Bonnet theorem, Trans. Amer. Math. Soc. 112 (1964), 508–535. (MR 29 #574.)
[3] On the 2nd cohomology group of a Kaehler manifold of positive curvature, Proc. Amer. Math. Soc. 16 (1965), 119–122. (MR 30 #2441.)
[4] On the topology of positively curved Kaehler manifolds II, Tôhoku Math. J. 17 (1965), 310–318. (MR 33 #3316b.)
[5] Rigidity of positively curved Kaehler manifolds, Proc. Nat. Acad. Sci. USA 54 (1965), 1037–1041. (MR 33 #3316a).
[6] A characterization of the Euclidean sphere, Bull. Amer. Math. Soc. 72 (1966), 122–124. (MR 32 #3011.)

BLANCHARD, A.
[1] Sur les variétés analytiques complexes, Ann. Sci. École Norm. Sup. 73 (1956), 157–202. (MR 19, 316.)

BLANUŠA, D.
[1] Eine isometrische und singularitätenfreie Einbettung des n-dimensionalen hyperbolischen Raumes im Hilbertschen Raum, Monatsch. Math. 57 (1953), 102–108. (MR 15, 61.)
[2] Über die Einbettung hyperbolischer Räume in euklidische Räume, Monatsch. Math. 59 (1955), 217–229. (MR 17, 188.)

BOCHNER, S.
[1] Vector fields and Ricci curvature, Bull. Amer. Math. Soc. 52 (1946), 776–797. (MR 8, 230.)
[2] Curvature in Hermitian metric, Bull. Amer. Math. Soc. 53 (1947), 179–195. (MR 8, 490.)
[3] On compact complex manifolds, J. Indian Math. Soc. 11 (1947), 1–21. (MR 9, 423.)
[4] Curvature and Betti numbers, Ann. of Math. 49 (1948), 379–390 (MR 9, 618); II. *ibid.* 50 (1949), 77–93 (MR 10, 571). (See Bochner-Yano [1] for corrections.)
[5] Euler-Poincaré characteristic for locally homogeneous and complex spaces, Ann. of Math. 51 (1950), 241–261. (MR 11, 617.)
[6] Tensor fields with finite bases, Ann. of Math. 53 (1951), 400–411. (MR 12, 750.)
[7] Tensor fields and Ricci curvature in Hermitian metric, Proc. Nat. Acad. Sci. USA. 37 (1951), 704–706. (MR 13, 385.)
[8] Vector fields on Riemannian spaces with boundary, Ann. Mat. Pura Appl. (4) 53 (1961), 57–62. (MR 28 #557.)

BOCHNER, S. and MONTGOMERY, D.
[1] Locally compact groups of differentiable transformations, Ann. of Math. 47 (1946), 639–653. (MR 8, 253.)
[2] Groups on analytic manifolds, Ann. of Math. 48 (1947), 659–669. (MR 9, 174.)
[3] Groups of differentiable and real or complex analytic transformations, Ann. of Math. 46 (1945), 685–694. (MR 7, 241.)

BOCHNER, S. and YANO, K.
[1] Tensor fields in non-symmetric connections, Ann. of Math. 56 (1952), 504–519. (MR 14, 904.)

BONAN, E.
[1] Structure presque quaternale sur une variété différentiable, C. R. Acad. Sci. Paris 260 (1965), 5445–5448. (MR 31 #5171.)
[2] Sur des variétés riemanniennes à groupe d'holonomie G_2 ou Spin (7), C. R. Acad. Sci. Paris 26 (1966), 127–129, (MR 33 #4855.)

[3] Structures presque quaternioniennes, C. R. Acad. Sci. Paris 258 (1964), 792–795. (MR 28 #4489.)
[4] Connexions presque quaternioniennes, C. R. Acad. Sci. Paris 258 (1964), 1696–1699. (MR 28 #4490.)
[5] Structures presque hermitiennes quaternioniennes, C. R. Acad. Sci. Paris 258 (1964), 1988–1991. (MR 28 #4491.)

BONNET, O.
[1] Sur quelques propriétés des lignes géodésiques, C. R. Acad. Sci. Paris 40 (1855), 1311–1313.
[2] Mémoire sur la théorie des surfaces applicables sur une surface donnée, J. École Polytechnique 42 (1867).

BOOTHBY, W.
[1] Some fundamental formulas for Hermitian manifolds with non-vanishing torsion, Amer. J. Math. 76 (1954), 509–534. (MR 15, 989.)
[2] Hermitian manifolds with zero curvature, Michigan Math. J. 5 (1958), 229–233. (MR 21 #2752.)
[3] Homogeneous complex contact manifolds, Proc. Symp. Pure Math. Vol. III, 144–154, Amer. Math. Soc., 1961. (MR 23 #A2173.)
[4] A note on homogeneous complex contact manifolds, Proc. Amer. Math. Soc. 13 (1962), 276–280. (MR 25 #582.)

BOOTHBY, W., KOBAYASHI, S., and WANG, H. C.
[1] A note on mappings and automorphisms of almost complex manifolds, Ann. of Math. 77 (1963), 329–334. (MR 26 #4284.)

BOOTHBY, W. and WANG, H. C.
[1] On contact manifolds, Ann. of Math. 68 (1958), 721–734. (MR 22 #3015.)

BOREL, A.
[1] Kählerian coset spaces of semi-simple Lie groups, Proc. Nat. Acad. Sci. USA 40 (1954), 1147-1151. (MR 17, 1108.)
[2]* Topology of Lie groups and characteristic classes, Bull. Amer. Math. Soc. 61 (1955), 397–432. (MR 17, 282.)
[3] On the curvature tensor of the Hermitian symmetric manifolds, Ann. of Math. 71 (1960), 508–521. (MR 22 #1923.)
[4] Sous-groupes compacts maximaux des groupes de Lie, Séminaire Bourbaki, Exposé 33, 1950.

BOREL, A. and HIRZEBRUCH, F.
[1] Characteristic classes and homogeneous spaces I, Amer. J. Math. 80 (1958), 458–538 (MR 21 #1586); II. ibid. 81 (1959), 315–382 (MR 22 #988); III. ibid. 82 (1960), 491–504 (MR 22 #11413.)

BOREL, A. and LICHNEROWICZ, A.
[1] Groupes d'holonomie des variétés Riemanniennes, C. R. Acad. Sci. Paris 234 (1952), 1835–1837. (MR 13, 986.)

[2] Espaces riemanniens et hermitiens symétriques, C. R. Acad. Sci. Paris 234 (1952), 2332–2334. (MR 13, 986.)

BOREL, A. and REMMERT, R.
[1] Über kompakte homogene Kählersche Mannigfaltigkeiten, Math. Ann. 145 (1961/62), 429–439. (MR 26 #3088.)

BOREL, A. and SERRE, J. P.
[1] Detérmination des p-puissances réduites de Steenrod dans la cohomologie des groupes classiques, Applications, C. R. Acad. Sci. Paris 233 (1951), 680–682. (MR 13, 319.)

BOTT, R.
[1] Non-degenerate critical manifolds, Ann. of Math. 60 (1954), 248–261. (MR 16, 276.)
[2] On manifolds all of whose geodesics are closed, Ann. of Math. 60 (1954). 375–382. (MR 17, 521.)
[3] An application of the Morse theory to the topology of Lie groups, Bull. Soc. Math. France 84 (1956), 251–281. (MR 19, 291.)
[4] Homogeneous vector bundles, Ann. of Math. 66 (1957), 203–248. (MR 19, 681.)
[5] Vector fields and characteristic numbers, Michigan Math. J. 14 (1967) 231–244. (MR 35 #2297.)

BOTT, R. and CHERN, S. S.
[1] Hermitian vector bundles and the equidistribution of the zeros of their holomorphic sections, Acta Math. 114 (1965), 71–112. (MR 32 #3070.)

BOTT, R. and SAMELSON, H.
[1] Applications of the theory of Morse to symmetric spaces, Amer. J. Math. 80 (1958), 964–1029 (MR 21 #4430); corrections: *ibid.* 83 (1961), 207–208 (MR 30 #589.)

BOURBAKI, N.
[1] Éléments de Mathématique, Groupes et Algèbres de Lie, Chapitre I, Hermann, Paris, 1960. (MR 24 #A2641.)

BOUZON, J.
[1] Structures presque-cocomplexes, Univ. et Polite. Torino Rend. Sem. Nat. 24 (1964/65), 53–123. (MR 33 #6560.)

BREMERMANN, H.
[1] Holomorphic continuation of the kernel function and the Bergman metric in several complex variables—Lectures on Functions of a Complex Variable, Univ. Michigan Press (1955), 349–383. (MR 17, 529.)

BRICKELL, F.
[1] Differentiable manifolds with an area measure, Canad. J. Math. 19 (1967) 540–549. (MR 35 #2253.)

BRIESKORN, E.
[1] Ein Satz über die komplexen Quadriken, Math. Ann. 154 (1964), 184–193. (MR 29 #5257.)

BURSTIN, C.
[1] Ein Beitrag zum Problem der Einbettung der Riemannschen Räume in Euklidischen Räume, Rec. Math. (Math. Sbornik) 38 (1931), 74–85.
[2] Beitrage der Verbiegung von Hyperflächen in euklidischen Räume, Rec. Math. (Math. Sbornik) 38 (1931), 86–93.

BUSEMANN, H.
[1] Similarities and differentiability, Tôhoku Math. J. 9 (1957), 56–67. (MR 20 #2772.)
[2] Convex Surfaces, Interscience Tracts #6, Interscience, New York, 1958. (MR 21 #3900.)
[3] Spaces with finite groups of motions, J. Math. Pure Appl. 37 (1958), 365–373. (MR 21 #913.)
[4] The Geometry of Geodesics, Academic Press, New York, 1955. (MR 17, 779.)

CALABI, E.
[1] Isometric imbedding of complex manifolds, Ann. of Math. 58 (1953), 1–23. (MR 15, 160.)
[2] On Kaehler manifolds with vanishing canonical class, Algebraic Geometry & Topology, Symposium in honor of S. Lefschetz, 78–89, Princeton, 1957. (MR 19, 62.)
[3] An extension of E. Hopf's maximum principle with an application to Riemannian geometry, Duke Math. J. 25 (1957), 45–56. (MR 19, 1056.)
[4] Improper affine hyperspheres of convex type and a generalization of a theorem of K. Jörgens, Michigan Math. J. 5 (1958), 105–126. (MR 21 #5219.)
[5] Construction and properties of some 6-dimensional almost complex manifolds, Trans. Amer. Math. Soc. 87 (1958), 407–438. (MR 24 #A558.)
[6] On compact, Riemannian manifolds with constant curvature, I, Proc. Symp. Pure Math. Vol. III, 155–180, Amer. Math. Soc. 1961. (MR 24 #A3612.)
[7] Minimal immersions of surfaces in Euclidean spheres, J. Differential Geometry 1 (1967), 111–126.

CALABI, E. and ECKMANN, B.
[1] A class of compact, complex manifolds which are not algebraic, Ann. of Math. 58 (1953), 494–500. (MR 15, 244.)

CALABI, E. and MARKUS, L.
[1] Relativistic space forms, Ann. of Math. 75 (1962), 63–76. (MR 24 #A3614.)

CALABI, E. and ROSENLICHT, M.
[1] Complex analytic manifolds without countable base, Proc. Amer. Math. Soc. 4 (1953), 335–340. (MR 15, 351.)

CALABI, E. and VESENTINI, E.
[1] Sur les variétés complexes compactes localement symétriques, Bull. Soc. Math. France 87 (1959), 311–317. (MR 22 #1922a.)
[2] On compact, locally symmetric Kaehler manifolds, Ann. of Math. 71 (1960) 472–507, (MR 22 #1922b.)

DO CARMO, M. P.
[1] The cohomology ring of certain Kaehlerian manifolds, Ann. of Math. 81 (1965), 1–14. (MR 30 #547.)

CARTAN, É.
[1] Sur les variétés à connexion affine et la théorie de la relativité généralisée, Ann. École Norm. Sup. 40 (1923), 325–412, 41 (1924) 1–25, and 42 (1925), 17–88.
[2] Les espaces à connexion conforme, Ann. Soc. Pol. Math. 2 (1923), 171–221.
[3] Sur un théorème fondamental de M. H. Weyl, J. Math. Pures Appl. 2 (1923), 167–192.
[4] Sur les variétés à connexion projective, Bull. Soc. Math. France 52 (1924), 205–241.
[5] La Géométrie des Espaces de Riemann, Mémorial des Sciences Math. Vol. 9, Gauthier-Villars, Paris, 1925.
[6] Les groupes d'holonomie des espaces généralisés, Acta Math. 48 (1926) 1–42.
[7] Sur une classe remarquable d'espaces de Riemann, Bull. Soc. Math. France 54 (1926), 214–264 and 55 (1927), 114–134.
[8] Leçons sur la Géométrie des Espaces de Riemann, Gauthier-Villars, Paris, 1928; 2nd ed., 1946.
[9] La Méthode du Repère Mobile, la Théorie de Groupes Continus et les Espaces Généralisés, Actualités Sci. et Ind. #194, Hermann, Paris, 1935.
[10] La déformation des hypersurfaces dans l'espace euclidien réel à n dimensions, Bull. Soc. Math. France 44 (1916), 65–99.
[11] Sur les variétés de courbure constante d'un espace euclidien ou non euclidien, Bull. Soc. Math. France 47 (1919), 125–160 and 48 (1920), 132–208.
[12] La géométrie des groupes de transformations, J. Math Pures Appl. 6 (1927), 1–119.
[13] La géométrie des groupes simples, Ann. Mat. Pura Appl. 4 (1927), 209–256 and 5 (1928), 253–260.
[14] Sur la possibilité de plonger un espace riemannien donné dans un espace euclidien, Ann. Soc. Pol. Math. 6 (1927), 1–17.

[15] Sur certaines formes riemanniennes remarquables des géométries à groupes fondamental simple, Ann. École Norm. Sup. 44 (1927), 345–467.
[16] Groupes simples clos et ouverts et géométrie riemannienne, J. Math. Pures Appl. 8 (1929), 1–33.
[17] Les groupes réels simples finis et continus, Ann. Sci. École Norm. Sup. 31 (1914), 263–355.
[18] Sur les domaines bornés de l'espace de n variables complexes, Abh. Math. Sem. Hamburg 11 (1935), 116–162.
[19] Les groupes de transformations continus, infinis, simples, Ann. École Norm. Sup. 26 (1909), 93–161.

CARTAN, É. and SCHOUTEN, J. A.
[1] On the geometry of the group manifold of simple and semi-simple groups, Proc. Amsterdam 29 (1926), 803–815.

CARTAN, H.
[1] Sur les groupes de transformations analytiques, Actualités Sci. et Ind. #198, Hermann, Paris, 1935.
[2] Notion d'algèbre différentielle; applications aux groupes de Lie et aux variétés où opère un groupe de Lie, Colloque de Topologie, Bruxelles (1950), 15–27. (MR 13, 107.)
[3] La transgression dans un groupe de Lie et dans un espace fibré principal, Colloque de Topologie, Bruxelles (1950), 57–71. (MR 13, 107.)

CARTAN, H. and EILENBERG, S.
[1] Homological Algebra, Princeton Univ. Press, 1956. (MR 17, 1040.)

CATTANEO-GASPARINI, I.
[1] Sur une classe de connexions linéaires à groupes d'holonomie isomorphes, C. R. Acad. Sci. Paris 246 (1958), 1145–1147. (MR 23 #A1324.)

CHARLAP, L. S.
[1] Compact flat Riemannian manifolds I, Ann. of Math. 81 (1965), 15–30. (MR 30 #543.)

CHARLAP, L. S. and VASQUEZ, A. T.
[1] Compact flat Riemannian manifolds II. The cohomology of Z_p-manifolds, Amer. J. Math. 87 (1965), 551–563. (MR 32 #6491.)

CHAVEL, J.
[1] Isotropic Jacobi fields, and Jacobi's equations on Riemannian homogeneous spaces, Comment. Math. Helv. 42 (1967) 237–248.
[2] On normal Riemannian homogeneous spaces of rank one, Bull. Amer. Math. Soc. 73 (1967), 477–481.

CHERN, S. S.
[1] Differential geometry of fiber bundles, Proc. Intern. Cong. Math. (1950), Vol. II, 397–411. (MR 13, 583.)

[2] Topics in Differential Geometry, Inst. for Adv. Study, Princeton, 1951. (MR 19, 764.)
[3] Pseudo-groupes continus infinis, Colloque de Géométrie Différentielle, Strasbourg (1953), 119–136. (MR 16, 112.)
[4] A simple intrinsic proof of the Gauss-Bonnet formula for closed Riemannian manifolds, Ann. of Math. 45 (1944), 747–752. (MR 6, 106.)
[5] On a theorem of algebra and its geometrical applications, J. Indian Math. Soc. 8 (1944), 29–36. (MR 6, 216).
[6] Integral formulas for the characteristic classes of sphere bundles, Proc. Nat. Acad. Sci. USA 30 (1944), 269–273. (MR 6, 106.)
[7] On the curvature integral in a Riemannian manifold, Ann. of Math. 46 (1945), 674–684. (MR 7, 328.)
[8] On Riemannian manifolds of four dimensions, Bull. Amer. Math. Soc. 51 (1945), 964–971. (MR 7, 216.)
[9] Some new characterizations of Euclidean sphere, Duke Math. J. 12 (1945), 279–290. (MR 7, 29.)
[10] Characteristic classes of Hermitian manifolds, Ann. of Math. 47 (1946), 85–121. (MR 7, 470.)
[11] Some new viewpoints in differential geometry in the large, Bull. Amer. Math. Soc. 52 (1946), 1–30. (MR 9, 101.)
[12] On the characteristic classes of Riemannian manifolds, Proc. Nat. Acad. Sci. USA 33 (1947), 78–82. (MR 8, 490.)
[13] Relations between Riemannian and Hermitian geometries, Duke Math. J. 20 (1953), 575–587. (MR 15, 743.)
[14] On curvature and characteristic classes of a Riemannian manifold, Abh. Math. Sem. Univ. Hamburg 20 (1955), 117–126. (MR 17, 783.)
[15]* La géométrie des sous-variétés d'un espace euclidien à plusieurs dimensions, Enseignement Math. 40 (1951–54), 26–46. (MR 16, 856.)
[16] An elementary proof of the existence of isothermal parameters on a surface, Proc. Amer. Math. Soc. 6 (1955), 771–782. (MR 17, 657.)
[17] On a generalization of Kaehler geometry, Algebraic Geometry Topology Symp. in honor of S. Lefschetz, Princeton Univ. Press 1957, 103–121. (MR 19, 314.)
[18] A proof of the uniqueness of Minkowski's problem for convex surfaces, Amer. J. Math. 79 (1957), 949–950. (MR 20 #2769.)
[19] Geometry of submanifolds in a complex projective space, Intern. Symp. Algebraic Topology, Mexico City 1958, 87–96. (MR 20 #6721.)
[20] Differentiable Manifolds, University of Chicago notes, also Textos Mat. No. 4, 1959, Recife, Brazil. (MR 24 #A3566.)

[21] Complex Manifolds, University of Chicago notes, also Textos de Mat. No. 5, 1959, Recife, Brazil. (MR 22 #1920.)
[22] Integral formulas for hypersurfaces in Euclidean space and their applications to uniqueness theorem, J. Math. Mech. 8 (1959), 947–955. (MR 22 #4997.)
[23] Pseudo-Riemannian geometry and Gauss-Bonnet formula, Ann. Acad. Brasil Ci. 35 (1963), 17–26. (MR 27 #5196.)
[24] Minimal surfaces in an Euclidean space of n dimensions, Symp. Differential and Combinatorial Topology in honor of M. Morse, 187–198, Princeton Univ. Press, 1965. (MR 31 #5156.)
[25] On the curvatures of a piece of hypersurface in Euclidean space, Abh. Math. Sem. Univ. Hamburg, 29 (1965), 77–91. (MR 32 #6376.)
[26]* The geometry of G-structures, Bull. Amer. Math. Soc. 72 (1966), 167–219. (MR 33 #661.)
[27] On Einstein hypersurfaces in a Kaehlerian manifold of constant holomorphic sectional curvature. J. Differential Geometry 1 (1967), 21–31.
[28] On the differential geometry of a piece of submanifold in Euclidean space, Proc. US-Japan Seminar in Differential Geometry, Kyoto, 1965, 17–21. (MR 35 #7268.)
[29] Curves and surfaces in Euclidean spaces, Studies in Global Geometry and Analysis, 16–65, Math. Assoc. Amer. 1967. (MR 35 #3610.)

CHERN, S. S., HANO, J., and HSIUNG, C. C.
[1] A uniqueness theorem on closed convex hypersurfaces, J. Math. Mech. 9 (1960), 85–88. (MR 22 #7084.)

CHERN, S. S. and HSIUNG, C. C.
[1] On the isometry of compact submanifolds in Euclidean space, Math. Ann. 149 (1962/63), 278–285. (MR 26 #5521.)

CHERN, S. S. and KUIPER, N.
[1] Some theorems on the isometric imbedding of compact Riemann manifolds in Euclidean space, Ann. of Math. 56 (1952), 422–430. (MR 14, 408.)

CHERN, S. S. and LASHOF, R. K.
[1] On the total curvature of immersed manifolds, Amer. J. Math. 79 (1957), 306–318 (MR 18, 927); II. Michigan Math. J. 5 (1958), 5–12 (MR 20 #4301).

CHERN, S. S. and SPANIER, E.
[1] A theorem on orientable surfaces in four dimensional space, Comment. Math. Helv. 25 (1951), 205–209. (MR 13, 492.)

CHEVALLEY, C.
[1] Theory of Lie Groups, Princeton Univ. Press, 1946. (MR 7, 412.)

[2] Invariants of finite groups generated by reflections, Amer. J. Math. 77 (1955), 778–782. (MR 17, 345.)

Chow, W. L.
[1] On compact complex analytic varieties, Amer. J. Math. 71 (1949), 893–914 MR 11, 389.) (See also the same journal 78 (1956), 898.)

Christoffel, E. B.
[1] Über die Transformation der homogenen Differentialausdrücke zweiten Grades, J. Reine Angew. Math. 70 (1869), 46–70.

Chu, H. and Kobayashi, S.
[1] The automorphism group of a geometric structure, Trans. Amer. Math. 113 (1963), 141–150. (MR 29 #1596.)

Clark, R. S.
[1] Special geometric object fields on a differentiable manifold, Tensor 13 (1963), 60–70. (MR 27 #5198.)

Clark, R. S. and Bruckheimer, M.
[1] Tensor structures on a differentiable manifold, Ann. Mat. Pura Appl. (4)54 (1961), 123–141. (MR 24 #A3596.)

Clifton, Y. H.
[1] On the completeness of Cartan connections, J. Math. Mech. 16 (1966), 569–576. (MR 34 #5017.)

Cohn, P. M.
[1] Lie Groups, Cambridge Tracts #46, Cambridge Univ. Press, New York, 1957. (MR 21 #2702.)

Cohn-Vossen, S.
[1] Zwei Sätze über Starrheit der Eiflächen, Nachr. Ges. Wiss. Göttingen, 1 (1927), 125–139.
[2] Unstarre geschlossene Flächen, Math. Ann. 102 (1929), 10–29.
[3] Kürzeste Wege und Totalkrümmung auf Flächen, Compositio Math. 2 (1935), 69–133.
[4] Existenz kürzester Wege, Compositio Math. 3 (1936), 441–452.

Couty, R.
[1] Sur les transformations de variétés riemanniennes et kähleriennes, Ann. Inst. Fourier (Grenoble) 9 (1959), 147–248. (MR 22 #12488.)
[2] Transformations projectives sur un espace d'Einstein complet, C. R. Acad. Sci. Paris (252 (1961), 109–1097. (MR 28 #569.)
[3] Transformations conformes. C.R. Acad. Sci. Paris 252 (1961), 3725, 3726. (MR 23 #A1332.)
[4] Transformations projectives des variétés presque kähleriennes, C.R. Acad. Sci. Paris 254 (1962), 4132–4134. (MR 28 #568.)

Crittenden, R.
[1] Minimum and conjugate points in symmetric spaces, Canad. J. Math. 14 (1962), 320–328. (MR 25 #533.)

[2] Covariant differentiation, Quart. J. Math. Oxford Ser. (2) 13 (1962), 285–298. (MR 26 #4283.)

VAN DANTZIG, D. and VAN DER WAERDEN, B. L.

[1] Über metrisch homogene Räume, Abh. Math. Sem. Univ. Hamburg 6 (1928), 374–376.

DAVIES, E. T. J.

[1] The first and second variations of the volume integral in Riemannian space, Quart. J. Math. Oxford 13 (1942), 58–64. (MR 4, 115.)

DIEUDONNÉ, J.

[1] Sur les espaces uniformes complets, Ann. Sci. École Norm. Sup. 56 (1939), 277–291. (MR 1, 220.)

DOLBEAULT, P.

[1] Formes différentielles et cohomologie sur une variété analytique complexe, Ann. of Math. 64 (1956), 83–130 (MR 18, 670); *ibid.* 65 (1957), 282–330 (MR 19, 171).

DOLBEAULT-LEMOINE, S.

[1] Sur la déformabilité des variétés plongées dans un espace de Riemann, Ann. Sci. École Norm. Sup. 73 (1956), 357–438. (MR 18, 819.)

[2] Réducibilité de variétés plongées dans un espace à courbure constante, Rend. Mat. Appl. (5)17 (1958), 453–474. (MR 21 #3009.)

DOMBROWSKI, P.

[1] On the geometry of the tangent bundles, J. Reine Angew. Math. 210 (1962), 73–88. (MR 25 #4463.)

[2] Krümmungsgrossen gleichungsdefinierter Untermannigfaltigkeiten Riemannscher Mannigfaltigkeiten, Jber. Deutsch. Math. Ver., to appear.

DOUGLIS, A. and NIRENBERG, L.

[1] Interior estimates for elliptic systems of partial differential equations, Comm. Pure Appl. Math. 8 (1955), 503–583 (MR 17, 743.)

DUBOIS, E.

[1] Beiträge zur Riemannschen Geometrie in Grossen, Comment. Math. Helv. 41 (1966/67), 30–50. (MR 34 #3497.)

DUSCHEK, A.

[1] Die Starrheit der Eiflächen, Montsch. Math. Phys. 36 (1929), 131–134.

[2] Zur geometrisihen Variationsrechnung, Math. Z. 40 (1936) 279–291.

VER EECKE, P.

[1] Sur les connexions d'éléments de contact, Topol. et Géom. Diff. (Sém. Ehresmann) Vol. 5, 70 pp. Inst. H. Poincaré, 1963. (MR 34 #6691.)

ECKMANN, B.
  [1] Sur les structures complexes et presque complexes, Colloque de Géométrie Différentielle, Strasbourg, 1953, 151–159. (MR 15, 649.)
  [2] Quelques propriétés globales des variétés kählériennes, C.R. Acad. Sci. 229 (1949), 577–579. (MR 11, 212.)
  [3] Complex analytic manifolds, Proc. Intern. Congr. Math. Cambridge, Mass. 1950, Vol. 2, 420–427. (MR 13, 574.)
  [4] Structures complexes et transformations infinitésimales, Convegno Intern. Geometria Differenziale, Italia, 1953, 176–184. (MR 16, 518.)
ECKMANN, B. and FRÖLICHER, A.
  [1] Sur l'intégrabilité de structures presques complexes, C.R. Acad. Sci. Paris 232 (1951), 2284–2286. (MR 13, 75.)
ECKMANN, B. and GUGGENHEIMER, H.
  [1] Formes différentielles et métrique hermitienne sans torsion I, C.R. Acad. Sci. Paris 229 (1949), 464–466 and II, 489–491. (MR 11, 212.)
  [2] Sur les variétés closes à métrique hermitienne sans torsion, C.R. Acad. Sci. Paris 229 (1949), 503–505. (MR 11, 212.)
EDWARDS, H. M.
  [1] A generalized Sturm theorem, Ann. of Math. 80 (1964), 22–57. (MR 29 #1652.)
EELLS, J., Jr.
  [1] A generalization of the Gauss-Bonnet theorem, Trans. Amer. Math. Soc. 92 (1959), 142–153. (MR 22 #1917.)
  [2]* A setting for global analysis, Bull. Amer. Math. Soc. 72 (1966), 751–807. (MR 34 #3590.)
EELLS, J., Jr. and SAMPSON, H.
  [1] Harmonic mappings of Riemannian manifolds, Amer. J. Math 86 (1964), 109–160. (MR 29 #1603.)
  [2] Énergie et déformations en géométrie différentielle, Ann. Inst. Fourier (Grenoble) 14 (1964), 61–69. (MR 30 #2529.)
  [3] Variational theory in fibre bundles, Proc. US-Japan Seminar in Differential Geometry, Kyoto, 1965, 22–33. (MR 35 # 7352.)
EFIMOV, N. V.
  [1] Generation of singularities on surfaces of negative curvature, Math. Sbornik 64 (106) (1964), 286–320. (MR 29 #5203.)
  [2]* Flächenverbiegung im Grossen, Akad. Verlag, Berlin, 1957. (MR 21 #4458.)
  [3] Qualitative problems of the theory of deformations of surfaces, Amer. Math. Soc. Transl. Ser. 1, #37. (original, MR 10, 324.)
EHRESMANN, C.
  [1] Sur la notion d'espace complet en géométrie différentielle, C.R. Acad. Sci. Paris 202 (1936), 2033.

[2] Les connexions infinitésimales dans un espace fibré différentiable, Colloque de Topologie, Bruxelles (1950), 29–55. (MR 13, 159.)
[3] Sur les espaces localement homogènes, Enseignement Math. 35 (1937), 317–333.
[4] Sur la théorie des espaces fibrés, Colloque de Topologie, Paris (1947), 3–15. (MR 11, 678.)
[5] Sur les variétés presque complexes, Proc. Intern. Congr. Math. Cambridge, Mass. 1950, Vol. 2, 412–419. (MR 13, 574.)
[6] Introduction à la théorie des structures infinitésimales et des pseudogroupes de Lie, Colloque de Géométrie Différentielle, Strasbourg, 1953, 97–110. (MR 16, 75.)
[7] Sur les pseudo-groupes de Lie de type fini, C.R. Acad. Sci. Paris 246 (1958), 360–362. (MR 21 #97.)

EISENHART, L. P.
[1]* Riemannian Geometry, Princeton Univ. Press 1949. (MR 11, 687.)
[2]* Non-Riemannian Geometry, Amer. Soc. Colloq. Publ. 8, 1927.

ELIASSON, H. I.
[1] Die Krümmung des Raumes $Sp(2)/SU(2)$ von Berger, Math. Ann. 164 (1966), 317–323. (MR 33 #4866.)
[2] Über die Anzahl geschlossener Geodätischen in gewissen Riemannschen Mannigfaltigkeiten, Math. Ann. 166 (1966), 119–147. (MR 34 #3498.)
[3] Geometry of manifolds of maps, J. Differential Geometry 1 (1967), 169–194.

FEEMAN, G. F. and HSIUNG, C. C.
[1] Characterizations of Riemannian n-spheres, Amer. J. Math. 81 (1959), 691–708. (MR 21 #5990.)

FELDMAN, E.
[1] The geometry of immersions, I, Bull. Amer. Math. Soc. 69 (1963), 693–698 (MR 27 #1967); Trans. Amer. Math. Soc. 120 (1965), 185–224 (MR 32 #3065.)
[2] The geometry of immersions, II, Bull. Amer. Math. Soc. 70 (1964), 600–607 (MR 29 #625); Trans. Amer. Math. Soc. 125 (1966), 181–215 (MR 34 #818).

FENCHEL, W.
[1] On total curvatures of Riemannian manifolds, J. London Math. Soc. 15 (1940), 15–22.

FENCHEL, W. and JESSEN, B.
[1] Mengenfunktionen und konvexe Körper, Danske Videns. Selskab. Math-Fysiske Medd. 16 (1938), 1–31.

FERUS, D.
[1] Über die absolute Totalkrümmung höher-dimensionaler Knoten. Math. Ann. 171 (1967), 81–86. (MR 35 #3682.)

FIALKOW, A.
- [1] Hypersurfaces of a space of constant curvature, Ann. of Math. 39 (1938), 762–785.

FIEBER, H.
- [1] Über eine Riemannsche Geometrie ohne Kürzeste, Arch. Math. 15 (1964), 388–391. (MR 30 #544.)

FLADT, K.
- [1] Über den Parallelismus von Levi-Civita in der Geometrie konstanten Krümmungsmasses, J. Reine Angew. Math. 201 (1959), 78–83. (MR 21 #6593.)

FLANDERS, H.
- [1] Development of an extended exterior differential calculus, Trans. Amer. Math. Soc. 75 (1953), 311–326. (MR 15, 161.)
- [2] Methods in affine connection theory, Pacific J. Math. 5 (1955), 391–431. (MR 17, 784.)
- [3] Remarks on mean curvature, J. London Math. Soc. 41 (1966), 364–366. (MR 33 #1818.)

FRANKEL, T. T.
- [1] Homology and flows on manifolds, Ann. of Math. 65 (1957), 331–339. (MR 19, 453.)
- [2] Fixed points and torsions on Kaehler manifolds, Ann. of Math. 70 (1959), 1–8. (MR 24 #A1730.)
- [3] Manifolds with positive curvature, Pacific J. Math. 11 (1961), 165–174. (MR 23 #A600.)
- [4] On the fundamental group of a compact minimal submanifold, Ann. of Math. 83 (1966), 68–73. (MR 32 #4637.)
- [5] On theorems of Hurwitz and Bochner, J. Math. Mech. 15 (1966), 373–377. (MR 33 #675.)
- [6] Critical submanifolds of the classical groups and Stiefel manifolds, Symposium on Differential and Combinatorial Topology in honor of M. Morse, Princeton Univ. Press, 1965, 37–53. (MR 33 #4952.)

FREUDENTHAL, H.
- [1] Die Topologie der Lieschen Gruppen als algebraisches Phänomen I, Ann. of Math. 42 (1941), 1051–1074. (MR 3, 198.)
- [2] Topologische Gruppen mit genügent vielen fastperiodischen Funktionen, Ann. of Math. 37 (1936), 57–77.
- [3] Neure Fassungen des Riemann-Helmholtzsche Raumproblem, Math. Z. 63 (1955/56), 374–405. (MR 18, 591.)
- [4] Zu den Weyl-Cartanschen Raumproblemen, Arch. Math. 11 (1960) 107–115. (MR 22 #5007.)
- [5] Clifford-Wolf-Isometrien symmetrischer Räume, Math. Ann. 150 (1963), 136–149. (MR 27 #693.)

[6] Das Helmholtz-Liesche Raumproblem bei indefiniter Metrik, Math. Ann. 156 (1964), 263–312. (MR 30 #5253.)

FRIEDMAN, A.
 [1] Local isometric imbeddings of Riemannian manifolds with indefinite metrics, J. Math. Mech. 10 (1961), 625–649. (MR 23 #A2845.)
 [2] Isometric embedding of Riemannian manifolds into Euclidean spaces, Rev. Modern Phys. 37 (1965), 201–203. (MR 31 #3983.)
 [3] Function-theoretic characterization of Einstein spaces and harmonic spaces, Trans. Amer. Math. Soc. 101 (1961), 240–258. (MR 24 #A1686.)

FROBENIUS, G.
 [1] Über die unzerlegbaren diskreten Bewegungsgruppen, Sitzungsber. König. Preuss. Acad. Wiss. Berlin 29 (1911), 654–665.

FRÖLICHER, A.
 [1] Zur Differentialgeometrie der komplexen Strukturen, Math. Ann. 129 (1955), 50–95. (MR 16, 857.)

FRÖLICHER, A. and NIJENHUIS, A.
 [1] Theory of vector valued differential forms I, Indag. Math. 18 (1956), 338–359 (MR 18, 569); II. *ibid*. 20 (1958), 414–429 (MR 21 #890).
 [2] Invariance of vector form operations under mappings, Comment Math. Helv. 34 (1960), 227–248. (MR 24 #2334.)
 [3] A theorem of stability of complex structures, Proc. Nat. Acad. Sci. USA 43 (1957), 239–241. (MR 18, 762.)

FUBINI, G.
 [1] Sulle metriche definite da una forma Hermitiana, Atti Instit. Veneto 6 (1903), 501–513.

FUJIMOTO, A.
 [1] On the structure tensor of G-structures, Mem. Coll. Sci. Univ. Kyoto Ser. A 18 (1960), 157–169. (MR 22 #8522.)
 [2] On decomposable symmetric affine spaces, J. Math. Soc. Japan, 9 (1957), 158–170. (MR 18, 932.)
 [3] On automorphisms of G-structures, J. Math. Kyoto Univ. 1 (1961/62), 1–20. (MR 25 #4459.)
 [4]* Theory of G-structures, A report in differential geometry (in Japanese), Vol. 1, 1966.

FUKS, B. A.
 [1] The Ricci curvature of the Bergman metric invariant with respect to biholomorphic mappings, Dokl. Akad. Nauk SSSR 167 (1966), 996–999 (Soviet Math. Dokl. 7 (1966), 525–529). (MR 33 #4954.)

GAFFNEY, M. P.
 [1] A special Stokes theorem for complete Riemannian manifolds, Ann. of Math. 60 (1954), 140–145. (MR 15, 986.)

[2] The heat equation method of Milgram and Rosenbloom for open Riemannian manifolds, Ann. of Math. 60 (1954), 458–466. (MR 16, 358.)
[3] Hilbert space methods in the theory of harmonic integrals, Trans. Amer. Math. Soc. 78 (1955), 426–444. (MR 16, 957.)

GANTMACHER, F. R.
[1] Canonical representations of automorphisms of a complex semi-simple Lie groups, Rec. Math. (Moscou) 5 (47) (1939), 101–146. (MR 1, 163.)
[2] On the classification of real simple Lie groups, Rec. Math. (Mat. Sbornik) 5 (47) (1939), 217–248. (MR 2, 5.)

GARABEDIAN, P. R. and SPENCER, D. C.
[1] A complex tensor calculus for Kaehler manifolds, Acta Math. 89 (1953), 279–331. (MR 16, 74.)

GARDNER, R.
[1] Invariant of Pfaffian systems, Trans. Amer. Math. Soc. 126 (1967) 514–533. (MR 35 #2233.)

GODEMENT, R.
[1] Topologie Algébrique et Théorie des Faisceaux, Actualités Sci. et Ind. #1252, Hermann, Paris, 1958. (MR 21 #1583.)

GOETZ, A.
[1] On induced connections, Fund. Math. 55 (1964), 149–174. (MR 21 #5194.)

GOLDBERG, J. N. and KERR, K. P.
[1] Some applications of the infinitesimal holonomy groups to the Petrov classification of Einstein spaces, J. Math. Phys. 2 (1961), 327–332. (MR 23 #A2836.)

GOLDBERG, S. I.
[1] Tensor fields and curvature in Hermitian manifolds with torsion, Ann. of Math. 63 (1956), 64–76. (MR 17, 787.)
[2] On characterizations of the Euclidean sphere, Indag. Math. 21 (1959), 384–390. (MR 22 #219.)
[3] Groups of transformations of Kaehler and almost Kaehler manifolds, Comment Math. Helv. 35 (1961), 35–46. (MR 23 #A606.)
[4] Conformal maps of almost Kaehlerian manifolds, Tôhoku Math. J. 13 (1961) 119–131. (MR 25 #3486.)
[5]* Curvature and Homology, Academic Press, New York, 1962. (MR 25 #2537.)

GOLDBERG, S. I. and KOBAYASHI, S.
[1] The conformal transformation group of a compact Riemannian manifold, Proc. Nat. Acad. Sci. USA 48 (1962), 25–26 (MR 25 #534); Amer. J. Math. 84 (1962), 170–174. (MR 25 #3481.)
[2] The conformal transformation group of a compact homogeneous

Riemannian manifold, Bull. Amer. Math. Soc. 68 (1962), 378–381. (MR 25 #3482.)

[3] On holomorphic bisectional curvature, J. Differential Geometry 1 (1967), 225–233.

GOTO, M.

[1] Faithful representations of Lie groups I, Math. Japon. 1 (1948), 1–13. (MR 10, 681.)

[2] On algegraic homogeneous spaces, Amer. J. Math. 76 (1954), 811–818. (MR 16, 568.)

GOTO, M. and SASAKI, S.

[1] Some theorems on holonomy groups of Riemannian manifolds, Trans. Amer. Math. Soc. 80 (1955), 148–158. (MR 17, 659.)

GRAUERT, H.

[1] Characterisierung der Holomorphiegebiete durch die vollständige Kählersche Metrik, Math. Ann. 131 (1956), 38–75. (MR 17, 1072.)

GRAUERT, H. and REMMERT, R.

[1] Über kompakte homogene komplexe Mannigfaltigkeiten, Arch. Math. 13 (1962), 498–507. (MR 26 #3089.)

GRAY, A.

[1] Minimal varieties and almost Hermitian submanifolds, Michigan Math. J. 12 (1965), 273–287. (MR 32 #1658.)

[2] Some examples of almost Hermitian manifolds, Illinois J. Math. 10 (1966), 353–366. (MR 32 #8289.)

[3] Pseudo-Riemannian almost product manifolds and submersions, J. Math. Mech. 16 (1967), 715–737. (MR 34 #5018.)

[4] Spaces of constancy of curvature operators, Proc. Amer. Math. Soc. 17 (1966), 897–902. (MR 33 #6550.)

GRAY, J. W.

[1] Some global properties of contact structures, Ann. of Math. 69 (1959), 421–450. (MR 22 #3016.)

GREEN, L. W.

[1] Surfaces without conjugate points, Trans. Amer. Math. Soc. 76 (1954), 529–546. (MR 16, 70.)

[2] A theorem of E. Hopf, Michigan Math. J. 5 (1958), 31–34. (MR 20 #4300.)

[3] A sphere characterization related to Blaschke's conjecture, Pacific J. Math. 10 (1960), 837–841. (MR 22 #7083.)

[4] Auf Wiedersehensflächen, Ann. of Math. 78 (1963), 289–299. (MR 27 #5206.)

GREUB, W. and TONDEUR, F.

[1] On sectional curvatures and characteristic of homogeneous spaces, Proc. Amer. Math. Soc. 17 (1966), 444–448. (MR 32 #6369.)

GRIFFITHS, P. A.
  [1] On a theorem of Chern, Illinois J. Math. 6 (1962), 468–479. (MR 26 #1905.)
  [2] Some geometric and analytic properties of homogeneous complex manifolds, Acta Math. 110 (1963), 115–155 (MR 26 #6993), and 157–208 (MR 27 #4249).
  [3] On the differential geometry of homogeneous vector bundles, Trans. Amer. Math. Soc. 109 (1963), 1–34. (MR 30 #2427.)
  [4] Hermitian differential geometry and the theory of positive and ample holomorphic vector bundles, J. Math. Mech, 14 (1965), 117–140. (MR 30 #1520.)

GROMOLL, D.
  [1] Differenzierbare Strukturen und Metriken positiver Krümmung auf Sphären, Math. Ann. 164 (1966), 353–371. (MR 33 #4940.)

GROMOLL, D. and MEYER, W. T.
  [1] On complete open manifolds of positive curvature, Ann. of Math., to appear.

GROSSMAN, N.
  [1] Two applications of the technique of length-decreasing variations, Proc. Amer. Math. Soc. 18 (1967) 327–333. (MR 35 #943.)

GROTEMEYER, K. P.
  [1] Über das Normalebündel differenzierbarber Mannigfaltigikeiten, Ann. Acad. Sci. Fenn. Ser. A I No. 336/15 (1963), 12 pp. (MR 29 #568.)
  [2] Zur eindeutige Bestimmung von Flächen durch die erste Fundamentalform, Math. Z. 55 (1952), 253–268. (MR 13, 984.)

GUGGENHEIMER, H.
  [1] Über komplex-analytischer Mannigfaltigkeiten mit Kählerscher Metrik, Comment Math. Helv. 25 (1951), 257–297. (MR 13, 781.)

GUILLEMIN, V. W.
  [1] The integrability problem for G-structures, Trans. Amer. Math. Soc. 116 (1965) 544–560. (MR 34 #3475.)

GUILLEMIN, V. W. and SINGER, I. M.
  [1] Differential equations and G-structures, Proc. US-Japan Seminar in Differential Geometry, Kyoto, 1965, 34–36. (MR 35 #4840.)

GUILLEMIN, V. W. and STERNBERG, S.
  [1] An algebraic model of transitive differential geometry, Bull. Amer. Math. Soc. 70 (1964), 16–47. (MR 30 #533.)
  [2] Sur les systèmes de formes différentielles, Ann. Inst. Fourier (Grenoble) 13 (1963), fasc. 2, 61–74, (MR 30 #2427.)
  [3] On differential systems II, Proc. Nat. Acad. Sci. USA 50 (1963), 994–995. (MR 30 #2428.)

GUNNING, R. C.
  [1] Connections for a class of pseudogroup structures, Proc. Conf. Complex Analysis (Minneapolis, 1964), 186–194, Springer, Berlin, 1965. (MR 31 #2684.)
GUNNING, R. C. and ROSSI, H.
  [1]* Analytic Functions of Several Complex Variables, Prentice-Hall, Englewood Cliffs, New Jersey, 1965. (MR 31 #4927.)
HADAMARD, J.
  [1] Les surfaces à courbures opposées et leurs lignes géodésiques, J. Math. Pures Appl. 4 (1898), 27–73.
HALL, M.
  [1] The Theory of Groups, Macmillan, New York, 1959. (MR 21 #1996.)
HANGAN, T.
  [1] Sur les connexions projectives, Rev. Math. Pures Appl. 3 (1958), 265–276. (MR 22 #952.)
  [2] Derivées de Lie et holonomie dans la théorie des connexions infinitésimales, Acad. Rep. Pop. Romîne Stud. Cerc. Mat. 11 (1960), 159–173. (MR 22 #5984.)
HANO, J-I.
  [1] On affine transformations of a Riemannian manifold, Nagoya Math. J. 9 (1955), 99–109. (MR 17, 891.)
  [2] On Kaehlerian homogeneous spaces of unimodular Lie groups, Amer. J. Math. 79 (1957), 885–900. (MR 20 #2477.)
  [3] On compact complex coset spaces of reductive Lie groups, Proc. Amer. Math. Soc. 15 (1964), 159–163. (MR 28 #1258.)
HANO, J-I. and KOBAYASHI, S.
  [1] A fibering of a class of homogeneous complex manifolds, Trans. Amer. Math. Soc. 94 (1960), 233–243. (MR 22 #5990.)
HANO, J-I. and MATSUSHIMA, Y.
  [1] Some studies of Kaehlerian homogeneous spaces, Nagoya Math. J. 11 (1957), 77–92. (MR 18, 934.)
HANO, J-I. and MORIMOTO, A.
  [1] Note on the group of affine transformations of an affinely connected manifold, Nagoya Math. J. 8 (1955), 71–81, (MR 16, 1053.)
HANO, J-I. and OZEKI, H.
  [1] On the holonomy groups of linear connections, Nagoya Math. J. 10 (1956), 97–100. (MR 18, 507.)
HANTZSCHE, W. and WENDT, W.
  [1] Dreidimensionale euklidische Raumformen, Math. Ann. 110 (1934), 593–611.
HARTMAN, P.
  [1] On isometric immersions in Euclidean space of manifolds with

non-negative sectional curvatures, Trans. Amer. Math. Soc. 115 (1965), 94–109. (MR 34 #1968.)

[2] On homotopic harmonic maps, Canad. J. Math. 19 (1967), 673–687. (MR 35 #4856.)

HARTMAN, P. and NIRENBERG, L.

[1] On spherical image maps whose Jacobians do not change sign, Amer. J. Math. 81 (1959), 901–920. (MR 23 #4106.)

HARTMAN, P. and WINTNER, A.

[1] On the embedding problem in differential geometry, Amer. J. Math. 72 (1950), 553–564. (MR 12, 51.)

[2] The fundamental equations of differential geometry, Amer. J. Math. 72 (1950), 757–774. (MR 12, 357.)

HATTORI, A.

[1] On 3-dimensional elliptic space forms (Japanese), Sugaku 12 (1960/61), 164–167. (MR 25 #2558.)

HAWLEY, N. S.

[1] Constant holomorphic curvature, Canad. J. Math. 5 (1953), 53–56. (MR 14, 690.)

[2] A theorem on compact complex manifolds, Ann. of Math. 52 (1950), 637–641. (MR 12, 603.)

HEINZ, E.

[1] On Weyl's embedding problem, J. Math. Mech. 11 (1962), 421–454. (MR 25 #2565.)

HELGASON, S.

[1] Some remarks on the exponential mapping for an affine connection, Math. Scand. 9 (1961), 129–146. (MR 24 #A1688.)

[2]* Differential Geometry and Symmetric Spaces, Academic Press, New York, 1962. (MR 26 #2986.)

[3] A duality in integral geometry on symmetric spaces, Proc. US-Japan Seminar in Differential Geometry, Kyoto, 1965, 37–56.

[4] On Riemannian curvature of homogeneous spaces, Proc. Amer. Math. Soc. 9 (1958), 831–838. (MR 21 #7523.)

[5] Totally geodesic spheres in compact symmetric spaces, Math. Ann. 165 (1966) 309–317. (MR 35 #938.)

HENRICH, C. J.

[1] Derivations on an arbitrary vector bundle, Trans. Amer. Math. Soc. 109 (1963), 411–419. (MR 27 #5195.)

HERGLOTZ, G.

[1] Über die Starrheit der Eiflächen, Abh. Math. Sem. Hamburg 15 (1943), 127–129. (MR 7, 322.)

Hermann, R.

[1] On the differential geometry of foliations, Ann. of Math. 72 (1960), 445–457; J. Math. Mech. 11 (1962), 305–315 (MR 25 #5524.)

[2] Compact homogeneous almost complex spaces of positive characteristic, Trans. Amer. Math. Soc. 83 (1956), 471–481. (MR 18, 762.)
[3] Geodesics of bounded symmetric domains, Comment. Math. Helv. 35 (1961), 1–8. (MR 23 #A2170.)
[4] Focal points of closed submanifolds of Riemannian spaces, Indag. Math. 25 (1963), 613–628. (MR 28 #1558.)
[5] Spherical compact hypersurfaces, J. Math. Mech. 13 (1964), 237–242. (MR 28 #3391.)
[6] An incomplete compact homogeneous Lorentz metric, J. Math. Mech. 13 (1964), 497–501. (MR 28 #5406.)
[7] Geometric aspects of potential theory in the symmetric bounded domains, Math. Ann. 148 (1962), 349–366 (MR 26 #6995); II. 151 (1963), 143–149 (MR 27 #5276); III. 153 (1964), 384–394 (MR 28 #3462).
[8] Convexity and pseudo-convexity for complex manifolds, J. Math. Mech. 13 (1964), 667–672 (MR 29 #5260); II. ibid. 1065–1070 (MR 30 #2531).
[9] Complex domains and homogeneous spaces, J. Math. Mech. 13 (1964), 243–247. (MR 28 #3463.)
[10] A sufficient condition that a map of Riemannian manifolds be a fiber bundle, Proc. Amer. Math. Soc. 11 (1960), 236–242. (MR 22 #3006.)
[11] Existence in the large of totally geodesic submanifolds of Riemannian spaces, Bull. Amer. Math. Soc. 66 (1960), 59–61. (MR 22 #8458.)
[12] Cartan connections and the equivalence problem for geometric structures, Contributions to Differential Equations 3 (1964), 199–248. (MR 29 #2741.)
[13] A Poisson Kernel for certain homogeneous spaces, Proc. Amer. Math. Soc. 12 (1961), 892–899. (MR 28 #4554.)
[14] Homogeneous Riemannian manifolds of non-positive sectional curvature, Indag. Math. 25 (1963), 47–56. (MR 27 #2940.)
[15] Vanishing theorems for homology of submanifolds, J. Math. Mech. 14 (1965), 479–483. (MR 33 #737.)
[16] Equivariance invariants for submanifolds of homogeneous spaces, Math. Ann. 158 (1965), 284–289. (MR 34 #3502.)
[17] Existence in the large of parallel homomorphisms, Trans. Amer. Math. Soc. 108 (1963), 170–183. (MR 27 #1905.)

HICKS, N.
[1] A theorem on affine connexions, Illinois J. Math. 3 (1959), 242–254. (MR 21 #6597.)
[2] An example concerning affine connexions, Proc. Amer. Math. Soc. 11 (1960), 952–956. (MR 22 #12480.)

[3] Connexion preserving, conformal and parallel maps, Michigan Math. J. 10 (1963), 295–302. (MR 27 #5202.)
[4] Linear perturbations of connexions, Michigan Math. J. 12 (1965), 389–397. (MR 32 #3012.)
[5] Submanifolds of semi-Riemannian manifolds, Rend. Circ. Mat. Palermo (2) 12 (1963), 137–149. (MR 29 #1601.)
[6] On the Ricci and Weingarten maps of a hypersurface, Proc. Amer. Math. Soc. 16 (1965), 491–493. (MR 31 #755.)
[7] Connexion preserving spray maps, Illinois J. Math. 10 (1966), 661–679. (MR 34 #5024.)

HILBERT, D.
[1] Über das Dirichletsche Prinzip, J. Reine Angew. Math. 129 (1905), 63–67.
[2] Über Flächen von konstanter Gausscher Krümmung, Trans. Amer. Math. Soc. 2 (1901), 87–99.

HIRONAKA, H.
[1] An example of a non-Kaehlerian complex analytic deformation of Kaehlerian complex structures, Ann. of Math. 75 (1962), 190–208. (MR 25 #2618.)

HIRZEBRUCH, F.
[1] Some problems on differentiable and complex manifolds, Ann. of Math. 60 (1954), 213–236. (MR 16, 518.)
[2] Automorphe Formen und der Satz von Riemann-Roch, Intern. Symp. Algebraic Topology, 129–144, Mexico City, 1958. (MR 21 #2058)
[3]* Topological Methods in Algebraic Geometry, third enlarged edition. Grundlehren der Math. Wissenschaften, 131, Springer-Verlag, New York, 1966. (MR 34 #2573.)

HIRZEBRUCH, F. and KODAIRA, K.
[1] On the complex projective spaces, J. Math. Pures Appl. 36 (1957), 201–216. (MR 19, 1077.)

HIRZEBRUCH, U.
[1] Über Jordan-Algebren und kompakten Riemannsche symmetrische Räume von Rang 1, Math. Z. 90 (1965), 339–354. (MR 32 #6371.)

HOCHSCHILD, G.
[1] The Structure of Lie Groups, Holden-Day, San Francisco, 1965. (MR 34 #7696.)

HODGE, W. V. D.
[1] The Theory and Applications of Harmonic Integrals, Cambridge Univ. Press, Cambridge, England, 1952. (MR 2, 296.)
[2] A special type of Kaehler manifolds, Proc. London Math. Soc. 1 (1951), 104–117. (MR 12, 848.)

[3] Structure problem for complex manifolds, Rend. Math. 11 (1952), 101–110. (MR 15, 649.)

HOPF, E.
[1] Closed surfaces without conjugate points, Proc. Nat. Acad. Sci. 34 (1948), 47–51. (MR 9, 378.)
[2] On S. Bernstein's theorem on surfaces $z(x,y)$ of non-positive curvature, Proc. Amer. Math. Soc. 1 (1950), 80–85. (MR 12, 13.)

HOPF, H.
[1] Zum Clifford-Kleinschen Raumproblem, Math. Ann. 95 (1925), 313–339.
[2] Über die Curvatura integra geschlossener Hyperflächen, Math. Ann. 95 (1925), 340–367.
[3] Géométrie infinitésimales et topologie, Enseignement Math. 30 (1931), 233–240.
[4] Differentialgeometrie und topologische Gestalt, Jber. Deutsch. Math. Ver. 41 (1932), 209–229.
[5] Zur Topologie der komplexen Mannigfaltigkeiten, Studies and Essays presented to R. Courant, Interscience, New York, 1948, 167–185. (MR 9, 298.)
[6] Über komplex-analytische Mannigfaltigkeiten, Rend. Mat. Univ. Roma 10 (1951), 169–182. (MR 13, 861.)
[7] Über Flächen mit einer Relation zwischen den Hauptkrümmungen, Math. Nachr. 4 (1951), 232–249. (MR 12, 634.)
[8] Zur Differentialgeometrie geschlossener Flächen im Euklidischen Raum, Conv. Int. Geometria Diff. Italia 1953, 45–54. Ed. Cremonese, Roma, 1954. (MR 16, 167.)

HOPF, H. and RINOW, W.
[1] Über den Begriff der vollständigen differentialgeometrischen Flächen, Comment. Math. Helv. 3 (1931), 209–225.
[2] Die topologischen Gestalten differentialgeometrisch verwandter Flächen, Math. Ann. 107 (1932), 113–123.

HOPF, H. and SAMELSON, H.
[1] Zum Beweis des Kongruenzsatzes für Eiflächen, Math. Z. 43 (1938), 749–766.

HOPF, H. and SCHILT, H.
[1] Über Isometrie und stetige Verbiegung von Flächen, Math. Ann. 116 (1938), 749–766.

HOPF, H. and VOSS, K.
[1] Ein Satz aus der Flächentheorie im Grossen, Arch. Math. 3 (1952), 187–192. (MR 14, 583.)

HÖRMANDER, L.
[1] An Introduction to Complex Analysis in Several Variables, Van Nostrand, Princeton, 1966. (MR 34 #2933.)

HSIANG, WU-YI
  [1] On the compact homogeneous minimal submanifolds, Proc. Nat. Acad. Sci. USA 56 (1966), 5-6. (MR 34 #5037.)
  [2] The natural metric on $SO(n)/SO(n-2)$ is the most symmetric metric, Bull. Amer. Math. Soc. 73 (1967), 55-58 (MR 35 #939).
  [3] On the bound of the dimensions of the isometry groups of all possible riemannian metrics on an exotic sphere, Ann. of Math. 85 (1967), 351-358. (MR 35 #4935.)
HSIUNG, C. C.
  [1] On differential geometry of hypersurfaces in the large, Trans. Amer. Math. Soc. 81 (1956), 243-252. (MR 17, 890.)
  [2] Some global theorems on hypersurfaces, Canad. J. Math. 9 (1957), 5-14. (MR 18, 818.)
  [3] A uniqueness theorem for Minkowski's problem for convex surfaces with boundary, Illinois J. Math. 2 (1958), 71-75. (MR 20 #2770.)
  [4] Some uniqueness theorems on compact Riemannian manifolds with boundary, Illinois J. Math. 4 (1960), 526-540. (MR 23 #A610.)
  [5] Curvature and homology of Riemannian manifolds with boundary, Math. Z. 82 (1963), 67-81. (MR 27 #2943.)
  [6] On the group of conformal transformations of a compact Riemannian manifold, Proc. Nat. Acad. Sci. USA 54 (1965), 1509-1513. (MR 32 #6372.)
  [7] Vector fields and infinitesimal transformations on Riemannian manifolds with boundary, Bull. Soc. Math. France 92 (1964), 411-434. (MR 31 #2693.)
  [8] On the congruence of hypersurfaces, Atti Accad. Naz. Incei Rend. Cl. Sci. Fis. Nat. Natur (8) 37 (1964), 258-266. (MR 33 #678.)
HU, S. T.
  [1] Homotopy Theory, Academic Press, New York, 1959. (MR 21 #5186.)
HUA, L. K.
  [1] Harmonic analysis of functions of several complex variables in the classical domains, Translations of Math. Monographs, Vol. 6, 1963, Amer. Math. Soc. (MR 30 #2162.)
HUSEMOLLER, D.
  [1] Fibre Bundles, McGraw-Hill, New York, 1966.
IGUSA, J.
  [1] On the structure of a certain class of Kähler manifolds, Amer. J. Math. 76 (1954), 669-678. (MR 16, 172.)
ISE, M.
  [1] Some properties of complex analytic vector bundles over compact complex homogeneous spaces, Osaka Math. J. 12 (1960), 217-252. (MR 23 #A2227.)

[2] On the geometry of Hopf manifolds, Osaka Math. J. 12 (1960), 387–402. (MR 23 #2228.)

[3] Generalized automorphic forms and certain holomorphic vector bundles, Amer. J. Math. 86 (1964), 70–108. (MR 29 #6502.)

ISHIHARA, S.

[1] Homogeneous Riemannian spaces of four dimensions, J. Math. Soc. Japan 7 (1955), 345–370. (MR 18, 599.)

[2] Groups of projective transformations and groups of conformal transformations, J. Math. Soc. Japan 9 (1957), 195–227. (MR 20 #311.)

IWAHORI, N.

[1] Fixed point theorem in Riemannian symmetric spaces, Summer Seminar in Diff. Geometry, Akakura, Japan, 1956 (in Japanese).

[2] Some remarks on tensor invariants of $O(n)$, $U(n)$, $Sp(n)$, J. Math. Soc. Japan 10 (1958), 145–160. (MR 23 #A1722.)

[3] On discrete reflection groups on symmetric spaces, Proc. US-Japan Seminar in Differential Geometry, Kyoto, Japan, 1965, 57–62. (MR 36 #830.)

[4] Theory of Lie Groups (in Japanese), Iwanami, Tokyo 1957.

IWAMOTO, H.

[1] Sur les espaces riemanniens symétriques I, Japan. J. Math. 19 (1948), 513–523. (MR 11, 399, 872.)

[2] On integral invariants and Betti numbers of symmetric Riemannian manifolds, Japan. J. Math. 1 (1949), 91–110 (MR 11, 377); II. 235–243 (MR 12, 122).

[3] On the structure of Riemannian spaces whose holonomy groups fix a null system, Tôhoku Math. J. 1 (1950), 109–135. (MR 12, 536.)

[4] On the relation between homological structure of Riemannian spaces and exact differential forms which are invariant under holonomy groups, Tôhoku Math. J. 3 (1951), 59–70. (MR 13, 75.)

IWASAWA, K.

[1] On some types of topological groups, Ann. of Math. 50 (1949), 507–558. (MR 10, 679.)

JACOBSON, N.

[1] Lie Algebras, Interscience Tracts #10, Interscience, New York, 1962. (MR 26 #1345.)

[2] Cayley numbers and normal simple Lie algebras of type G, Duke Math. J. 5 (1939), 775–783. (MR 1, 100.)

JANET, M.

[1] Sur la possibilité de plonger un espace riemannien donné dans un espace euclidien, Ann. Soc. Math. Pol. 5 (1926), 38–43.

KÄHLER, E.

[1] Über eine bemerkenswerte Hermitesche Metrik, Abh. Math. Sem. Univ. Hamburg 9 (1933), 173–186.

KAMBER, F. and TONDEUR, P.
 [1] Flat bundles and characteristic classes of group representations, Amer. J. Math. 89 (1967), 857–886.
KANEYUKI, S. and NAGANO, T.
 [1] On the first Betti numbers of compact quotient spaces of complex semi-simple Lie groups by discrete subgroups, Sci. Papers Coll. Gen. Ed. Univ. Tokyo 12 (1962), 1–11. (MR 26 #2547.)
 [2] On certain quadratic forms related to symmetric Riemannian spaces, Osaka Math. J. 14 (1962), 241–252. (MR 28 #2564.)
KARRER, G.
 [1] Einfürhrung von Spinoren auf Riemannschen Mannigfaltigkeiten, Ann. Acad. Sci. Fenn Ser. A. I No. 336/5 (1963), pp. 16. (MR 29 #1595.)
KASHIWABARA, S.
 [1] On the reducibility of an affinely connected manifold, Tôhoku Math. J. 8 (1956), 13–28. (MR 18, 332.)
 [2] The decomposition of a differentiable manifold and its applications, Tôhoku Math. J. 11 (1959), 43–53. (MR 21 #5998.)
 [3] The structure of a Riemannian manifold admitting a parallel field of one-dimensional tangent vector subspaces, Tôhoku Math. J. 11 (1959), 327–350. (MR 22 #4076.)
KATSURADA, Y.
 [1] Generalized Minkowski formulas for closed hypersuperfaces in Riemann space, Ann. Mat. Pura Appl. 57 (1962), 283–293. (MR 26 #2989.)
 [2] On a certain property of closed hypersurfaces in an Einstein space, Comment. Math. Helv. 38 (1964), 165–171. (MR 28 #4483.)
 [3] On the isoperimetric problem in a Riemann space, Comment. Math. Helv. 41 (1966/67), 18–29. (MR 34 #5032.)
 [4] Some congruence theorems for closed hypersurfaces in Riemann spaces (Part I: Method based on Stokes' theorem), Comment. Math. Helv. 43 (1968), 176–194.
 [5] On a piece of hypersurface in a Riemann manifold with mean curvature bounded away from zero, Trans. Amer. Soc. 129 (1967) 447–457.
KAUP, W.
 [1] Reele Transformationsgruppe und invariante Metriken auf komplexen Räumen, Invent. Math. 3 (1967), 43–70.
KELLEY, J. L.
 [1] General Topology, Van Nostrand, Princeton, 1955. (MR 16, 1136.)
KERBRAT, Y.
 [1] Sous-variétés complexes de $\mathbf{C}^n$, C.R. Acad. Sci. Paris 262 (1966), 1171–1174. (MR 33 #4871.)

KERVAIRE, M.
  [1] Courbure intégrale généralisée et homotopie, Math. Ann. 131 (1956), 219–252. (MR 19, 160.)
KILLING, W.
  [1] Die nicht-Euklidischen Raumformen in analytische Behandlung, Teubner, Leibzig, 1885.
  [2] Über die Clifford-Kleinschen Raumformen, Math. Ann. 39 (1891), 257–278.
  [3] Über die Grundlagen der Geometrie, J. Reine Angew. Math. 109 (1892), 121–186.
  [4] Einführung in die Grundlagen der Geometrie, Paderborn, 1893.
KIRCHHOFF, A.
  [1] Sur l'existence de certains champs tensoriels sur les sphères à n dimensions, C.R. Acad. Sci. Paris 225 (1947), 1258–1260. (MR 9, 298.)
  [2] Beiträge zur topologischen linearen Algebra, Compositio Math. 11 (1953), 1–36. (MR 15, 500.)
KLEIN, F.
  [1] Vergleichende Betrachtungen über neuere geometrische Forschungen, Math. Ann. 43 (1893), 63–100.
  [2] Vorlesungen über nicht-Euklidische Geometrie, Springer, Berlin, 1928.
  [3] Zur nicht-Euklidischen Geometrie, Math. Ann. 37 (1890), 544–572.
KLINGENBERG, W.
  [1] Eine Kennzeichnung der Riemannschen sowie der Hermiteschen Mannigfaltigkeiten, Math. Z. 70 (1959), 300–309. (MR 21 #3891.)
  [2] Contributions to Riemannian geometry in the large, Ann. of Math. 69 (1959), 654–666. (MR 21 #4445.)
  [3] Über kompakte Riemannsche Mannigfaltigkeiten, Math. Ann. 137 (1959), 351–361. (MR 21 #4446.)
  [4] Neue Ergebnisse über konvexe Flächen, Comment. Math. Helv. 34 (1960), 17–36. (MR 22 #4996.)
  [5] Über Riemannsche Mannigfaltigkeiten mit positiver Krümmung, Comment. Math. Helv. 35 (1961), 47–54. (MR 25 #2559.)
  [6] Über Riemannsche Mannigfaltigkeiten mit nach oben beschränkter Krümmung Ann. Mat. Pura Appl. 60 (1962), 49–59. (MR 28 #556.)
  [7] Neue Methoden und Ergebnisse in der Riemannschen Geometrie, Jber. Deut. Math. Ver. 66 (1963/64), 85–94. (MR 29 #2748.)
  [8] Manifolds with restricted conjugate locus, Ann. of Math. 78 (1963), 527–547 (MR 28 #2506); II. 80 (1964), 330–339. (MR 29 #2749.)
  [9] The triangle theorem in the Riemannian geometry, Textos de Mat. No. 14, Recife, Brazil, 1964. (MR 32 #4634.)

[10] On the number of closed geodesics on a riemannian manifold, Bull. Amer. Math. Soc. 70 (1964), 279–282 (MR 29 #1604.)
[11] The theorem of three closed geodesics, Bull. Amer. Math. Soc. 71 (1965), 601–605. (MR 31 #1637.)
[12] Closed geodesics on Riemannian manifolds, Rev. Roum. Math. Pures Appl. 10 (1965), 1099–1104. (MR 34 #761.)

KLOTZ, T. and OSSERMAN, R.
 [1] Complete surfaces in E^3 with constant mean curvature, Comment. Math. Helv. 41 (1966/67), 313–318. (MR 35 #2213.)

KNEBELMAN, M. S. and YANO, K.
 [1] On homothetic mappings of Riemann spaces, Proc. Amer. Math. Soc. 12 (1961), 300–303. (MR 22 #11340.)

KOBAYASHI, E. T.
 [1] Integration of subspaces derived from a linear transformation field, Canad. J. Math. 13 (1961), 273–282. (MR 23 #A2818.)
 [2] A remark on Nijenhuis tensor, Pacific J. Math. 12 (1962), 963–977; correction, ibid. 1467. (MR 27 #678.)
 [3] A remark on the existence of a G-structure, Proc. Amer. Math. Soc. 16 (1965), 1329–1331. (MR 35 #2235.)

KOBAYASHI, S.
 [1] Le groupe de transformations qui laissent invariant un parallélisme, Colloque de Topologie, Ehresmann, Strasbourg, 1954. (MR 19, 576.)
 [2] Espaces à connexion de Cartan complets, Proc. Japan Acad. Sci. 30 (1954), 709–710. (MR 16, 1053.)
 [3] Espaces à connexions affines et riemanniennes symétriques, Nagoya Math. J. 9 (1955), 25–37. (MR 17, 891.)
 [4] A theorem on the affine transformation group of a Riemannian manifold, Nagoya Math. J. 9 (1955), 39–41. (MR 17, 892.)
 [5] Induced connections and imbedded Riemannian spaces, Nagoya Math. J. 10 (1956), 15–25. (MR 18, 332.)
 [6] Theory of connections, Ann. Mat. Pura Appl. 43 (1957), 119–194. (MR 20 #2760.)
 [7] Une remarque sur la connexion affine symétrique, Proc. Japan Acad. Sci. 31 (1956), 14–15. (MR 16, 1151.)
 [8] Holonomy groups of hypersurfaces, Nagoya Math. J. 10 (1956), 9–14. (MR 18, 503.)
 [9] Canonical connections and Pontrjagin classes, Nagoya Math. J. 11 (1957), 93–109. (MR 19, 313.)
 [10] Principal fibre bundles with the 1-dimensional toroidal group, Tôhoku Math. J. 8 (1956), 29–45. (MR 18, 328.)
 [11] Fixed points of isometries, Nagoya Math. J. 13 (1958), 63–68. (MR 21 #2276.)

[12] Compact homogeneous hypersurfaces, Trans. Amer. Math. Soc. 88 (1958) 137–143. (MR 20 #2768.)
[13] Remarks on complex contact manifolds, Proc. Amer. Math. Soc. 10 (1959), 164–167. (MR 22 #1925.)
[14] Geometry of bounded domains, Trans. Amer. Math. Soc. 92 (1959), 267–290. (MR 22 #3017.)
[15] On the automorphism group of a certain class of algebraic manifolds, Tôhoku Math. J. 11 (1959), 184–190. (MR 22 #3014.)
[16] Riemannian manifolds without conjugate points, Ann. Mat. Pura Appl. 53 (1961), 149–155. (MR 27 #692.)
[17] On compact Kähler manifolds with positive Ricci tensor, Ann. of Math. 74 (1961), 570–574. (MR 24 #A2922.)
[18] On characteristic classes defined by connections, Tôhoku Math. J. 13 (1961), 381–385. (MR 25 #1512.)
[19] Homogeneous Riemannian manifolds of negative curvature, Tôhoku Math. J. 14 (1962), 413–415. (MR 26 #5525.)
[20] Topology of positively pinched Kähler manifolds, Tôhoku Math. J. 15 (1963), 121–139. (MR 27 #4185.)
[21] Frame bundles of higher order contact, Proc. Symp. Pure Math. Vol. 3, Amer. Math. Soc. 1961, 186–193. (MR 23 #A4104.)
[22] On complete Bergman metrics, Proc. Amer. Math. Soc. 13 (1962), 511–513. (MR 25 #5192.)
[23] Imbeddings of homogeneous spaces with minimum total curvature, Tôhoku Math. J. 19 (1967), 63–74. (MR 35 #3592.)
[24] Intrinsic metrics on complex manifolds, Bull. Amer. Math. Soc. 73 (1967), 347–349. (MR 35 #1046.)
[25] Isometric imbeddings of compact symmetric spaces, Tôhoku Math. J., 20 (1968), 21–25.
[26] Invariant distances on complex manifolds and holomorphic mappings, J. of Math. Soc. Japan 19 (1967), 460–480.
[27] On conjugate and cut loci, Studies in Global Geometry and Analysis, Math. Assoc. Amer., 1967, 96–122. (MR 35 #3603.)
[28] Hypersurfaces of complex projective space with constant scalar curvature, J. Differential Geometry 1 (1967), 369–370.

KOBAYASHI, S. and EELLS, J. Jr.
[1] Problems in differential geometry, Proc. US-Japan Seminar in Differential Geometry, Kyoto, 1965, 167–177.

KOBAYASHI, S. and NAGANO, T.
[1] On projective connections, J. Math. Mech. 13 (1964), 215–236. (MR 28 #2501.)
[2] On a fundamental theorem of Weyl-Cartan on G-structures, J. Math. Soc. Japan 17 (1965), 84–101. (MR 33 #663.)
[3] On filtered Lie algebras and geometric structures I. J. Math. Mech.

13 (1964), 875–908 (MR 29 #5961); II. 14 (1965), 513–522 (MR 32 #2512); III. *ibid.* 679–706 (MR 32 #5803); IV. 15 (1966), 163–175 (MR 33 #4189); V. *ibid.* 315–328 (MR 33 #4189).
[4] A report on filtered Lie algebras, Proc. US-Japan Seminar in Differential Geometry, Kyoto, 1965, 63–70. (MR 35 #6722.)

KOBAYASHI, S. and NOMIZU, K.
[1] On automorphisms of a Kählerian structure, Nagoya Math. J. 11 (1957), 115–124. (MR 20 #4004.)

KOEBE, P.
[1] Riemannsche Mannigfaltigkeiten und nichteuklidische Raumformen, Bericht Preuss. Akad. Wiss. (1927), 164–196; (1928), 345–442; (1929), 413–457; (1930), 304–364, 505–541; (1931), 506–534; (1932), 249–284.

KODAIRA, K.
[1] Harmonic fields in Riemannian manifolds, Ann. of Math. 50 (1949), 587–665. (MR 11, 108.)
[2] On differential geometric method in the theory of analytic stacks, Proc. Nat. Acad. Sci. USA 39 (1953), 1263–1273. (MR 16, 618.)
[3] On Kähler varieties of restricted type, Ann. of Math. 60 (1954), 28–48. (MR 16, 252.)
[4] Complex structures on $S^1 \times S^3$, Proc. Nat. Acad. Sci. USA 55 (1966), 240–243. (MR 33 #4955.)

KODAIRA, K. and SPENCER, D. C.
[1] Groups of complex line bundles over compact Kähler manifolds, Proc. Nat. Acad. Sci. 39 (1953), 868–872. (MR 16, 75.)
[2] On the variation of almost complex structure, Algebraic Geometry, Topology, Symposium in honor of S. Lefschetz, 139–150, Princeton Univ. Press, 1957. (MR 19, 578.)

KOECHER, M.
[1] Die Geodätischen von Positivitätsbereichen, Math. Ann. 135 (1958), 192–202. (MR 21 #2749.)

KOH, S. S.
[1] On affine symmetric spaces, Trans. Amer. Math. Soc. 119 (1965), 291–309. (MR 32 #1643.)

KOHN, J. J.
[1] Harmonic integrals on strongly pseudo-convex manifolds, Ann. of Math. 78 (1963), 112–148 (MR 27 #2999); II. 79 (1964), 450–472 (MR 34 #8010).

KOSMANN, Y.
[1] Dérivées de Lie des spineurs, C. R. Acad. Sci. Paris, Sér. A-B 262 (1966), A 289–292. (MR 34 #723); Applications, *ibid.* A394–397. (MR 34 #724.)

KOSTANT, B.
- [1] Holonomy and the Lie algebra of infinitesimal motions of a Riemannian manifold, Trans. Amer. Math. Soc. 80 (1955), 528–542. (MR 18, 930.)
- [2] On differential geometry and homogeneous spaces, Proc. Nat. Acad. Sci. 42 (1958), 258–261, 354–357. (MR 19, 454.)
- [3] On invariant skew tensors, Proc. Nat. Acad. Sci. USA 42 (1956), 148–151. (MR 17, 1128.)
- [4] On holonomy and homogeneous spaces, Nagoya Math. J. 12 (1957), 31–54. (MR 21 #6003.)
- [5] A characterization of invariant affine connections, Nagoya Math. J. 16 (1960), 35–50. (MR 22 #1863.)

KOSZUL, J. L.
- [1] Homologie et cohomologie des algèbres de Lie, Bull. Soc. Math. France, 78 (1950), 65–127. (MR 12, 120.)
- [2] Sur la forme hermitienne canonique des espaces homogènes complexes, Canad. J. Math. 7 (1955), 562–576. (MR 17, 1109.)
- [3] Multiplicateurs et classes caractéristiques, Trans. Amer. Math. Soc. 89 (1958), 256–266. (MR 20 #6099.)
- [4] Domaines bornés homogènes et orbites de groupes de transformations affines, Bull. Soc. Math. France 89 (1961), 515–533. (MR 26 #3090.)
- [5] Variétés localement plates et convexité, Osaka J. Math. 2 (1965), 285–290. (MR 33 #4849.)
- [6] Exposés sur les Espaces Homogènes Symétriques, Publ. Soc. Mat. Sao Paulo, 1959. (MR 24 #A2640.)
- [7] Lectures on Fibre Bundles and Differential Geometry, Tata Inst. Fund. Research, Bombay, 1960.
- [8] Sur un type d'algèbre différentielles en rapport avec la transgression, Colloque de Topologie (Espaces Fibrés), Bruxelles, 1950, 73–81. (MR 13, 109.)
- [9] Ouverts convexes homogènes des espaces affines, Math. Z. 79 (1962), 254–259. (MR 27 #774.)

KRAINES, V. Y.
- [1] Topology of quaternionic manifolds, Bull. Amer. Math. Soc. 71 (1965), 526–527 (MR 33 #6653); same title, Trans. Amer. Math. Soc. 122 (1966), 357–367. (MR 33 #738.)

KUBOTA, T.
- [1] Über die Eibereiche im n-dimensionalen Raum, Science Reports Tôhoku Imp. Univ. 14 (1925), 399–402.

KUIPER, N. H.
- [1] Sur les surfaces localement affines, Colloque de Géométrie Différentielle, Strasbourg, 1953, 79–88.

[2] Einstein spaces and connections I, II, Indag. Math. 12 (1950), 505–521. (MR 12, 636.)
[3] On conformally flat spaces in the large, Ann. of Math. 50 (1949), 916–924. (MR 11, 133.)
[4] On compact conformally Euclidean spaces of dimension >2, Ann. of Math. 52 (1950), 478–490. (MR 12, 283.)
[5] On C^1-isometric imbeddings I, Indag. Math. 17 (1955), 545–556; II, 683–689. (MR 17, 782.)
[6] Groups of motions of order $\frac{1}{2}n(n-1)+1$ in Riemannian n-space, Indag. Math. 17 (1955), 313–318. (MR 18, 232.)
[7] Immersions with minimal absolute curvature, Colloque de Géométrie Différentielle Globale, Bruxelles, 1958, 75–87. (MR 23 #A608.)
[8] On surfaces in Euclidean three-space, Bull. Soc. Math. Belg. 12 (1960), 5–22. (MR 23 #A609.)
[9] Convex immersions of closed surfaces in E^3. Non-orientable closed surfaces in E^3 with minimal total absolute Gauss-curvature, Comment. Math. Helv. 35 (1961), 85–92. (MR 23 #A2175.)
[10] On convex maps, Nieuw. Arch. Wisk. 10 (1962), 147–164. (MR 26 #3076.)
[11] Der Satz von Gauss-Bonnet für Abbildungen in E^N und damit verwandte Probleme, Jber. Deut. Math. Ver. 69 (1967), 77–88.

KUIPER, N. H. and YANO, K.
[1] Two algebraic theorems with applications, Indag. Math. 18 (1956), 319–328. (MR 18, 5.)

KULKARNI, R. S.
[1] Curvature and metric, Thesis, Harvard Univ. 1967.

KURITA, M.
[1] On the vector in homogeneous spaces, Nagoya Math. J. 5 (1953), 1–33. (MR 15, 469.)
[2] On the holonomy group of the conformally flat Riemannian manifold, Nagoya Math. J. 9 (1955), 161–172. (MR 17, 528.)
[3] A note on conformal mappings of certain Riemannian manifolds, Nagoya Math. J. 21 (1962), 111–114. (MR 26 #718.)
[4] Tensor fields and their parallelism, Nagoya Math. J. 18 (1961), 133–151. (MR 25 #3458.)

LAUGWITZ, D.
[1] Differential and Riemannian Geometry, Academic Press, New York, 1965. (MR 30 #2406.)
[2] Über die Erweiterung der Tensoranalysis auf Mannigfaltigkeiten unendlicher Dimension, Tensor 13 (1963), 295–304. (MR 28 #3396.)
[3] Differentialgeometrie in Vektorraum, unter besonderer Berücksichtigung der unendlichdimensionalen Räumen, Frier. Vieweg. Sohn, Braunschweig, 1965. (MR 32 #406.)

[4] Über eine Vermutung von Hermann Weyl zum Raumproblem, Arch. Math. 9 (1958), 128–133. (MR 21 #2267.)

[5] Einige differentialgeometrische Charakterisierungen der Quadriken, Ann. Mat. Pura Appl. (4) 55 (1961), 307–314. (MR 25 #1498.)

LEDGER, A. J.

[1] Harmonic homogeneous spaces of Lie groups, J. London Math. Soc. 29 (1954), 345–347. (MR 15, 986.)

[2] Symmetric harmonic spaces, J. London Math. Soc. 32 (1957), 53–56. (MR 18, 761.)

LEDGER, A. J. and YANO, K.

[1] Linear connections on tangent bundles, J. London Math. Soc. 39 (1964), 495–500. (MR 29 #6440.)

[2] The tangent bundle of a locally symmetric space, J. London Math. Soc. 40 (1965), 487–492. (MR 34 #3480.)

LEGRAND, G.

[1] Étude d'une généralisation des structures presque complexes sur les variétés différentiables, Rend. Circ. Mat. Palermo 7 (1958), 323–354; ibid. 8 (1959), 5–48. (MR 28 #1570.)

LEHMANN, D.

[1] Extensions à courbure nulle d'une connexion, C.R. Acad. Sci. Paris 258 (1964), 4903–4906. (MR 29 #547.)

[2] Remarques sur la connexion canonique d'une variété de Stiefel, C.R. Acad. Sci. Paris 259 (1964), 2754–2757. (MR 30 #534.)

[3] Remarques sur les fibrés C^∞ de base compacte, admettant une connexion à groupe d'holonomie fini, C.R. Acad. Sci. Paris 263 (1966), 348–351. (MR 34 #8407.)

[4] Remarques sur l'indice d'un espace fibré vectoriel complexe différentiable, C.R. Acad. Sci. Paris 263 (1966), 470–473. (MR 34 #6797.)

LEHMANN-LEJEUNE, J.

[1] Intégrabilité des G-structures définies par une 1-forme 0-déformable à valeurs dans le fibre tangent, Ann. Inst. Fourier (Grenoble) 16 (1966), fasc. 2, 329–387. (MR 35 #3586.)

LEICHTWEISS, K.

[1] Zur Riemannschen Geometrie in Grassmannschen Mannigfaltigkeiten, Math. Z. 76 (1961), 334–366. (MR 23 #A4102.)

[2] Über eine Art von Krümmungsinvarianten beliebiger Untermannigfaltigkeiten des n-dimensionalen euklidischen Raumes, Abh. Math. Sem. Univ. Hamburg 26 (1963/64), 155–190. (MR 28 #3371.)

LELONG-FERRAND, J.

[1] Quelques propriétés des groupes de transformations infinitésimales d'une variété riemannienne, Bull. Soc. Math. Belg. 8 (1956), 15–30. (MR 19, 168.)

[2] Sur les groupes à un paramètre de transformations des variétés différentiables, J. Math. Pure Appl. 37(3) (1958), 269–278. (MR 20 #4874.)
[3] Sur l'application des méthodes de Hilbert à l'étude des transformations infinitésimales d'une variété différentiable, Bull. Soc. Math. Belg. 9 (1957), 59–73. (MR 21 #2274.)

LERAY, J.
[1] Sur l'homologie des groupes de Lie, ..., Colloque de Topologie, Bruxelles, 1950, 101–115. (MR 12, 802.)

LEVI-CIVITA, T.
[1] Nozione di parallelismo in una varieta qualunque e consequente specificazione geometrica della curvature Riemanniana, Rend. Palermo 42 (1917), 73–205.

LEVI-CIVITA, T. and RICCI, G.
[1] Méthodes de calcul différentiel absolu et leurs applications, Math. Ann. 54 (1901), 125–201.

LEWY, H.
[1] On differential geometry in the large, Trans. Amer. Math. Soc. 45 (1938), 258–290.

LIBERMANN, P.
[1] Sur le problème d'equivalence de certaines structures infinitésimales, Ann. Mat. Pura Appl. 36 (1954), 27–120. (MR 16, 520.)
[2] Sur les structures presque complexes et autres structures infinitésimales régulières, Bull. Soc. Math. France 83 (1955), 195–224. (MR 18, 143.)
[3] Sur les automorphismes infinitésimaux des structures symplectiques et des structures contactes, Colloque Géométrie Différentielle Globale, Bruxelles, 1958, 37–59. (MR 22 #9919.)
[4] Sur la géométrie des prolongements des espaces fibrés vectoriels, Ann. Inst. Fourier (Grenoble) 14 (1964), fasc. 1, 145–172. (MR 29 #5190.)
[5] On sprays and higher order connections, Proc. Nat. Acad. Sci. USA 49 (1963), 459–462. (MR 27 #6207.)
[6] Pseudogroupes infinitésimaux attachés aux pseudogroupes de Lie, Bull. Soc. Math. France 87 (1959), 409–425. (MR 23 #A607.)
[7] Sur quelques exemples de structures pfaffiennes et presque cosymplectiques, Ann. Mat. Pur Appl. (4) 60 (1962), 153–172. (MR 28 #565.)

LICHNEROWICZ, A.
[1] Espaces homogènes kählériens, Colloque Géométrie Différentielle, Strasbourg, 1953, 171–201. (MR 16, 519.)
[2]* Théorie Globale des Connexions et des Groupes d'Holonomie, Ed. Cremonese, Rome, 1955. (MR 19, 453.)

[3]* Géométrie des Groupes de Transformations, Dunod, Paris, 1958. (MR 23 #A1329.)
[4] Sur les espaces riemanniens complètement harmoniques, Bull. Soc. Math. France 72 (1944), 146–168. (MR 7, 80.)
[5] Quelques théorèmes de géométrie différentielle globale, Comment. Math. Helv. 22 (1949), 271–301. (MR 10, 571.)
[6] Courbure, nombres de Betti et espaces symétriques, Proc. Intern. Congr. Math. Cambridge, Mass. 1950, Vol. 2, 216–223. (MR 13, 492.)
[7] Équations de Laplace et espaces harmoniques, Premier Colloque Équations Derivées Partielles, Louvain, 1953, 9–23. (MR 15, 898.)
[8] Isométries et transformations analytiques d'une variété kählérienne compacte, Bull. Soc. Math. France 87 (1959), 427–437. (MR 22 #5012.)
[9] Laplacien sur une variété riemannienne et spineurs, Atti. Acad. Nat. Lincei Rend. Cl. Sci. Fis Mat. Nat. 33 (1962), 187–191. (MR 27 #5203.)
[10] Transformations des variétés à connexion linéaire et des variétés riemanniennes, Enseignement Math. 8 (1962), 1–15. (MR 26 #717.)
[11] Théorèmes de réductivité sur des algèbres d'automorphismes, Rend. Mat. Appl. 22 (1963), 197–244. (MR 28 #544.)
[12] Sur les transformations conformes d'une variété riemannienne compacte, C.R. Acad. Sci. Paris 259 (1964), 697–700. (MR 29 #4007.)
[13] Opérateurs différentiels invariants sur un espace homogène, Ann. Sci. École Norm Sup. 81 (1964), 341–385. (MR 32 #4628.)
[14] Variétés complexes et tenseur de Bergmann, Ann. Inst. Fourier (Grenoble) 15 (1965), 345–407. (MR 33 #744.)
[15] Sur certaines variétés kählériennes compactes, C.R. Acad. Sci. Paris 263 (1966), 570–575. (MR 34 #6700.)

LIE, S. and ENGEL, F.
[1] Theorie der Transformationsgruppen, 3 Volumes, B. G. Teubner, Leipzig, 1888–1893.

LIEBMANN, H.
[1] Über Flächen von konstanter Gausscher Krümmung, Trans. Amer. Math. Soc. 2 (1901), 87–99.

LITTLE, J. A.
[1] On singularities of submanifolds of higher dimensional spaces, Thesis, University of Minnesota, 1968.

LÖBELL, F.
[1] Beispiele geschlossener dreidimensionaler Clifford-Kleinscher Räume negativer Krümmung, Berichte Sachs Akad. Wiss. (Leibziger Berichte) 83 (1931), 167–174.

Loos, O.
[1] Spiegelungsräume und homogene symmetrische Räume, Math. Z. 99 (1967) 141–170. (MR 35 #3608.)

McKean, H. P. and Singer, I. M.
[1] Curvature and the eigenvalues of the Laplacian, J. Differential Geometry 1 (1967) 43–70. (MR 36 #828.)

Malcev, A.
[1] On a class of homogeneous spaces, Amer. Math. Soc. Transl. No. 39. (1951). (MR 12, 589.)

Malgrange, B.
[1] Théorème de Frobenius complexe, Séminaire Bourbaki, (1957/58), Exposé 166.

Maltz, R.
[1] The nullity spaces of the curvature operator, Cahiers de Topologie et Géométrie Différentielle VIII (1966), 1–20. (MR 35 #940.)

Martinelli, E.
[1] Varietà a struttura quaternionale generalizzata, Atti. Accad. Naz. Lincei Rend. Cl. Sci. Fis. Mat. Nat. 26 (1959), 353–362. (MR 26 #1835.)
[2] Modello metrico reale dello spazio proiettivo quaternionale, Ann. Mat. Pura Appl. 49 (1960), 78–89. (MR 26 #1836.)

Massey, W. S.
[1] Surfaces of Gaussian curvature zero in Euclidean 3-space, Tôhoku Math. J. 14 (1962), 73–79. (MR 25 #2527.)

Matsushima, Y.
[1] On the discrete subgroups and homogeneous spaces of nilpotent Lie groups, Nagoya Math. J. 2 (1951), 95–110. (MR 12, 802.)
[2] Un théorème sur les espaces homogènes complexes, C.R. Acad. Sci. Paris 241 (1955), 785–787. (MR 17, 410.)
[3] Sur la structure du groupe d'homéomorphismes analytiques d'une certaine variété kählérienne, Nagoya Math. J. 11 (1957), 145–150. (MR 20 #995.)
[4] Sur les espaces homogènes kählériens d'un groupe de Lie réductif, Nagoya Math. J. 11 (1957), 53–60. (MR 19, 315.)
[5] Fibrés holomorphes sur un tore complexe, Nagoya Math. J. 14 (1959), 1–24. (MR 21 #1403.)
[6] Espaces homogènes de Stein des groupes de Lie complexes, Nagoya Math. J. 16 (1960), 205–218 (MR 22 #739); II. 18 (1961), 153–164 (MR 23 #A244).
[7] Sur certaines variétés homogènes complexes, Nagoya Math. J. 18 (1961), 1–12. (MR 25 #2147.)
[8] On the first Betti number of compact quotient spaces of higher

dimensional symmetric spaces, Ann. of Math. 75 (1962), 312–330. (MR 28 #1629.)

[9] On Betti numbers of compact, locally symmetric Riemannian manifolds, Osaka Math. J. 14 (1962), 1–20. (MR 25 #4549.)

MATSUSHIMA, Y. and MORIMOTO, A.

[1] Sur certains espaces fibrés holomorphes sur une variété de Stein, Bull. Soc. Math. France 88 (1960), 137–155. (MR 23 #A1061.)

MATSUSHIMA, Y. and MURAKAMI, S.

[1] On vector bundle valued harmonic forms and automorphic forms on symmetric riemannian manifolds, Ann. of Math. 78 (1963), 365–416. (MR 27 #2997.)

[2] On certain cohomology groups attached to Hermitian symmetric spaces, Osaka J. Math. 2 (1965), 1–35. (MR 32 #1728.) Same title, Proc. US-Japan Seminar in Differential Geometry, Kyoto, 1965, 82–94.

MAUTNER, F. I.

[1] Geodesic flows on symmetric Riemann spaces, Ann. of Math. 65 (1957), 416–431. (MR 18, 929.)

MILNOR, J. W.

[1] Groups which act on S^n without fixed points, Amer. J. Math. 79 (1957), 623–630. (MR 19, 761.)

[2] On manifolds homeomorphic to the 7-sphere, Ann. of Math. 64 (1956), 394–405. (MR 18, 498.)

[3] Lectures on Morse Theory, Ann. Math. Studies No. 51, Princeton Univ. Press, Princeton, New Jersey 1963. (MR 29 #634.)

[4] Lectures on Characteristic Classes, Notes from Princeton Univ. 1958.

[5] Topology from the Differentiable Viewpoint, Univ. Virginia Press, 1965.

[6] On the existence of a connection with curvature zero, Comment. Math. Helv. 32 (1958), 215–223. (MR 20 #2020.)

[7] Eigenvalues of the Laplace operator on certain manifolds, Proc. Nat. Acad. Sci. USA 51 (1964), 542. (MR 28 #5403.)

MINAGAWA, T. and RADO, T.

[1] On the infinitesimal rigidity of surfaces, Osaka Math. J. 4 (1952), 241–285. (MR 14, 794.)

MINKOWSKI, H.

[1] Diskontinuitätsbereich für arithmetische Äquivalenz, J. Reine Angew. Math. 129 (1905), 220–274.

MOLINO, P.

[1] Champs d'éléments sur un espace fibré principal différentiable, Ann. Inst. Fourier (Grenoble) 14 (1964), 163–219. (MR 31 #1630; also, MR 23 #A1326.)

MONTGOMERY, D.
  [1] Simply connected homogeneous spaces, Proc. Amer. Math. Soc. 1 (1950), 467–469. (MR 12, 242.)
MONTGOMERY, D. and SAMELSON, H.
  [1] Transformation groups of spheres, Ann. of Math. 44 (1943), 454–470. (MR 5, 60.)
MONTGOMERY, D. and ZIPPIN, L.
  [1] Transformation Groups, Interscience Tracts #1, Interscience, New York, 1955. (MR 17, 383.)
MORIMOTO, A.
  [1] Structures complexes invariantes sur les groupes de Lie semi-simples, C.R. Acad. Sci. Paris 242 (1956), 1101–1103. (MR 18, 583.)
  [2] Sur la classification des espaces fibrés vectoriels holomorphes admettant des connexions holomorphes sur un tore complexe, Nagoya Math. J. 15 (1959), 83–154. (MR 22 #239.)
  [3] Sur le groupe d'automorphismes d'un espace fibré principal analytique complexe, Nagoya Math. J. 13 (1958), 157–178. (MR 20 #2474.)
  [4] On normal almost contact structures, J. Math. Soc. Japan 15 (1963), 420–436. (MR 29 #548.)
MORIMOTO, A. and NAGANO, T.
  [1] On pseudo-conformal transformations of hypersurfaces, J. Math. Soc. Japan 15 (1963), 289–300. (MR 27 #5275.)
MORITA, K.
  [1] On the kernel functions of symmetric domains, Sci. Rep. Tokyo Kyoiku Daigaku, Sect. A5 (1956), 190–212. (MR 19, 541.)
MORSE, M.
  [1] The Calculus of Variations in the Large, Amer. Math. Soc., New York, 1934.
MURAKAMI, S.
  [1] On the automorphisms of a real semi-simple Lie algebra, J. Math. Soc. Japan 4 (1952), 103–133 (MR 14, 531); corrections, *ibid.* 5 (1953), 105–112 (MR 15, 500).
  [2] Algebraic study of fundamental characteristic classes of sphere bundles, Osaka Math. J. 8 (1956), 187–224. (MR 18, 667.)
  [3] Sur certains espaces fibrés principaux différentiables et holomorphes, Nagoya Math. J. 15 (1959), 171–199. (MR 22 #1927.)
  [4] Sur certains espaces fibrés principaux holomorphes admettant des connexions holomorphes, Osaka Math. J. 11 (1959), 43–62. (MR 22 #958.)
  [5] Sur certains espaces fibrés principaux holomorphes dont le groupe est abélian connexe, Osaka Math. J. 13 (1961), 143–167. (MR 26 #788.)

[6] Sur la classification des algèbres de Lie réelles et simples, Osaka J. Math. 2 (1965), 291–307. (MR 34 #1460.)

MYERS, S. B.
[1] Riemannian manifolds in the large, Duke Math. J. 1 (1935), 39–49.
[2] Connection between differential geometry and topology, Duke Math. J. 1 (1935), 376–391; ibid. 2 (1936), 95–102.
[3] Riemannian manifolds with positive mean curvature, Duke Math. J. 8 (1941), 401–404. (MR 3, 18.)
[4] Curvature of closed hypersurfaces and non-existence of closed minimal hypersurfaces, Trans. Amer. Math. Soc. 71 (1951), 211–217. (MR 13, 492.)

MYERS, S. B. and STEENROD, N.
[1] The group of isometries of a Riemannian manifold, Ann. of Math. 40 (1939), 400–416.

NACHBIN, L.
[1] The Haar Integral, Van Nostrand, Princeton, New Jersey, 1965. (MR 31 #271.)

NAGANO, T.
[1] The conformal transformation on a space with parallel Ricci tensor, J. Math. Soc. Japan 11 (1959), 10–14. (MR 23 #A1330.)
[2] On fibred Riemannian manifolds, Sci. Papers Coll. Gen. Ed. Univ. Tokyo 10 (1960), 17–27. (MR 28 #560.)
[3] On conformal transformations of Riemann spaces, J. Math. Soc. Japan 10 (1958), 79–93. (MR 22 #1859.)
[4] Sur des hypersurfaces et quelques groupes d'isométries d'un espace riemannien, Tôhoku Math. J. 10 (1958), 242–252. (MR 21 #344.)
[5] Transformation groups with $(n-1)$-dimensional orbits on non-compact manifolds, Nagoya Math. J. 14 (1959), 25–38. (MR 21 #3513.)
[6] Compact homogeneous spaces and the first Betti number, J. Math. Soc. Japan 11 (1959), 4–9. (MR 21 #2706.)
[7] On some compact transformation groups on spheres, Sci. Papers Coll. Gen. Ed. Univ. Tokyo 9 (1959), 213–218. (MR 22 #4799.)
[8] Homogeneous sphere bundles and the isotropic Riemann manifolds, Nagoya Math. J. 15 (1959), 29–55. (MR 21 #7522.)
[9] On the minimum eigenvalues of the Laplacians in Riemannian manifolds, Sci. Papers Coll. Gen. Ed. Univ. Tokyo 11 (1961), 177–182. (MR 26 #1830.)
[10] Transformation groups on compact symmetric spaces, Trans. Amer. Math. Soc. 118 (1965), 428–453. (MR 32 #419.)
[11] A problem on the existence of an Einstein metric, J. Math. Soc. Japan 19 (1967), 30–31. (MR 34 #5030.)

NAGANO, T. and TAKAHASHI, T.
[1] Homogeneous hypersurfaces in Euclidean spaces, J. Math. Soc. Japan 12 (1960), 1–7. (MR 22 #5008.)

NAKAI, M.
[1] Algebras of some differentiable functions on Riemannian manifolds, Japan. J. Math. 29 (1959), 60–69. (MR 22 #11337.)

NAKANO, S.
[1] On complex analytic vector bundles, J. Math. Soc. Japan 7 (1955), 1–12. (MR 17, 409.)

NARASIMHAN, R.
[1] Imbeddings of holomorphically complete complex spaces, Amer. J. Math. 82 (1960), 917–934. (MR 26 #6438.)

NARASIMHAN, M. S. and RAMANAN, S.
[1] Existence of universal connections, Amer. J. Math. 83 (1961), 563–572 (MR 24 #A3597); II. 85 (1963), 223–231. (MR 27 #1904.)

NASH, J. F.
[1] C^1-isometric imbeddings, Ann. of Math. 60 (1954), 383–396. (MR 16, 515.)
[2] The imbedding problem for Riemannian manifolds, Ann. of Math. 63 (1956), 20–63 (MR 17, 782.)

NEWLANDER, A. and NIRENBERG, L.
[1] Complex analytic coordinates in almost complex manifolds, Ann. of Math. 65 (1957), 391–404. (MR 19, 577.)

NGUYEN-VAN HAI
[1] Sur le groupe d'automorphismes de la variété affine des connexions linéaires invariantes d'un espace homogène, C.R. Acad. Sci. Paris 258 (1964), 3952–3955. (MR 29 #1599.)
[2] Conditions nécessaires et suffisantes pour qu'un espace homogène admette une connexion linéaire invariante, C.R. Acad. Sci. Paris 259 (1964), 49–52. (MR 29 #3955.)
[3] Un type de connexion linéaire invariante sur un espace homogène, C.R. Acad. Sci. Paris 259 (1964), 2065–2068. (MR 29 #6439.)
[4] Relations entre les diverses obstructions relatives à l'existence d'une connexion linéaire invariante sur un espace homogène, C.R. Acad. Sci. Paris 260 (1965), 45–48. (MR 31 #689.)
[5] Construction de l'algèbre de Lie des transformations infinitésimales affines sur un espace homogène à connexion linéaire invariante, C. R. Acad. Sci. Paris 263 (1966), 876–879. (MR 34 #6685.)

NICKERSON, H. K.
[1] On complex form of the Poincaré lemma, Proc. Amer. Math. Soc. 9 (1958), 183–188. (MR 20 #2473.)
[2] On differential operators and connections, Trans. Amer. Math. Soc. 99 (1961), 509–539. (MR 28 #5411.)

NIJENHUIS, A.
  [1] X_{n-1}-forming sets of eigenvectors, Indag. Math. 13 (1951), 200–212. (MR 13, 281.)
  [2] On the holonomy group of linear connections IA, IB, Indag. Math. 15 (1953), 233–249 (MR 16, 171); II, 16 (1954), 17–25 (MR 16, 172).
  [3] A note on infinitesimal holonomy groups, Nagoya Math. J. 12 (1957), 145–147. (MR 21 #1626.)
  [4] A note on hyperconvexity in Riemannian manifolds, Canad. J. Math. 11 (1959), 576–582. (MR 22 #214.)
  [5] Jacobi-type identities for bilinear differential concomitants of certain tensor fields I, II, Nederl. Akad. Wetensch. Proc. Ser. A 58 (1955), 390–403. (MR 17, 661.)
  [6] Geometric aspects of formal differential operators on tensor fields, Proc. Intern. Congr. Math. 1958 (Edinburgh), 463–469. (MR 30 #531.)

NIJENHUIS, A. and WOOLF, W. B.
  [1] Some integration problems in almost complex manifolds, Ann. of Math. 77 (1963), 424–483. (MR 26 #6992.)

NIRENBERG, L.
  [1] The Weyl and Minkowski problems in the differential geometry in the large, Comm. Pure Appl. Math. 6 (1953), 337–394. (MR 15, 347.)
  [2] A complex Frobenius theorem, Seminar Analytic Functions, Princeton Univ. Press (1957), Vol. 1, 172–189.
  [3] Elementary remarks on surfaces with curvature of fixed sign, Proc. Symp. Pure Math. Vol III, 181–185, Amer. Math. Soc., 1961. (MR 23 #2843.)
  [4] Rigidity of a class of closed surfaces, Nonlinear problems, Proc. Symp. Madison Wisc. 1962, 177–193. (MR 27 #697.)

NITSCHE, J. C. C.
  [1]* On recent results in the theory of minimal surfaces, Bull. Amer. Math. Soc. 71 (1965), 195–270. (MR 30 #4200.)

NOMIZU, K.
  [1] On the group of affine transformations of an affinely connected manifold, Proc. Amer. Math. Soc. 4 (1953), 816–823. (MR 15, 468.)
  [2] Invariant affine connections on homogeneous spaces, Amer. J. Math. 76 (1954), 33–65. (MR 15, 468.)
  [3] Studies on Riemannian homogeneous spaces, Nagoya Math. J. 9 (1955), 43–56. (MR 17, 891.)
  [4] Reduction theorem for connections and its application to the problem of isotropy and holonomy groups of a Riemannian manifold, Nagoya Math. J. 9 (1955), 57–66. (MR 17, 891.)

- [5] Un théorème sur les groupes d'holonomie, Nagoya Math. J. 10 (1956), 101–103. (MR 18, 489.)
- [6] On infinitesimal holonomy and isotropy groups, Nagoya Math. J. 11 (1957), 111–114. (MR 18, 931.)
- [7] Lie Groups and Differential Geometry, Publ. Math. Soc. Japan, #2, 1956. (MR 18, 821.)
- [8] On local and global existence of Killing vector fields, Ann. of Math. 72 (1960), 105–120. (MR 22 #9938.)
- [9] Sur les algèbres de Lie de générateurs de Killing et l'homogénéité d'une variété riemannienne, Osaka Math. J. 14 (1962), 45–51. (MR 25 #4467.)
- [10] Holonomy, Ricci tensor and Killing vector fields, Proc. Amer. Math. Soc. 12 (1961), 594–597. (MR 24 #A1099.)
- [11] Recent development in the theory of connections and holonomy groups, Advances in Mathematics, Academic Press, New York, 1961, fasc. 1, 1–49. (MR 25 #5473.)
- [12] On hypersurfaces satisfying a certain condition on the curvature tensor, Tôhoku Math. J. 20 (1968), 46–59.

NOMIZU, K. and OZEKI, H.
- [1] The existence of complete Riemannian metrics Proc. Amer., Math. Soc. 12 (1961), 889–891. (MR 24 #A3610.)
- [2] On the degree of differentiability of curves used in the definition of the holonomy group, Bull. Amer. Math. Soc. 68 (1962), 74–75. (MR 24 #A2916.)
- [3] A theorem on curvature tensor fields, Proc. Nat. Acad. Sci. USA 48 (1962), 206–207. (MR 24 #A2347.)

NOMIZU, K. and SMYTH, B.
- [1] Differential geometry of complex hypersurfaces II, J. Math. Soc. Japan 20 (1968), 498–521.

NOMIZU, K. and YANO, K.
- [1] Une démonstration simple d'un théorème sur le groupe d'holonomie affine d'un espace de Riemann, C.R. Acad. Sci. Paris 258 (1964), 5334–5335. (MR 29 #562.)
- [2] On infinitesimal transformations preserving the curvature tensor field and its covariant differentials, Ann. Inst. Fourier (Grenoble) 14 (1964), 227–236. (MR 30 #4227.)
- [3] Some results related to the equivalence problem in Riemannian geometry, Proc. US-Japan Seminar in Differential Geometry, Kyoto, Japan, 1965, 95–100.
- [4] Same title as [3], Math. Z. 97 (1967), 29–37.

NOWACKI, W.
- [1] Die euklidischen, dreidimensionalen geschlossenen und offenen Raumformen, Comment. Math. Helv. 7 (1934), 81–93.

OBATA, M.
[1] Affine transformations in an almost complex manifold with a natural affine connection, J. Math. Soc. Japan 8 (1956), 345–362. (MR 18, 822.)
[2] On subgroups of the orthogonal groups, Trans. Amer. Math. Soc. 87 (1958), 347–358. (MR 20 #1711.)
[3] Certain conditions for a Riemannian manifold to be isometric with a sphere, J. Math. Soc. Japan 14 (1962), 333–340. (MR 25 #5479.)
[4] Conformal transformations of compact Riemannian manifolds, Illinois J. Math. 6 (1962), 292–295. (MR 25 #1507.)
[5] Riemannian manifolds admitting a solution of a certain system of differential equations, Proc. US-Japan Seminar in Differential Geometry, Kyoto, Japan, 1965, 101–114. (MR 35 #7263.)
[6] Quelques inégalités intégrales sur une variété riemanniene compacte, C.R. Acad. Sci. Paris, Sér. A-B 264 (1967), A123–125. (MR 35 #2246.)

OBATA, M. and YANO, K.
[1] Sur le groupe de transformations conformes d'une variété de Riemann dont le scalaire de courbure est constant, C.R. Acad. Sci. Paris 260 (1965), 2698–2700. (MR 31 #697.)

OCHIAI, T.
[1] On the automorphism group of a G-structure, J. Math. Soc. Japan 18 (1966), 189–193. (MR 33 #3224.)

OLIVIER, R.
[1] Die Existenz geschlossener Geodätischen auf kompakten Mannigfaltigkeiten, Comment. Math. Helv. 35 (1961), 146–152. (MR 24 #A2349.)
[2] Über die Dehnung von Sphärenabbildungen, Invent. Math. 1 (1966), 380–390, (MR 34 #3500.)

OMORI, H.
[1] Isometric immersions of Riemannian manifolds, J. Math. Soc. Japan, 19 (1967), 205–214. (MR 35 #6101.)

O'NEILL, B.
[1] An algebraic criterion for immersion, Pacific J. Math. 9 (1959), 1239–1247. (MR 23 #A3537.)
[2] Immersion of manifolds of non-positive curvature, Proc. Amer. Math. Soc. 11 (1960), 132–134. (MR 22 #8459.)
[3] Umbilics of constant curvature immersions, Duke Math. J. 32 (1965), 149–159. (MR 31 #5181.)
[4] Isotropic and Kaehler immersions, Canad. J. Math. 17 (1965), 907–915. (MR 32 #1654.)

[5] Isometric immersion of flat Riemannian manifolds in Euclidean space, Michigan Math. J. 9 (1962), 199–205. (MR 27 #2941.)
[6] Isometric immersions which preserve curvature operators, Proc. Amer. Math. Soc. 13 (1962), 759–763. (MR 26 #721.)
[7] The fundamental equations of a submersion, Michigan Math. J. 13 (1966), 459–469. (MR 34 #751.)

O'NEILL, B. and STIEL, E.
[1] Isometric immersions of constant curvature manifolds, Michigan Math. J. 10 (1963), 335–339. (MR 28 #1554.)

OSBORN, H.
[1] The Morse index theorem, Proc. Amer. Math. Soc. 18 (1967), 759–762 (MR 35 #3704); correction, in the same journal, 19 (1968).

OTSUKI, T.
[1] On the existence of solutions of a system of quadratic equations and its geometrical application, Proc. Japan Acad. 29 (1953), 99–100. (MR 15, 647.)
[2] Isometric imbedding of Riemannian manifolds in a Riemannian manifold, J. Math. Soc. Japan 6 (1954), 221–234. (MR 16, 747.)
[3] Note on the isometric imbedding of compact Riemannian manifolds in Euclidean spaces, Math. J. Okayama Univ. 5 (1956), 95–102. (MR 18, 66.)
[5] On focal elements and the spheres, Tôhoku Math. J. 17 (1965), 285–304. (MR 33 #676.)
[6] Integral formulas for hypersurfaces in a Riemannian manifold and their applications, Tôhoku Math. J. (2) 17 (1965), 335–348. (MR 33 #3233.)
[7] On the total curvature of surfaces in Euclidean spaces, Japan J. Math. 35 (1966), 61–71. (MR 34 #692.)
[8] Surfaces in the 4-dimensional Euclidean space isometric to a sphere, Kōdai Math. Sem Rep. 18 (1966), 101–115. (MR 34 #3503.)

OZEKI, H.
[1] Infinitesimal holonomy groups of bundle connections, Nagoya Math. J. 10 (1956), 105–124. (MR 18, 232.)
[2] Chern classes of projective modules, Nagoya Math. J. 23 (1963), 121–152. (MR 29, 3522.)

PALAIS, R. S.
[1] A Global Formulation of the Lie Theory of Transformation Groups, Mem. Amer. Math. Soc. #22, 1957.
[2] On the differentiability of isometries, Proc. Amer. Math. Soc. 8 (1957), 805–807. (MR 19, 451.)
[3] Morse theory on Hilbert manifolds, Topology 2 (1963), 299–340. (MR 28 #1633.)

PAPY, G.
  [1] Sur la définition intrinsèque des vecteurs tangents à une variété de classe C^r lorsque $1 \leq r < \infty$, C. R. Acad. Sci. Paris 242 (1956), 1573–1575. (MR 17, 892.)

PASIENCIER, S.
  [1] Homogeneous almost complex spaces of positive characteristic, Proc. Amer. Math. Soc. 15 (1964), 471–481. (MR 30 #2444.)

PATTERSON, L. N.
  [1] On the index theorem, Amer. J. Math. 85 (1963), 271–297. (MR 27 #5204.)

PETERS, K.
  [1] Über holomorphe und meromorphe Abbildungen gewisser kompakter komplexer Mannigfaltigkeiten, Arch. Math. 15 (1964), 222–231. (MR 29 #4071.)

PETROW, A. S.
  [1]* Einstein-Räume, Akad. Verlag, Berlin, 1964. (MR 28 #5792.)

POGORELOW, A. W.
  [1] Einige Untersuchungen zur Riemannschen Geometrie im Grossen, VEB Deutscher Verlag der Wissenschaften, Berlin, 1960. (MR 22 #5946.)
  [2] Some results on surface theory in the large, Advances in Math. 1, fasc. 2 (1964), 191–264. (MR 31 #2695.)
  [3] Continuous maps of bounded variations, Dokl. Akad. Nauk SSSR, 111 (1956), 757–759. (MR 19, 309).
  [4] An extension of Gauss' theorem on the spherical representation of surface of bounded exterior curvature, Dokl. Akad. Nauk SSSR, 111 (1956), 945–947. (MR 19, 309.)

POHL, W. F.
  [1] Differential geometry of higher order, Topology, 1 (1962), 169–211. (MR 27 #4242.)
  [2] Extrinsic complex projective geometry, Proc. Conf. Complex Analysis (Minneapolis, 1964), 18–29, Springer, Berlin, 1965. (MR 30 #5335.)
  [3] Connexions in differential geometry of higher order, Trans. Amer. Math. Soc. 125 (1966), 310–325. (MR 34 #3477.)

POINCARÉ, H.
  [1] Sur les lignes géodésiques des surfaces convexes, Trans. Amer. Math. Soc. 5 (1905), 237–274.

PONTRJAGIN, L.
  [1] Topological Groups, Princeton Univ. Press, 1939. (MR 19, 867.)
  [2] Some topological invariants of closed Riemannian manifolds, Amer. Math. Soc. Transl. No. 49, 1951. (MR 12, 848.)

POSTNIKOV, M. M.
[1] The Variational Theory of Geodesics, translated from the Russian, W. B. Saunders Co., Philadelphia-London, 1967. (MR 35 #937.)

PREISMANN, A.
[1] Quelques propriétés globales des espaces de Riemann, Comment. Math. Helv. 15 (1942), 175–216. (MR 6, 20.)

PYATETZKI-SHAPIRO, I.
[1] On a problem proposed by E. Cartan, Dokl. Akad. Nauk. SSSR. 124 (1959), 272–273. (MR 21 #728.)
[2]* Géométrie des Domaines Classiques et Théorie des Fonctions Automorphes, Dunod, Paris, 1966. (MR 33 #5949.)

RAGHUNATHAN, M. S.
[1] Deformations of linear connections and Riemannian manifolds, J. Math. Mech. 13 (1966), 97–123 (MR 28 #4484); addendum. 1043–1045 (MR 29 #5201).

RAŠEVSKIĬ, P. K.
[1] On the geometry of homogeneous spaces, Dokl. Akad. Nauk, SSSR (N. S.) 80 (1951), 169–171. (MR 13, 383.)
[2] On the geometry of homogeneous spaces, Trudy Sem. Vektor Tenzor Analiz., 9 (1952), 49–74. (MR 14, 795.)

RAUCH, H. E.
[1] A contribution to differential geometry in the large, Ann. of Math. 54 (1951), 38–55. (MR 13, 159.)
[2] The global study of geodesics in symmetric and nearly symmetric Riemannian manifolds, Comment. Math. Helv. 35 (1961), 111–125. (MR 24 #A3604.)
[3] Geodesics, symmetric spaces and differential geometry in the large, Comment. Math. Helv. 27 (1953), 294–320. (MR 15, 744.)
[4] Geodesics and Curvature in Differential Geometry in the Large, Yeshiva Univ., N.Y., 1959.
[5] Geodesics and Jacobi equations on homogeneous Riemannian manifolds, Proc. US-Japan Seminar in Differential Geometry, Kyoto, Japan, 1965, 115–127.

REEB, G.
[1] Sur certaines propriétés topologiques des variétés feuilletées, Actualités Sci. Ind. 1183 (91–154), Hermann, Paris, 1952. (MR 14, 113.)

REINHART, B.
[1] Foliated manifolds with bundle-like metrics, Ann. of Math. 69 (1959), 119–131. (MR 21 #6004.)
[2] Harmonic integrals on almost product manifolds, Trans. Amer. Math. Soc. 88 (1958), 243–276. (MR 21 #3687.)

[3] Harmonic integrals on foliated manifolds, Amer. J. Math. 81 (1959), 529–536. (MR 21 #6005.)

DE RHAM, G.
 [1] Sur la réductibilité d'un espace de Riemann, Comment. Math. Helv. 26 (1952), 328–344. (MR 14, 584.)
 [2] Variétés Différentiables, Hermann, Paris, 1955. (MR 16, 957.)

RIEMANN, B.
 [1] Über die Hypothesen, welche der Geometrie zugrunde liegen, Ges. Werke.

RINOW, W.
 [1]* Die Innere Geometrie der Metrischen Räume, Springer, Berlin, 1961. (MR 23 #A1290.)
 [2] Über Zusammenhange zwischen der Differentialgeometrie im Grossen und im Kleinen, Math. Z. 35 (1932), 512–528.

RODRIGUES, A. A. M.
 [1] Characteristic classes of complex homogeneous spaces, Bol. Soc. Mat. Sao Paulo 10 (1955), 67–86. (MR 22 #4082.)

RÖHRL, H.
 [1]* Holomorphic fibre bundles over Riemann surfaces, Bull. Amer. Math. Soc. 68 (1960), 125–160, (MR 26 #4373.)

ROSENTHAL, A.
 [1] Riemannian manifolds of constant nullity, Michigan Math. J. 14 (1967), 469–480. (MR 36 #3282.)

RUH, E. A.
 [1] On the automorphism group of a G-structure, Comment. Math. Helv. 39 (1964), 189–264. (MR 31 #1631.)

RUSE, H. S., WALKER, A. G., and WILLMORE, T. J.
 [1]* Harmonic Spaces, Ed. Cremonese, Rome, 1961. (MR 25 #5456.)

SACKSTEDER, R.
 [1] On hypersurfaces with no negative sectional curvatures, Amer. J. Math. 82 (1960), 609–630. (MR 22 #7087.)
 [2] The rigidity of hypersurfaces, J. Math. Mech. 11 (1962), 929–939. (MR 26 #1833.)

SAGLE, A.
 [1] On anti-commutative algebras and homogeneous spaces, J. Math. Mech. 16 (1967), 1381–1394.
 [2] A note on simple anti-commutative algebras obtained from reductive homogeneous spaces, Nagoya Math. J. 31 (1968), 105–124.

SAGLE, A. and WINTER, D. J.
 [1] On homogeneous spaces and reductive subalgebras of simple Lie algebras, Trans. Amer. Math. Soc. 128 (1967), 142–147.

SAITO, M.
 [1] Sur certains groupes de Lie résolubles, Sci. Papers Coll. Gen. Ed.

Univ. Tokyo 7 (1957), 1–11 (MR 20 #3931); II. 157–168 (MR 20 #3932).

SAMELSON, H.
[1]* Topology of Lie groups, Bull. Amer. Math. Soc. 58 (1952), 2–37. (MR 13 #533.)
[2] On curvature and characteristic of homogeneous spaces, Michigan Math. J. 5 (1958), 13–18. (MR 21 #2277.)

SAMPSON, H.
[1] A note on automorphic varieties, Proc. Nat. Acad. Sci. USA. 38 (1952), 895–898. (MR 14, 633.)

SASAKI, S.
[1] On the differential geometry of tangent bundles of Riemannian manifolds, Tôhoku Math. J. 10 (1958), 338–345 (MR 22 #3007); II, 14 (1962) 146–155, (MR 26 #2987).
[2] A global formulation of the fundamental theorem of the theory of surfaces in three dimensional Euclidean space, Nagoya Math. J. 13 (1958), 69–82. (MR 20 #3565.)
[3]* Almost Contact Manifolds, Part I, Math. Inst. Tôhoku Univ. 1965. Part II, 1967, Part III, 1968.
[4] On almost contact manifolds, Proc. US-Japan Seminar in Differential Geometry, Kyoto, 1965, 128–136. (MR 35 #6075.)
[5] On differentiable manifolds with certain structures which are closely related to almost contact structures, Tôhoku Math. J. 12 (1960), 459–476 (MR 23 #A591); II (with Hatakeyama, Y.) 13 (1961), 281–294 (MR 25 #1513.)

SATAKE, I.
[1] The Gauss-Bonnet theorem for V-manifolds, J. Math. Soc. Japan 9 (1957), 464–492. (MR 20 #2022.)
[2] On representations and compactifications of symmetric Riemannian spaces, Ann. of Math. 71 (1960), 77–110. (MR 22 #9546.)

SCHMID, J.
[1] Zum Begriff des linearen Zusammenhanges, Monatsch. Math. 68 (1964), 326–367. (MR 30 #1471.)

SCHOENBERG, I. J.
[1] Some applications of the calculus of variations to Riemannian geometry, Ann. of Math. 33 (1932), 485–495.

SCHOUTEN, J. A.
[1] Über unitäre Geometrie, Proc. Kon. Ned. Akad. 32 (1929), 457–465.
[2]* Ricci-Calculus, 2nd ed., Springer, Berlin, 1954. (MR 16, 521.)

SCHOUTEN, J. A. and VAN DANTZIG, D.
[1] Über unitäre Geometrie, Math. Ann. 103 (1930), 319–346.

SCHOUTEN, J. A. and STRUIK, D. J.
  [1] On some properties of general manifolds relating to Einstein's theory of gravitation, Amer. J. Math. 43 (1921), 213–216.
SCHUR, F.
  [1] Über den Zusammenhang der Räume konstanten Krümmungsmasses mit den projektiven Räumen, Math. Ann. 27 (1886), 537–567.
SCHWARTZ, M-H.
  [1] Connexions adaptées à une sous-variété, An. Acad. Brasil Ci. 34 (1962), 427–444. (MR 27 #4172.)
SEIFERT, H. and THRELFALL, W.
  [1] Topologische Untersuchung der Diskontinuitätsbereiche endlicher Bewegungsgruppen des dreidimensionalen sphärischen Raumes, Math. Ann. 104 (1930), 1–70; 107 (1933), 543–586.
SHIODA, T.
  [1] On algebraic varieties uniformizable by bounded domains, Proc. Jap. Acad. 39 (1963), 617–619. (MR 28 #5448.)
SIBUYA, Y.
  [1] Note on real matrices and linear dynamical systems with periodic coefficients, J. Math. Anal. Appl. 1 (1960), 363–372. (MR 23 #A2480.)
SIEGEL, C. L.
  [1] Einheiten quadratische Formen, Abh. Math. Sem. Hamburg 13 (1940), 209–239.
SIERPINSKI, W.
  [1] Sur les espaces métriques localement séparables, Fund. Math. 21 (1933), 107–113.
SIMONS, J.
  [1] On transitivity of holonomy systems, Ann. of Math. 76 (1962), 213–234. (MR 26 #5520.)
  [2] A note on minimal varieties, Bull. Amer. Math. Soc. 73 (1967), 491–495. (MR 34 #8291.)
  [3] Minimal varieties in Riemannian manifolds, Ann. of Math. 88 (1968), 62–105.
SINGER, I. M.
  [1] Geometric interpretation of a special connection, Pacific J. Math. 9 (1959), 585–590. (MR 22 #1926.)
  [2] Infinitesimally homogeneous spaces, Comm. Pure Appl. Math. 13 (1960), 685–697. (MR 24 #A1100.)
SINGER, I. M. and STERNBERG, S.
  [1] On the infinite groups of Lie and Cartan, I, Ann. Inst. Fourier (Grenoble) 15 (1965), 1–114. (MR 36 #911.)

SINGH, K. D.
    [1] Sous-espaces d'une variété kählérienne, Bull. Sci. Math. (2) 81 (1957), 21–29. (MR 19, 577.)

ŚLEBODZIŃSKI, A.
    [1] Formes Extérieurs et Leurs Applications, 2 volumes, Warzawa, 1954 and 1963. (MR 16, 1082; 29 #2738.)
    [2] Sur les prolongements d'une connexion linéaire, Bull. Acad. Pol. Sci. Ser. Math. Astr. Phys. 8 (1960), 145–150. (MR 22 #7147.)

SMITH, J. W.
    [1] Lorentz structures on the plane, Trans. Amer. Math. Soc. 95 (1960), 226–237. (MR 22 #8454.)

SMITH, P. A.
    [1] Fixed point theorem for periodic transformations, Amer. J. Math. 63 (1941), 1–8. (MR 2, 179.)

SMYTH, B.
    [1] Differential geometry of complex hypersurfaces, Ann. of Math. 85 (1967), 246–266. (MR 34 #6697.)
    [2] Homogeneous complex hypersurfaces, J. Math. Soc. Japan 20 (1968), 643–647.

STEENROD, N.
    [1] Topology of Fibre Bundles, Princeton Univ. Press, 1951. (MR 12, 522.)

STERNBERG, S.
    [1] Lectures on Differential Geometry, Prentice-Hall, Englewood Cliffs, New Jersey, 1964. (MR 23 #1797.)

STIEL, E.
    [1] Isometric immersions of manifolds of nonnegative constant sectional curvature, Pacific J. Math. 15 (1965), 1415–1419, (MR 32 #6375.)
    [2] On immersions with singular second fundamental form operators, Proc. Amer. Math. Soc. 17 (1966), 699–702, (MR 33 #6557.)
    [3] Immersions into manifolds of constant negative curvature, Proc. Amer. Math. Soc. 18 (1967), 713–715. (MR 35 #3609.)

STOKER, J. J.
    [1] Developable surfaces in the large, Comm. Pure Appl. Math. 14 (1961), 627–635, (MR 25 #493.)
    [2] On the form of complete surfaces in three-dimensional space for which $K \leq -c^2$ or $K \geq c^2$, Studies in Math. Analysis and Related Topics, pp. 377–387. Stanford Univ. Press, 1962. (MR 26 #2991.)

STUDY, E.
    [1] Kürzeste Wege im komplexen Gebiete, Math. Ann. 60 (1905), 321–377.

SÜSS, W.
    [1] Zur relativen Differentialgeometrie V. Tôhoku Math. J. 30 (1929), 202–209.

SYNGE, J. L.
[1] The first and second variations of length in Riemannian space, Proc. London Math. Soc. 25 (1926), 247–264.
[2] On the connectivity of spaces of positive curvature, Quart. J. Math. (Oxford Ser.) 7 (1936), 316–320.

SZYBIAK, A.
[1] On the general connections and continuity of the operator ∇, Bull. Acad. Polon. Sci. Sér. Sci. Math. Astronom. Phys. 13 (1965), 665–667. (MR 33 #4845.)
[2] On prolongation of fibre structures and connections of higher order, Bull. Acad. Pol. Sci. Ser. Sci. Math. Astronom. Phys. 13 (1965), 661–664. (MR 33 #1810.)

TAFT, E. J.
[1] Invariant Wedderburn factors, Illinois J. Math. 1 (1957), 565–573. (MR 20 #4586.)
[2] Invariant Levi factors, Michigan Math. J. 9 (1962), 65–68. (MR 24 #A1973.)

TAI, S. S.
[1] On minimum imbeddings of compact symmetric spaces of rank one, J. Differential Geometry 2 (1968), 55–66.

TAKAHASHI, T.
[1] Correction to "The index theorem in Riemannian geometry" by W. Ambrose, Ann. of Math. 80 (1964), 538–541. (MR 30 #545.)
[2] Minimal immersions of Riemannian manifolds, J. Math. Soc. Japan 18 (1966), 380–385. (MR 33 #6551.)
[3] Hypersurface with parallel Ricci tensor in a space of constant holomorphic sectional curvature, J. Math. Soc. Japan 19 (1967), 199–204. (MR 35 #2252.)

TAKEUCHI, M.
[1] On Pontrjagin classes of compact symmetric spaces, J. Fac. Sci. Univ. Tokyo Sect. I, 9 (1962), 313–328. (MR 26 #2548.)
[2] On the fundamental group and the group of isometries of a symmetric space, J. Fac. Sci. Univ. Tokyo Sect. I, 10 (1964), 88–123. (MR 30 #1217.)
[3] Cell decompositions and Morse equalities on certain symmetric spaces, J. Fac. Sci. Univ. Tokyo Sect. I, 12 (1965), 81–192.

TAKEUCHI, M. and KOBAYASHI, S.
[1] Minimal imbeddings of R-spaces, J. Differential Geometry 2 (1968), 203–215.

TAKIZAWA, S.
[1] On Cartan connections and their torsions, Mem. Coll. Sci. Univ. Kyoto Ser. A 29 (1955), 199–217. (MR 21 #3017.)

[2] On Stiefel characteristic classes of a Riemannian manifold, Mem. Coll. Sci. Univ. Kyoto Ser. A 28 (1953), 1–10. (MR 15, 646.)
[3] On the primary difference of two frame functions in a Riemannian manifold, Mem. Coll. Sci. Univ. Kyoto Ser. A 28 (1953), 11–14. (MR 15, 647.)
[4] Some remarks on invariant forms of a sphere bundle with connections Mem. Coll. Sci. Univ. Kyoto Ser. A 29 (1955), 193–198. (MR 21 #867.)
[5] On the induced connexions, Mem. Coll. Sci. Univ. Kyoto Ser. A 30 (1957), 105–118. (MR 20 #317.)
[6] On soudures of differentiable fibre bundles, J. Math. Kyoto Univ. 2 (1962/63), 237–275. (MR 27 #5197.)
[7] On contact structures of real and complex manifolds, Tôhoku Math. J. 15 (1963), 227–252. (MR 28 #566.)

TANAKA, N.
[1] Conformal connections and conformal transformations, Trans. Amer. Math. Soc. 92 (1959), 168–190. (MR 23 #A1331.)
[2] Projective connections and projective transformations, Nagoya Math. J. 11 (1957), 1–24. (MR 21 #3899.)
[3] On the pseudo-conformal geometry of hypersurfaces of the space of n complex variables, J. Math. Soc. Japan 14 (1962), 397–429. (MR 26 #3086.)
[4] On the equivalence problem associated with a certain class of homogeneous spaces, J. Math. Soc. Japan 17 (1965), 103–139. (MR 32 #6358.)

TANNO, S.
[1] Strongly curvature-preserving transformations of pseudo-Riemannian manifolds, Tôhoku Math. J. 19 (1967), 245–250. (MR 36 #3284.)

TASHIRO, Y.
[1] Complete Riemannian manifolds and some vector fields, Trans. Amer. Math. Soc. 117 (1965), 251–275. (MR 30 #4229.)

TELEMAN, G.
[1] Sur les connexions infinitésimales qu'on peut définir dans les structures fibrées différentiables de base donnée, Ann. Mat. Pura Appl. 62 (1963), 379–412. (MR 29 #1597.)
[2] Sur une classe d'espaces riemanniens symétriques, Rev. Math. Pures Appl. 2 (1957), 445–470. (MR 21 #1631.)
[3] Sur les variétés de Grassmann, Bull. Math. Soc. Sci. Math. Phys. Rep. Popul. Roumanie 2 (50) (1958), 202–224. (MR 22 #11343.)
[4] Sur les groupes des mouvements d'un espace de Riemann, J. Math. Soc. Japan 15 (1963), 134–158. (MR 27 #6219.)

THOMAS, T. Y.
  [1] Riemannian spaces of class one and their characterizations, Acta Math. 67 (1935), 169–211.
  [2] Imbedding theorems in differential geometry, Bull. Amer. Math. Soc. 45 (1939), 841–850. (MR 1, 88).
  [3] Extract from a letter by E. Cartan concerning my note: On closed spaces of constant mean curvature, Amer. J. Math. 59 (1937), 793–794.
  [4] On a class of existence theorems in differential geometry, Bull. Amer. Math. Soc. 40 (1934), 721–728.
  [5] On closed spaces of constant mean curvature, Amer. J. Math. 58 (1936), 702–704.
  [6] On the variation of curvature in Riemann spaces of constant mean curvature, Ann. Mat. Pura Appl. 13 (1934/35), 227–238.

THORPE, J. A.
  [1] Sectional curvatures and characteristic classes, Ann. of Math. 80 (1964), 429–443. (MR 30 #546.)
  [2] Parallelizability and flat manifolds, Proc. Amer. Math. Soc. 16 (1965), 138–142. (MR 30 #1473.)
  [3] On the curvatures of Riemannian manifolds, Illinois J. Math. 10 (1966), 412–417. (MR 33 #4868.)

TITS, J.
  [1] Sur certaines classes d'espaces homogènes de groupes de Lie, Acad. Roy. Belg. Cl. Sci. Mem. Coll. 29 (1955), fasc. 268 pp. (MR 7, 874.)
  [2] Espaces homogènes complexes compacts, Comment. Math. Helv. 37 (1962/63), 111–120. (MR 27 #4248.)

TOMPKINS, C.
  [1] Isometric imbedding of flat manifolds in Euclidean space, Duke Math. J. 5 (1939), 58–61.
  [2] A flat Klein bottle isometrically imbedded in Euclidean 4-space, Bull. Amer. Math. Soc. 47 (1941), 508. (MR 2, 301.)

TONDEUR, P.
  [1] Ein invariantes Vektorraumfeld auf einem reduktiven, lokal-symmetrischen homogenen Raum ist involutorisch, Math. Z. 85 (1964), 382–384. (MR 31 #6181.)
  [2] Champs invariants de p-plans sur un espace homogène, C.R. Acad. Sci. Paris 259 (1964), 4473–4475. (MR 30 #2445.)
  [3] Invariant subbundle of the tangent bundle of a reductive homogeneous space, Math. Z. 89 (1965), 420–421. (MR 32 #410.)

TOPONOGOV, V. A.
  [1] Riemannian spaces with curvature bounded below, Uspehi Math. Nauk 14 (1959), 87–130. (MR 21 #2278.)

[2] Evaluation of the length of closed geodesics on convex surfaces, Dokl. Akad. Nauk. SSSR 124 (1959), 282–284. (MR 21 #850.)
[3] Dependence between curvature and topological structure of Riemannian spaces of even dimensions, Dokl. Akad. Nauk. SSSR. 133 (1960), 1031–1033. Soviet Math. Dokl. 1 (1961). 943–945. (MR 25 #A3476.)
[4] A bound for the length of a closed geodesic in a compact Riemannian space of positive curvature, Dokl. Akad. Nauk. SSSR. 154 (1964), 1047–1049. (MR 28 #3392.)
[5] The metric structure of Riemannian spaces of non-negative curvature containing straight lines, Sibirsk Mat. Z. 5 (1964), 1358–1369. (MR 32 #3017.)

TSUKAMOTO, Y.
 [1] On Kaehlerian manifolds with positive holomorphic sectional curvature, Proc. Japan Acad. 33 (1957), 333–335. (MR 19, 880.)
 [2] On a theorem of A. D. Alexandrov, Mem. Fac. Sci. Kyushu Univ. Ser. A 15 (1962), 83–89. (MR 25 #2562.)
 [3] On Riemannian manifolds with positive curvature, Mem. Fac. Sci. Kyushu Univ. A 15 (1962), 90–96. (MR 25 #2561.)
 [4] A proof of Berger's theorem, Mem. Fac. Sci. Kyushu Univ. Ser. A 17 (1963), 168–175. (MR 28 #4486.)
 [5] On certain Riemannian manifolds of positive curvature, Tôhoku Math. J. 18 (1966), 44–49. (MR 34 #8343.)
 [6] Closed geodesics on certain Riemannian manifolds of positive curvature, Tôhoku Math. J. 18 (1966), 138–143. (MR 34 #3501.)
 [7]* Curvature, geodesics and topology of Riemannian manifolds, A report in differential geometry (in Japanese), Vol. 2, 1966.

VAISMAN, IZU
 [1] Sur quelques formules du calcul de Ricci global, Comment. Math. Helv. 41 (1966/67), 73–87. (MR 34 #6688.)

VEBLEN, O. and WHITEHEAD, J. H. C.
 [1] The Foundations of Differential Geometry, Cambridge Tracts No. 29, Cambridge Univ. Press, 1932.

VINBERG, E. B.
 [1] On invariant linear connections, Dokl. Akad. Nauk. SSSR. 128 (1959), 653–654. (MR 22 #956.)
 [2] Invariant linear connections in a homogeneous space, Trudy Moscow Mat. Obsc. 9 (1960), 191–210. (MR 31 #690.)

VINCENT, G.
 [1] Les groupes linéaires finis sans points fixes, Comment. Math. Helv. 20 (1947), 117–171. (MR 9, 131.)

Voss, K.
[1] Differentialgeometrie geschlossener Flächen im Euklidischen Raum, Jber. Deutsch. Math. Verein., 63 (1960/61), 117–136. (MR 23 #A4107.)
[2] Einige differentialgeometrische Kongruentzsätze für geschlossene Flächen und Hyperflächen, Math. Ann. 131 (1956), 180–218. (MR 18, 229.)
[3] Über geschlossene Weingartensche Flächen, Math. Ann. 138 (1959), 42–54. (MR 21 #7536.)

Vranceanu, G.
[1] Leçons de Géométrie Différentielle, 3 volumes, Gauthier-Villars, Paris, 1964.
[2] Tenseurs harmoniques et groupes de mouvement d'un espace de Riemann, Comment. Math. Helv. 33 (1959), 161–173. (MR 21 #5999.)

Vranceanu, G., Hangan, T., and Teleman, C.
[1] Recherches de géométrie différentielle en Roumaine, Rev. Roumaine Math. Pures Appl. 11 (1966) 1147–1156. (MR 35 #882.)

Wakakuwa, H.
[1] On affinely connected manifolds with homogeneous holonomy group $GL(n; Q) \times T^1$; Tôhoku Math. J. 11 (1959), 364–375. (MR 22 #959.)
[2] Remarks on 4-dimensional differentiable manifolds, Tôhoku Math. J. 16 (1964), 154–172. (MR 29 #6449.)

Walker, A. G.
[1] On Ruse's spaces of recurrent curvature, Proc. London Math. Soc. (2) 52 (1950), 36–64. (MR 12, 283.)
[2] The fibring of Riemannian manifolds, Proc. London Math. Soc. (3) 3 (1953), 1–19. (MR 15, 159.)
[3] Connexions for parallel distribution in the large, Quart. J. Math. (Oxford) (2) 6 (1955), 301–308 (MR 19, 312); II, 9 (1958), 221–231 (MR 20 #6135).
[4] Distributions and global connexions, Colloque Géométrie Différentielle Globale, Bruxelles, 1958, 63–74. (MR 23 #A590.)
[5] Dérivation torsionelle et second torsion pour une structure presque complexe, C.R. Acad. Sci. Paris 245 (1957), 1213–1215. (MR 19, 680.)
[6] Note on metrisable Lie groups and algebras, Calcutta Math. Soc. Golden Jubilee Commemoration Volume (1958/59), Part I, 185–192, Calcutta Math. Soc., 1963. (MR 29 #567.)
[7] Almost-product structures, Proc. Symp. Pure Math. III (1961), 94–100. (MR 23 #A1314.)

WALLACH, N.
- [1] A classification of involutive automorphisms of compact simple Lie algebras up to inner equivalence, Thesis, Washington Univ., 1966.
- [2] On maximal subsystems of root systems, Canad. J. Math., 20 (1968), 555–574.

WANG, H. C.
- [1] On invariant connections over a principal fibre bundle, Nagoya Math. J. 13 (1958), 1–19. (MR 21 #6001.)
- [2] Homogeneous spaces with non-vanishing Euler characteristics, Ann. of Math. 50 (1949), 925–953. (MR 11, 326.)
- [3] Two point homogeneous spaces, Ann. of Math. 55 (1952), 177–191. (MR 13, 863.)
- [4] Closed manifolds with homogeneous complex structure, Amer. J. Math. 76 (1954), 1–32. (MR 16, 518.)
- [5] Complex parallisable manifolds, Proc. Amer. Math. Soc. 5 (1954), 771–776. (MR 17, 531.)

WARNER, F. W.
- [1] The conjugate locus of a Riemannian manifold, Amer. J. Math. 87 (1965), 575–604. (MR 34 #8344.)
- [2] Extensions of the Rauch comparison theorem to submanifolds, Trans. Amer. Math. Soc. 122 (1966), 341–356. (MR 34 #759.)
- [3] Conjugate loci of constant order, Ann. of Math. 86 (1967), 192–212. (MR 35 #4857.)

WASSERMAN, A.
- [1] Morse theory for G-manifolds, Bull. Amer. Math. Soc. 71 (1965), 384–388. (MR 30 #4271.)

WATANABE, S.
- [1] Sur les formes spatiales de Clifford-Klein, Japan. J. Math. 8 (1931), 65–102.
- [2] Formes spatiales de l'espace elliptique, Japan. J. Math. 9 (1932), 117–134.

WEIL, A.
- [1] Sur un théorème de de Rham, Comment. Math. Helv. 26 (1952), 119–145. (MR 14, 307.)
- [2] Introduction à l'Étude des Variétés Kähleriennes, Hermann, Paris, 1958. (MR 22 #1921)
- [3] On discrete subgroups of Lie groups, I, Ann. of Math. 72 (1960), 369–384 (MR 25 #1241); II. ibid. 75 (1962), 578–602 (MR 25 #1242).

WEINSTEIN, A. D.
- [1] The cut locus and conjugate locus of a Riemannian manifold, Ann. of Math. 87 (1968), 29–41.

[2] On the homotopy type of positively-pinched manifolds, Arch. Math. 18 (1967) 523–524.
[3] A fixed point theorem for positively curved manifolds, J. Math. Mech. 18 (1968), 149–153.

WEST, A.
[1] The projection of infinitesimal connection, J. London Math. Soc. 40 (1965), 551–564. (MR 31 #2686.)

WEYL, H.
[1] Reine Infinitesimalgeometrie, Math. Z. 2 (1918), 384–411.
[2] Zur Infinitesimalgeometrie; Einordnung der projektiven und konformen Auffassung, Göttingen Nachrichten (1912), 99–112.
[3] Raum, Zeit, Materie, Springer, 1918.
[4] On the foundations of general infinitesimal geometry, Bull. Amer. Math. Soc. 35 (1929), 716–725.
[5] Die Einzigartigkeit der Pythagoreischen Massbestimmung, Math. Z. 12 (1922), 114–146.
[6] Zur Infinitesimalgeometerie; p-dimensionale Fläche im n-dimensionalen Raum, Math. Z. 12 (1922), 154–160.
[7] On the volume of tubes, Amer. J. Math. 61 (1939), 461–472.
[8] Über die Bestimmung einer geschlossenen konvexen Fläche durch ihr Linienelement, Vierteljarsch. Naturforsch. Ges. Zürich 61 (1916), 40–72.
[9] Über die Starrheit der Eiflächen und konvexen Polyeder, Preuss. Akad. Wiss. Sitzungsber. (1917), 250–266.

WHITEHEAD, J. H. C.
[1] Convex regions in the geometry of paths, Quart. J. Math. 3 (1932), 33–42, 226–227.
[2] On the covering of a complete space by the geodesics through a point, Ann. of Math. 36 (1935), 679–704.
[3] Affine spaces of paths which are symmetric about each point, Math. Z. 35 (1932), 644–659.

WHITMAN, A. P. and CONLON, L.
[1] A note on holonomy, Proc. Amer. Math. Soc. 16 (1965), 1046–1051. (MR 32 #1641.)

WILLMORE, T. J.
[1] Parallel distributions on manifolds, Proc. London Math. Soc. (3) 6 (1956), 191–204. (MR 19, 455.)
[2] Connexions for systems of parallel distributions, Quart. J. Math. (2) 7 (1956), 269–276. (MR 20 #4299.)
[3] Connexions associated with foliated structures, Ann. Inst. Fourier (Grenoble) 14 (1964), 43–47. (MR 30 #538.)
[4] Systems of parallel distributions, J. London Math. Soc. 32 (1957), 153–156. (MR 19, 455.)

[5] Derivations and vector 1-forms, J. London Math. Soc. 35 (1960), 425–432. (MR 23 #2821.)
[6] Generalized torsional derivation, Atti Acad. Naz. Lincei Rend. Cl. Sci. Fis. Mat. Nat. 26 (1959), 649–653. (MR 22 #960.)
[7] The Poincaré Lemma associated with certain derivations, J. London Math. Soc. 37 (1962), 345–350. (MR 25 #2540.)
[8] An Introduction to Differential Geometry, Oxford Univ. Press, 1959. (MR 28 #2482.)

WILSON, J. P.
[1] The total absolute curvature of an immersed manifold, J. London Math. Soc. 40 (1965), 362–366. (MR 31 #698.)

WOLF, J. A.
[1] Sur la classification des variétés riemanniennes homogènes à courbure constante, C.R. Acad. Sci. Paris 250 (1960), 3443–3445. (MR 22 #5948.)
[2] The manifolds covered by a Riemannian homogeneous manifold Amer. J. Math. 82 (1960), 661–688. (MR 23 #A246.)
[3] Homogeneous manifolds of constant curvature, Comment. Math. Helv. 36 (1961), 112–147. (MR 24 #A2921.)
[4] The Clifford-Klein space forms of indefinite metric, Ann. of Math. 75 (1962), 77–80. (MR 24 #A3613.)
[5] Vincent's conjecture of Clifford translations of the sphere, Comment. Math. Helv. 36 (1961), 33–41. (MR 25 #532.)
[6] Discrete groups, symmetric spaces, and global holonomy, Amer. J. Math. 84 (1962), 527–542. (MR 26 #5523.)
[7] Spaces of Constant Curvature, McGraw-Hill, New York, 1967. (MR 36 #829.)
[8] Complex homogeneous contact manifolds and quaternionic symmetric spaces, J. Math. Mech. 14 (1965), 1033–1047. (MR 32 #3020.)
[9] Exotic metrics on immersed surfaces, Proc. Amer. Math. Soc. 17 (1966), 871–877. (MR 33 #3235.)
[10] On locally symmetric spaces of non-negative curvature and certain other locally homogeneous spaces, Comment. Math. Helv. 37 (1963), 266–295. (MR 27 #4178.)
[11] Geodesic spheres in Grassmann manifolds, Illinois J. Math. 7 (1963), 425–446. (MR 27 #6220.)
[12] Elliptic spaces in Grassmann manifolds, Illinois J. Math. 7 (1963), 447–462. (MR 27 #6221.)
[13] Homogeneous manifolds of zero curvature, Trans. Amer. Math. Soc. 104 (1962), 462–469 (MR 25 #3474); Errata, 106 (1963), 540 (MR 30 #1474).

[14] Isotropic manifolds of indefinite metric, Comment. Math. Helv. 39 (1964), 21–64 (MR 31 #3987.)
[15] Curvature in nilpotent Lie groups, Proc. Amer. Math. Soc. 15 (1964), 271–274. (MR 28 #5405.)
[16] Homogeneity and bounded isometries in manifolds of negative curvature, Illinois J. Math. 8 (1964), 14–18. (MR 29 #565.)
[17] Differentiable fibre spaces and mappings compatible with Riemannian metrics, Michigan Math. J. 11 (1964), 65–70. (MR 28 #2502.)
[18] Locally symmetric homogeneous spaces, Comment. Math. Helv. 37 (1962/63), 65–101. (MR 26 #5522.)
[19] Space forms of Grassmann manifolds, Canad. J. Math. 15 (1963), 193–205. (MR 26 #5524.)

WOLF, J. A. and GRIFFITHS, P. A.
[1] Complete maps and differentiable coverings, Michigan Math. J. 10 (1963), 253–255. (MR 27 #4179.)

WONG, Y. C.
[1] Recurrent tensors on a linearly connected differentiable manifold, Trans. Amer. Math. Soc. 99 (1961), 325–341. (MR 22 #12485.)
[2] Existence of linear connections with respect to which given tensor fields are parallel or recurrent, Nagoya Math. J. 24 (1964), 67–108. (MR 30 #4222.)
[3] Two dimensional linear connexions with zero torsion and recurrent curvature, Monatsch. Math. 68 (1964), 175–184. (MR 30 #4221.)
[4] Differential geometry of Grassmann manifolds, Proc. Nat. Acad. Sci. USA 57 (1967), 589–594. (MR 35 #7266.)
[5] Linear connections and quasi-connections on a differentiable manifold, Tôhoku Math. J. 14 (1962), 48–63. (MR 25 #2547.)
[6] Linear connections with zero torsion and recurrent curvature, Trans. Amer. Math. Soc. 102 (1962), 471–506, (MR 24 #A3601)
[7] Conjugate loci in Grassmann manifolds, Bull. Amer. Math. Soc. 74 (1968), 240–245. (MR 36 #3285)

WU, H.
[1] On the de Rham decomposition theorem, Illinois J. Math. 8 (1964), 291–311. (MR 28 #4488.)
[2] Decomposition of Riemannian manifolds, Bull. Amer. Math. Soc. 70 (1964), 610–617. (MR 29 #566.)
[3] Holonomy groups of indefinite metrics, Pacific J. Math. 20 (1967), 351–392. (MR 35 #3606.)
[4] Two remarks on sprays, J. Math. Mech. 14 (1965), 873–879. (MR 31 #5174.)
[5] Normal families of holomorphic mappings, Acta Math., 119 (1967), 193–233.

WU, W. T.
  [1] Sur les classes caractéristiques de structures fibrés sphériques, Actualités Sci. Ind. #1183, Hermann, Paris, 1952, 5–89. (MR 14, 1112.)

YAMABE, H.
  [1] On an arcwise connected subgroup of a Lie group, Osaka Math. J. 2 (1950), 13–14. (MR 12, 158.)
  [2] On a deformation of Riemannian structures on compact manifolds, Osaka Math. J. 12 (1960), 21–37. (MR 23 #A2847.)

YANO, K.
  [1] On harmonic and Killing vectors, Ann. of Math. 55 (1952), 38–45. (MR 13, 689.)
  [2]* The Theory of Lie Derivatives and its Applications, North-Holland Amsterdam, 1957. (MR 19, 576.)
  [3] Quelques remarques sur les variétés à structure presque complexe, Bull. Soc. Math. France 83 (1955), 57–80. (MR 17, 662.)
  [4] Sur un théorème de M. Matsushima, Nagoya Math. J. 12 (1957), 147–150. (MR 20 #2476.)
  [5] Harmonic and Killing tensor fields in Riemannian spaces with boundary, J. Math. Soc. Japan 10 (1958), 430–437. (MR 21 #3012.)
  [6] Harmonic and Killing vector fields in compact orientable Riemannian spaces with boundary, Ann. of Math. 69 (1959), 588–597. (MR 21 #3887.)
  [7] Some integral formulas and their applications, Michigan Math. J. 5 (1958), 68–73. (MR 21 #1860.)
  [8]* Differential Geometry on Complex and Almost Complex Spaces, Pergamon Press, New York, 1965. (MR 32 #4635.)
  [9] Closed hypersurfaces with constant mean curvature in a Riemannian manifold, J. Math. Soc. Japan 17 (1965), 333–340. (MR 32 #1657.)
  [10] On Riemannian manifolds with constant scalar curvature admitting a conformal transformation group, Proc. Nat. Acad. Sci. USA 55 (1966), 472–476. (MR 33 #677.)
  [11] On a structure defined by a tensor field f of type $(1, 1)$ satisfying $f^3 + f = 0$, Tensor, 14 (1963), 99–109. (MR 28 #2513.)
  [12] Notes on hypersurfaces in a Riemannian manifold, Canad. J. Math. 19 (1967), 439–446.

YANO, K. and BOCHNER, S.
  [1]* Curvature and Betti Numbers, Annals of Math. Studies, No. 32 Princeton Univ. Press, 1953. (MR 15, 989.)

YANO, K. and ISHIHARA, S.
  [1] Structure defined by f satisfying $f^3 + f = 0$, Proc. US-Japan Seminar in Differential Geometry, Kyoto, 1965, 153–166. (MR 36 #815.)

[2] Differential geometry in tangent bundle, Kōdai Math. Sem. Rep. 18 (1966), 271–292. (MR 34 #8322.)

YANO, K. and KOBAYASHI, S.
[1] Prolongations of tensor fields and connections to tangent bundles, J. Math. Soc. Japan 18 (1966), 194–210 (MR 33 #1816); II, 236–246 (MR 34 #743); III. 19 (1967), 486–488.

YANO, K. and MOGI, I.
[1] On real representations of Kaehlerian manifolds, Ann. of Math. 61 (1955), 170–189. (MR 16, 859.)

YANO, K. and NAGANO, T.
[1] Einstein spaces admitting a one-parameter group of conformal transformations, Ann. of Math. 69 (1959), 451–461. (MR 21 #345.)
[2] Some theorems on projective and conformal transformations, Indag. Math. 19 (1957), 451–458. (MR 22 #1861.)
[3] The de Rham decomposition, isometries and affine transformations in Riemannian spaces, Japanese J. Math. 29 (1959), 173–184. (MR 22 #11341.)
[4] On geodesic vector fields in a compact orientable Riemannian space, Comment. Math. Helv. 35 (1961), 55–64. (MR 23 #A2164.)

YANO, K. and TAKAHASHI, T.
[1] Some remarks on Einstein spaces and spaces of constant curvature, J. Math. Soc. Japan 12 (1960), 89–96. (MR 22 #5009.)

YOSHIDA, K.
[1] A theorem concerning the semi-simple Lie groups, Tôhoku Math. J. 43, Part II (1937), 81–84.

ZASSENHAUS, H.
[1] Beweis eines Satzes über diskrete Gruppe, Abh. Math. Sem. Univ. Hamburg 12 (1938), 298–312.
[2] Über endliche Fastkörper, Abh. Math. Sem. Univ. Hamburg 11 (1936), 187–220.

Summary of Basic Notations

We summarize only those basic notations that are used most frequently throughout Volumes I and II.

1. Σ_i, $\Sigma_{i,j}$, ..., etc. stand for the summation taken over i or $i, j, \ldots$, where the range of indices is generally clear from the context. $\mathfrak{S}$ denotes the cyclic sum, e.g. $\mathfrak{S} R(X, Y)Z = R(X, Y)Z + R(Y, Z)X + R(Z, X)Y$.

2. **R** and **C** denote the real and complex number fields, respectively.

$\mathbf{R}^n$: vector space of n-tuples of real numbers $(x^1, \ldots, x^n)$
$\mathbf{C}^n$: vector space of n-tuples of complex numbers $(z^1, \ldots, z^n)$
(x, y): standard inner product $\Sigma_i x^i y^i$ in $\mathbf{R}^n$ ($\Sigma_i x^i \bar{y}^i$ in $\mathbf{C}^n$)
$GL(n; \mathbf{R})$: general linear group acting on $\mathbf{R}^n$
$\mathfrak{gl}(n; \mathbf{R})$: Lie algebra of $GL(n; \mathbf{R})$
$GL(n; \mathbf{C})$: general linear group acting on $\mathbf{C}^n$
$\mathfrak{gl}(n; \mathbf{C})$: Lie algebra of $GL(n; \mathbf{C})$
$O(p, q)$: orthogonal group for $-\Sigma_{i=1}^p (x^i)^2 + \Sigma_{i=p+1}^{p+q} (x^i)^2$ on $\mathbf{R}^{p+q}$
$O(n) = O(0, n)$: orthogonal group for the standard inner product in $\mathbf{R}^n$
$\mathfrak{o}(p, q)$: Lie algebra of $O(p, q)$
$\mathfrak{o}(n)$: Lie algebra of $O(n)$
$U(p, q)$: unitary group for $-\Sigma_{i=1}^p x^i \bar{x}^i + \Sigma_{i=p+1}^{p+q} x^i \bar{x}^i$ on $\mathbf{C}^{p+q}$
$U(n) = U(0, n)$: unitary group for the standard inner product in $\mathbf{C}^n$
$\mathfrak{u}(p, q)$: Lie algebra of $U(p, q)$
$\mathfrak{u}(n)$: Lie algebra of $U(n)$
$\mathbf{T}_s^r(V)$: tensor space of type (r, s) over a vector space V
$\mathbf{T}(V)$: tensor algebra over V
V^c: complexification of a real vector space V
A^n: space $\mathbf{R}^n$ regarded as an affine space
$A(n; \mathbf{R})$: group of affine transformations of A^n
$\mathfrak{a}(n; \mathbf{R})$: Lie algebra of $A(n; \mathbf{R})$

$G(n, p)$: Grassmann manifold of n-planes in $\mathbf{R}^{n+p}$
$\tilde{G}(n, p)$: Grassmann manifold of oriented n-planes in $\mathbf{R}^{n+p}$
$V(n, p)$: Stiefel manifold of n-frames in $\mathbf{R}^{n+p}$
$G_{p,q}(\mathbf{C})$: complex Grassmann manifold of p-planes in $\mathbf{C}^{p+q}$
$D_{p,q}$: space of complex matrices Z with q rows and p columns such that $I_p - {}^t\bar{Z}Z$ is positive-definite
$P_n(\mathbf{C}) = G_{n,1}(\mathbf{C})$: complex projective space
$D_n = D_{n,1}$: unit disk in $\mathbf{C}^n$

3. M denotes an n-dimensional differentiable manifold.
$T_x(M)$: tangent space of M at x
$\mathfrak{F}(M)$: algebra of differentiable functions on M
$\mathfrak{X}(M)$: Lie algebra of vector fields on M
$\mathfrak{T}(M)$: algebra of tensor fields on M
$\mathfrak{D}(M)$: algebra of differential forms on M
$T(M)$: tangent bundle of M
$L(M)$: bundle of linear frames of M
$O(M)$: bundle of orthonormal frames of M (with respect to a given Riemannian metric g)
$\theta = (\theta^i)$: canonical 1-form on $L(M)$ or $O(M)$
$A(M)$: bundle of affine frames of M
$T_s^r(M)$: tensor bundle of type (r, s) of M
f_*: differential of a differentiable mapping f
$f^*\omega$: the transform of a differential form ω by f
$\dot{x}_t$: tangent vector of a curve x_t at the point x_t
L_X: Lie differentiation with respect to a vector field X

4. For a Lie group G, G^0 denotes the identity component and $\mathfrak{g}$ the Lie algebra of G.
L_a: left translation by $a \in G$
R_a: right translation by $a \in G$
ad a: inner automorphism by $a \in G$; also adjoint representation in $\mathfrak{g}$
$P(M, G)$: principal fibre bundle over M with structure group G
A^*: fundamental vector field corresponding to $A \in \mathfrak{g}$
ω: connection form
Ω: curvature form
$E(M, F, G, P)$: bundle associated to $P(M, G)$ with fibre F

5. For an affine (linear) connection on M,
    $\Theta = (\Theta^i)$: torsion form
    Γ^i_{jk}: Christoffel symbols
    $\Psi(x)$: linear holonomy group at $x \in M$
    $\Phi(x)$: affine holonomy group at $x \in M$
    ∇_X: covariant differentiation with respect to a vector (field) X
    R: curvature tensor field (with components R^i_{jkl}), giving rise to curvature transformations $R(X, Y)$. Also, the Riemannian curvature tensor:

    $$R(X, Y, Z, W) = g(R(Z, W)Y, X)$$

    T: torsion tensor field (with components T^i_{jk})
    S: Ricci tensor field (with components $R_{ij} = \Sigma_k R^k_{jki}$)
    $K(p)$: sectional curvature for a 2-plane p in $T_x(M)$, where M is a Riemannian manifold
    $\mathfrak{A}(M)$: group of all affine transformations of M
    $\mathfrak{a}(M)$: Lie algebra of all infinitesimal affine transformations of M
    $\mathfrak{I}(M)$: group of all isometries of a Riemannian manifold M
    $\mathfrak{i}(M)$: Lie algebra of all infinitesimal isometries

6. For a manifold M immersed in a Riemannian manifold N,
    $T_x(M)^\perp$: normal space to M at x
    $T(M)^\perp$: normal bundle of M
    $O(N, M)$: bundle of adapted frames
    $\mathfrak{X}(M)^\perp$: set of vector fields normal to M
    $\alpha(X, Y)$: second fundamental form defined as a mapping

    $$T_x(M) \times T_x(M) \to T_x(M)^\perp$$

    A_ξ: endomorphism such that $g(A_\xi X, Y) = g(\alpha(X, Y), \xi)$, where $\xi \in T_x(M)^\perp$
    $h^i(X, Y)$: symmetric bilinear forms on $T_x(M)$ defined by

    $$\alpha(X, Y) = \sum h^i(X, Y) \xi_i,$$

    where $\xi_1, \ldots, \xi_p$ is an orthonormal basis of $T_x(M)^\perp$

    η: mean curvature normal $= \dfrac{1}{n} \Sigma_i (\text{trace } A_{\xi_i}) \xi_i$, $n = \dim M$

K_n: Gaussian curvature of a hypersurface M ($n = \dim M$)
$t(x)$: type number

7. For a manifold M with an almost complex structure J,
    $T_x^c(M)$: complex tangent space at x
    $T_x^{1,0}$: space of tangent vectors of type $(1, 0)$
    $T_x^{0,1}$: space of tangent vectors of type $(0, 1)$
    $\mathfrak{C}(M)$: set of complex r-forms
    $\mathfrak{C}^{p,q}(M)$: set of complex forms of degree (p, q)
    $C(M)$: bundle of complex linear frames with structure group $GL(n: \mathbf{C})$
    $U(M)$: bundle of unitary frames (with respect to a Hermitian metric) with structure group $U(n)$

8. For a homogeneous space G/H,
    $\tilde{H}$: linear isotropy group
    $\mathfrak{g} = \mathfrak{m} + \mathfrak{h}$: decomposition of Lie algebra such that
    $$[\mathfrak{m}, \mathfrak{h}] \subset \mathfrak{m},$$
    $X_\mathfrak{m}$ and $X_\mathfrak{h}$ denoting the $\mathfrak{m}$-component and the $\mathfrak{h}$-component of $X \in \mathfrak{g}$, respectively
    B: Killing-Cartan form
    σ: involutive automorphism of G which defines a symmetric space (G, H, σ)
    $(\mathfrak{g}, \mathfrak{h}, \sigma)$: symmetric Lie algebra
    $(\mathfrak{g}^*, \mathfrak{h}, \sigma^*)$: dual of symmetric Lie algebra $(\mathfrak{g}, \mathfrak{h}, \sigma)$

9. For a vector bundle E over M,
    $c_i(E)$: i-th Chern class
    $c(E)$: total Chern class
    $ch(E)$: Chern character
    $p_k(E)$: k-th Pontrjagin class
    $\chi(E)$: Euler class

Index for Volumes I and II

(Note: For example, I-201 refers to p. 201 of Volume I, whereas 201 refers to p. 201 of Volume II.)

Abelian
  Lie algebra, 325
  variety, 131
Absolute parallelism, I-122
Adapted frame, 2, 54
Adjoint representation, I-40
Affine
  connection, I-129
    generalized, I-127
    invariant, 375
    invariant by parallelism, 194
    rigid, 376
  frame, I-126
  holonomy group, I-130, 331
  locally symmetric, 222
  mapping, I-225
  parallel displacement, I-130
  parameter, I-138
  symmetric, 223
  space, I-125
    tangent, I-125
  transformation, I-125, I-226
    infinitesimal, I-230
Allowable imbedding, 52
Almost
  cocomplex structure, 383
  complex connection, 143
  complex manifold, 121
    natural orientation of, 121
  complex mapping, 122, 127
  complex structure, 121
    conjugate, 123
    integrability of, 124, 321
    integrable, 124, 321
    invariant, 216
    on spheres, 138-140
    torsion of, 123

Almost
  effective action, 187
  Hamiltonian manifold, 149
  Hermitian manifold, 147
  Kaehler manifold, 149
  product structure, 384
  symplectic manifold, 149
Alternation, I-28
Analytic continuation, I-254
Arc-length, I-157
Atlas, I-2
  complete, I-2
Augmented index, 89
Automorphism of
  an almost complex structure, 127
  a connection, I-81
  a G-structure, I-307, 186, 333
  a Lie algebra, I-40
  a Lie group, I-40
Auto-parallel submanifold, 53

Bergman metric, 163
Bianchi's identities, I-78, I-121, I-135
Bonnet's theorem, 78
Bounded domain, 162, 375
  symmetric, 263
Bundle
  associated, I-55
  holonomy, I-85
  homomorphism of, I-53
  induced, I-60
  normal, 3
  of adapted frames, 2, 54
  of affine frames, I-126
  of complex linear frames, 141
  of normal frames, 2
  of orthonormal frames, I-60

Bundle
  of unitary frames, 152
  principal fibre, I-50
  reduced, I-53
  sub-, I-53
  tangent, I-56
  tensor, I-56
  trivial, I-50
  vector, I-113
    Hermitian, 178
    orientable, oriented, 314
    Riemannian, 315

C-space, 138, 373
canonical
  complex structure, 115
  connection, I-110, I-301, 7, 230
    on a symmetric space, 230
  decomposition (= de Rham decomposition), I-185, I-192, 171, 246, 263, 331
  decomposition of a symmetric Lie algebra, 226
  flat connection, I-92
  form on $L(M)$, I-118
  Hermitian form, 374
  invariant connection, I-110, I-301, 192, 230
  invariant Riemannian metric, I-155
  linear connection, I-302
  metric, I-155
  1-form on a group, I-41
  parameter of a geodesic, I-162
Cayley numbers, 139
Center of gravity, 109
Characteristic class, 293
Chart, I-2
Chern character, 311
Chern class, 305
Christoffel's symbols (Γ^i_{jk}), I-141
Clifford translation, 105
Codazzi, equation of, 25–26
Compact-open topology, I-46
Compact real form, 291
Compact type
  Lie algebra of, 204, 252, 329
  symmetric Lie algebra of, 252

Compact type
  symmetric space of, 252, 256
Comparison theorem of Rauch, 76
Complete
  linear connection, I-134
  Riemannian manifold, I-172
  Riemannian metric, I-172
  vector field, I-13
Complex
  affine (locally) symmetric, 259
  conjugation, 116
  contact form, 385
  differential form, 124, 125
    holomorphic, 129
  Grassmann manifold, 133, 160, 286
  homogeneous space, 220, 373
  hyperbolic space, 282
  hypersurface, 378, 379
  Lie algebra, 120, 329
  Lie group, 130
  linear frame, 141
  manifold, I-3, 121
  parallelizable, 132, 373
  projective space, 134, 159, 273
  quadric, 278, 378
  structure (on a vector space), 114
  structure (on a manifold), 122
  submanifold, 164, 175, 378
  tangent space, 124
  tangent vector, 124
    of type (1, 0) or (0, 1), 125
  torus, 131, 159
Complexification
  of a Lie algebra, 329
  of a vector space, 116
Components
  of a linear connection, I-141
  of a 1-form, I-6
  of a tensor (field), I-21, I-26
  of a vector (field), I-5
Conformal transformation, I-309
  infinitesimal, I-309
Conjugate
  almost complex structure, 123
  complex manifold, 123
  point, 67, 71
    multiplicity of, 88

Connection, I-63
  affine, I-129
  canonical, I-110, I-301, 7, 230
  canonical flat, I-92
  canonical invariant, I-110, I-301, 192, 230
  canonical linear connection, I-302
  flat, I-92
  form, I-64
  generalized affine, I-127
  Hermitian, 178
  in normal bundle, 4, 15
  induced, I-82
  invariant, I-81, I-103, 376
    by parallelism, I-262, 194
  Levi-Civita, I-158
  linear, I-119
  metric, I-117, I-158
  $(-)$-, 198
  natural torsion-free, 197
  normal invariant, 208
  $(+)$-, 198
  Riemannian, I-158
  torsion-free, 332
  universal, I-290, 332
  (0)-, 199
Constant curvature, I-202
  space of, I-202, I-204, 24, 71, 264, 268
  surface of, 343
Constant holomorphic sectional curvature, 168
  space of, 134, 159, 169, 282
Contact form, 381
  complex, 385
Contact structure, 381
  almost, 382
  complex, 385
Contraction, I-22
Contravariant tensor (space), I-20
Convex
  hypersurface, 40
    strictly, 40
  neighborhood, I-149, I-166
Coordinate neighborhood, I-3
Covariant
  derivative, I-114, I-115, I-122
  differential, I-124

Covariant
  differentiation, I-115, I-116, I-123
  tensor (space), I-20
Covector, I-6
Covering space, I-61
Critical point, 362
  index of, 362
  non-degenerate, 362
Cross section, I-57
  adapted to a normal coordinate system, I-257
Cubic neighborhood, I-3
Curvature, I-132
  constant, I-202
  form, I-77
  Gaussian (Gauss-Kronecker), 33
  holomorphic bisectional, 372
  holomorphic sectional, 168
  Kaehlerian sectional, 369
  mean, 33
  operator, 367
  principal, 32
  recurrent, I-305
  scalar, I-294
  sectional, I-202
  tensor (field), I-132, I-145
    Riemannian, I-201
  total, 362
  transformation, I-133
Cut locus, 100
Cut point, 96
Cylinder, I-223
  Euclidean, I-210
  twisted, I-223

Degree (p, q), 125
Derivation
  of $\mathfrak{D}(M)$, I-33
  of $\mathfrak{X}(M)$, I-30
  of tensor algebra, I-25
Derived series of a Lie algebra, 325
Descending central series of a Lie algebra, 325
Development, I-131
Diffeomorphism, I-9
Differential
  covariant, I-124

Differential
  form, I-6, I-7
    complex, 124, 125
    holomorhpic, 129
    of a function, I-8
    of a mapping, I-8
Direct product of symmetric spaces, 228
Direct sum of symmetric Lie algebras, 228
Discontinuous group, I-44
  properly, I-43
Distance function, I-157
Distribution, I-10
  involutive, I-10
Divergence, I-281, 337
Dual symmetric Lie algebra, 253

Effective action of a group, I-42, 187
  almost, 187
Effective symmetric Lie algebra, 226
Effective symmetric space, 225
  almost, 225
Einstein
  hypersurface, 36, 378
  manifold, I-294, 336, 341
Elliptic
  linear Lie algebra, 334
  space form, I-209, 264
n-frame, 6
Equation of
  Codazzi, 25, 26, 47
  Gauss, 23, 26
  Jacobi, 63
Equivalence problem, I-256, 357
Equivariant isometric imbedding, 356
Euclidean
  cylinder, I-210
  locally, I-197, I-209, I-210
  metric, I-154
  motion, I-215
  subspace, I-218
  tangent space, I-193
  torus, I-210
Euler class, 314
Exponential mapping, I-39, I-140, I-147

Exterior
  covariant derivative, I-77
  covariant differentiation, I-77
  derivative, I-7, I-36
  differentiation, I-7, I-36

Fibre, I-55
  bundle, principal, I-50
  metric, I-116
  transitive, I-106
Finite type, linear Lie algebra of, 333
First normal space, 353
Flat
  affine connection, I-209
  connection, I-92
    canonical, I-92
  linear connection, I-210
  Riemannian manifold, I-209, I-210
Form
  curvature, I-77
  1-form, I-6
  r-form, I-7
  tensorial, I-75
    pseudo-, I-75
  torsion, I-120
Frame
  adapted, 2, 54
  affine, I-126
  complex linear, 141
  linear, I-55
  normal, 2
  orthonormal, I-60
  unitary, 152
Free action of a group, I-42
Frobenius, theorem of, I-10, 323
Fubini-Study metric, 160, 274
Fundamental
  theorem for hypersurfaces, 47
  2-form of a Hermitian manifold, 147
  vector field, I-51

Gauss
  equation of, 23, 26, 47
  formula, 15, 18
  spherical map of, 9, 18
  theorema egregium of, 33

Gauss-Bonnet theorem, 318, 358
Gaussian (Gauss-Kronecker) curvature, 33
Geodesic, I-131, I-146
  minimizing, I-166
  totally, I-180, 54
Gradient, 337
Grassmann manifold, 6, 9, 271
  complex, 133, 160, 286
  of oriented p-planes, 9, 272
  oriented, 272
Green's theorem, I-281
G-structure, I-288, 332, 333

Hamiltonian manifold, 149
  almost, 149
Harmonic function, 339
Hermitian
  connection, 178
  inner product, 118
  locally symmetric, 259
  manifold, 147
  metric, 146
  symmetric, 259
  vector bundle, 178
Holomorphic, I-2
  bisectional curvature, 372
  form, 129
  sectional curvature, 168
  transformation, 336
  vector field, 129
Holonomy bundle, I-85
Holonomy group, I-71, I-72
  affine, I-130
  homogeneous, I-130
  infinitesimal, I-96, I-151
  linear, I-130
  local, I-94, I-151
  of a Kaehler manifold, 173
  of a submanifold, 355
  restricted, I-71, I-72
Holonomy theorem, I-85
Homogeneous
  complex manifold, 220, 373
  coordinate system, 134
  holonomy group, I-130

Homogeneous
  Kaehler manifold, 374, 376
  Riemannian manifold, I-155, I-176, 200, 208, 211, 376
  space, I-43
    complex, 220
    Kaehlerian, 374, 376
    naturally reductive, 202
    reductive, 190, 376
    symmetric, I-301, 225
    strongly curvature, 357
Homomorphism of
  fibre bundle, I-53
  symmetric Lie algebra, 227
  symmetric space, 227
Homothetic, 289
  transformation, I-242, I-309
Hopf manifold, 137
Hyperbolic space form, I-209, 268
  complex, 282
Hypersurface, 5
  Codazzi equation of, 26, 30
  complex, 378
  Einstein, 36
  fundamental theorem for, 47
  Gauss equation of, 23, 24, 30
  Gaussian curvature of, 33
  in a Euclidean space, 17, 29
  mean curvature of, 33
  principal curvature of, 32
  Ricci tensor of, 35
  rigidity of, 45
  second fundamental form of, 13
  spherical map of, 9
  type number of, 42

Imbedding, I-9, I-53
  isometric, I-161, 354
  equivariant, 356
Immersion, I-9
  isometric, I-161, 354
  Kaehlerian, 164
  minimal (in mean curvature), 376
  minimal (in total curvature), 363
Indefinite Riemannian metric, I-155
  invariant, 200

Index, 89
  augmented, 89
  form, 81
  of a critical point, 362
  of nullity, 347
  of relative nullity, 348
  theorem (of Morse), 89
Induced
  bundle, I-60
  connection, I-82
  Riemannian metric, I-154
Infinite type, linear Lie algebra of, 333
Infinitesimal
  affine transformation, I-230
  automorphism of
    an almost complex structure, 127
    a G-structure, 186
  holonomy group, I-96, I-151
  isometry, I-237
  variation of a geodesic, 63
Inhomogeneous coordinate system, 134
Inner product, I-24
Integrability conditions of almost
    complex structure, 125, 145, 321,
    324
Integrable almost complex structure,
    124, 321, 324
Integral
  curve, I-12
  manifold, I-10
Interior product, I-35
Invariant
  affine connection, 375
  almost complex structure, 216
  by parallelism, I-262, 194
  connection, I-81, I-103, 375
  indefinite Riemannian metric, 200
  polynomial, 293, 298
  Riemannian metric, I-154
Involutive
  distribution, I-10
  Lie algebra, 225
    orthogonal, 246
Irreducible
  group of Euclidean motions, I-218
  Riemannian manifold, I-179
  symmetric Lie algebra, 252

Irreducible
  weakly, 331
Isometric, I-161
  imbedding, I-161, 354, 355, 356, 379
    equivariant, 356
  immersion, I-161, 354, 355
Isometry, I-46, I-161, I-236, 335
  infinitesimal, I-237
Isotropy
  group, linear, I-154, 187
  subgroup, I-49

Jacobi
  equation, 63
  field, 63, 68

Kaehler
  manifold, 149
    almost, 149
    pseudo, 149
    non-degenerate, 175
  metric, 149
Kaehlerian
  homogeneous, 374, 376
  pinching, 369
  sectional curvature, 369
Killing-Cartan form, I-155, 252, 325
  of $\mathfrak{o}(n+1)$, 266
  of $\mathfrak{o}(n, 1)$, 270
  of a symmetric Lie algebra, 250
Killing vector field, I-237
Klein bottle, I-223

Laplacian, 338
Lasso, I-73, I-184, I-284
Leibniz's formula, I-11
Length function, 79
Levi-Civita connection, I-158
Levi
  decomposition, 238, 327
  subalgebra, 327
Lie algebra
  abelian, 325
  complex, 120, 329
  dual symmetric, 253
  effective symmetric, 226
  involutive, 225

Lie algebra
  irreducible symmetric, 252
  nilpotent, 325
  of (non) compact type, 204, 252, 329
  orthogonal symmetric, 246
  reductive, 326
  semi-simple, 325
  simple, 325
  solvable, 325
  symmetric, 225, 238
Lie derivative, I-29
Lie differentiation, I-29
Lie group, I-38
  complex, 130
Lie subgroup, I-39
Lie transformation group, I-41
Lie triple system, 237
Lift, I-64, I-68, I-88
  horizontal, I-64, I-68, I-88
  natural, I-230
Linear
  connection, I-119
  frame, I-55
    complex, 141
  holonomy group, I-130
  isotropy group, I-154, 187
  isotropy representation, 187
Local
  basis of a distribution, I-10
  coordinate system, I-3
Locally
  affine, I-210
  Euclidean, I-197, I-209, I-210
  symmetric, I-303, 222, 243, 259
Lorentz group, 268
Lorentz manifold (metric), I-292, I-297

Manifold, I-2, I-3
  complex (analytic), I-3, 121
  differentiable, I-2, I-3
  Hermitian, 147
  Kaehler, 149
  oriented, orientable, I-3
  real analytic, I-2
  sub-, I-9
  symplectic, 149
Maurer-Cartan, equations of, I-41

Maximal nilpotent ideal, 325
Mean curvature, 33
  constant, 346
  normal, 34, 340, 341
Metric connection, I-117, I-158
Minimal immersion
  (in mean curvature), 379
  (in total curvature), 363
Minimal submanifold, 34, 340, 342, 379
Minimum point, 96
($-$)-connection, 199
Möbius band, I-223
Multiplicity
  of a conjugate point, 88
  of an index form, 89

Natural lift of a vector field, I-230
Natural torsionfree connection, 197
Naturally reductive homogeneous
  space, 202
Nilpotent Lie algebra, 325
Non-compact type
  Lie algebra of, 204, 252
  symmetric Lie algebra of, 252
  symmetric space of, 252, 256
Non-degenerate Kaehler manifold, 173
Non-positive curvature, space of, 29,
  70, 102, 109
Non-prolongeable, I-178
Normal
  bundle, 3
  coordinate system, I-148, I-162
  frame, 2
  invariant connection, 208
  space, 2
Nullity
  index of, 347
  index of relative, 348
  of a bilinear form, 89

1-parameter
  group of transformations, I-12
  subgroup, I-39
Orbit, I-12
Orientable, oriented, vector bundle, 314
Orientation, I-3
  natural, 121

Orthonormal frame, I-60
Orthogonal symmetric Lie algebra, 246
 of (non) compact type, 252
Ovaloid, 346

Paracompact, I-58
Parallel
 cross section, I-88
 displacement, I-70, I-87, I-88
  affine, I-130
 tensor field, I-124
Parallelism
 absolute, I-122
 complex, 132, 373
 invariant by, I-262, 194
Partition of unity, I-272
Pinched, pinching, 364, 369
 Holomorphic, 369
 Kaehlerian, 369
$(+)$-connection, 198
Point field, I-131
Polynomial function, 298
 invariant, 293, 298
Pontrjagin class, 312
Positive curvature, space of, 74, 78, 88, 364
Positive Ricci tensor, space of, 74, 88
Principal curvature, 32
Principal direction, 32
Projectable vector field, 218
Projection, covering, I-50
Projective space, I-52, 134, 159
Prolongation (of linear Lie algebra), 333
Properly discontinuous, I-43
Pseudo-conformal transformation, 335
Pseudo-group of transformations, I-1, I-2
Pseudo-Kaehler manifold, 149
Pseudo-tensorial form, I-75

Quotient space, I-43, I-44

Radical of a Lie algebra, 238, 325
Radon measure, 108
Rank of a mapping, I-8

Rauch, comparison theorem of, 76
Real form of a complex Lie algebra, 291, 330
Real projective space, I-52
Real representation
 of $GL(n; \mathbf{C})$, 115
 of $SL(n; \mathbf{C})$, 151
Recurrent
 curvature, I-305
 tensor, I-304
Reduced bundle, I-53
Reducible
 connection, I-81, I-83
 Riemannian manifold, I-179
 structure group, I-53
Reduction
 of connection, I-81, I-83
 of structure group, I-53
Reduction theorem, I-83
Reductive
 homogeneous space, 190
  naturally, 202
 Lie algebra, 326
 subalgebra, 326
Restriction of tensor field, 57, 58
de Rham decomposition, I-185, I-192, 171, 246, 263, 331
Ricci form, 153, 183
Ricci tensor, I-248, I-292, 35
 of a Kaehler manifold, 149
Riemannain
 connection, I-158
 curvature tensor, I-201
 homogeneous space, I-155, I-176, 200, 208, 211, 376
 locally symmetric, 232, 243
 manifold, I-60, I-154
 metric, I-27, I-154, I-155
  canonical invariant, I-155
  indefinite, I-155
  induced, I-154
  invariant, I-154
 symmetric, 232, 243
 vector bundle, 315
Rigid, 45
 affine connection, 376

Rigidity, 349
  theorem, 43, 46, 343, 353

Scalar curvature, I-294
Schur, theorem of, I-202
  Kaehlerian analogue of, 168
Second fundamental form, 13, 20
  of a complex hypersurface, 175
Sectional curvature, I-202
  holomorphic, 168
  Kaehlerian, 369
Segment, I-168
Semi-simple Lie algebra, 325
Simple Lie algebra, 325
Simple covering, I-168
Skew-derivation, I-33
Solvable Lie algebra, 325
Space form, I-209
Sphere theorem, 366
Spherical map of Gauss, 9, 18, 358
Standard horizontal vector field, I-119
Stiefel manifold, 6
Strongly curvature preserving, 357
Strongly curvature homogeneous, 357
Structure
  constants, I-41
  equations, I-77, I-78, I-118, I-120, I-129
  group, I-50
Subbundle, I-53
Submanifold, I-9, 1
  auto-parallel, 53
  complex, 164, 175, 378
  totally geodesic, I-180, 54, 234
Symmetric
  affine (locally), 222, 223
  complex affine (locally), 259
  Hermitian (locally), 259
  (homogeneous) space, 225
  Lie algebra, 225, 238
    dual, 253
    effective, 226
    irreducible, 252
    orthogonal, 246
  locally, I-303, 222, 232
  Riemannian (locally), I-302, 232, 243

Symmetric
  space, 225
    effective, 225
    almost effective, 225
  subspace, 227
Symmetrization, I-28
Symmetry, I-301, 222, 225
Symplectic manifold, 149
  almost, 149
  homogeneous,
Symplectic structure on $T^*(M)$, 165
Synge's formula, 87

Tangent
  affine space, I-125
  bundle, I-56
  space, I-5
    complex, 124
  vector, I-4
Tensor
  algebra, I-22, I-24
  bundle, I-56
  complex, 124
  contravariant, I-20
  covariant, I-20
  field, I-26
  product, I-17
  space, I-20, I-21
Tensorial form, I-75
  pseudo-, I-75
Torsion
  form of an affine connection, I-120
  of an almost complex structure, 123
  of two tensor fields of type $(1, 1)$, I-38
  tensor (field) of an affine connection, I-132, I-145
  translation, I-132
Torsion-free connection, 332
  natural, 197
Torus, I-62
  Complex, 131, 154
  Euclidean, I-210
  twisted, I-223
Total curvature, 362
Total differential, I-6
  of the length function, 79

Totally geodesic submanifold, I-180, 54
    of a symmetric space, 234, 237
Transformation, I-9
Transition functions, I-51
Transvection, 236
Trivial fibre bundle, I-51
Twisted
    cylinder, I-223
    torus, I-223
Type
    ad G, I-77
    of tensor, I-21
    (1, 0), complex vector of, 125
    (0, 1), complex vector of, 125
Type number, 42, 349

Umbilic (umbilical point), 30
Unitary frame, 152
Universal factorization property, I-17

Variation of a geodesic, 63
    infinitesimal, 63

Vector, I-4
    bundle, I-113
        orientable, (oriented), 314
        Riemannian, 315
    field, I-5
        holomorphic, 129
Vertical
    component, I-63
    subspace, I-63, I-87
    vector, I-63
Volume element, I-281

Weakly irreducible, 331
Weil homomorphism, 297
Weingarten's formula, 15
Weyl, theorem of, 204, 291, 330
Weyl group, 305
Whitney sum, 306
    formula, 306, 315

(0)-connection, 199

Errata for
Foundations for Differential Geometry, Volume I

p. 9   Line 10 from bottom: "independent of p" should read "constant in a neighborhood of M in M'
p. 16 Line 2 from bottom: "constantvector" should read "constant vector"
p. 21 About the middle of the page:

$$\bar{e}_i = \sum_j A_i^j e_j \text{ should read } e_i = \sum_j \overline{A_i^j} \bar{e}_j$$

$$\bar{e}^i = \sum_j B_j^i e^j \text{ should read } e^i = \sum_j \overline{B_j^i} \bar{e}^j$$

p. 22 Line 3 from bottom: $K_{j_1 \cdots j_s}^{i_1 \cdots k \cdots i_r}$ should read $K_{j_1 \cdots k \cdots j_{s-1}}^{i_1 \cdots k \cdots i_{r-1}}$
p. 35 Line 11 from bottom: "partitions" should read "permutations"
p. 46 Line 10 from bottom: $d(\phi_n(b), \phi_n(b))$ should read $d(\phi_n(b), \phi_N(b))$
p. 53 Line 7: $f''(a)$ should read $f''(a')$
p. 53 Line 12 through line 14 should read as follows: "or *injection* if the induced mapping $f : M' \to M$ is an imbedding and if $f : G' \to G$ is a monomorphism. By identifying P' with"
p. 64 Line 7: $\omega((R_a)_* X)$ should read $(R_a^* \omega)(X)$
p. 84 Line 1: "a horizontal curve" should read "horizontal curves"
p. 89 Line 2 from bottom: $-2\omega(X_i^*, X_j^*)$ should read $-2\Omega(X_i^*, X_j^*)$
p. 106 In the proof of Proposition 11.4, the last 6 lines beginning with "On the other hand, ..." should read as follows: Now note that the mapping $X' \in \mathfrak{k} \to \tilde{X} \in \mathfrak{X}(P)$ is induced by the action of K on P to the left and hence satisfies the condition $[\widetilde{X}, \widetilde{Y}] = -[\tilde{X}, \tilde{Y}]$ (in contrast to the situation in Proposition 4.1, p. 42, where the group acts on the right so that we have a Lie algebra homomorphism). Thus we have

$$\omega_{u_0}([\tilde{X}, \tilde{Y}]) = \omega_{u_0}(-\widetilde{[X, Y]}) = -\Lambda([X, Y])$$

so that

$$2\Omega_{u_0} = [\Lambda(X), \Lambda(Y)] - \Lambda([X, Y])$$

p. 111 Line 5: $[\tilde{X}, \tilde{Y}]$ should read $[\tilde{X}, Y]$
p. 118 Line 10: "i-th column" should read "j-th column"   Line 11: "j-th row" should read "i-th row"
p. 128 Line 14 from bottom: $= \text{ad}(\bar{a})(A)$ should be $= \text{ad}(\bar{a})(A)$ (for both equations)
p. 131 Line 4 of Proposition 4.1: C_t should read C_t
p. 135 Line 3: $= \Theta(\ldots)$ should be $= 2\Theta(\ldots)$ $T(X, Z)$ should be $T(Y, Z)$
p. 136 Line 11 from bottom: p. 611 should read pp. 61–62
p. 142 Line 12: Insert "the (i, j)-component of" before ψ_{UV}
p. 149 Line 3 of Corollary 8.6 should read $d\omega = (-1)^r A(\nabla \omega)$ for $\omega \in \mathfrak{D}^r(M)$; also in the proof
p. 150 Line 10: $= \sum_{j,k} \cdots$ should be $= 2 \sum_{j,k} \cdots$
p. 158 Line 4 from bottom: "definining" should be "defining"
p. 164 Line 8: R_{jkl}^i should be $\bar{R}_{jkl}^i$
p. 177 in the diagram: the vertical map into $N(x; 2r)$ should be called p_*, not p
p. 178 Line 11: vector at X^* should be vector at x^*
p. 181 Line 5: $y \in M$ should read $y \in M'$
p. 194 Line 3 from bottom: $\partial/\partial x^j$ should be $\partial/\partial x^i$
p. 205 Line 10 from bottom: insert "one-to-one" before "linear"
p. 208 Line 6 from bottom: Change "By Proposition 2.4," to "It follows that"
p. 209 Line 3 from bottom: $A(M)$ should be $A(U)$
p. 227 Line 14 from bottom: $\tilde{f}^* \omega$ should read $(\tilde{f}^{-1})^* \omega$
p. 235 Equation in (2) of Proposition 2.6: $+R(X, Y)$ should be $-R(X, Y)$

p. 239 Line 6: In $[X, B]$ and $[X, B']$, X should read $\tilde{X}$
p. 244 Line 5 from bottom: "infinitesiaml" should be "infinitesimal"
p. 245 Line 8: A_Z should read A_X
p. 247 Line 10: "respect this" should be "respect to this"
p. 253 Line 12: "or" should read "for"
p. 255 Line 2 from bottom: Insert "of the same dimension" at the end
p. 256 Delete the six lines of Remark
p. 258 Line 10: $\hat{R}^i_{jkl}$ should be $\hat{R}^i_{jlm}$
p. 280 Line 12: $A \in \mathfrak{g}$ should read $A \in \bar{\mathfrak{g}}$
p. 283 Line 7: $dv(X_1, \ldots, X_n) = 1$ should read $dv(X_1, \ldots, X_n) = 1/n!$
p. 283 Lines 9 and 8 from bottom: The equation should read as follows:

$$dv(\partial/\partial x^1, \ldots, \partial/\partial x^n) = \sum_{i_1,\ldots,i_n} \varepsilon\, C_1^{i_1} \cdots C_n^{i_n} dv(X_{i_1}, \ldots, X_{i_n})$$

$$= \sum_{i_1,\ldots,i_n} \varepsilon\, C_1^{i_1} \cdots C_n^{i_n} dv(X_1, \ldots, X_n)$$

$$= \det(C_i^k)\frac{1}{n!} = \frac{1}{n!}\sqrt{G},$$

p. 289 Line 3 from bottom: "there exists" should read "there exist"
p. 293 Line 1 of Proposition 2: "then i" should read "then it"
p. 306 Line 4 from bottom: insert "invariant" after "structure"
p. 308 Lines 11-9 from bottom should read as follows: If we write $M = \tilde{M}/\Gamma$, where Γ is a discrete subgroup of $I(\tilde{M})$, then Γ must commute with the identity component $I^0(\tilde{M})$ of $I(\tilde{M})$. If $\tilde{M} = \mathbf{R}^n$, then $I^0(\tilde{M})$ is the group of proper motions and only the identity transformation commutes with $I^0(\tilde{M})$. If $\tilde{M} = S^n$, then $I^0(\tilde{M}) = SO(n + 1)$ and only $\pm I$ commutes with $SO(n + 1)$. If $\tilde{M}$ is a simply connected hyperbolic space, then $I(\tilde{M}) = O(1, n)$ and the identity element is the only element of $I(\tilde{M})$ that commutes with $I^0(\tilde{M})$.
p. 308 Line 8 from bottom: insert ", then" between S^n and "for".
p. 308 Line 7 from bottom: delete the comma "," between "x'" and "there"
p. 309 Line 19: insert "of Chapter VI" between "3.3" and "are"
p. 310 Line 12: $\mathfrak{J}_0(M)$ should read $\mathfrak{J}^0(M)$
p. 311 Line 7: x_0 should be x^0
p. 315 Berger, M. [1]: (1953) should read (1955)
p. 318 Line 1: "Momorimoto" should be "Morimoto"
p. 318 Line 3 from bottom: "Forchungen" should be "Forschungen"
p. 319 Line 17: "Lie algebra in" should be "Lie algebra of"
p. 321 Line 2 from bottom: "dreidmensionalen" should read "dreidimensionalen"
p. 322 Line 4 from bottom: "Einornung" should read "Einordnung"
p. 322 Line 2 from bottom: (1912) should be (1921)

Errata for
Foundations for Differential Geometry, Volume II

p. 31 Line 17: $x + \xi_x$ should read $\lambda x + \xi_x$
p. 37 Equation (1): λ_2 should read λ^2
p. 45 Equation (I): $\Gamma^k_{ij} e^k$ should read $\Gamma^k_{ij} e_k$
p. 48 Equation (I_p): Σ^n_{k+1} should read $\Sigma^n_{k=1}$
p. 53 Line 2: $t < t_1$ should be $t \leq t_1$
p. 55 Line 4 below Prop. 8.2: M should read N
p. 64 Line 4: X_x should read X_{x_t}

p. 67   Line 3 of Theorem 1.4: After x, insert "along a geodesic"
p. 67   Line 1 from bottom: After "x and y", insert "without being identically 0"
p. 71   Line 1 from bottom: I^b should be I_a^b
p. 72   Line 7: $a \geq t \geq b$ should be $a \leq t \leq b$
p. 74   Line 10 from bottom: $\sin \pi \frac{(t-a)}{(t-b)}$ should read $\sin \pi \frac{(t-a)}{(b-a)}$
p. 76   Line 2 of Theorem 4.1: $a \leq t \geq b$ should read $a \leq t \leq b$
p. 81   Line 3 of Theorem 5.4: The equation $x^\perp = X - (1/r)g(X, \dot{\tau})\dot{\tau}$ should read $X^\perp = X - (1/r^2)g(X, \dot{\tau})\dot{\tau}$
p. 81   Line 3 of Theorem 5.5: $\Sigma_{j=1}^{k}$ should read $\Sigma_{j=1}^{h}$
p. 90   Line 6: "and x_{a_i}" should read "at x_{a_i}"
p. 101 Line 4 of Example 7.1: "north" should read "south"
p. 115 Line 9: $(y^1, \ldots, y^n, -x^1, \ldots, -x^n)$ should read $(-y^1, \ldots, -y^n, x^1, \ldots, x^n)$
p. 115 Line 12: $\begin{pmatrix} 0 & I_n \\ -I_n & 0 \end{pmatrix}$ should read $\begin{pmatrix} 0 & -I_n \\ I_n & 0 \end{pmatrix}$
p. 115 Line 6 from bottom: Same as above
p. 115 Line 2 from bottom: $\begin{pmatrix} A & B \\ -B & A \end{pmatrix}$ should read $\begin{pmatrix} A & -B \\ B & A \end{pmatrix}$
p. 137 Line 9 from bottom: **x** should read $\times$
p. 157 Equation (22): $\Gamma^\gamma_{\alpha\gamma}/\partial \bar{z}^\beta$ should read $\partial \Gamma^\gamma_{\alpha\gamma}/\partial \bar{z}^\beta$
p. 162 Just above Example 6.6: The minus sign in the middle of the numerator should be the plus sign
p. 169 Line 4 from bottom: The minus sign in the middle of the numerator should be the plus sign
p. 179 Line 4: $\dim (T_u(Q) \cap J(Q))$ should read $\dim (T_u(Q) \cap J(T_uQ))$
p. 183 Line 6: $2iR$ should read $2i\tilde{R}$
p. 183 Line 8: Proposition 4.11 should read Proposition 4.9
p. 188 Line 2 of Corollary 1.3: $\tilde{X}$ should read Λ
p. 271 Line 10 from bottom: $\mathbf{R}^{n+1}$ should read $\mathbf{R}^n$
p. 283 Line 13: $T'_z(M)$ should read $T_z(M)$
p. 318 Line just above Theorem 5.1: $(-1)^p$ should read 1
p. 334 Line 5: "the theorem that" should read "the theorem on"
p. 338 Line 15: formula (ii): $(\operatorname{grad} f) \cdot f$ should read $2(\operatorname{grad} f) \cdot f$
p. 338 Next line: $g(df, df)$ should read $2g(df, df)$
p. 341 Line 10 from bottom: curvature 1 should read curvature $n(n-1)$
p. 341 Line 7 from bottom: Obata [1] should read Obata [3], [5]
p. 345 Line 2: $\nabla_Y Y = -bY$ should read $\nabla_Y Y = -bX$
p. 351 Line 2: $f(x) = y_\beta$ should read $f(x) = -z_\beta$. Also, $y_\beta \in \operatorname{range} f$ should read $z_\beta \in \operatorname{range} f$
p. 351 Line 3: $z_\beta \in \operatorname{range} f$ should read $y_\beta \in \operatorname{range} f$
p. 351 Line 3 below Lemma 2: $\Sigma_{j=1}^{k}$ should read $\Sigma_{j=1}^{n}$
p. 379 Line 12: $-\|\alpha(X, X)\|^2$ should read $-2\|\alpha(X, X)\|^2$
p. 383 Line 10: $T_k^0 = u(\mathbf{R})$ should read $T_x^0 = u(\mathbf{R})$
p. 404 Dombrowski, P. [2]: The journal should be replaced with "Math. Nachr. 38(1968) 133-180 (MR 39 #7536)"
p. 412 Hadamard, J. [1] should be replaced with "Sur certaines propriétés des trajectoires en Dynamique, J. Math. Pure Appl. (5)3(1897), 331-388"
p. 437 Line 12: 19(1968) should read 20(1969), 337-338
p. 446 Thomas, T.Y. [1]: (1935) should read (1936)
p. 446 After Thomas, T.Y. [6], add [7] Algebraic determination of the second fundamental form of a surface by its mean curvature, Bull. Amer. Math. Soc. 51(1945), 390-399
p. 454 Line 5: #1816 should read #1814